Light Absorption in Sea Water

ATMOSPHERIC AND OCEANOGRAPHIC SCIENCES LIBRARY

VOLUME 33

The titles published in this series are listed at the end of this volume.

Light Absorption in Sea Water

Bogdan Woźniak[1,2] and Jerzy Dera[1]

[1]*Institute of Oceanology*
Polish Academy of Science
Powstańców Warszawy 55
81-706 Sopot, Poland

[2]*Institute of Physics*
Pomeranian Academy
Arciszewskiego 22
76-200 Słupsk, Poland

Springer

Bogdan Woźniak
Institute of Oceanology
Polish Academy of Sciences
Powstańców Warszawy 55
81-712 Sopot, Poland
and
Institute of Physics
Pomeranian Academy
Arciszewskiego 22
76-200 Słupsk, Poland

Jerzy Dera
Institute of Oceanology
Polish Academy of Sciences
Powstańców Warszawy 55
81-712 Sopot, Poland

ISBN-13: 978-1-4419-2149-9
eISBN-10: 0-387-49560-6
eISBN-13: 978-0-387-49560-6

Printed on acid-free paper.

9 8 7 6 5 4 3 2 1

springer.com

Contents

1 Introduction: Absorption of Sunlight in the Ocean

1.1 Inflow and Absorption of Sunlight in the Ocean

The absorption of light by the oceans is a fundamental process in the Earth's harvesting of the vast resources of solar radiation and its conversion into other forms of energy. Of course, light is also absorbed by the atmosphere and the continents, but the scale of the process there is very much smaller than in the oceans, inasmuch as the absorption capability of the atmosphere is lower and the surface area of the land is only about one-third of that of the oceans. At any instant, half the Earth's surface is illuminated by a beam of solar rays (Figure 1.1) which, at this distance from the sun and given the comparative size of the Earth, are practically parallel. Their angle of divergence is that through which we see the Sun from the Earth; that is, $\Delta\varphi \approx 6.8 \times 10^{-5}$ radians, or c. 0.004 of a degree.

Taking into consideration the daily cycles of the Earth's revolutions about its axis and the unevenness of its surface, the mean insolation of the Earth amounts to c. 342 W m^{-2}, that is, around one quarter of the solar constant (S = 1365 W m^{-2}; Wilson (1993)), because the surface area of the Earth is four times as large as its cross-section (Schneider 1992). If we allow for a 26% loss of this radiation due to reflection in the atmosphere and a further 19% due to absorption in the atmosphere (Harrison et al. 1993, Trenberth 1992), the mean insolation of the ocean is 55% of the original value, or c.188 W m^{-2}. In other words, this is the time- and space-averaged downward irradiance of the sea surface. Furthermore, if we bear in mind that c. 6% of the light reaching the sea is reflected from the surface and within the water itself—this is the mean albedo of the sea (Payne 1979)—we have an average flux of radiant energy equal to c. 177 W that is constantly being absorbed in the water column beneath every square meter of the sea surface and converted to other forms of energy.

If we take the World Ocean to have a total surface area of 361 million km^2, it is unceasingly absorbing a solar radiation flux of c. 6.4×10^{10} MW. This energy is consumed in warming and evaporating the waters of the ocean, warming the atmosphere by conduction, the latent heat of evaporation, and

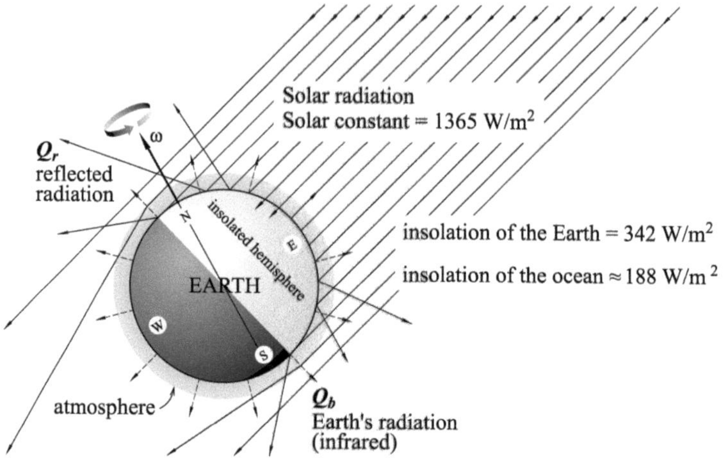

FIGURE 1.1. Solar light fluxes and the Earth. Solar constant: the mean solar radiation flux incident on a plane at right angles to the radiation reaching the top of the atmosphere; insolation of the Earth: the average component at the top of the atmosphere of the above flux (solar constant) perpendicularly incident at every spot on the Earth's surface during the day; insolation of the ocean (and land areas, i.e., the combined surface area of the ocean and land): the insolation of the Earth minus the averaged amount of radiation lost in the atmosphere. (Adapted from Dera (2003).)

the thermal radiation of the sea surface, as well as setting (and maintaining) the masses of water in motion. Finally, less than 1% of this energy drives the photosynthesis of organic matter and photochemical reactions in the sea water (e.g., Dera 2003).

From the point of view of quantum mechanics, this flux of absorbed energy means that an average of 10^{20} collisions between photons and the component molecules of sea water and the absorption of the former by the latter takes place under every single square meter of ocean during every single second. The effect of every elementary collision of a photon with a particle of matter is strictly dependent on the energy of a photon $E_\nu = h\nu$ (where $h = 6.62517 \times 10^{-34}$ Js, Planck's constant and ν is the frequency of vibrations of a photon, equal to the reciprocal of the period of vibrations $T = 1/\nu$, which is linked to the wavelength of the light λ and its velocity c by the relationship $c = \nu\lambda$). The most likely effects of these collisions, during which photons of the appropriate energy are absorbed, are shown in Table 1.1.

We see in the table how very different the energies of photons are from different frequency bands of the light spectrum and also how large the extent to which the effects of their absorption vary, from the ionization of atoms by the far ultraviolet (UV) to the translation motions induced by the far infrared (IR). It thus becomes essential to know what the spectrum of light incident on the sea is and the changes it is subject to as it penetrates ever deeper into

TABLE 1.1. Relationships between the length of an electromagnetic wave λ (in a vacuum), the frequency of vibrations ν, the energy of a photon E_ν, the color of the light, and the probable effect of the absorption of a photon.

Wave-length λ [nm]	Frequency of vibrations ν [Hz]	Energy of a photon E_ν [10^{-19}J]	Energy of a photon E_ν [eV]	Number of photons per J of energy [10^{18}J^{-1}]	Color of light (type of waves)	Probable effect of the absorption of a photon by a molecule
-1-	-2-	-3-	-4-	-5-	-6-	-7-
100	3.001×10^{15}	19.88	12.41	0.503		Ionization,
150	2.000×10^{15}	13.25	8.27	0.756	Far	electronic
200	1.500×10^{15}	9.93	6.20	1.006	ultraviolet	excitation
250	1.199×10^{15}	7.95	4.96	1.258	Ultraviolet	
300	9.993×10^{14}	6.62	3.87	1.510		
350	8.566×10^{14}	5.68	3.54	1.761	Violet	
400	7.495×10^{14}	4.97	3.10	2.012	Blue	
450	6.662×10^{14}	4.41	2.76	2.268	Green	
500	5.996×10^{14}	3.97	2.48	2.519	Greenish-	Vibrational -
550	5.451×10^{14}	3.61	2.25	2.770	Yellow	rotational
						excitation
600	4.997×10^{14}	3.31	2.07	3.021	Orange	
650	4.612×10^{14}	3.06	1.91	3.268		
700	4.283×10^{14}	2.84	1.77	3.521	Red	
750	3.997×10^{14}	2.65	1.65	3.774		
800	3.748×10^{14}	2.48	1.55	4.032		
1,000	2.998×10^{14}	1.99	1.24	5.025	Infrared	
2,000	1.499×10^{14}	0.99	0.62	10.10		
3,000	9.993×10^{13}	0.66	0.41	15.15		
5,000	6.038×10^{13}	0.40	0.25	25.00	Far infrared	
10,000	3.019×10^{13}	0.20	0.13	50.00		Rotational
20,000	1.509×10^{13}	0.10	0.06	100.0		excitation,
50,000	6.032×10^{12}	0.04	0.03	250.0		translations

Adapted from Dera (2003).

the water. Spectra of the irradiance of the sea surface by sunlight and the changes it undergoes with increasing depth in different seas are illustrated in Figure 1.2. Numerous studies have shown that, under average conditions, c. 50% of the sunlight incident on the sea surface consists of IR bands, some 45% of visible radiation and only around 5% of UV. The figure also shows how quickly the visible light spectrum narrows down with depth in the sea: bluish-green light ($\lambda \approx 450$ nm) penetrates the farthest in optically clear oceanic waters, whereas greenish-yellow light ($\lambda \approx 550$ nm) is the most penetrating in sea waters containing large amounts of organic substances. We can also see from Figure 1.2a that light reaching a depth of barely one meter contains practically no more IR, which means that IR radiation is very strongly absorbed by sea water and that all its energy entering the sea is absorbed in the very thin surface layer.

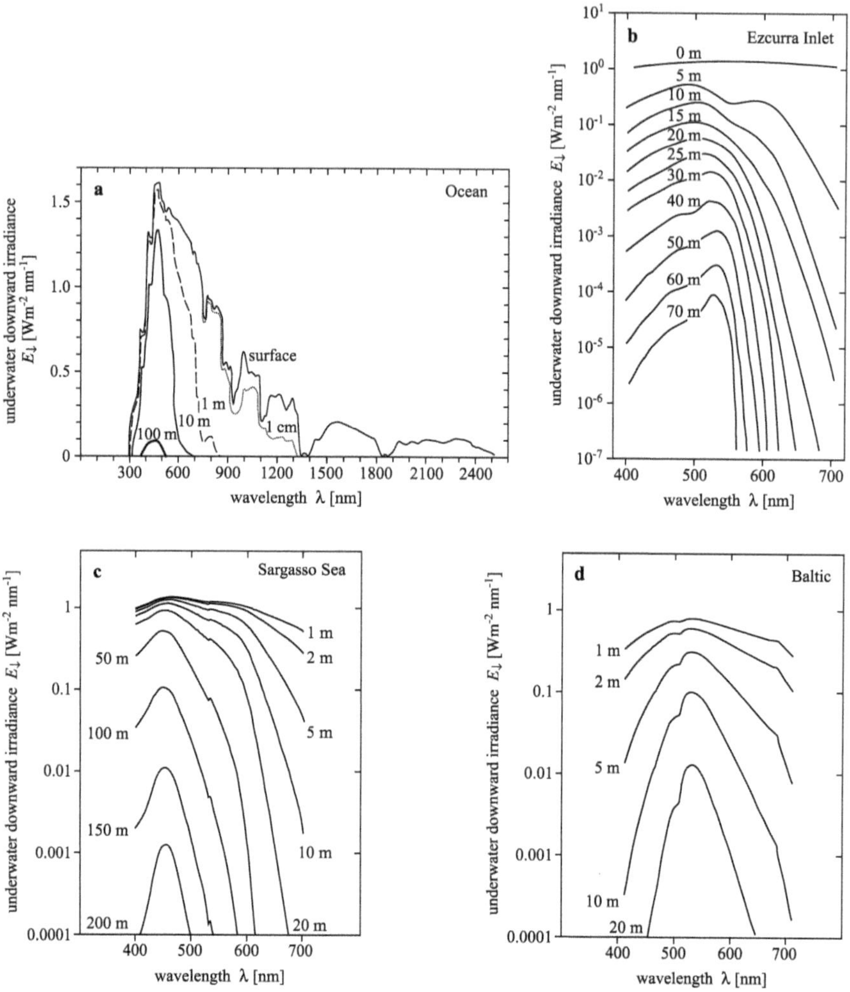

FIGURE 1.2. Spectra of daily downward irradiance in sea waters at various depths: (a) in the ocean over a broad spectral range (Jerlov 1976); (b) in the Ezcurra Inlet, Antarctica, at around noon in January (from measurements by Woźniak et al. during the second Antarctic Expedition of the Polish Academy of Sciences (Dera 1980); (c) in the clear oligotrophic waters of the Sargasso Sea; (d) in the eutrophic waters of the Baltic Sea (drawn on the basis of data from IO PAN Sopot.) Reproduced from J. Dera, *Marine Physics*, 2nd ed., updated and supplemented, 2003 (in Polish), with the kind permission of PWN, Warszawa.

Water molecules play the most important part in the absorption of solar light energy in the ocean, not only because of their amount (>96% of all the molecules contained in sea water), but also because of their absorption properties: light in the IR band is very strongly absorbed (Chapter 2). Even so, there is a multitude of other substances in sea water that are also capable of absorbing this energy. The complexity of sea water as a substance means that its optical properties are essentially different from those of pure water. Sea water contains numerous dissolved mineral salts and organic substances, suspensions of solid organic and inorganic particles, including various live microorganisms, and also gas bubbles and oil droplets. Many of these components participate directly in the interactions with solar radiation in that they absorb or scatter photons. Many also participate indirectly by fulfilling diverse geochemical and biological functions, for instance, in photosynthesis, which regulates the circulation of matter in marine ecosystems, and in doing so, affects the concentrations of most of the optically active components of sea water. The resources and the concentrations of the most important sea water components of the World Ocean are listed in Table 2.17, which also gives the principal optical and biological functions of these components.

The occurrence in sea water of suspended particles as well as other inhomogeneities, such as gas bubbles, oil droplets, and turbulence, means that from the optical standpoint it is a turbid medium, a light absorber and scatterer, the optical properties of which vary together with changes in the composition and concentration of the components and depend on the physical conditions prevailing at any given time (Jerlov 1976, Morel and Prieur 1977, Shifrin 1983b,1988, Højerslev 1986, Kirk 1994, Spinard et al. 1994, Mobley 1994, Stramski et al. 2001, Dera 1992, 2003).

Because of this powerful interaction between the molecules and the large number of components in sea water, we observe in the optical spectrum not discrete spectral absorption lines but broad absorption bands. Overlapping to various degrees and in different regions of the spectrum, the absorption bands actually form an absorption continuum with local maxima and minima. The absolute principal minimum of electromagnetic wave absorption in sea water lies in the visible region, as is the case with pure water (Figures 1.3 and 2.11). As we see in Figure 1.3, there is a great variety for different seas (e.g., a large absorption for the Baltic). The huge number of sea water components and the continuous nature of its absorption spectrum preclude any meaningful discussion of the optical properties of each component in turn. Nevertheless, it is possible to distinguish groups of components that are especially actively involved in the absorption of the solar radiation entering the sea water.

Generally speaking, we have to examine the following groups of components, which differ distinctly in their optical properties: water molecules and their associated forms, sea salt ions, dissolved organic matter (DOM), live phytoplankton and suspended phytoplanktonlike particles, and other particulate organic matter (POM; organic detritus, zooplankton, and zooplanktonlike particles), suspended mineral particles, and other components such as oil droplets and gas bubbles. It is thus the aim of the present volume to describe and assess the current state of knowledge of the absorption properties of these optically significant substances and the part they play in the interaction with light in sea water.

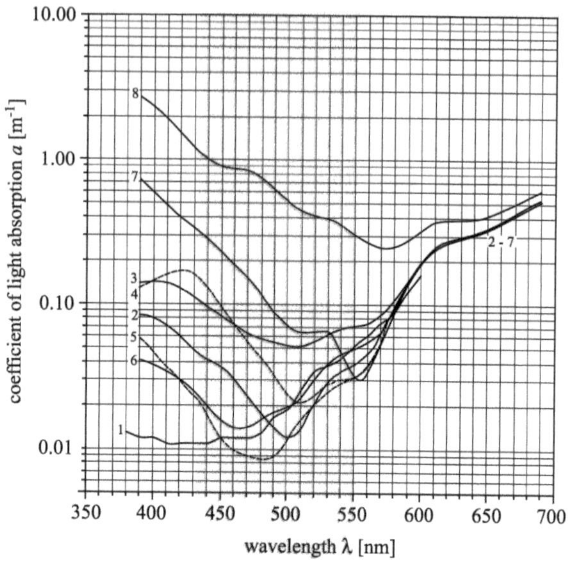

FIGURE 1.3. Spectra of light absorption in the visible range measured in different sea waters, curves: 1. Central Pacific; a very clear, deep-sea water (data from Pelevin and Rostovtseva (2001)); 2. Pacific Ocean, northern zone of subtropical convergence, depth 85 m; 3. Pacific Ocean, North Equatorial Current, depth 200 m; 4. Pacific Ocean, South Equatorial Current, depth 10 m; 5. Pacific Ocean, Tonga Trench, depth 10,000 m; 6. Atlantic Ocean, Sargasso Sea surface water; 7. Baltic Sea, Gotland Deep surface water; 8. Baltic Sea, Gulf of Riga surface water; plots 2 to 8, data selected from Kopelevitch et al. (1974); expedition of the Soviet research vessel "Dmitrii Mendeleyev" to the oceans (1971) and other cruises in the Baltic Sea (1970).

1.2 Case 1 and Case 2 Waters

In addition to water molecules and salt ions, most of the other substances interacting to a significant extent with light in the open waters of oceans have their origin in the photosynthesis of organic matter in phytoplankton cells. The organic matter produced in this process supplies a whole trophic chain of marine organisms with nutritional energy, and the functioning of the ecosystem in turn gives rise to a range of dissolved and suspended substances: the metabolic and decay products of these organisms. The photosynthetic process is strictly dependent on the concentration of chlorophyll a in phytoplankton cells. Hence, the concentration of a mixture of light-absorbing substances in such waters is correlated with the chlorophyll a concentration, which can serve both as an index of the trophicity (fertility) of waters and as a parameter of their optical properties. Waters in which the functioning of the local marine ecosystem is practically the only source of substances affecting the optical properties of waters—these substances are thus autogenic—have come to be known as *Case 1 waters* (Morel and Prieur 1977).

Unlike such waters, which are unaffected by the inflow of various substances from rivers, shores, or other external sources, many marine basins, for example, the Baltic Sea, do contain waters that are affected by such inflows. Waters in which allogenic substances (entering from outside the local ecosystem, not produced by it) play a significant part in the interactions with light are referred to as *Case 2 waters* (Morel and Prieur 1977). In these latter waters the dependence of the absorption properties and other optical properties on the chlorophyll *a* concentration is more complex and has to be determined empirically with the aid of statistical methods, separately for particular marine regions and different seasons of the year. The classification of sea waters in accordance with this criterion has become standard in the oceanographic literature. The authors of this concept have stated that more than 98% of oceanic waters are Case 1 waters. This division is, however, merely an approximation of reality, as with many other such approximations necessary for modeling nature and investigating prevailing phenomena. In actual fact, ocean waters always contain certain admixtures not derived from the local ecosystem, for example, windborne sands from the Sahara, volcanic ash and other atmospheric dust, and traces of inflows from the great rivers carried along by the ocean currents. Nonetheless, these admixtures are usually of secondary importance where the optical properties of Case 1 open ocean waters are concerned.

1.3 The Light Absorption Coefficient and Its Components in Sea Water

An exact description and the measurement of light absorption in a medium requires the radiant energy transfer equation to be invoked (Preisendorfer 1961). When passing through a medium, light is not only absorbed but also scattered, and, unlike the classical Bouguer–Lambert law of absorption, the radiant energy transfer equation takes both processes into account. In the simplest case, where only one single, thin, parallel, time-invariable beam of light travels a distance r through a homogeneous medium that itself contains no sources of light, this equation can be written as

$$\frac{dL}{dr} = -cL, \tag{1.1}$$

where $L \equiv L(r,\lambda)$, the radiance of this light beam; λ is the wavelength of the light; $c \equiv c(\lambda)$, the total volumetric coefficient of light attenuation (for definitions, see Dera (2003, 1992), Højerslev (1986), and Jerlov (1976)).

The dimension of the coefficient of light attenuation is [m^{-1}] and comprises the sum of the coefficients of absorption $a(\lambda)$ and scattering $b(\lambda)$:

$$c(\lambda) = a(\lambda) + b(\lambda). \tag{1.2}$$

In the sea these coefficients are also functions of the coordinates of a given water body and are time-variable. They are the inherent optical properties of the medium (in this case, sea water) and thus characterize only the nature of the medium and are independent of the light's intensity. The correct way of measuring these coefficients, even if different techniques are applied, should lead to identical results. Later on, we cover in detail the light absorption coefficient $a(\lambda)$ of sea water, which characterizes the absorption properties of this water and its components. A general description of these and other optical properties of sea water can be found in earlier monographs (Jerlov 1976, Mobley 1994, Dera 2003). What the literature lacks, however, is a compact description of the optical absorption properties of sea water in conjunction with a detailed analysis of the properties and the contribution to absorption of the several groups of its components, as well as some individual components that play a particularly important role in certain types of seas. It is the aim of our book to fill this gap and to acquaint the reader with the state of the art in this field.

Determining the optical absorption properties of sea water components has become imperative, particularly in view of the ever-wider applications of remotely sensing, satellite techniques for monitoring the state of the marine environment (Gordon et al. 1988, Karabashev et al. 2002, Arst 2003, Zaneveld et al. 2005a,b). One of the fundamental formulas in the remote sensing of the concentrations of selected sea water components is the one linking the light absorption coefficient $a(\lambda)$ with the remote reflectance of radiance in the sea $R_{rs}(\lambda)$ recorded by a satellite (Gordon et al. 1988):

$$R_{rs}(\lambda) = \frac{L_u(\lambda)}{E_d(\lambda)} \approx C \frac{b_b(\lambda)}{a(\lambda) + b_b(\lambda)}, \tag{1.3}$$

where L_u is the radiance emerging vertically upwards from the sea measured just below the sea surface, E_d is the downward irradiance just below the sea surface, b_b is the light backscattering coefficient; that is, in scattering angles from 90° to 180°, C is a constant coefficient determined empirically for the given conditions of an investigation. This formula also takes account of the significant part played by the light-scattering coefficient, but the correlations of the absorption coefficients with the reflectance enable one to detect the presence, and even calculate the concentrations, of certain substances in the sea on the basis the remotely measured reflectance R_{rs} (Sathyendranath et al. 1994, 2001, Platt et al. 1995, Olszewski and Darecki 1999, Lee et al. 1994; Gordon 2002). A well-known application of the reflectance of the sea surface radiance and its correlation with the absorption coefficient of sea water components is the remote sensing of the chlorophyll concentration and hence of primary production in the ocean (Platt and Sathyendranath 1993a,b; Sathyendranath et al. 1989, Morel 1991, Antoine et al. 1996, Woźniak et al. 2004, see also Figure 6.13).

The energy of the light absorbed in unit volume (energy density) of water dP/dV at depth z in the sea depends only on the inflow of this light (from all directions) and on the coefficient of light absorption in this volume of water. This is described by the Law of Conservation of Radiant Energy, well known in marine optics, which emerges, for example, from Gershun's equation (1958):

$$\frac{dP}{dV} = -aE_0 \quad \left[\mathrm{Wm}^{-3}\right], \tag{1.4}$$

where E_0 is the scalar irradiance at the point of investigation. This equation is exact in the situation where there are no internal light sources in the volume of water under study (for details, see Gordon (2002) or Dera (2003)).

Because the concentrations of the various substances dissolved and suspended in sea water are small, light absorption by the various components or groups of components of sea water is assumed to be additive. This means that the overall absorption coefficient of sea water $a(\lambda)$ is the sum of the partial coefficients: these express the contribution of various groups of components in the water to the resultant absorption. We can ascribe a separate absorption coefficient to each of these component groups, so that their sum total will express the overall coefficient for sea water:

$$\left.\begin{aligned} a = a_w(\lambda) + a_s(\lambda) + a_{DOM}(\lambda) + a_p(\lambda) + a_r(\lambda) \\ \text{and } a_p(\lambda) = a_{POM}(\lambda) + a_{PIM}(\lambda),\ a_{POM}(\lambda) = a_{pl}(\lambda) + a_{Od}(\lambda) \end{aligned}\right\} \tag{1.5a}$$

or

$$\left.\begin{aligned} a = a_w(\lambda) + a_s(\lambda) + a_{DOM}(\lambda) + a_{pl}(\lambda) + a_{NAP}(\lambda) + a_r(\lambda) \\ \text{and } a_{NAP}(\lambda) = a_{Od}(\lambda) + a_{PIM}(\lambda) \end{aligned}\right\}, \tag{1.5b}$$

where the subscripts to the respective partial coefficients denote the following: w is water; s is sea salt; DOM is dissolved organic matter; p is suspended particulate matter; r is the remaining substances, including oil droplets, gas bubbles, and others not covered by further subscripts; POM is particulate organic matter; PIM is particulate inorganic matter (suspended minerals); pl is live phytoplankton and bacteria (a_{pl} is often identified with the total absorption coefficient of all the phytoplankton pigments); Od is organic detritus (together with nonalgal living organismus as zooplankton); NAP is nonalgal particles. The dimension of the light absorption coefficients is [m^{-1}].

The partial absorption coefficient related to the mass concentration of a given absorber in sea water is called the *mass-specific light absorption coefficient* of that absorber, and is usually denoted $a_j^* = a_j/C_j$ (where the subscripts j denote the absorber in question and C_j is the mass concentration of the absorber in the water). The dimension of the mass-specific light absorption coefficients of an absorber is [m^{-1}[(mg of the jth absorber) $\mathrm{m}^{-3}]^{-1}$] ≡ [m^2(mg of the jth absorber)$^{-1}$].

In the case of phytoplankton the absorption coefficient is often related to the mass concentration of the principal phytoplankton pigment, that is, to the total chlorophyll *a* concentration in sea water C_a (the sum of chlorophyll *a* + pheo.). Denoted $a^*_{pl} = a_{pl}/C_a$, it is referred to as the *chlorophyll-specific light absorption coefficient* of phytoplankton. Its dimension is $[m^{-1}[(\text{mg tot. chl } a)\ m^{-3}]^{-1}] \equiv [m^2(\text{mg tot. chl } a)^{-1}]$.

In line with the partial absorption coefficients we have distinguished here, the several chapters in this book describe and analyze the absorption properties of these various groups of sea water components. We devote the greatest amount of space to the components most strongly differentiating sea and ocean waters from the optical point of view: the water itself, the dissolved organic substances, and the suspended particles of diverse provenance. We place special emphasis on the role of phytoplankton pigments, which are particularly strong and important absorbers of visible light in the sea.

2 Light Absorption by Water Molecules and Inorganic Substances Dissolved in Sea Water

The principal absorber of light and other electromagnetic radiation in seas and oceans is, of course, water as a chemical substance. Above all, this is due to the absolute numerical superiority of H_2O molecules over the molecules of all other substances contained in sea water: for every 100 H_2O molecules there are only 3–4 molecules of other substances, chiefly sea salt, but also dissolved organic substances, and the numerous suspended particles of mineral and organic matter, including phytoplankton cells and other live organisms. Although small in quantity, these various substances contained in sea water very significantly differentiate marine areas from the optical point of view. Water itself is a very strong absorber of electromagnetic radiation in the infrared (IR) region; thanks to this property it plays a prominent and indispensable role for life on Earth in that the ocean absorbs solar IR radiation and converts it into heat. This process leads directly to the warming of the ocean's surface waters, causing them to evaporate, and to the heating and circulation of the atmospheric air.

In this chapter we describe the mechanisms and spectra of light absorption by small molecules—principally water molecules—as well as the spectral absorption properties of liquid water, ice, and water vapor, and of the components of sea salt and other mineral substances dissolved in the water. Because of its exceptional significance in Nature, we give the water itself pride of place, describing its absorption spectra with respect not only to visible light but also to a wide spectrum of electromagnetic radiation, from the extremely short γ- and X-rays to long radio waves.

2.1 Light Absorption Spectra of Small Molecules such as Water: Physical Principles

As we stated in Chapter 1, the light absorption spectra of matter in its various states are determined by quantum changes in the atomic and molecular energies of that matter as a result of its having absorbed photons. Among these quantum processes we must include the energetic electronic transitions in atoms, and also the electronic, vibrational, and rotational transitions in molecules, all of which in fact take place simultaneously. We can explain these quantum changes of energy with respect to a molecule most simply if we examine the energy states of a free molecule.

The total energy of a freely moving molecule E_M (Λ,υ,J) consists of the temperature-dependent energy of its translational motion E_{TR}, the energy of its rotation about its various axes of symmetry $E_{ROT}(J)$, the vibrational energy of its atoms around its equilibrium position $E_{VIB}(\upsilon)$, and the energy of its electrons $E_E(\Lambda)$, where J, υ, Λ, are respective quantum numbers. We can thus write down this total energy as the sum of these component energies:

$$E_M\,(\Lambda,\upsilon,J) = E_E(\Lambda) + E_{VIB}(\upsilon) + E_{ROT}(J) + E_{TR}. \qquad (2.1)$$

The first three—quantized—energy constituents on the right-hand side of Equation (2.1) depend on the extent to which the molecule is excited. In other words, they take discrete values, strictly defined by the given quantum (energy) state of the molecule. The energy states of a simple molecule (e.g., H_2O, CO_2) are defined by the rotational energy (the quantum number or numbers J), the vibrational energy (the quantum number or numbers υ), and the electronic energy (given by the quantum number Λ), defining the absolute value of the projection of the angular momentum on to the molecule's axis.

The absorption or emission by a molecule of a photon of energy $E_{h\nu} = h\nu$ invariably involves its transition from one quantum state, described by quantum numbers Λ, υ, J, to another quantum state, described by Λ', υ', J' (in the general case all three quantum numbers change) in accordance with the allowed quantum mechanical selection rules. The energy of this photon is then equal to the difference between the molecule's energies in these two states and can be written as

$$E_{h\nu} = E_M(\Lambda,\upsilon,J) - E_M(\Lambda',\upsilon',J') = \Delta E_E(\Lambda\rightarrow\Lambda') + \Delta E_{VIB}(\upsilon\rightarrow\upsilon') + \Delta E_{ROT}(J\rightarrow J') + \Delta E_{TR}. \qquad (2.2)$$

So it is the sum of the increments (or losses) of energy of the several components in these two states (initial and final), that is, the increments of electronic (ΔE_E), vibrational (E_{VIB}), rotational (E_{ROT}), and translational (E_{TR}) energy. The increments of these first three energy components of the molecule following the absorption of a photon give rise to the three principal types of absorption band, which we discuss presently. The transition of a molecule from one electronic state to another (the selection rule for such transitions is $\Delta\Lambda = 0, \pm1$) usually brings about changes in the vibrational state (selection rule: $\Delta\upsilon = \pm1, \pm2, \ldots$) and the rotational state of the molecule (selection rule: $\Delta J = 0, \pm1$) as well. A transition during which all three types of molecular energy change as a result of the absorption (or emission) of energy gives rise to an *electronic-vibrational-rotational* spectrum of the absorption of electromagnetic wave energy. This is a band spectrum with a highly complex structure consisting of very many spectral lines forming two or three branches denoted R($\Delta J = 1$), P($\Delta J = -1$), and Q($\Delta J = 0$). The lines of these branches lie in the ultraviolet and visible regions of the electromagnetic waves spectrum (in the case of water only in the ultraviolet).

In certain electronic states, a molecule may, as a result of allowed vibrational-rotational energy transitions, change only its vibrational and rotational energy. Transitions of this kind give rise to vibrational-rotational absorption spectra as a result of the absorption (or emission) of relatively low-energy quanta, that is, from the infrared region of electromagnetic waves.

If a molecule is not symmetrical or has a dipole moment other than zero (like the water molecule, for example), then as a result of the absorption (or emission) of energy quanta from the microwave or radiowave ranges, its rotational energy may change without the electronic or vibrational states being affected. What we then have is a rotational absorption band.

This simplified picture of a molecule's internal energy changes reflects the structural complexity of a molecular spectrum of the absorption (or emission) of electromagnetic wave energy. In this chapter we present the theoretical foundations underlying this process, which are essential for understanding the structure of the light absorption spectra of water molecules; it also helps in understanding the absorption band structures of other small molecules. A more detailed treatment of the subject can be found in the numerous monographs on molecular physics, spectroscopy, and quantum mechanics and chemistry, for example, Barrow (1969), Herzberg (1950, 1992), Banwell (1985), Hollas (1992), Haken and Wolf (1995, 1998, 1996, 2002), Kowalczyk (2000), and Linne 2002, to mention but a few. The following works, dealing specifically with the various properties of water, in particular the interaction of water molecules with electromagnetic radiation, are also deserving of attention: Eisenberg and Kauzmann (1969), Lemus (2004), and Chaplin (2006), again, to mention just three. Moreover, Bernath (2002) provides a detailed review of the subject literature, and in the course of this chapter we cite yet other works.

2.1.1 *Vibrational-Rotational Absorption Spectra*

Water molecules appear to be the most important ones involved in the process of solar energy absorption because of their untold numbers in the ocean and atmosphere, as well as their crucial optical properties. Of particular significance in Nature is the very strong absorption by water molecules of infrared radiation (IR), as a result of transitions between the vibrational-rotational energy states in these molecules. According to our estimates, this IR absorption by water molecules is equivalent to some 60% of the total solar radiation energy absorbed in the Earth's epigeosphere (i.e., around 70% of the energy absorbed in the atmosphere and some 50% of the energy absorbed in the sea).

How the water molecule interacts with electromagnetic radiation depends closely on its physical properties. Important in this respect are its geometrical parameters (the positions of the atoms vis-à-vis one another and the configuration of electrons), as well as its dynamic, electrical, and magnetic properties. Table 2.1 lists many of these properties that are directly related to the optical properties of water. We refer frequently to this table in the present (2.1) and the next section (2.2).

As we can see in Table 2.1 (items (1) to (3)), the triatomic molecule of water (H_2O) has a nonlinear structure: the respective distances between the atoms of oxygen and hydrogen and between the two hydrogen atoms are $\bar{d}_{OH} \approx 9.57$ 10^{-11} m and $\bar{d}_{HH} \approx 1.54\ 10^{-10}$ m (see also Figure 2.4a). These distances are the ones prevailing in the equilibrium state, when the angle HOH $\bar{\alpha}_{HOH} \approx$ 104.5°. These parameters define the geometry of the water molecule, in which a rotation through an angle π or 2π around the axis of rotation in the planes

TABLE 2.1. Selected physical properties of the water molecule $^1H_2{}^{16}O$, governing its interaction with electromagnetic radiation or associated with its optical properties.

No.	Name or symbol (explanation)	Value: In common units	Value: In SI units
-1-	-2-	-3-	-4-
	Geometrical Parameters		
1	Mean OH bond length in the ground state $\bar{d}_{OH}$	0.9572 (±0.0003) Å	9.572 (±0.003) × 10^{-11}m
2	Mean HOH bond angle in the ground state $\bar{\alpha}_{HOH}$	104.52° (± 0.05°)	1.82 rad (± 8.43 × 10^{-4} rad)
3	Mean distance between H atoms in the ground state $\bar{d}_{HH}$	c. 1.54 Å	c. 1.54 × 10^{-10} m
	Dynamic Parameters		
4	Inert molecular weight m_{H_2O}	2.9907243 × 10^{-23} g	2.9907243 × 10^{-26} kg
5	Moments of inertia in the ground state(*): $I^y_{H_2O}$	2.9376 × 10^{-40} g cm^{-2}	2.9376 × 10^{-39} kg m^{-2}
6	$I^z_{H_2O}$	1.959 × 10^{-40} g cm^{-2}	1.959 × 10^{-39} kg m^{-2}
7	$I^x_{H_2O}$	1.0220 × 10^{-40} g cm^{-2}	1.0220 × 10^{-39} kg m^{-2}
	Electric and Magnetic Properties		
8	Relative permittivity (dielectric constants) ε	Gas: 1.0059 (100°C, 101.325 kPa) Liquid: 87.9 (0°C), 78.4 (25°C), 55.6 (100°C) Ice Ih: 99 (–20°C), 171 (–120°C)	
9	Relative polarizability α/ε_0 (where α = polarizability, ε_0 = permittivity of a vacuum)	1.44 × 10^{-10} m^3	
10	Dipole moment in the equilibrium state p_p	Gas: 1.854 D (debye) Liquid (27°C): 2.95 D Ice Ih: 3.09 D	6.18 × 10^{-30} A s m 9.84 × 10^{-30} A s m 10.31 × 10^{-30} A s m
11	Volume magnetic susceptibility $\chi = \mu - 1$ (where μ - relative magnetic permeability) at 20°C	–7.19 × 10^{-7} [dimensionless] (cgs convention)	–9.04 × 10^{-6} [dimensionless]

TABLE 2.1. Selected physical properties of the water molecule $^{1}H_2{}^{16}O$, governing its interaction with electromagnetic radiation or associated with its optical properties.—Cont'd.

No.	Name or symbol (explanation)	Value: In common units	Value: In SI units
-1-	*-2-*	*-3-*	*-4-*
	Characteristic Molecular Energies		
12	Bond energy between the constituent atoms in the molecule at temperature 0K	–9.511 eV	-1.52×10^{-18} J
12a	As above, at temperature 25°C	–10.09 eV	-1.62×10^{-18} J
13	Ground state vibrational energy	+0.574 eV	9.20×10^{-20} J
14	Energy of electronic bonds (the difference (12) – (13))	–10.085 eV	-1.62×10^{-18} J
15	The sum of the energies of the discrete constituent atoms in the ground state	–2070.46 eV	-3.32×10^{-16} J
16	The total energy of the molecule at temperature 0 K (sum of (14) + (15))	–2080.55 eV	-3.33×10^{-16} J
16a	The kinetic energy input (equal to-(16))	+2080.55 eV	3.35×10^{-16} J
16b	The potential energy input (equal to 2(16) – (17))	–4411.3 eV	-7.07×10^{-16} J
17	Nuclear repulsion energy	+250.2 eV	4.01×10^{-17} J
18	Electronic excitation energy at light wavelength λ = 124 nm	c. 10.0 eV	c. 1.60×10^{-18} J
19	Ionization potential first (I)	12.62 eV	2.02×10^{-18} J
20	Second (II)	14.73 eV	2.36×10^{-18}
21	Third (III)	16.2 ±0.3 eV	2.60×10^{-18} J $\pm 4.81 \times 10^{-20}$ J
22	Fourth (IV)	18.0 ± 0.3 eV	2.88×10^{-18} J $\pm 4.81 \times 10^{-20}$ J
23	Total H–O bond energy at temperature 0K (equal to 1/2 (12))	–4.756 eV	-7.62×10^{-19} J
24	Dissociation energy of H–O bond at temperature 0 K	4.40 eV	7.05×10^{-19} J
25	Dissociation energy of H–OH bond at temperature 0 K (equal to-(12) – (24))	5.11 eV	8.19×10^{-19} J
26	Energy of the lowest vibrational transition	0.198 eV	3.17×10^{-20} J
27	Typical energy of a rotational transition	0.005 eV	8.01×10^{-22} J
28	Change of internal energy per molecule during the formation of water vapor at boiling point	0.39 eV	6.25×10^{-20} J
29	Change of internal energy per molecule during the formation of type Ih ice at temperature 0°C	–0.06 eV	-9.61×10^{-21} J
30	Change of internal energy per molecule during the transition from type Ih ice to type II ice	–0.0007 eV	-1.12×10^{-22} J

(Continued)

TABLE 2.1. Selected physical properties of the water molecule $^1H_2{}^{16}O$, governing its interaction with electromagnetic radiation or associated with its optical properties.—Cont'd.

No.	Name or symbol (explanation)	Value: In common units	Value: In SI units
-1-	*-2-*	*-3-*	*-4-*
	$^1H_2{}^{16}O$ Occurrence on the Background of Molar Isotopic Composition of Natural Waters (according to VSMOW[a])		
		Percent of total mass	
31	$H_2{}^{16}O$	99.7317	
32	$H_2{}^{17}O$	0.0372	
33	$H_2{}^{18}O$	0.199983	
34	$HD^{16}O$	0.031069	
35	$HD^{17}O$	0.0000116	
36	$HD^{18}O$	0.0000623	
37	$D_2{}^{16}O$	0.0000026	
38	$HT^{16}O$	variable trace	
39	$T_2{}^{16}O$	~ 0	

After various authors, cited in Eisenberg and Kauzman (1969), Haken and Wolf (1995,1998), Dera (2003), and Chaplin (2006); among others.

* Moment of inertia: (y) = relative to the *y*-axis passing through the center of mass of the molecule and perpendicular to the HOH plane; (z) = relative to the axis bisecting the angle α_{HOH}; (x) = relative to the axis perpendicular to the (*y,z*) plane and passing through the center of mass.

[a] The Vienna Standard Mean Ocean Water (VSMOW) for a general number of hydrogen atoms contains 99.984426% atoms of ^1H, 0.015574% atoms of ^2H (D), and 18.5×10^{-16}% atoms of ^3H (T), and for a general number of oxygen atoms contains 99.76206% atoms of ^{16}O, 0.03790% atoms of ^{17}O and 0.20004% atoms of ^{18}O (see, e.g., Chaplin (2006)).

of symmetry (the plane of the molecule, and the plane perpendicular to it, passing through the oxygen atom) do not affect its configuration. Determined with respect to these axes of rotation, the three moments of inertia of the water molecule are different (see items (5)–(7) in Table 2.1). It is for these reasons that we replace the description of the rotational motion of the water molecule by a quantum description of the motion of an asymmetrical top. This approach makes it very much easier to describe the purely rotational spectrum of the vibrations of the water molecule.

The asymmetrical structure of the water molecule, which can be likened to a three-dimensional anharmonic vibrator, also affects its vibrational states. It is responsible for the fact that water molecules have a high, permanent dipole moment: $p_p = 1.854$ debye in the equilibrium state (see item (10) in Table 2.1). The value of this dipole moment changes due to the interaction of interatomic forces, which alter the individual interatomic distances in the molecule and also the angle α_{HOH} between the OH bonds. The effect of these interactions manifests itself as vibrations of the atoms around their equilibrium positions in the molecule's electric field; generally characteristic of a particular molecule, this effect depends on its structure. The number *f* of normal vibrations of a molecule depends on the number of its internal degrees of freedom *N*:

f = 3N–5 for linear molecules.

f = 3N–6 for nonlinear molecules.

In the water molecule (N = 3) there are thus three "modes" of normal vibrations, described by three vibrational quantum numbers: υ_1 (mode I), υ_2 (mode II), and υ_3 (mode III); see Figure 2.1. As this figure shows, the vibrational modes characteristic of the water molecule differ. The excited vibrations of mode I are symmetric stretch vibrations, whereas those of mode II are deformation vibrations, causing the molecule to bend. Mode III vibrations are asymmetric stretch vibrations.

We describe the vibrational energy states of the water molecule by stating the values of all three of its vibrational quantum numbers: υ_1, υ_2, and υ_3. In the ground (not excited) vibrational state the values of all three quantum numbers are zero, so that we can write this state as (0,0,0). The vibrational

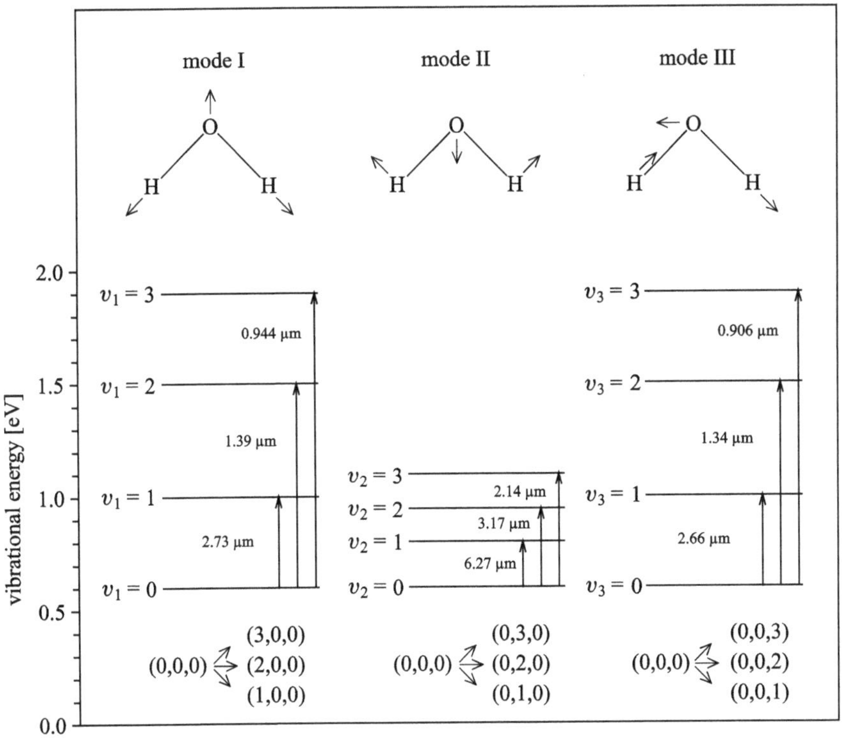

FIGURE 2.1. Normal vibration modes of a water molecule and its characteristic vibrational energy levels in the ground state, and in selected vibrational excited states described by quantum numbers υ_1, υ_2, and υ_3. The figure also gives the approximate wavelengths of light [μm] absorbed by such a molecule during its transition from the ground state to these excited states. The exact values of the vibrational energies shown in the figure are given in Table 2.2.

TABLE 2.2. Vibrational energies of the water molecule E_{VIB} in the ground state and selected simple excited states.[a]

Quantum number υ_1 or υ_2 or υ_3	Mode I (υ_1) E_{VIB} [eV]	Mode II (υ_2) E_{VIB} [eV]	Mode III (υ_3) E_{VIB} [eV]
-1-	-2-	-3-	-4-
0	0.574	0.574	0.574
1	1.03	0.772	1.04
2	1.47	0.965	1.5
3	1.89	1.15	1.94
4	2.34	1.33	2.38
5	2.74	1.51	2.80
6	3.12	1.67	3.21
7	3.55	1.83	3.61

Calculated from data in Eisenberg and Kauzmann (1969) and Lemus (2004).
[a] That is, for the vibrations of one mode.

energy of the water molecule in this state is $E_{VIB} = 0.574$ eV (i.e., 9.20×10^{-20} J). In the excited states of the molecule, its vibrational energy is greater, every higher level of excitation being described by the corresponding higher value of the vibrational quantum number. This is exemplified in Table 2.2, which gives the vibrational energies of the water molecule in various "simple" excited states, that is, when it is vibrating in only one of its modes. The data in Table 2.2 show that these energies vary for different modes. The most highly energetic ones are the asymmetric stretch vibrations, that is, mode III. Only slightly less energetic are the symmetric stretch vibrations (mode I). Finally, the vibrations with the lowest energy (less than the previous two types by c. 30–60%, depending on the level of excitation) are the mode II deformation vibrations. These differences exert a fundamental influence on the position of the light absorption spectral bands associated with the molecule's different vibrational modes (see below), these bands arising as a result of transitions between lower and higher vibrational-rotational energy levels. These transitions can be divided into a number of groups, which we now briefly discuss.

Fundamental Transitions

Transitions from the ground state to the first excited state in a given mode are recorded in the light absorption spectrum as fundamental absorption bands, which can be written as follows:

$$(0,0,0) \rightarrow (1,0,0) \text{ mode I}$$

$$(0,0,0) \rightarrow (0,1,0) \text{ mode II}$$

$$(0,0,0) \rightarrow (0,0,1) \text{ mode III}$$

If we know the vibrational energy of the ground state E_{VIB1} and the energies of the first excited state E_{VIB2} for each of the three vibrational modes (see Table 2.2), we can easily work out the positions in the spectrum of these three fundamental absorption bands for the water molecule from the differences between these energies.[1] In the spectrum, the fundamental light absorption bands for the vibrational modes I, II, and III of the water molecule lie in the vicinity of the wavelengths $\lambda = 2.73$ μm, 6.27 μm, and 2.66 μm, respectively. As we can see, all three bands lie in the IR region, whereby the longest-wave band (i.e., with the lowest energy of absorbed photons) is due to mode II vibrations. In contrast, the changes in the energy states of the water molecule performing mode I or mode III vibrations cause radiation of a much shorter wavelength, that is, much higher-energy photons to be absorbed (or emitted).

In addition to these three fundamental bands of light absorption or emission, corresponding to molecular transitions from the ground state to the first excited state, or back from the latter state to the former (i.e., transitions for which the condition $\Delta\upsilon_1 = \pm1$, $\Delta\upsilon_2 = \pm1$, or $\Delta\upsilon_3 = \pm1$ is satisfied), the absorption spectrum of the water molecule reveals a whole series of further bands due to energy transitions between various vibrational levels. For analyzing their origin and characteristic features, it is convenient to distinguish four categories of such transitions, namely:

- Harmonic transitions (also known as overtones) from the ground state
- Combination transitions from the ground state
- Harmonic transitions between excited states only
- Combination transitions between excited states only

We now proceed to discuss the meanings of these concepts and the positions of the absorption bands to which they give rise.

Harmonic and Combination Transitions from the Ground State

In the water molecule, inasmuch as it is anharmonic, further transitions are allowed from the ground state to higher excited states in a given mode (i.e., as a result of which one of the three vibrational quantum numbers changes: $\Delta\upsilon_1$ or $\Delta\upsilon_2$ or $\Delta\upsilon_3 > 1$. As a consequence of these transitions, harmonic absorption bands are formed, also known as overtones. We can write these transitions as above, but with higher values of the quantum numbers $\upsilon'_1, \upsilon'_2, \upsilon'_3 = 2, 3, \ldots$; for example, (0,0,0) → (0,2,0) denotes a mode II transition from the ground state to the first state of harmonic vibrations The quantum mechanical selection rules also allow transitions in the water molecule from the ground state to narrower vibrational energy states with a simultaneous change of more than one vibrational quantum number. An example of such a change is (0,0,0) → (3,1,1), which means that on absorbing a photon, the molecule passes from the ground state to an excited state that is a mixture of

---

[1] The wavelength of a photon absorbed (or emitted) by a molecule is given by the obvious relationship $\lambda = hc/(E_{VIB2} - E_{VIB1})$, derived from Equation (2.2), where E_{VIB1} is the ground state energy (i.e., (0,0,0); E_{VIB2} is the excited state energy (i.e., in this particular case (1,0,0), (0,1,0), or (0,0,1)).

all three vibrational modes, because all three quantum numbers have changed by $\Delta \upsilon_1 = 3$, $\Delta \upsilon_2 = 1$, $\Delta \upsilon_3 = 1$. Transitions of this type give rise to so-called combination absorption bands.

There may be a large number of such harmonic and combination absorption bands. In practice we find several tens of them in any spectrum (Eisenberg and Kauzmann 1969, Lemus 2004). The more important of these bands, recorded experimentally, are listed in Table 2.3 together with the definition of

TABLE 2.3. Vibrational bands of light absorption by H_2O molecules excited from the ground state.

Quantum numbers of excited states $\upsilon_1, \upsilon_2, \upsilon_3$	Wavelength [μm]	Quantum numbers of excited states $\upsilon_1, \upsilon_2, \upsilon_3$	Wavelength [μm]
-1-	*-2-*	*-1-*	*-2-*
0,1,0	6.27	1,2,2	0.733
0,2,0	3.17	2,2,1	0.732
1,0,0	2.73	1,7,0	0.732
0,0,1	2.66	2,0,2	0.723
0,3,0	2.14	3,0,1	0.723
1,1,0	1.91	0,7,1	0.723
0,1,1	1.88	1,2,2	0.719
0,4,0	1.63	0,2,3	0.711
1,2,0	1.48	4,0,0	0.703
0,2,1	1.46	1,0,3	0.698
2,0,0	1.39	0,0,4	0.688
1,0,1	1.38	1,5,1	0.683
0,0,2	1.34	1,3,2	0.662
0,5,0	1.33	2,3,1	0.661
1,3,0	1.21	2,1,2	0.652
0,3,1	1.19	3,1,1	0.652
2,1,0	1.14	0,3,3	0.644
1,1,1	1.14	4,1,0	0.635
0,6,0	1.13	1,1,3	0.632
0,1,2	1.11	3,2,1	0.594
0,4,1	1.02	2,2,2	0.594
2,2,0	0.972	3,0,2	0.592
1,2,1	0.968	2,0,3	0.592
0,2,2	0.950	4,2,0	0.580
3,0,0	0.944	1,2,3	0.578
2,0,1	0.942	5,0,0	0.573
1,0,2	0.920	4,0,1	0.572
0,0,3	0.906	1,0,4	0.563
1,3,1	0.847	3,3,1	0.547
1,1,2	0.824	3,1,2	0.544
2,1,1	0.823	2,1,3	0.544
1,1,2	0.806	4,1,1	0.527
0,1,3	0.796	3,0,3	0.506
2,4,0	0.757	5,0,1	0.487
1,4,1	0.754	3,1,3	0.471
0,4,2	0.744	4,0,3	0.444

- Based on data gleaned from Lemus (2004).

the type of transition. We should bear in mind, however, that the intensities of the spectral lines in bands corresponding to harmonic vibrations and also in combination bands, are several orders of magnitude less than the intensities of the fundamental lines.

As we can see from Table 2.3, these harmonic and combination absorption bands of the water molecule lie in the near-infrared region of the spectrum, and some of them encroach into the visible region. This means that such photons are absorbed in the transitions, whose energies exceed those of the photons absorbed during fundamental transitions. Absorption of these photons is reflected in the spectrum by the fundamental absorption bands in the vicinity of wavelengths 2.73 μm (mode I), 6.27 μm (mode II), and 2.66 μm (mode III). These last wavelengths, characteristic of the fundamental bands of light absorption by the water molecule, constitute the longwave boundaries (at the longwave end of the spectrum) of the sets of wavelengths corresponding to all the possible absorption bands due to molecular transitions from the ground state to diverse excited states.

On the other hand, at the shortwave end, there is no such boundary due to vibrational transitions. Theoretically, however, we could expect there to be, in this region of the spectrum, shortwave boundaries connected with molecular dissociation, separating the several vibrational-rotational absorption spectra from the continuum absorption spectrum. These boundaries are delineated by the wavelengths of photons, whose absorption (associated with the transition from the ground vibrational state to an excited state with quantum numbers taking infinitely large values) causes the water molecule to dissociate, or more precisely, causes first the H–OH bond, then the H–O bond, to break down (see Table 2.1, items 24 and 25). These are waves of length c. 0.28 μm breaking the H–O bond (dissociation energy c. 4.4 eV) and c. 0.24 μm for the H–OH bond (dissociation energy c. 5.11 eV). On the longwave side of these boundaries the absorption spectrum should consist of separate bands, corresponding to the excitation of successively higher vibrational states. On the shortwave side, however, it should be a continuum, inasmuch as there can be no question of any quantized, discrete energy levels being present in such a configuration. Quite simply, the excess energy of the absorbed photon, beyond that required to dissociate a molecule, may be converted into the kinetic energy (of any value) of these dissociated fragments of the molecule.

Theoretically, therefore, we can also expect there to be vibrational-rotational absorption bands due to transitions of the water molecule to very high vibrational energy states, lying not only in the visible region of electromagnetic waves, but also in the ultraviolet. Empirically, however, such absorption bands are not recorded for water. This is very likely because they are of very low intensity, because the probability of energy transitions of the molecule decreases sharply as the energy of the photons increases. Direct photodissociation of the water molecule only by vibrational-rotational excitation, in the absence of electronic excitation, is so very unlikely as to be practically impossible. Being a single-photon process, it is forbidden by the selection rules for vibrational transitions, which only allow transitions involving a small change in the vibrational quantum numbers. It is also forbidden by the Franck–Condon principle (see, e.g., Barrow (1969) and Haken and Wolf (1995)). Photodissociation is, however, possible as a result of the absorption of high-energy photons, which give rise to transitions between the electronic energy states of the molecule. We return to the problem of photodissociation in Section 2.1.2.

Harmonic and Combination Transitions Between Excited States

Apart from the aforementioned fundamental vibrational transitions and their overtones and combination transitions from the ground state, whose absorption bands lie in the near-IR and visible regions of the spectrum—$\lambda \leq 6.27$ µm (i.e., the fundamental band of mode II)—the water molecule can, as a result of vibrational transitions, also absorb radiation of a longer wavelength, in the region of $\lambda > 6.27$ µm. The differences in vibrational energy between successive, ever higher vibrational states diminish with the increasing quantum numbers characterizing these states (see, for instance, the successive vibrational energies given in Table 2.2). Therefore, if vibrational transitions, both harmonic and combination, are going to take place solely between excited states, they can be induced not only by photons with energies higher (shorter wavelengths) than those required for transitions from the ground state (because the quantum numbers of ground states are sufficiently low and those of the final states sufficiently high), but also by lower-energy photons (i.e., longer wavelengths;[2] because the quantum numbers of the initial and final states are sufficiently high and the differences $\Delta \upsilon$ sufficiently low).

In practice, however, transitions between excited states are far less probable than fundamental transitions or overtones and combination transitions from the ground state. Hence, the absorption bands due to transitions between excited states are far less intense. This is because under normal illumination conditions (e.g., when the sea is illuminated by daylight) the number of molecules not excited (or in very low states of excitation) far exceeds the number of highly excited molecules. So the probability of "coming across" (and absorbing a photon from) a highly excited molecule is many orders of magnitude lower than that of coming across (and absorbing a photon from) a molecule in the ground or a low excited state.

Transitions Between the Vibrational States of Different Isotopes of Water

We should also mention that in natural aquatic environments, or other environments containing water in different states of aggregation, in addition to the

---

[2] For example, the following simple transitions for the case of mode II: (0,1,0) → (0,2,0), and (0,2,0) → (0,3,0), and so on, are due to the absorption of photons of 6.42 µm, 6.57 µm, and so on, that is, of a longer wavelength than the fundamental band at 6.27 µm.

TABLE 2.4. Positions of light absorption bands of isotopic variants of water molecules on excitation from the ground state.

Quantum numbers of excited states	Wavelength λ [μm]					
$\upsilon_1, \upsilon_2, \upsilon_3$	$H_2{}^{16}O$	$H_2{}^{17}O$	$H_2{}^{18}O$	$HD^{16}O$	$D_2{}^{16}O$	$T_2{}^{16}O$
-1-	*-2-*	*-3-*	*-4-*	*-5-*	*-6-*	*-7-*
0,1,0	6.27	6.28	6.30	7.13	8.49	10.05
1,0,0	2.73	2.74	2.74	3.67	3.47	4.48
0,0,1	2.66	2.67	2.67	2.70	3.59	4.23
0,2,0	3.17	—	—	3.59	—	—
0,1,1	1.88	—	—	1.96	2.53	—
0,2,1	1.45	—	—	1.55	1.96	—
1,0,1	1.38	—	—	1.56	1.86	—
1,1,1	1.28	—	—	—	1.53	—
2,0,1	0.942	—	—	—	1.27	—

Based on data from: Eisenberg and Kauzmann (1969) and Chaplin (2006) (http://www.lsbu.ac.uk/water/vibrat.html).

overwhelming numbers of water molecules H_2O consisting of the most common isotopes of hydrogen and oxygen (^1H, ^{16}O), there are, although in very much smaller quantities, also water molecules made up of the heavier isotopes of hydrogen and oxygen (e.g., $^1H_2{}^{17}O$; $^1H_2{}^{18}O$; $^1H^2D^{16}O$; $^2D_2{}^{16}O$; $^3T_2{}^{16}O$). It turns out that the parameters of such molecules differ dynamically from those of ordinary water $^1H_2{}^{16}O$. Consequently, the excited state vibrational energies of these heavier variants of water are generally lower. This quite substantially increases the length of the light waves emitted or absorbed as a result of transitions between their vibrational states. Examples of such transitions can be found in Table 2.4.

A Simple Analytical Description of the Vibrational States of the Water Molecule

An exact definition of the vibrational energies of the various possible states of the molecule, and hence the parameters (energy, wave number, wavelength) of the photons absorbed or emitted as a result of changes in these states, demands time-consuming model quantum-mechanical calculations,[3] or else complex experimental procedures. As Benedict et al. (1956) demonstrated (see also Eisenberg and Kauzmann (1969)), however, there is a straightforward empirical formula for water and its isotopic variants, which describes with good accuracy these vibrational energies in the ground state and in a

[3] An example of such calculations of arbitrary vibrational states of the $H_2{}^{16}O$ molecule are those by Lemus (2004) using a model he derived himself. In his work he gives a vibrational description of $H_2{}^{16}O$ in terms of Morse local oscillators for both bending and stretching degrees of freedom.

whole range of excited states. This applies to simple excited states (when only one mode of vibration is excited), and also to mixed states (when various arbitrary configurations of two or all three modes are excited) with low or medium states of excitation, such as may occur under natural conditions. This formula makes the vibrational energy of a water molecule E_{VIB} dependent on the quantum numbers υ_1 υ_2 υ_3 and takes the form:

$$E_{VIB}(\upsilon_1,\upsilon_2,\upsilon_3) = hc\left[\sum_{i=1}^{3}\omega_i\left(\upsilon_1+\frac{1}{2}\right) + \sum_{i=1}^{3}\sum_{k\geq i}^{3} x_{i,k}\left(\upsilon_i+\frac{1}{2}\right)\left(\upsilon_k+\frac{1}{2}\right)\right], \qquad (2.3)$$

where: h is Planck's constant; c is the velocity of light in a vacuum; i is the number of the vibrational mode; k is a natural number taking values from i to 3; and ω_i [cm^{-1}] and $x_{i,k}$ [cm^{-1}] are empirical constants whose values for the water molecule and selected variants of it are given in Table 2.5.

The terms ω_i appearing in expression (2.1.3) have the dimension [cm^{-1}] and describe the component frequencies (more precisely, the wave numbers) of the harmonic vibrations of the bonds in the water molecule, whereas the terms $x_{i,k}$ (expressed in [cm^{-1}]) are anharmonic constants describing the effect (on the resultant frequency of vibrations) of deviations of a real anharmonic vibrator from the oscillations of a harmonic vibrator. The above formula may be very useful in practice, as it allows us to define simply the energies of the various possible vibrational states of the water molecule, and hence, also the energies, frequencies, and wavelengths of the photons absorbed or emitted during transitions of the molecule between these states.

TABLE 2.5. Values of the constants appearing in Equation (2.3) describing the vibrational state energies of water molecules H_2O, D_2O, and HDO.

Molecule	Constant harmonic and anharmonic frequencies [cm^{-1}]								
	ω_1	ω_2	ω_3	x_{11}	x_{22}	x_{33}	x_{12}	x_{13}	x_{23}
-1-	-2-	-3-	-4-	-5-	-6-	-7-	-8-	-9-	-10-
H_2O	3832.17	1648.47	3942.53	−42.576	−16.813	−47.566	−15.933	−165.824	−20.332
HDO	2824.32	1440.21	3889.84	−43.36	−11.77	−82.88	−8.60	−13.14	−20.08
D_2O	2763.80	1206.39	2888.78	−22.58	−9.18	−26.15	−7.58	−87.15	−10.61

Based on Benedict et al. (1956).

Rotational Transitions

Because molecular rotations are not a factor that noticeably affects the resultant absorption spectra of liquid water in the sea, we deal with this particular problem here rather briefly.

As we said at the beginning of this section, transitions between the various vibration states of a molecule are accompanied by changes in its rotational states. If we use a spectrophotometer with sufficient resolving power to record IR absorption spectra of molecules in the gas state (i.e., separate molecules), we shall see that these spectra are not smooth, but take the form of bands with an extremely intricate structure. Resulting from changes in the rotational states of molecules, these bands appear both on the longwave side of the central line corresponding to a purely vibrational transition, that is, when the rotational energy decreases (the P($\Delta J = -1$) branch), and also on the shortwave side of this line, that is, when the rotational energy increases following the absorption of these waves (the R($\Delta J = +1$) branch). It is often the case that the lines of these bands are even more intense than the central line, corresponding to a purely vibrational transition, with no change in the rotational energy (band Q ($\Delta J = 0$). Figure 2.2 illustrates such examples for water vapor in the atmosphere. In condensed phases (liquid water, ice), however, where the molecules are acted upon by the forces of their interaction, the rotational energy levels are subject to line broadening. Hence in liquid water, the rotational structure in vibrational absorption bands is scarcely visible, if at all.

The limiting cases of transitions between the vibrational-rotational energy states of molecules are those where the vibrational quantum numbers do not change; only the rotational states of the molecules do so. Then we have a purely rotational transition. Because the fundamental rotational energies of molecules are usually c. 100 times smaller than their vibrational energies, such rotational transitions involve the absorption or emission of low-energy

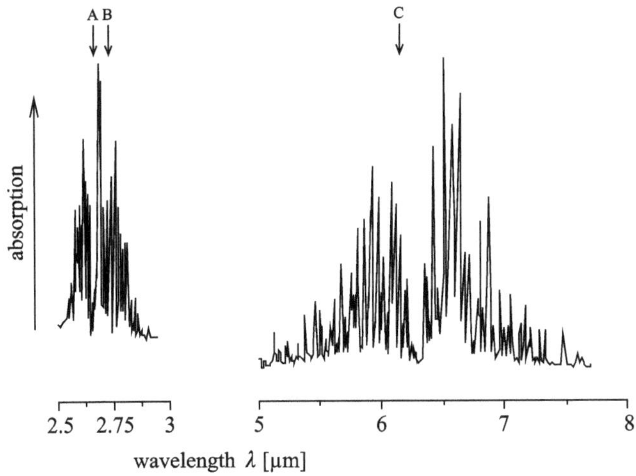

FIGURE 2.2. Absorption spectrum of water vapor corresponding to the three principal vibrational transitions and showing part of the rotational structure: A, 2.66 μm band, corresponding to mode III, that is, (0,0,0) → (0,0,1); B, 2.73 μm band, corresponding to mode I, that is, (0,0,0) → (1,0,0); C, 6.27 μm band, corresponding to mode II, that is, (0,0,0) → (0,1,0). (Adapted from Banwell (1985).)

radiation. The photon energies required to induce such transitions in water molecules are of the order of $10^{-2} - 10^{-3}$ eV, (i.e., light in the far IR). Hence, the rotational bands of light absorption for these molecules lie in the far infrared, microwave, and radiowave regions of the spectrum. The most intense absorption lines of these bonds appear in the 50 μm region (Eisenberg and Kauzmann 1969). Also characteristic are the absorption bands of quanta of wavelengths 27.9 μm and 118.6 μm, and in the vicinity of 1.64 mm and 1.35 cm (Shifrin 1983b, 1988).

Vibrational-Rotational Absorption Spectra of the Water Molecule

The large number of vibrational-rotational energy transitions in the water molecule allowed by the quantum-mechanical selection rules means that the light absorption spectrum of this molecule over the whole wavelength range of IR and microwaves is an extremely complex one, consisting of many bands of different intensities and widths. This is exemplified in Figure 2.3 by the spectrum of the specific light absorption coefficient of water vapor in the atmosphere, which we defined approximately[4] on the basis of empirical data and information gleaned from the works of the various authors cited in the caption to this figure. The absorption of radiation in rarefied water vapor is practically of the same nature as absorption by discrete molecules of water.

As we can see in Figure 2.3, the most intense bands in the near-IR region are the three fundamental absorption bands corresponding to vibrational-rotational transitions of the water molecule from the ground state to the first excited state in the relevant vibrational mode (see the band marked X and 6.3μ). The most intense and the widest of these three bands is the one corresponding to vibrational-rotational transitions in vibrational mode II, deforming the molecules. The center of this band lies around the wavelength $\lambda = 6.27$ μm. This band possesses a fine structure (not visible on the spectrum), in which there appear a large number of spectral lines varying in width and intensity. These lines, including the weakest ones in this intense band, can be identified in the Earth's atmosphere, even in the so-called "windows" in the absorption spectrum of the atmosphere (Kondratev 1969).

---

[4] Owing to the fine structure of the absorption bands of gases in the IR, which consist of narrow, almost monochromatic, natural absorption lines and narrow "absorptionless intervals" between them, the exact empirical determination of the so-called "logarithmic" coefficients of absorption (the ones applied inter alia in marine optics) is complicated and can be achieved only to a greater or lesser approximation. This is because the widths of these natural absorption lines and "absorption less intervals" in the absorption spectrum are usually much smaller than those of the spectral detection intervals, which are a consequence of the resolving powers of spectrophotometers. With respect to water vapor, these questions are discussed by Kondratev (1969) and Bird and Riordan (1986), among others.

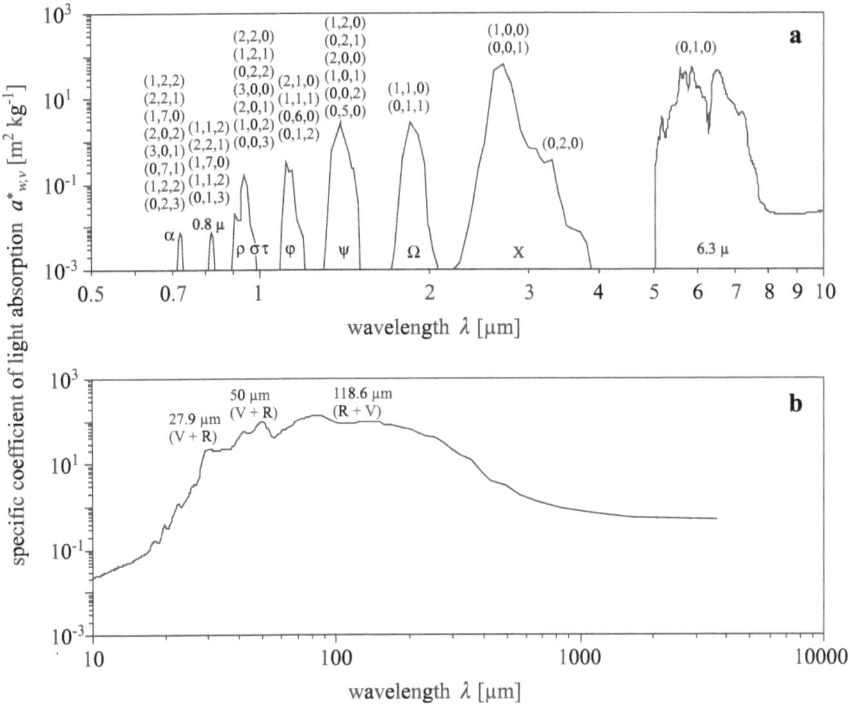

FIGURE 2.3. Spectrum of the approximate specific coefficient of IR and microwave absorption for atmospheric water vapor $a^*_{w,v}$; the details of the rotational structures (as visible on Figure 2.2) have been omitted. (a) standard (atmospheric optical) codes for the various bands are given below the plot; the excitations responsible for absorption (from the ground state) of the various vibrational states (υ_1, υ_2, υ_3) are stated in parentheses above the relevant bands; (b) (V + R) denotes the absorption peaks due to the overlapping of vibrational-rotational and rotational bands; (R + V) denotes the peaks due mainly to rotational transitions. This spectrum was plotted on the basis of empirical data and other information taken from Yamamoto and Onishi (1952), Shifrin (1983b, 1988), Kondratev (1969), Bird and Riordan (1986).

Somewhat less intense and narrower than the previous band is the fundamental absorption band formed as a result of vibrational-rotational transitions in mode III of the water molecule. The center of this band lies in the vicinity of wavelength $\lambda = 2.66$ μm. Near this band we find the fundamental absorption band of the water molecule in the vicinity of $\lambda = 2.73$ μm, corresponding to vibrational-rotational transitions in mode I. These last two bands ((0,0,1) and (1,0,0)), and the harmonic band (0,2,0) in the vicinity of $\lambda = 3.17$ μm, partially overlap, together forming one very wide absorption band of the water molecule in the wavelength range from $\lambda = 2.3$ μm to $\lambda = 3.9$ μm (marked X on the spectrum). This band as a whole is the most characteristic one of the water molecule and plays an important part in Nature in the absorption of near-IR radiation (Hollas 1992).

In the near-IR the water molecule also absorbs radiation of wavelengths $\lambda \approx 0.81$ μm, $\lambda \approx 0.94$ μm, $\lambda \approx 1.13$ μm, $\lambda \approx 1.38$ μm, and $\lambda \approx 1.88$ μm (see the bands denoted by the symbols 0.8μ, ρστ, φ, Ψ, and Ω in Figure 2.3a). The absorption of energy quanta of these wavelengths involves energy transitions of the water molecule, in which, according to the selection rules, the values of at least two vibrational quantum numbers change simultaneously (e.g., the band for $\lambda = 1.88$ μm is associated mainly with the transition $(0,0,0) \rightarrow (0,1,1)$. The widths and intensities of these last absorption bands vary, but they are many times smaller than the three fundamental bands we discussed earlier. Close by this last-mentioned group of bands we find narrow, low-intensity absorption bands corresponding to higher harmonic vibrations of the water molecule (overtones) as well as further allowed combination bands, that is, a combination of the aforementioned vibrational modes of the water molecule and their higher harmonics. Meriting our particular attention are those combination bands lying in the visible part of the spectrum ($\lambda < 0.8$ μm). They form a single, quite strong absorption band in the atmosphere, denoted on the spectrum by the symbol α, with an absorption maximum at $\lambda \approx 718$ nm. This is the only clear visible band of light absorption by the water molecule in the visible range of electromagnetic waves.

We now move on to the absorption of light in the somewhat farther infrared (see Figure 2.3b). Here we find the slight absorption by the water molecule of energy quanta from the wavelength range $12 < \lambda < 20$ μm, which corresponds to higher harmonic vibrations in modes I and II, and is characteristically very variable in intensity, this increasing with the wavelength of the absorbed light in this spectral region. In the wavelength range 20 μm $< \lambda <$ 1 mm the absorption spectrum is practically a continuum, but a few bands of greater intensity, among them, of wavelengths $\lambda = 27.9$ μm, $\lambda = 50$ μm, and $\lambda = 118.6$ μm, do stand out. The continuum absorption spectrum of the water molecule in this range is due to the superposition of vibrational-rotational bands with the spectral lines induced by purely rotational energy transitions. Nevertheless, the two strongest microwave absorption bands due to rotational transitions in the water molecule lie in the vicinity of the wavelengths $\lambda = 1.64$ mm and $\lambda = 13.48$ mm (Shifrin 1983b, 1988) (these last are not shown in Figure 2.3b).

2.1.2 *Electronic Absorption Spectra*

Molecules can absorb or emit radiation not only as a result of changes in their rotational and vibrational energies, as we have just been discussing, but also in consequence of changes in their electronic configurations, and hence, their electronic energy. Energy changes caused by a transition from one electronic state to another are usually quite large, corresponding to the energies of photons in the ultraviolet region, and in the case of the large unsaturated molecules, in the visible region as well. We discuss the formation of absorption spectra of such large molecules in Chapter 3; for the present, we focus our attention on the electronic spectra of the water molecule.

We should expect—as we stated at the beginning of this chapter—that alongside electronic transitions involving such large changes in energy, there would be changes in vibrational energy (much smaller) and changes in rotational energy (even smaller). Indeed, the example of water shows us that electronic transitions do in fact lead to the formation, not of single absorption or emission lines, but of entire complexes of electron-vibrational absorption and emission bands with a highly intricate fine structure.

The energies of the electronic states of a molecule depend on its electronic structure as determined by the electronic structure of its constituent atoms. In order better to illustrate this problem, we now introduce the concepts of atomic orbitals (AO) and molecular orbitals (MO). These orbitals are useful in defining the energy of individual electrons in atoms and molecules. Again, the changes in these electronic energies, specified by the quantum rules of selection, determine the energies of photons absorbed or emitted by given molecules; in other words, they determine which wavelengths of electromagnetic radiation will correspond to the photon energies of the transitions.

The Concept of an Orbital

The quantum-mechanical magnitude characterizing the state of an electron in an atom or molecule is the single-electron wave function $\varphi_e(x,y,z,s)$, called the spin orbital. For the sake of simplicity, we can represent it as the product of the configurational function $\psi_e(x,y,z)$, dependent only on the coordinates (symbols: x, y, z) of its position in space, and the spin (symbol s) function $\xi(s)$, dependent only on the electron's spin s:

$$\varphi_e(x,y,z,s) = \psi_e(x,y,z) \cdot \xi(s). \tag{2.4}$$

The configurational wave function $\psi_e(x,y,z)$ can be determined for any electron in an atom or molecule by assuming the "single-electron approximation" (discussed in greater detail in Section 3.1) and solving the spinless Schrödinger equation.[5] The energy of an electron at every point in space is proportional to $|\psi_e|^2$ and defines the spatial distribution of the electronic charge. For a better picture of the localization of an electron, it is convenient to use the concept of orbitals: AO when we are dealing with the electrons in an atom, and MO if the electrons are those in a molecule. We usually assume an orbital to be that volume of space in which the majority (e.g., 95%) of the electronic charge is concentrated, or to put it another way, that there is a 95% probability

[5] For complex molecules these solutions can be exceedingly time-consuming, which is why we often use suitable approximate methods, some of which are mentioned in due course.

of finding the electron in this space. Orbitals conceived in this way have various shapes and sizes and localizations within the confines of the atom or molecule, depending on the type of atoms or molecular bonds and on the configuration of electrons. In Chapter 3, we discuss many such electronic orbitals typical of atoms and complex molecules, that is to say, the many and diverse configurations of electrons in atoms and molecules absorbing light. Here we begin by presenting first the atomic orbitals of hydrogen and oxygen, and then the orbitals of the water molecule formed from these atoms.

Atomic Orbitals of Hydrogen and Oxygen; Molecular Orbitals of Water

Before describing the electronic states of the water molecule, let us recall briefly the main features and conventions applied to the description of electronic states in atoms. For this we use atomic orbitals, whereby every such orbital is unequivocally defined by the values of the following quantum numbers, which can be zero or an integer.

1. Principal quantum number $n = 1, 2, 3, 4, \ldots$, which is at the same time the number of the orbit or electronic shell.
2. Orbital quantum number $l = 0, 1, 2, \ldots, (n - 1)$, whose increasing values are denoted by the letters *s*, *p*, *d*, ... It is the quantum number of the orbital angular momentum (OAM) of the electron that defines the scalar value of this angular momentum in accordance with the expression: OAM $= \sqrt{l(l + 1)}\, h/2\pi$ (*h* is Planck's constant); to the various letters *l* are attributed subshells (suborbits) denoted successively by the letters *s*, *p*, *d*, ...
3. Magnetic quantum number $m = -l, -l + 1, \ldots, 0, \ldots, +l$, which defines the scalar values of the projections of this orbital angular momentum on to $(2l + 1)$ distinct directions, equal $mh/2\pi$.

However, the energy state of every electron in an atom is characterized unequivocally by four quantum numbers. The first three are the ones just mentioned, *n*, *l*, *m*, characterizing the type of orbital, whereas the fourth quantum number of an electron is its spin number, or simply spin, which can take one of two values: +1/2 or −1/2. This is in agreement with the Pauli exclusion principle in quantum mechanics, according to which each electron in an atom (and also in a molecule) must differ from all the other electrons in the value of at least one of the four quantum numbers describing its state. Likewise, each atomic (and also molecular) orbital may contain at most two electrons with different spins. Notice too, that the four quantum numbers (*n*, *l*, *m*, *s*) so defined unequivocally determine the value of the electronic energy (mentioned earlier) described by the quantum number Λ, which defines the absolute value of the projection of the orbital angular momentum on to the axis of the molecule.

In accordance with this convention, the electronic configurations of the hydrogen and oxygen atoms forming the water molecule in the ground state can be written as:

Hydrogen: $1s^1$;
Oxygen: $[1s^2]$, $2s^2$, $2p_x^2$, $2p_y^1$, $2p_z^1$.

So every configurational code begins with a number equal to the principal quantum number, that is, the number of the orbit. The letter in position two of the code denotes the subshell; that is, it defines the orbital quantum number ($l = 0$ for *s*, $l = 1$ for *p*). Information about the magnetic quantum number is also given here. In the case of the *s* subshell (i.e., 1*s* for hydrogen, and 1*s* and 2*s* for oxygen) there is no extra information, because in this case $l = 0$, and the magnetic quantum number takes the only possible value of $(2l + 1) = 1$. But in the case of the 2*p* subshell (i.e., for $l = 1$) for oxygen, there are $(2l + 1) = 3$ possibilities, represented by the relevant components of the projection: $2p_x(m = 0)$, $2p_y$ $(m = -1)$ and $2p_z(m = +1)$. Finally, the superscript on the right-hand side of the various orbital codes indicates the number of electrons in these subshells (at most equal to two electrons with different spins). Filled inner shells, that is, shells containing their full complement of electrons, are customarily enclosed in square brackets (see the first shell for oxygen, $[1s^2]$).

The approximate shapes and orientations in space of the atomic orbitals of hydrogen and oxygen are illustrated in Figure 2.4b,c. In the same figure (Figure 2.4d,e,f) there is a schematic representation of how the atomic orbitals of the valence electrons of the two hydrogen atoms and the oxygen atom combine to form (mainly) the molecular orbitals of the water molecule. Similar information is given in Table 2.6.

The table specifies the components of the atomic orbitals of hydrogen and oxygen (column 1), the applied symbols[6] (column 3) and selected features (columns 2,4,5). The energy relationships between these orbitals of the water molecule, which can be determined by various quantum-mechanical methods,[7] are illustrated in approximate form in Figure 2.5 (detailed values of the potential energy of electrons in the various molecular orbitals of water are analyzed further and presented in Table 2.7).

In accordance with the information and conventions given in Figure 2.4 and in Table 2.6, we can present the electron configuration of the H_2O molecule in the ground state (in the same way as above for the hydrogen and

[6] For clarity's sake, we should add that the system of symbols used here to describe molecular orbitals is one of many. In the spectroscopic literature we can find many other ways of denoting energy states and molecular orbitals, depending on the quantum-mechanical calculation method or the type of spectroscopic analysis.
[7] One such simple method, described in relation to the H_2O molecule, by Haken and Wolf (1995) among others, involves the application of group theory, which describes the action of so-called symmetry operators on the wave functions of electrons. Another very efficient method is the one using RHF (the restricted Hartree–Fock wave function) (see, e.g., Chaplin (2006)).

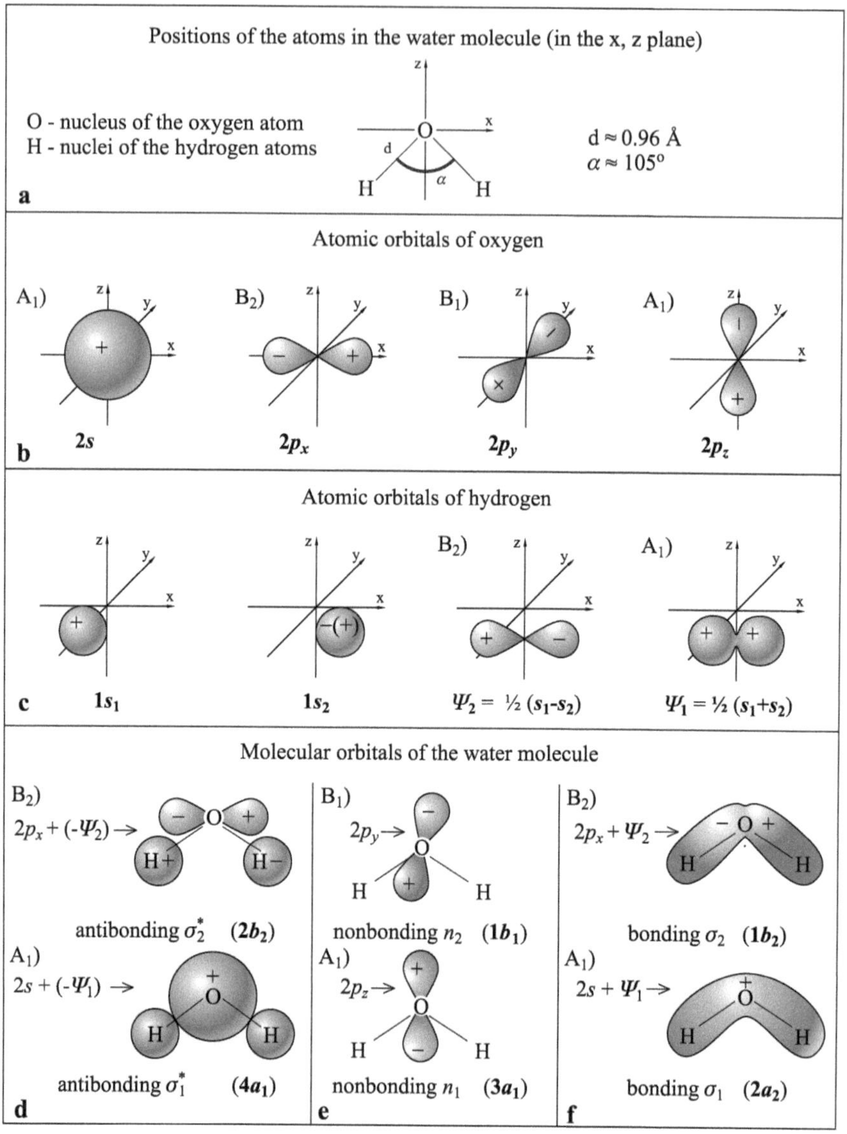

FIGURE 2.4. Sketches of the atomic orbitals of hydrogen and oxygen, and the molecular orbitals of the water molecule (explanation in the text).

oxygen atoms) in the form of the following sequence of orbitals of increasing energy,

$$-\text{water:}\underbrace{[1a_1^2],\ \ [2a_1^2,1b_2^2,3a_1^2,1b_1^2]}_{HOMO},\ \ \underbrace{4a_1 2b_2\ldots}_{LUMO}.$$

TABLE 2.6. List of symbols and features of the molecular orbitals of H_2O (MO) and the input orbitals of hydrogen and oxygen forming them (input AO).

		MO			
Input AO	Symmetry types	Symbols	Functions[a]	Occupations in ground state[b]	
-1-	-2-	-3-	-4-	-5-	
1*s* Hydrogen and $2p_x$ Oxygen	B_2	$2b_2$	σ_2*	LUMO	
1*s* Hydrogen and 2*s* Oxygen	A_1	$4a_1$	σ_1*		
$2p_z$	B_1	$1b_1$	n_2	HOMO	} Valence orbitals
$2p_y$	A_1	$3a_1$	n_1		} Valence orbitals
1*s* Hydrogen and $2p_x$ Oxygen	B_2	$1b_2$	σ_2		} Valence orbitals
1*s* Hydrogen and 2*s* Oxygen	A_1	$2a_1$	σ_1		} Valence orbitals
1*s* Oxygen	A_1	$1a_1$	n_0		

[a] Division according to the bonding ability of the atoms in the molecule (as used later in this book to describe the functions of electrons in large organic molecules): bonding (σ), nonbonding (n), and antibonding (σ*);
[b] Division in terms of the occupation of the various electronic orbitals: highest occupied molecular orbitals (HOMO) or lowest unoccupied molecular orbitals (LUMO).

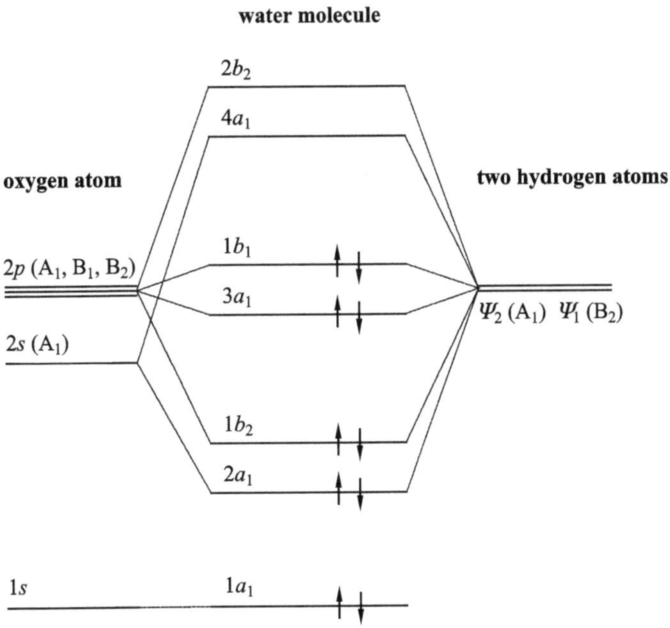

FIGURE 2.5. The energy levels of a water molecule and its constituent atoms: one atom of oxygen and two atoms of hydrogen. The pairs of vertical arrows on the energy diagram of the water molecule illustrate the distribution of electron pairs (with opposite spins) in this molecule in its ground state. Ψ_1 and Ψ_2 denote the electron wave function of the two hydrogen atoms: $\Psi_1 = 1/2(s_1 + s_2)$ and $\Psi_2 = 1/2(s_1 - s_2)$. The symbols in parentheses next to the energy levels denote the relevant types of symmetry (see text).

TABLE 2.7. Energies of the electrons in a water molecule [eV] described by different molecular orbitals and defined by various authors.*

		Orbital	Gas	Liquid	Ice	Orbital	Water: Dimer H_2O–H_2O	Ions: H_3O^+–H_2O C_s Symmetry	Ions: H_3^+–H_2O C_2 Symmetry	Ions: HO^-–H_2O
-1-		-2-	-3-	-4-	-5-	-6-	-7-	-8-	-9-	-10-
A	LUMO	$3b_2$	+28.0[1]							
		$2b_2$	+8.0[1]							
						$4a_1$ proton donor	+7.3[1]	+0.6[1]	+0.5[1]	
		$4a_1$	+6.0[1]			$4a_1$ proton acceptor	+5.3[1]	–0.9[1]	–0.7[1]	+12.6[1]
B	HOMO	$1b_1$	–12.6[2] –14.0[1]	–11.16[8]	–12.3[4,5] –11.8[6]	$1b_1$ proton donor	–12.7[1]	–21.0[1]	–21.9[1]	–3.2[1]
						$1b_1$ proton acceptor $3a_1$ proton donor	–14.1[1]	–22.6[1]	–21.9	–3.2[1]
		$3a_1$	–14.84[8] –14.80[2] –15.0[1]	–13.50[8]	–14.2[4,5]	$3a_1$ proton donor $1b_1$ proton acceptor	–14.8[1]	–23.4[1]	–23.1[1]	–6.0[1]
						Major $3a_1$ proton acceptor Minor $1b_2$ proton donor	–16.3[1]	–26.9[1]	–27.8[1]	–7.0[1]
		$1b_2$	–18.78[8] –18.60[2,8] –19.0[1]	–17.34[8]	–17.6[4,5] –18.0[7]	Major $1b_2$ proton donor Minor $3a_1$ proton acceptor	–18.7[1]	–28.7[1]	–27.8[1]	–7.0[1]
						$1b_1$ proton acceptor	–20.2[1]	–29.0[1]	–28.8[1]	–11.3[1]
		$2a_1$	–32.62[8] –32.60[2] –37.0[1]	–30.90[8]	–31.0[6]	$2a_1$ proton donor $2a_1$ proton acceptor	–35.8[1] –37.4[1]	–44.0[1] –46.3[1]	–44.6[1] –45.5[1]	–25.1[1] –28.7[1]
		$1a_1$	–559[1]	(–559[1])	(–559[1])	0-atom 1*s* proton donor 0-atom 1*s* proton acceptor	–558.4[1] –560.1[1]	–566.7[1] –568.6[1]	–567.6[1] –567.6[1]	–549.0[1] –551.6[1]
C	Rydberg states	$3pb_1$	–2.43[3]							
		$4sa_1$	–2.00[3]							
		$3pa_1$	–2.62[3]							
		$3sa_1$	–5.18[3]							

* References: (1) Chaplin 2006; (2) Banna et al. 1986; (3) based on data in Mota et al. 2005; (4) Henderson 2002; (5) Krischok et al. 2001; (6) Shibaguchi et al.

The first component of each designation stands for the consecutive number of the orbital belonging to a given type of molecular symmetry[8] (A_1 or B_1 or B_2; see column 2 in Table 2.6). The letter in position two of the designation (a_1 or b_1 or b_2) with the subscript repeats the symbol for the type of symmetry. The remaining denotations are as for atomic orbitals (see above).

As we can see, we can distinguish two groups of orbitals: HOMO, the highest-energy occupied molecular orbitals, which in the ground state of the water molecule are occupied by electrons, and LUMO, the lowest unoccupied molecular orbitals, which in the ground state are unoccupied. They come about as the combination of different O and H atomic orbitals (see Figures 2.4 and 2.5), they have various properties, and fulfill various functions (see Column 4 in Table 2.6). The first of these HOMOs, $1a_1$, a nonbonding orbital (n_0), arises from the inner atomic orbital of oxygen ($1s$) and in fact duplicates it (i.e., the symmetry of the electronic charge distributions is practically spherical with respect to the oxygen nucleus). The next HOMO, $2a_1$, is a contribution from the $2s$ orbital of the oxygen atom (and only partially a contribution from the 1s atomic orbitals of hydrogen, or to be exact, their functions Ψ_1; see Figure 2.4c), respectively, and so is approximately spherical. The next three HOMOs—$1b_2$, $3a_1$, and $1b_1$—are orthogonal around the oxygen atom and without obvious sp^3 hybridization[9] characteristics. But the highest-energy HOMO, $1b_1$, is predominantly p_z in character, with no contribution from the hydrogen 1s orbital, and contributes mainly to the so-called "lone pair" effects. It thus has the character of a nonbonding orbital (n_2). Orbitals $2a_1$, $1b_1$, and $3a_1$ all contribute to the O–H bonds; $3a_1$, however, formed as it is mainly from the p_y orbital of oxygen, is only weakly bonding, so we can assume it to be nonbonding (n_1). In contrast, the O–H bond is based mainly on two σ-type bonding orbitals, that is, on orbital $1b_2$, which is strongly bonding (σ_2), and on orbital $2a_1$, which is formally bonding (σ_1), but weakly so.

In addition to these five HOMOs, the next, energetically higher orbitals $4a_1$ and $2b_2$, mentioned in Table 2.6, and also the orbitals of even higher energies, such as $3b_2$, not accounted for in this table, are unoccupied in the ground state, and are therefore classified as LUMOs. They are antibonding orbitals (σ^*-type). $4a_1$ and $2b_2$ are O–H antibonding orbitals. They have the greatest electron densities around the O atom, whereas orbital $3b_2$ has the greatest electron density around the H atom.

---

[8] Representations of A_1, B_1, B_2, and others linked with this type of molecular symmetry are described in the monographs by Barrow (1969) and Haken and Wolf (1995), among others.

[9] By hybridization we mean the situation when a molecular orbital arises when different types of atomic orbital mix, in this particular case the $2p$ orbital of oxygen and the $1s$ orbital of hydrogen.

We should add that we can predict these HOMOs and LUMOs for water fairly easily from the description of electronic wave functions if we avail ourselves of symmetry operators and group theory (see, e.g., Haken and Wolf (1995)). The set of these orbitals is a kind of fundamental base set. But the diversity and complex electronic nature of the water molecule are such that there are yet further series of interactive electronic states (molecular orbitals) resembling the main ones. We address this particular question a little later.

Energy States of Electrons

Tables 2.7A and B, columns 3, 4, and 5, list the electronic energy values characteristic of all three states of aggregation and described by HOMO and LUMO orbitals. These values were determined by various authors, usually with a method employing RHF (restricted Hartree–Fock wave function; see, e.g., Chaplin (2006)). Obviously, the energies of all bonding and nonbonding electrons in HOMO are negative; those in LUMO are positive.

It is worth noting that in a natural environment, apart from single molecules of water, not interacting among themselves, various kinds of supramolecular structures can occur (Eisenberg and Kauzmann 1969, Dera 2003, Chaplin 2006). These larger structures can be electrically neutral (dimers, polymers, crystalline elements), but also electrically charged (ions and their combinations). This applies in larger or smaller measure to water in all its three states: gaseous water (water vapor), liquid water, and ice. It is also the reason for the formation of a complex series of electronic states, quantitatively very different from the states of the single molecule mentioned earlier. Some of these states and their characteristic energy values are also given in Table 2.7, columns 6 to 10.

It is clear from these data that the energy differences between these main HOMO and LUMO electronic energy states of the water molecule and its supramolecular structures are very great: of the order of tens or hundreds of eV. The least of these differences is almost 20 eV. This implies that the absorption or emission of radiation of very short wavelengths, usually shorter or very much shorter than 100 nm, is responsible for electronic transitions between these states. However, as we show in due course, water absorbs, as a result of electronic transitions, not only very short waves (i.e., high-energy photons) but also waves from the $\lambda > 100$ nm range. The absorption of these longer waves may be the result of transitions between the principal energy states, such as are listed in Tables 2.7A and B. We explain below the question of absorption due to electronic transitions of photons from the $\lambda > 100$ nm range.

Interactive Orbitals and Rydberg States

As we stated above, in addition to the main HOMO and LUMO electronic energy states listed in Tables 2.7A and B, the diversity and complex electronic nature of the water molecule means that it is further characterized by a whole series of interactive electronic states (molecular orbitals) similar to the main ones, for example, $3a_1$-like orbitals, $4a_1$-like orbitals, and so on. Among these, the so-called Rydberg states (and their corresponding orbitals) of electrons play a particularly important part in the formation of electronic absorption spectra of water (see, e.g., Haken and Wolf (1996, 2002) and Mota et al. (2005)). They occur in both atoms and molecules, and usually refer to a few (one or two) of the outermost valence electrons. We recall that this is connected with the change in the electronic configuration of an atom (both isolated and as a component of a molecule).

These are highly excited electronic states, described by orbitals whose dimensions are large in comparison with those of the core of the atom or molecule in the ground state. This means that when one of these outermost electrons is excited to a very high energy level, it enters a spatially extensive orbit, that is, an orbital that is situated much farther away from the core than the orbitals of all the other electrons. This excited electron then acts from this great distance upon the atomic or molecular core, consisting of the nucleus (or nuclei) and all the other electrons, as if it were practically a point charge, equal to 1 (+1e), that is, the same charge as that of the hydrogen nucleus. As long as the excited electron does not approach the atomic or molecular core too closely, it behaves as though it belonged to a hydrogen atom. Hence, the behavior of "Rydberg" atoms and molecules is in many respects similar to that of highly excited hydrogen atoms. In particular, the relationship between the electronic energy E_n and the number n (the principal quantum number) of the Rydberg orbit resembles, after correcting for the so-called quantum defect, that for the hydrogen atom and takes the form:

$$E_n = hcR_y \frac{z_e}{(n - \delta)^2} - E_i, \qquad (2.5)$$

where z_e is the core charge ($z_e = 1$), E_i is the ionization energy of a given electron, R_y is 109677.5810 cm^{-1} the Rydberg constant, δ is the quantum defect resulting from the penetration of the Rydberg orbital into the core.

The codes used to describe Rydberg orbitals are also similar to those used for atomic orbitals. First we write the principal quantum number n (i.e., the number of the Rydberg orbit), then the denotation of the subshell ($s, p, d, \ldots$), and finally the type of electron. So for water we have the following.

$n\,s\,a_1$ ($n = 1,2, \ldots$), $n\,p\,a_1$ ($n = 1,2, \ldots$) with respect to the orbital of a valence electron of symmetry A_1;

$n\,s\,b_1$ ($n = 1,2, \ldots$), $n\,p\,b_1$ ($n = 1,2, \ldots$) with respect to the orbital of a valence electron of symmetry B_1.

Studies have shown that in the case of the water molecule, the following four Rydberg states most probably exist (see, e.g., Mota et al. (2005)): $3sa_1$, $3pa_1$, $4sa_1$ and $3pb_1$.

The values of the quantum defect δ appearing in Equation (2.5) depend on two quantum numbers: n (the number of the orbit) and 1 (the orbital quantum number, equivalent to the type of subshell: s, p, d, etc.) and are usually determined empirically. Mota et al. (2005) obtained these values for the above-mentioned frequent states of the water molecule: $\delta(3sa_1) = 1.31$, $\delta(3pa_1) = 0.72$, $\delta(4sa_1) = 1.39$, and $\delta(3pb_1) = 0.65$. The electronic energies characteristic of these Rydberg states of the water molecule, calculated on this basis using Equation (2.5), are given in Table 2.7C. In these calculations it was assumed that the ionization energy with respect to the individual electrons is respectively equal to (see Table 2.1, items 19 and 20) minus the bonding energy of electron $3a_1$, which is c. 14.73 eV for states from the series nsa_1 and npa_1, and minus the bonding energy of electron $1b_1$, which is c. 12.62 eV for states from the series npb_1.

As we can see from the energies of electrons in Rydberg states (Table 2.7C) and in HOMO states $3a_1$ and $1b_1$ (Table 2.7B), the energy differences between these two types of orbitals are relatively small and may explain the formation of absorption (emission) spectra in the $\lambda > 100$ nm range. Transitions between these two types of states do not, however, explain the absorption of higher-energy light quanta ($\lambda < 100$ nm): these are absorbed as a result of the photodissociation, photoionization, and photolysis of water (see below).

Electronic Absorption Spectra of the Water Molecule

The direct empirical study of the electronic absorption spectra of water is an extremely complicated matter, and is, moreover, encumbered with considerable error. This is because these spectra are located in the far ultraviolet, a region difficult to access with classical spectroscopic techniques. The results of measurements in the form of photoabsorption spectra obtained for water vapor by means of synchrotron radiation (Mota et al. 2005), or determined indirectly from spectral data for liquid water on the real and imaginary parts of the dielectric function measured by X-ray scattering (Hayashi et al. 2000) are illustrated in Figures 2.6a–d. The figures show that the electronic photoabsorption spectrum consists of:

- One principal, very wide, practically continuous band[10] with a maximum at c. 65 nm (other authors give a slightly different position for this maximum), which, as research has confirmed, is characteristic of water in all three states of matter.
- A number of distinct discrete features (see the positions of bands A, B, and C in Figure 2.6) located in the spectral region around $\lambda > 100$ nm, which are especially conspicuous for water vapor.

---

[10] Except for certain discontinuities in the region c. 80 nm, which have not been examined in detail; they may well be involved in the photoionization of the electron $3a_1$.

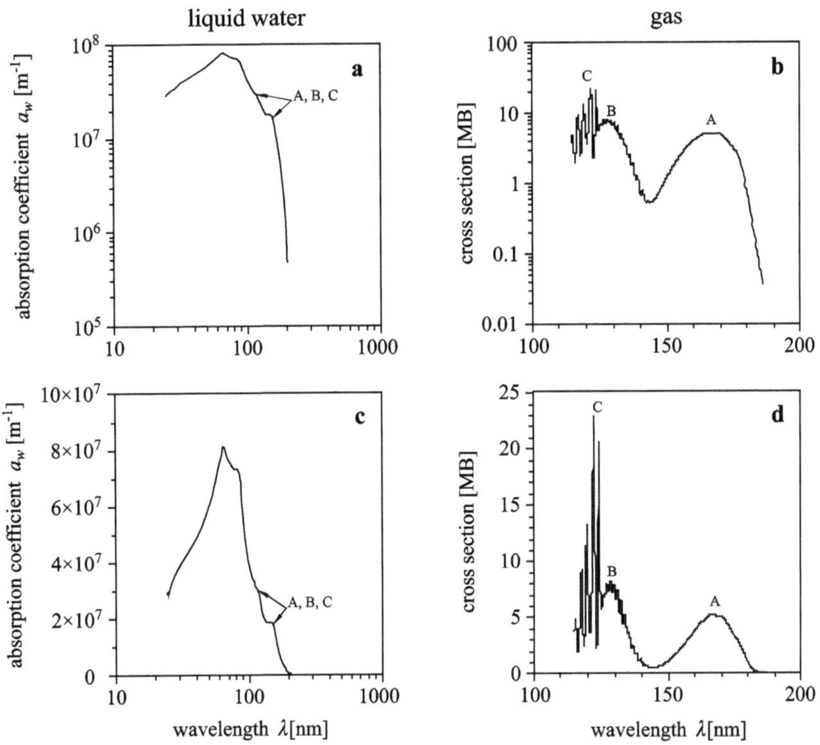

FIGURE 2.6. Electronic UV absorption spectra of water: absorption coefficient spectra of liquid water on a logarithmic scale (a) and a linear scale (c), determined from spectral data of the real and imaginary parts of the dielectric function measured by inelastic X-ray scattering (by Hayashi et al. (2000)); spectral absorption cross-section of water vapor on a logarithmic scale (b) and a linear scale (d), determined using synchrotron radiation (based on Mota et al. (2005)). MB = Mega Barn = 10^{-20} m^2 per 1 molecule.

The reason for the occurrence of this wide, almost continuous, principal photoabsorption band is as follows. It consists mainly of superposed absorption spectra of high-energy photons causing photoionization and photodissociation, which are continuous in the case of energies higher than the ionization and dissociation energies. In the shortwave region of the electromagnetic spectrum (γ and X-radiation), we have continuous absorption of photons photolyzing water.

There is no doubt that further discrete subbands, corresponding to electronic transitions between (suitably distant) discrete energy levels, are also components of this band. Figures 2.6a and c show, however, that these discrete features are not particularly conspicuous (apart from the distinct discontinuity around 80 nm mentioned in footnote 10) in this wide continuous band, which is similar for water in all its states of matter. The huge diversity and spectral differentiation of these discrete features, as well as their vast numbers, are further reasons for their inconspicuousness.

Overlapping this wide continuous band are several, more subtle features that are characteristic of the water molecule, mostly in the gas state (see Figures 2.6 b and d).

A: A relatively wide subband in the 145–180 nm region with a maximum at c. 166.5 nm
B: A relatively wide subband in the 125–145 nm region with a maximum at c. 128 nm
C: A set of narrow bands in the 115 nm–c. 125 nm region

All these three features of the light absorption spectra were obtained empirically by means of synchrotron radiation and analyzed in detail (see Table 2.8) by Mota et al. (2005). Among other things, they showed that subband A (145–180 nm) is the result of the $1b_1 \rightarrow 4a_1$-like orbital transition, which has been shown to dissociate water into OH + H. This band has an electronic-vibrational structure, even though it is very broad and not particularly conspicuous. The positions of these fine-structural features of subband A in the absorption spectrum, along with the more important results of this vibrational analysis are given in Table 2.8A and illustrated in Figure 2.7. This band is also partially overlapped by a Rydberg structure (among others, absorption corresponding to the $1b_1 \rightarrow 3sa_1$ transition; see below).

Subband B (125–145 nm), also involved in photodissociation, is generated by the excitation of electron $3a_1$ and to a lesser extent by the excitation of electron $1b_1$ to the Rydberg level $3sa_1$; these are therefore the $3a_1 \rightarrow 3sa_1$ and $1b_1 \rightarrow 3sa_1$ transitions, that is, from the Rydberg series. Unlike subband A, this one has a very distinct vibrational structure (see Figure 2.8), the most significant features of which are listed in Table 2.8B.

Finally, the set of narrow bands C (from 115 nm to c. 125 nm) is an extension of the system of Rydberg transitions in bands A and B and corresponds to the following two systems of these transitions:

- Rydberg series converging to the lowest ionic ground state, which correspond to electronic transitions from the $1b_1$ state to Rydberg states of types nsa_1 (mainly for $n = 3$ and $n = 4$), npa_1 (mainly for $n = 3$), and npb_1 (mainly for $n = 3$)
- Rydberg series converging to the second ionic ground state, which come into being as a result of electrons $3a_1$ being excited to Rydberg states $3sa_1$

Both systems in these series have an intricate vibrational structure (Figure 2.9), associated mainly with mode II vibrational transitions, that is, bending, and to a lesser extent with mode I transitions or stretching. The positions of the absorption maxima due to these electronic transitions, together with the various changes in the vibrational states accompanying them, are given in Tables 2.8C and D. Both of these Rydberg series have a natural boundary on

TABLE 2.8. The principal structural elements in the absorption spectrum of a water molecule in the range from c. 110 to c. 180 nm.

A. Some energies and wavelengths of photons absorbed as a result of $1b_1 \rightarrow 4a_1$ type transitions with simultaneous changes in vibrational quantum numbers ($\Delta\upsilon_1$, $\Delta\upsilon_2$, $\Delta\upsilon_3$,) in subband A

No.	Assignment	Energy [eV]	λ [μm]
-1-	*-2-*	*-3-*	*-4-*
1	0,0,0	7.069	0.175
2	0,1,0	7.263	0.171
3	0,2,0 1,0,0	7.464	0.166
4	0,3 0	7.668	0.162
5	0,4,0 2,0,0	7.872	0.158
6	0,5,0	8.067	0.1537
7	0,6,0 3,0,0	8.260	0.150
8	0,7,0	8.463	0.147
9	4,0,0	8.604	0.144
10	0,8,0	8.658	0.143

B. Some energies and wavelengths of photons absorbed as a result of $1a_1 \rightarrow 3sa_1$ type transitions with simultaneous changes in vibrational quantum numbers ($\Delta\upsilon_1$, $\Delta\upsilon_2$, $\Delta\upsilon_3$,) in subband B

No.	Assignment	Energy [eV]	λ [μm]
-1-	*-2-*	*-3-*	*-4-*
1	0,0,0	8.598	0.144
2	0,1,0	8.658	0.143
3	0,2,0	8.775	0.141
4	0,3,0	8.875	0.140
5	0,4,0	8.978	0.138
6	0,5,0	9.083	0.137
7	0,6,0	9.198	0.135
8	0,7,0	9.294	0.133
9	0,8,0	9.393	0.132
10	0,9,0	9.479	0.131
11	0,10,0	9.574	0.130
12	0,11,0	9.671	0.128
13	0,12,0	9.770	0.127
14	0,13,0	9.864	0.126
15	0,14,0	9.995	0.124

C. Some energies and wavelengths of photons absorbed as a result of transitions of electron $1b_1$ to selected Rydberg levels with simultaneous changes in vibrational quantum numbers ($\Delta\upsilon_1$, $\Delta\upsilon_2$, $\Delta\upsilon_3$,) (band set C)

No.	Assignment	Energy [eV]	λ [μm]
-1-	*-2-*	*-3-*	*-4-*
1	$3s\ a_1$	7.464	0.166
2	$4s\ a_1$	10.624	0.117
. . .	. . .	. . .	. . .
1	$3p\ a_1$+(0,0,0)	10.011	0.124
2	$3p\ a_1$+(0,1,0)	10.179	0.122
3	$3p\ a_1$+(0,2,0)	10.354	0.120
4	$3p\ a_1$+(1,0,0)	10.401	0.119
5	$3p\ a_1$+(0,3,0)	10.476	0.118
6	$3p\ a_1$+(1,1,0)	10.556	0.117
7	$3p\ a_1$+(1,2,0)	10.721	0.116
. . .	. . .	. . .	. . .
1	$3p\ b_1$+(0,0,0)	10.163	0.122
2	$3p\ b_1$+(0,1,0)	10.360	0.120
3	$3p\ b_1$+(0,2,0)	10.556	0.117
4	$3p\ b_1$+(1,0,0)	10.574	0.117
5	$3p\ b_1$+(0,3,0)	10.763	0.115
6	$3p\ b_1$+(1,1,0)	10.777	0.115

D. Some energies and wavelengths of photons absorbed as a result of transitions of electron $3a_1$ to selected Rydberg levels with simultaneous changes in vibrational quantum numbers ($\Delta\upsilon_1$, $\Delta\upsilon_2$, $\Delta\upsilon_3$,) (band set C)

No.	Assignment	Energy [eV]	λ [μm]
-1-	*-2-*	*-3-*	*-4-*
1	$3s\ a_1$+(0,0,0)	9.991	0.124
2	$3s\ a_1$+(0,1,0)	10.142	0.122
3	$3s\ a_1$ +(0,2,0)	10.320	0.120
4	$3s\ a_1$+(1,0,0)	10.384	0.119
5	$3s\ a_1$+(0,3,0)	10.458	0.119
6	$3s\ a_1$+(1,1,0)	10.516	0.118
7	$3s\ a_1$+(0,2,0)	10.777	0.115

[a] Selected results of analyses; taken from Mota et al. (2005).

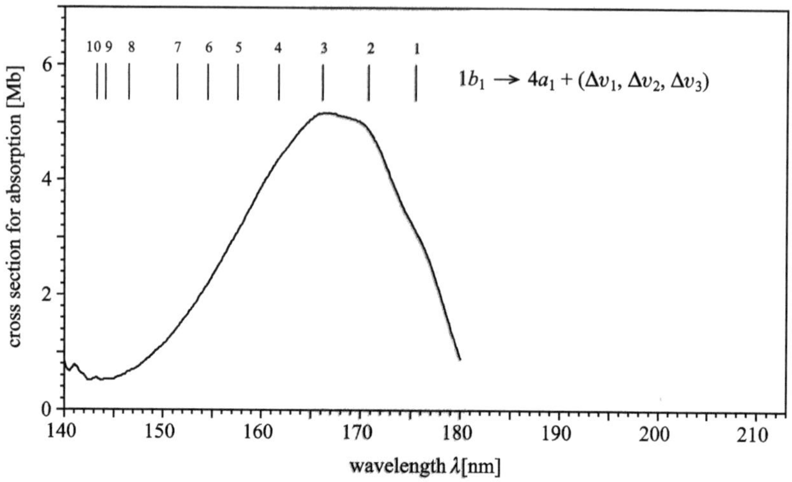

FIGURE 2.7. H_2O molecule photoabsorption spectrum in the A-band (from 145 to 180 nm) with some vibrational series labeled (numbers correspond to Table 2.8A). (Based on data from Mota et al. (2005).)

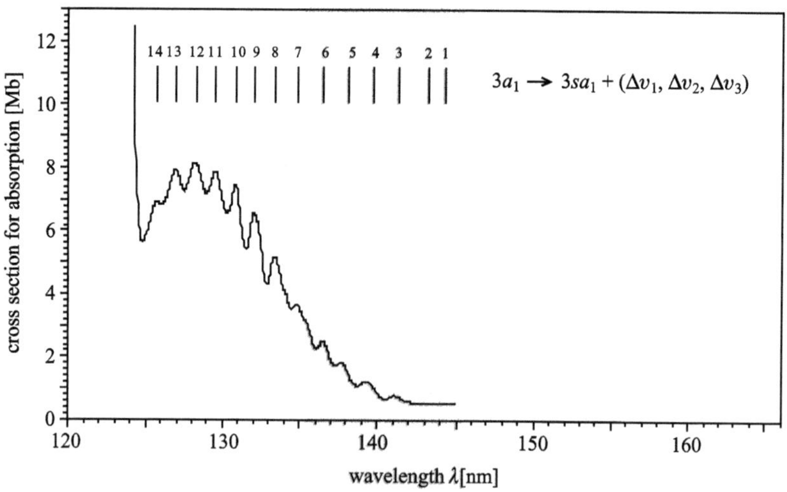

FIGURE 2.8. H_2O molecule photoabsorption spectrum in the B-band (from 125 to 145 nm) with some vibrational series labeled (numbers correspond to Table 2.8B). (Based on data from Mota et al. (2005).)

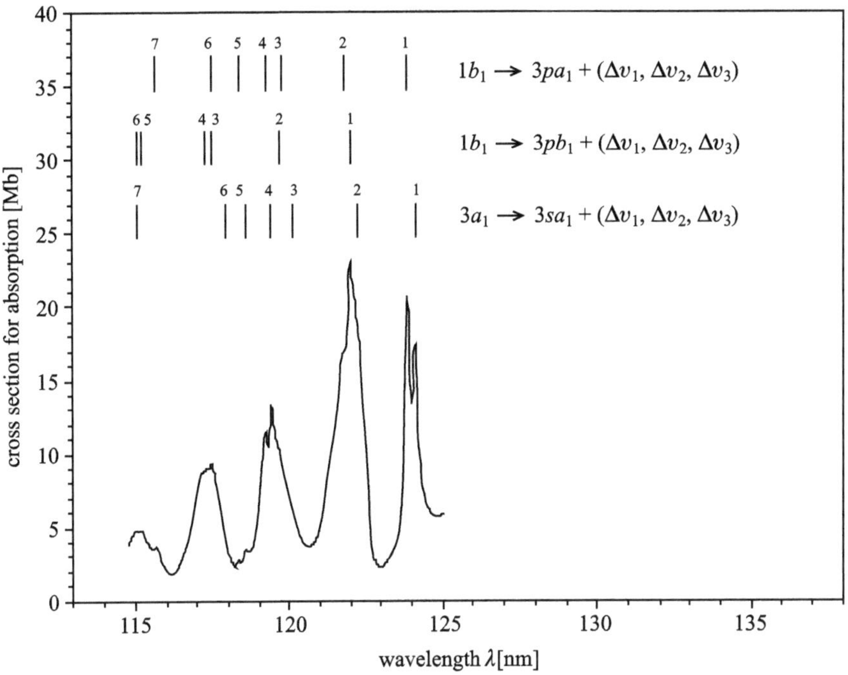

FIGURE 2.9. H_2O molecule photo-absorption spectrum in band set C (from 115 to 125 nm) with Rydberg series labeled and some vibrational excitation modes designated (numbers correspond to Tables 2.8C and D). (Based on data from Mota et al. (2005).)

the shortwave side of the spectrum. The maximum photon energies due to this boundary are equal to the energies of the ionizations that detach individual valence electrons from the molecule:

- Electron $1b_1$, whose ionization energy (minus the bond energy) is c. 12.62 eV (see Table 2.1 item 19; after Lias (2005)), that is, $\lambda \approx 98.3$ nm for the first of these Rydberg series types.
- Electron $3a_1$, with an ionization energy of c. 14.73 eV (see Table 2.1 item 20; after Diereksen et al. (1982)); that is, $\lambda \approx 84.2$ nm for the second of these Rydberg series types.

Clearly, these ionizations can also be caused by the absorption of higher-energy photons (from the shortwave range) and this excess energy would be converted, for example, into the kinetic energy of the released electrons. Such transitions lead to the formation of a continuous spectrum, and that is why a wide band of light absorption by water in this spectral range is observed.

2.2 The Absorption of Light and Other Electromagnetic Radiation in Pure Liquid Water and Ice

Our aim in this section is to characterize the electromagnetic absorption spectra of water in its condensed states, that is, the liquid state (liquid water) and the solid state (ice). A complete explanation of the formation and structure of the absorption spectra of water $a_w(\lambda)$ and of ice $a_{w,ice}(\lambda)$ is not possible on the basis of only what we know from Section 2.1. There we discussed the physical mechanisms of the interaction between electromagnetic radiation and freely moving H_2O molecules not interacting with each other. But these mechanisms provide for a satisfactory explanation of the absorption of such radiation only by water vapor, in which intermolecular interactions are negligible. They are not sufficient for an adequate description of the absorption properties of liquid water and ice, which are substantially modified by these interactions. Moreover, the mechanisms described in Section 2.1 do not explain the absorption of shortwave radiation in the $\lambda < 20$ nm range, that is, very-high-energy UV photons, and X- and γ-radiation quanta. Neither do they explain the absorption of longwave electromagnetic radiation such as radio waves. Section 2.2.1, therefore, extends the material of Section 2.1 by describing the physical principles underpinning the formation of spectra of the absorption of radiation by water in its condensed states.

Sections 2.2.2 and 2.2.3 contain detailed descriptions of the radiation absorption spectra of liquid water and ice. In order to produce them, we analyzed more than 30 empirical spectra of the absorption of radiation by liquid water (both distilled and very clear natural water in seas and lakes)[11] $a_w(\lambda)$ and ice[12] $a_{w,ice}(\lambda)$ in different spectral ranges. On the basis of this analysis and also reviewing the information on the interaction of electromagnetic radiation with water, available in numerous publications, we compiled a list of the spectral features of the absorption of this radiation in water in all its three states (water vapor, liquid water, and ice). Covering a very wide range of

---

[11] Spectra of $a_w(\lambda)$ or the relevant data enabling them to be determined are available in these publications: Clarke and James (1939), Le Grand (1939), Shuleykin (1959, 1968), Smoluchowski (1908), Ivanov (1975), Zolotarev et al. (1969), Wieliczka et al. (1989), Tam et al. (1979), Sullivan (1963), Sogandares and Fry (1997), Smith and Baker (1981), Shifrin 1988, Segelstein (1981), Quickenden and Irvin (1980), Pope and Fry (1997), Pope (1993), Palmer and Williams (1974), Morel and Prieur (1977), Kopelevitch (1976), Hale and Querry (1973), Buiteveld et al. (1994), Bricaud et al. (1995), Boivin et al. (1986), Baker and Smith (1982), Pelevin and Rostovtseva (2001), Hayashi et al. (2000), and Mota et al. (2005).

[12] Spectra of $a_{w,ice}(\lambda)$ or the relevant data enabling them to be determined are available in these publications: Warren (1984), Perovich and Govoni 1991, Grenfell 1983, 1991, Perovich et al. 1986, Grenfell and Perovich (1981, 1986), Grenfell and Maykut (1977), Irvine and Pollack (1968), and Kou et al. (1993).

radiation wavelengths—from vacuum UV to microwaves—this list is set out in Table 2.9. This gives the positions of the most important absorption bands (columns 3 to 5), together with an assignment of their origin (column 2). The information provided by this table is very useful when discussing the radiation absorption spectra of liquid water and ice, and we make reference to it many times in this section.

2.2.1 *Physical Mechanisms of Absorption*

We could, in principle, explain the light absorption spectrum of liquid water and ice in terms of the same mechanisms described for water vapor. However, we would have to significantly modify and extend that description in order to take account of the fact that individual liquid water molecules are not independent of each other but interact through intermolecular forces to form groups of molecules held together by hydrogen bonds—these $(H_2O)_n$-type polymers are known as clusters—and other various ionized structures (see, e.g., Horne (1969), Chaplin (2006), and the papers cited therein). Furthermore, in the ice crystal, every H_2O molecule is hydrogen-bonded to its neighbors (see Eisenberg and Kauzmann (1969), Chaplin (2006), and the papers cited therein). As a result, we have a whole range of intermolecular interactions, which strongly and in various ways disturb or modify the configurations of the excited energy states of the molecules. In some aspects, then, the absorption spectra of liquid water and ice do resemble those of free molecules or water vapor, but in many others they diverge quite substantially from the latter. Figure 2.10 illustrates absorption spectra of water over a wide range of wavelengths for all three of its states: liquid water at room temperature $a_w(\lambda)$, a monocrystal of ice at different temperatures $a_{w,ice}(\lambda)$, and the equivalent quantity of atmospheric water vapor[13] $a_{w,v}(\lambda)$.

This figure shows (but see also Figures 2.11 and 2.14) that the absorption spectra of liquid water and ice in the long-wavelength range (VIS and longer) exhibit a series of vibrational-rotational absorption bands as for water vapor, but in distinctly modified form. In contrast to the spectra of water vapor, then, these are *modified* vibrational-rotational absorption bands of water and ice molecules. Apart from the latter, there are also *condensed* phase absorption bands, which are not present in the water vapor spectra. Finally, the absorption of radio waves in liquid water is fairly strong, stronger than in ice or water vapor. We now briefly discuss the origin of these three characteristic features of the absorption spectra of water in its condensed states. To end this section, we also outline the absorption of shortwave radiation in water,

---

[13] The equivalent absorption coefficients for water vapor $a_{w,v}(\lambda)$, were determined on the basis of known approximate specific absorption coefficients for water vapor $a^*_{w,v}(\lambda)$ (see Figure 2.3) from the relationship $a_{w,v}(\lambda) = a^*_{w,v}(\lambda)\ C_{w,v}$, where $C_{w,v}$ is the equivalent (for liquid water) concentration of water vapor, equal to 1000 kg m^{-3}.

TABLE 2.9. The more important spectral features of the absorption of electromagnetic radiation in water vapor, liquid water, and ice (type Ih).

No.	Assignment	Water vapor[a]	Liquid water[a]	Ice (type Ih)[a]
-1-	-2-	-3-	-4-	-5-
1	T_B - intermolecular bend (translation)	—	~200 μm weak wide band	166 μm intense band
2	T_S - intermolecular stretch	—	~50–55 μm very intense band (local maximum)	44 μm very intense band (distinct local maximum)
3	Rotation	~ 118.6, ~70, ~50 μm three very intense bands (absolute maximum in longwave range)	—	—
4	Rotation (for water vapor) L_1 - librations (for liquid water and ice)	27.9 μm intense band (distinct local maximum)	25 μm weak band	~20 μm weak band
5	Rotation (for water vapor) L_2 - librations (for liquid water and ice)	From 16 to 23 μm three weak bands	15 μm very intense wideband (from 11 to 33 μm) (distinct local maximum)	11.9 μm very wide band (distinct local maximum)
6	Vibrational υ_2 - bend (fundamental band Mode II)	6.27 μm several very intense bands in the region from 5 to 7 μm (local maximum)	6.08 μm intense band with a half-width of c. 0.30 μm (local maximum)	6.06 μm weak wide band
7	Vibrational υ_2 - bend + L_2 and associated band		4.65 μm intense band with a half-width of c. 1.3 μm	4.41 μm weak wide band
8	Vibrational 2 υ_2 - bend (overtone) (0,0,0) → (0,2,0)	3.17 μm intense absorption band		
9	Vibrational υ_1 - symmetrical stretch (fundamental band Mode I)	2.73 μm weak band; 2.68 μm so-called X-band	3.05 μm very intense band; A very intense band with a half-width of c. 0.60 μm combined with a maximum in the 2.90–3.00 μm region (absolute maximum in longwave range)	3.24 μm very intense, structurally complex wideband
10	Vibrational υ_3 - asymmetrical stretch (fundamental band Mode III)	2.66 μm very intense band (strong local maximum)	2.87 μm very intense band	3.08 μm (absolute maximum in longwave range)

11	Vibrational combination transitions (0,0,0) → (a,1,b); a + b = 1	1.88 μm intense band in the 1.76–1.98 μm region so-called Ω - band	1.94 μm intense band (local maximum)	~2.00 μm weak band (local maximum)
12	Vibrational combination transitions (0,0,0) → (a,0,b); a + b = 2	1.38 μm intense band in the 1.32–1.50 μm region so-called Ψ - band	~1.44 μm intense band (local maximum)	1.50 μm weak band (local maximum)
13	Vibrational combination transitions (0,0,0) → (a,1,b); a + b = 2	1.13 μm intense band in the 1.10–1.17 μm region so-called Φ - band	1.20 μm weak band (bulge)	1.27 μm weak band (local maximum)
14	Vibrational combination transitions (0,0,0) → (a,0,b); a + b = 3	0.935 μm intense band in the 0.92–0.98 μm region so-called. ρστ- band	0.970 μm weak band (local maximum)	1.04 μm very weak band
15	Vibrational combination transitions (0,0,0) → (a,1,b); a + b = 3	0.810 μm intense band in the 0.79–0.84 μm region so-called 0.8μ - band	0.836 μm very weak band	~0.85 μm very weak band
16	Vibrational combination transitions (0,0,0) → (a,0,b); a + b = 4	0.718 μm intense band in the 0.70–0.74 μm region so-called α - band	0.739 μm weak band (small local maximum)	—
17	Vibrational combination transitions (0,0,0) → (a,0,b); a + b = 5	—	0.606 μm very weak band (bulge)	—
18	Vibrational combination transitions (0,0,0) → (a,0,b); a + b = 6	—	0.514 μm very weak band	—
19	Electronic transition: $1b_1 \rightarrow 4a_1$ like orbital	0.1665 μm (distinct local maximum)	~0.150 μm (distinct local maximum)	~0.143 μm (distinct local maximum)
20	Electronic transitions: $3a_1 \rightarrow 3sa_1$ and $1b_1 \rightarrow 3sa_1$	0.128 μm (distinct local maximum)	—	~0.121 μm weak band
21	Electronic transitions: Rydberg series	0.115–0.125 μm (set of narrow bands)	—	—
22	Different electronic transitions, photo-ionization, photodissociation	0.065 μm intense wide band (absolute maximum)	0.056–0.086 μm intense wide band (absolute maximum)	0.073 μm intense wide band (absolute maximum)

[a] Note: in columns 3–5 wavelengths are expressed in μm.

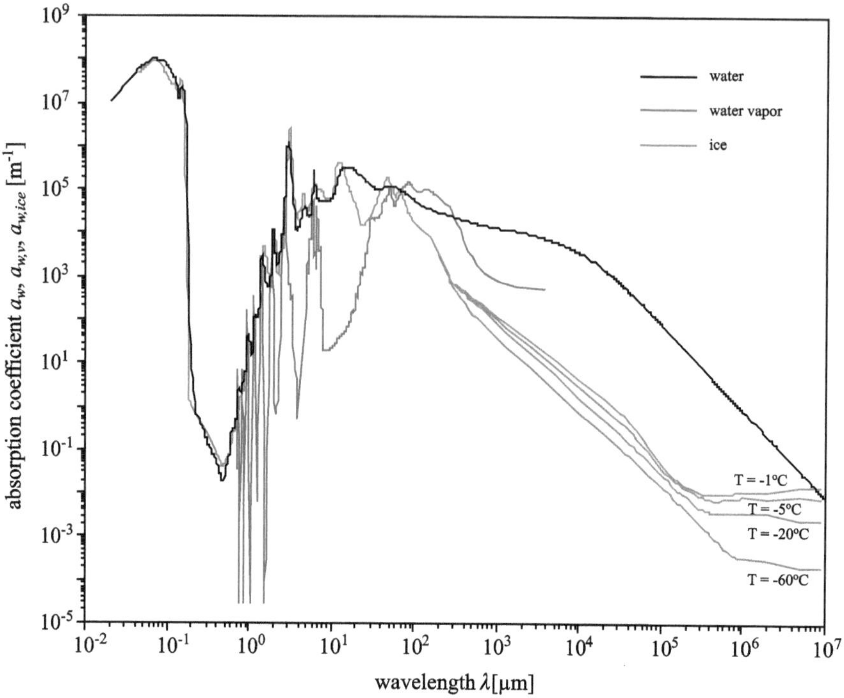

FIGURE 2.10. Absorption spectra in wide range of wavelengths (from UV to short radiowaves): liquid water at room temperature (based on data from Segelstein (1981)); ice crystals at various temperatures (based on data from Warren (1984)); atmospheric water vapor (based on Figure 2.3). (See Colour Plate 1)

that is, the absorption of high-energy UV photons and of X- and γ-radiation, which was not covered in Section 2.1.

Modified Vibrational-Rotational Absorption Bands in Liquid Water and Ice

One would expect that rotations of separate H_2O molecules in the condensed states are to a large extent retarded and sometimes brought to a standstill by the forces acting on the molecules in supramolecular structures and, in theory at least, the resulting vibrational-rotational absorption band structure ought to be less intricate. But this is not so; quite the opposite: in fact, the band structure of condensed phases is very much more complex than that of free molecules. This is because intermolecular interactions give rise to a whole series of vibrational states of the molecules for the same vibrational quantum numbers. These states correspond to a large number of energy levels so close to each other that they form almost a continuum. Hence, the absorption spectra of these condensed phases, unlike the fine structure of absorption spectra of free H_2O molecules, are practically continuum spectra. As a result these intermolecular interactions, the absorption bands in liquid water and ice, corresponding to diverse vibrational-rotational transitions of the H_2O molecule, are much broadened in contrast to those in water vapor. Moreover, the positions of the wave spectra peaks are shifted in comparison with those of the analogous bands for free single molecules; see the examples in Table 2.10. These shifts are usually in the longwave direction, and in ice are greater than in liquid water. The exception is the fundamental mode II band (bending), which in liquid water and ice is shifted toward the short waves relative to water vapor. The relations between the intensities of the individual bands also vary. For example, according to Eisenberg and Kauzmann (1969), in the gas state, the vibrations involve combinations of symmetric stretch (υ_1, mode I), asymmetric stretch (υ_3, mode III), and bending (υ_2, mode II), of the covalent bonds with absorption intensity $a(\upsilon_1):a(\upsilon_2): a(\upsilon_3) = 0.07:1.47:1.00$. This means that the fundamental mode II band of water in the gas state is the most intense and the fundamental mode I band is the least intense.

However, in the condensed states of water, this situation is radically different. In liquid water the most intense absorption is due to the fundamental mode III band (asymmetric stretch), and the least intense is the mode II band (bending). In liquid water the ratio of these intensities is $a(\upsilon_1):a(\upsilon_2):a(\upsilon_3) = 0.87:0.33:1.00$. For ice these relations are similar to those for liquid water. None of these intricacies characterizing the different positions and intensities of the vibrational-rotational absorption bands in the condensed states as opposed to water vapor are strictly constant; they depend to varying degrees on temperature and pressure, among other things. Detailed treatments of these questions can be found in Eisenberg and Kauzmann (1969), Walrafen (2004), Raichlin et al. (2004), Yakovenko et al. (2002), Worley and Klotz (1966), Symons (2001), Segtnan et al. (2001), Brubach et al. (2005), and elsewhere.

Absorption Bands of Condensed Phases

As we have already mentioned, there appear in the condensed states of water completely new phenomena and interactions between supramolecular structures and electromagnetic radiation that we did not discuss in Section 2.1. Consequently, the absorption spectra of liquid water and ice exhibit a number of features not present in the absorption spectra of water vapor, which may be partially, indirectly, or not at all due to electronic, vibrational, or rotational transitions. The most significant of these features are:

- *Librations* of hydrogen bonds (restricted rotations, or rocking motions). They form two absorption bands: a minor L_1 band with a peak in liquid water at $\lambda \approx 25$ μm and a major L_2 band with a maximum in liquid water at $\lambda \approx 15$ μm. Furthermore, L_1-type librations in combination with the

TABLE 2.10. Positions of the fundamental vibrational-rotational bands of light absorption by water in its various states.

Quantum numbers of excited states	Wavelength λ [nm]		
$\upsilon_1, \upsilon_2, \upsilon_3$	Water vapor	Liquid water	Ice
-1-	*-2-*	*-3-*	*-4-*
0,1,0	6.27	6.08	6. 06
1,0,0	2.73	3.05	3.24
0,0,1	2.66	2.87	3.08

Based on data from Chaplin (2006), Eisenberg and Kauzmann (1969), and Venyaminov and Prendergast (1997).

principal vibrational transition, that is, (0,0,0) → (0,1,0), produce an absorption band peaking in liquid water at around 4.65 μm. The intensities of these three bands increase with rising temperature, that is, in liquid water together with the falling number of molecules held in clusters by hydrogen bonds. These bands are also characteristic of ice, but are shifted towards shorter waves (see Table 2.10 and also Table 2.9 items 4, 5, and 7). These librations are discussed in greater detail in, for example, Zelsmann (1995), and Eisenberg and Kauzmann (1969).

- *Cluster vibrations*, or translational vibrations: these involve the stretching (T_S) or bending (T_B) of intermolecular hydrogen bonds (O–H . . . O). Intermolecular stretching is responsible for the absorption band centered around 55 μm in liquid water and around 44 μm in ice (see Table 2.9, item 2). Intermolecular bending, however, gives rise to an absorption band at c. 200 μm in liquid water and at c. 166 μm in ice (see Table 2.9 item 1). The higher the temperature is, the less intense the intermolecular stretching bands, but the more intense are the intermolecular bending bands. Cluster vibrations are accorded detailed treatment in Gaiduk and Vij (2001), Brubach et al. (2005), Vij (2003), and Dunn et al. (2006), among others.
- *Molecular vibrations in the crystal lattice* are associated with many absorption bands around $\lambda > 9.5$ μm. These bands are weak in the absorption spectrum of liquid water, but are very distinct, and practically characteristic only of ice. They are due to molecular vibrations in the crystal lattice transmitted by hydrogen bonds as a result of translational ($\lambda > 20$ μm) and rotational (9.5 μm $< \lambda <$ 20 μm) vibrations of the molecules. The reader can find this topic covered in greater detail in Hobbs (1974) and Warren (1984).

Apart from these three features, contributing to the complexity of the absorption properties of liquid water and ice, there are others that affect these properties to a lesser extent. The interested reader is referred to the relevant literature, for example, Johnson (2000), Tsai and Wu (2005), Padró and Marti (2003), Olander and Rice (1972), and Woods and Wiedemann (2004).

Absorption of Radio Waves

This process is typical of liquid water: water vapor and ice absorb radio waves only weakly. We know that electromagnetic radiation of wavelength around $\lambda > 2$ cm (i.e., long microwaves and radiowaves) interacts strongly with the dipoles of H_2O molecules, and that the energy of these interactions is dissipated in the work done shifting these dipoles in the electric field of the wave (see, e.g., Dera (1992)). Quantitatively, then, these interactions depend on the medium in which the radio waves propagate, and in particular on the medium's permittivity, magnetic permeability, and specific electrical conductivity. This problem with regard to aquatic media is discussed in greater depth by Jackson (1975). This author shows that in dipole–electric field interactions energy is absorbed from those frequency ranges in which the period of vibrations is close to the relaxation time of the H_2O dipole: dipole–electric field interactions are therefore relaxation processes. The relaxation time of these dipoles is governed by the forces of intermolecular interactions, which depend strictly on the temperature and pressure of the medium. Although strongly attenuated as a result of friction with adjacent molecules in the medium, these vibrations are maintained by the absorption of radiowaves: the higher their frequency, the stronger is the absorption. This is why radio communication, especially in the shorter wavebands, is difficult in fresh waters and well-nigh impossible in the salt waters of the sea. The radiowave absorption spectra of fresh and salt waters are illustrated in Figure 2.13, curves 2 and 3 (see below).

Absorption of Shortwave Radiation (High-Energy UV, X, γ)

In this book we do not analyze in any great detail the differences between the absorption properties of the three states of water with respect to UV light and radiation of even smaller wavelengths (see Section 2.1.2). Nevertheless, we should mention that these properties are not as well differentiated as those of water regarding visible and longwave radiation. This applies in particular to the very short wave range, where $\lambda < 10$ nm (i.e., mainly X and γ radiation). The absorption properties of these very short waves are determined more by the physical properties of atoms than of molecules. In the spectral interval from around $\lambda < 10$ nm right across to the shortwave range, corresponding to the high-energy quanta giving rise to the photoelectric effect (i.e., $\lambda \approx 23.5$ Å), the absorption properties of water have not, to our knowledge, been investigated empirically.

In contrast, in the wavelength range $\lambda < 23.5$ Å, the absorption spectrum of water in any of its states is very well known, having been studied empirically by the methods of nuclear physics. In this region of the spectrum, high-energy photons of X and γ radiation interact with individual atoms (and their nuclei), successively (with increasing photon energy) giving rise to the photoelectric effect, Compton effect, electron–positron pair formation, and other high-energy processes up to and including intranuclear processes, which were investigated by nuclear physicists some 40 years ago (see, e.g., the monographs by Aglintsev (1961/1957), Jaeger (1962/1959), Ciborowski (1962), Adamczewski (1965), and others). The absorption coefficient, the wavelength relationship for water in this high-energy spectral range, in which the coefficient is independent of the state of matter of water, is depicted graphically in Figure 2.13, curve 1 (see below). In this spectral region the absorption of radiation falls as the wavelength of the radiation shortens. Detailed information on these interactions of high-energy photons with matter can be found in the classical monographs on nuclear physics and radiation chemistry (see e.g. Wishart and Nocera 1998).

A slightly greater differentiation in the radiation absorption spectra of water in its various states is, however, undoubtedly characteristic as regards UV absorption in the c. 115–180 nm range (see Section 2.1), where the absorption bands are due to Rydberg energy states of the water molecule. The intensity of these absorption bands is distinctly greater for water vapor than for water in its condensed states.

2.2.2 The Absorption of Electromagnetic Radiation in Pure Liquid Water

What we mean here by pure water is a chemically pure substance that consists of water molecules occurring under the natural conditions of Planet Earth: these molecules contain different isotopes of hydrogen and oxygen in their structure. In Nature, the usual structure of the water molecule is $H_2{}^{16}O$; the most common of its isotopic variants are $H_2{}^{18}O$, $H_2{}^{17}O$, and the $HD^{16}O$ heavy water molecule. Occurring in a ratio of roughly 2:0.4:0.3, these last three molecules have a joint concentration in pure water of no more than 0.3% (±0.1%). Natural water also contains even heavier variants of heavy water ($HD^{17}O$, $HD^{18}O$, $D_2{}^{16}O$, $HT^{16}O$, $T_2{}^{16}O$), but only in trace amounts. The proportions in sea waters of the different isotopes of hydrogen and oxygen, and of the several isotopic variants of water are given in Table 2.1 (items 31–39) in accordance with the VSMOW (Vienna Standard Mean Ocean Water) convention.

The electromagnetic radiation absorption spectra of these isotopic variants of pure water differ among themselves, and all differ from the spectra of $H_2{}^{16}O$ water molecules, as numerous authors have shown (Barret and Mansell 1960, Eisenberg and Kauzmann 1969, Janca et al. 2003, Greenwood and Earnshaw 1997, Chaplin 2006). These differences are due to the fact that the molecules of these isotopic variants, with their different masses, all have slightly different geometrical structures and characteristic dimensions. As a result, their dynamic, electric, and magnetic properties (e.g., moments of inertia or dipole moments) also differ, hence the diversity in their energy states. The individual transitions between the energy states of these isotopic variants of water therefore correspond to the absorption of photons of different energies (i.e., different wavelengths).

This is exemplified by the positions of the radiation absorption bands set out in Table 2.4. We see there that the wavelengths of the photons absorbed as a result of these transitions differ for the different isotopic variants of molecules: in the case of water molecules, the wavelengths increase as the molecular mass does so. This differentiation also holds good with respect to absorption due to other types of energy transition—electronic, rotational, and other transitions characteristic of condensed phases (librations, transitional vibrations, and dipole–electric field interactions)—and not just to the positions of the bands and lines in the absorption spectrum, but also to their intensities. Nevertheless, this diversity in the absorption properties of the isotopic variants of water does not significantly affect the total coefficient of radiation absorption $a_w(\lambda)$ of the pure liquid water that fills the oceans and other basins of the Earth. This is because the overall concentration in Nature of this admixture of the isotopic variants of water is generally more than a thousandfold smaller than the concentration of $H_2{}^{16}O$ water. That is why, in our further description of the absorption properties of pure water, we concern ourselves only with the properties of water formed from $H_2{}^{16}O$ molecules.

As we have said, it emerges from the physical principles underlying the formation of the absorption spectra of free H_2O molecules (Section 2.1), and of their aggregations in condensed states (Section 2.2.1), that liquid water (and also water vapor and ice) absorbs ultraviolet strongly and shortwave radiation even more strongly, and on the other hand absorbs infrared radiation strongly and the longer electromagnetic wavelengths even more so. The consequence, therefore, of the arrangement of energy levels specific to the water molecule and its constituent atoms of hydrogen and oxygen is that its radiation absorption spectrum generally consists of two very wide regions of strong absorption—one on the shortwave side and the other on the longwave side of the spectrum—with a relatively narrow absorption minimum in the visible region lying in between (see Figures 2.10, 2.12, and 2.13). This "window" in the spectrum, which has a very high transmittance, lies in the VIS region, and the maximum transmittance, that is, the absolute minimum coefficient of absorption of liquid water $a_w(\lambda)$, lies at a wavelength of c. $\lambda \approx 415$ nm. That, in a nutshell, is how we can characterize the relationships of $a_w(\lambda)$, that is, the light absorption spectrum of water.

A more detailed analysis of the structure of this spectrum of $a_w(\lambda)$ with respect to its several spectral intervals will bring its complexity into focus: this is the next step in our story. To begin with, we describe the fundamental IR absorption bands (spectral range 2.3–8 μm), the absorption of near-IR ($\lambda < 2.3$ μm) and visible light, the absorption of longwave radiation ($\lambda > 8$ μm), and the absorption of UV, X-, and γ-radiation. Then we go on to review the empirical light absorption spectra of liquid water on the basis of the material available in the worldwide subject literature. Finally, we briefly discuss the influence that the absorption of radiation by water has had on the biological evolution of life on our planet.

Fundamental Infrared Absorption Bands (2.3 μm < λ < 8 μm)

The IR, microwave, and radiowave absorption spectrum of liquid water $a_w(\lambda)$ is compared with that of water vapor $a_{w,v}(\lambda)$ in Figure 2.11. This shows that the absorption bands due to fundamental vibrational-rotational transitions (modes I, II, and III, i.e., (0,0,0) → (1,0,0); (0,1,0); (0,0,1)) in pure liquid water molecules are, as a result of these molecular interactions, much broadened in comparison with those in water vapor. In liquid water they are also shifted towards the long waves (except mode II transitions) in comparison with the bands for free molecules in water vapor; see the right-hand part of the plot in Figure 2.11a, and also Table 2.9, items 6–10. This line broadening

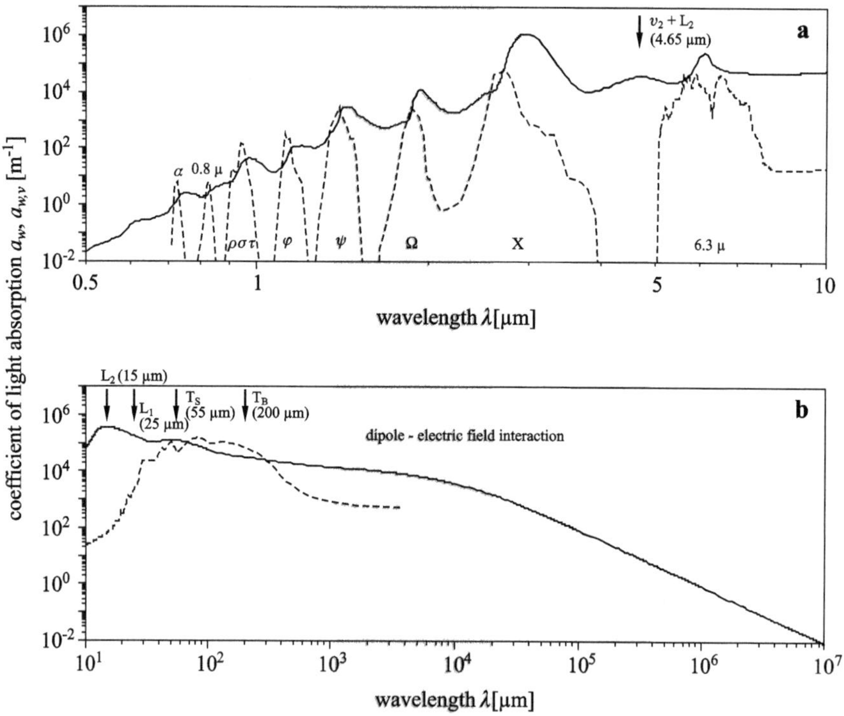

FIGURE 2.11. Visible light-short radio wave absorption spectra of: liquid water – continuous curve (after Segelstein (1981)) and atmospheric water vapor – dashed curve (defined as in the captions to Figs. 2.3 and 2.10);. Explanations: (a) the symbols are the individual water vapor absorption band codes commonly used in atmospheric optics; (a) and (b) the arrows give the positions of the absorption band peaks characteristic of liquid water but not present in the water vapor spectra (T_B, T_S – bands due to transitional vibrations, L_1, L_2 – bands due to librations, $v_2 + L_1$ - combination bending band (v_2) in L_2 libration modes; the heading "dipole-electric field interaction" denotes the spectral area of radio wave absorption).

and shift are of such an extent that the absorption bands of liquid water now coalesce to form an absorption continuum.

The strongest absorption band on the longwave side of the spectrum, starting from visible light, lies in the $2.3 < \lambda < 8$ μm wavelength range. Corresponding to the so-called X-band for water vapor, this band is formed in liquid water by combined fundamental mode I and mode III transitions, and peaks at $\lambda = 2.9$–3.0 μm (see Table 2.9, items 9 and 10). It is worth drawing attention to the fact that absorption in this 2.3–8 μm range is more intense, much more intense even, than the corresponding absorption in water vapor. Within this same wavelength range we also observe additional electromagnetic radiation absorption bands that do not occur in the absorption spectrum of free single molecules in water vapor. The most significant of these bands is a wide intense one peaking at $\lambda = 4.65$ μm: it is a combination band of bending (υ_2) and librations (mode L_2; Table 2.9, item 7). The existence of this band is the reason why in the spectral interval between 4 and 5 μm the absorption of radiation by liquid water is on average around 10^4 times greater than the absorption of an equivalent quantity of water vapor comprised of free, noninteracting H_2O molecules.

Absorption of Near Infrared and Visible Light (0.4 μm < λ < 2.3 μm)

In the wavelength range $\lambda < 2.3$ μm there is a whole series of broad, overlapping absorption bands of pure liquid water, the intensities of which decrease with increasing photon energy (i.e., with shortening wavelength); see Figure 2.11a (left-hand side of the plot) and also Table 2.9, items 11–18; notice that these bands are less intense than the fundamental absorption bands we mentioned earlier. As in the case of discrete H_2O molecules, these bands are due to vibrational-rotational harmonic and combination transitions from the ground state, specified earlier in Figure 2.3a, but they are wider and usually shifted in the direction of the longer wavelengths. Again, as in the case of such molecules, some of these broadened bands lie in the visible range of the spectrum, although there are more of them in the VIS range in the case of liquid water than of water vapor. These include two discernible, although quite weak bands at c. 606 nm and c. 515 nm (See Table 2.9, items 17 and 18). Notice also, that as in the case of the fundamental absorption bands, these harmonic and combination absorption bands of liquid water are generally more intense than those of water vapor. Exceptional here are certain absorption bands of water vapor, principally those designated in the accepted notation of atmospheric optics as α, 0.8μ, ρστ, and φ; see Figure 2.11a. These relatively narrow absorption bands of discrete molecules are characterized by greater intensities than the absorption coefficients of liquid water, although only in narrow ranges around their peaks.

Absorption of Longwave Radiation (λ > 8 μm)

The longwave absorption spectra ($\lambda > 8$ μm) in liquid water and water vapor in this spectral region differ from each other the most (see Figure 2.11b and the right-hand side of Figure 2.11a; see also Table 2.9, items 1–5). In this region then, free molecules absorb quite strongly, at a level comparable to that of liquid water only in the c. 50 μm < λ < 300 μm range. The complex absorption bands here are due to vibrational-rotational transitions between various excited states, superimposed on which are bands due to purely rotational transitions.

On the longwave side of this range (at c. λ > 300 μm), light absorption by free molecules in water vapor is more than one order of magnitude smaller than that by liquid water, even though water vapor is also a strong absorber of short radiowaves. On the shortwave side of this range (c. 8 μm < λ < 50 μm) the differences in absorption of water vapor and liquid water are even greater: the absorption coefficients of the latter are three to four orders of magnitude larger than the former. Generally speaking, the absorption coefficients of liquid water are very high right across the IR and microwave regions and do not begin to decrease visibly until we reach the long radiowave region.

Responsible for this absorption are the condensed phase bands (discussed earlier in Section 2.2.1), and not the rotational or vibrational-rotational transitions typical of free H_2O molecules. In liquid water they are librations L_1 and L_2 with peaks at c. 15 μm and c. 25 μm, and the translational vibrations T_S and T_B with peaks at 55 μm and c. 200μm. At even longer wavelengths (microwaves and radiowaves), this absorption in liquid water is caused by dipole–electric field interactions of the H_2O molecules, which we described above (see, e.g., Dera (1992, 2003)). Having such absorption properties, liquid water strongly absorbs electromagnetic radiation across the entire IR, microwave, and radiowave range: pure water is therefore practically opaque to these wavelengths. This extremely important physical property of water plays a fundamental part in its warming by solar radiation and in its evaporation from the surface layer of the oceans, and hence in the overall processes shaping climate and life on the Earth (see, e.g., Trenberth (1992)).

The Absorption of Ultraviolet, X-, and γ-Radiation

As we mentioned in our description of the shortwave region of the electromagnetic spectrum (UV), the absorption of UV light in pure liquid water (and also by water vapor and ice) is due to electronic vibrational-rotational transitions in H_2O molecules, and also to their dissociation and ionization. These processes give rise to certain spectral features of this absorption (see Section 2.1.2, Figures 2.6–2.9, and Table 2.9, items 19–22), namely, a wide absorption band peaking at c. 65 nm and a set of narrower, intricately structured absorption bands lying between c. 115 nm and c. 180 nm. These latter are very strong bands (see, e.g., Figure 2.10), the most intense in the entire electromagnetic spectrum, with absorption coefficients of up to c. 10^8 m^{-1} (at c. 65 nm). These bands are characteristic of water in all its three states, although there is no doubt that the bands in the 115–180 nm range, brought about by Rydberg state changes in the electronic states of the water molecule, are more intense for free molecules (water vapor) than for the condensed phases of water.

Notice that this strong absorption concerns very short wave UV radiation. In contrast, the band intensity for the absorption of longer UV waves ($\lambda >$ 180 nm) plunges by more than nine orders of magnitude (i.e., more than a billion times!), and in the near-UV and the shortwave part of the VIS region oscillates between 10^0 and 10^{-2} m^{-1}. Hence the depth of penetration of VIS radiation in pure sea water (e.g., the Sargasso Sea) is relatively great: c. 100 m. In practical terms, it is only at 100 m and more beneath the surface that the irradiance due visible light drops to 1% of its value at the sea surface. The transmittance of this light is therefore also greater than that of shortwave X and gamma (γ) radiation in the range above $\lambda > 10^{-4}$ Å (see Section 2.2.1). Evidence to support this claim is provided by Figure 2.13, which compares X- and γ-radiation absorption spectra as determined by the methods of nuclear physics (plot 1 in this figure) with the VIS absorption spectrum (see the position of the minimum of plot 2 in this figure).

Empirical Light Absorption Spectra of Liquid Water: A Review

The experimental determination of the light absorption spectra of liquid water is in practice very difficult, both because of the technical problems involved in the direct or indirect measurement of absorption coefficients and the impossibility of obtaining ideally pure water for our research (these questions are discussed more broadly, e.g., by Shifrin (1988) and Dera (2003)). In such investigations it is particularly important to eliminate suspended particles from the water samples to be examined. Light scattering by the tiny quantity of these particles present even in filtered distilled water and in the purest ocean waters (of the order of 1000 particles with diameters in excess of 1 μm (see, e.g., Jonasz and Fournier (1999) and Dera (2003)) can lead to major errors in the measurement of absorption. This is especially significant in studies of VIS and near-UV absorption, where the light absorption coefficients of structurally pure liquid water are very small; they may in fact be smaller than or comparable to the light-scattering coefficients of the suspended particles. Hence, as we show in due course, the absorption of VIS and near-UV light in liquid water measured by various authors are encumbered with serious errors and are frequently very different.

In view of these difficulties, practically the only reliable source of data on light absorption in pure water was for many years the work by Clarke and James (1939). From the light attenuation coefficients that they measured in distilled water $c_w(\lambda)$ in the $375 < \lambda < 800$ nm wavelength range (Table 2.11, column 2) the light absorption coefficients of water $a_w(\lambda)$ in this spectral interval can be determined indirectly with the aid of the relationship (1.2) given in Chapter 1. For this, we also need the relevant light-scattering coefficients in water, $b_w(\lambda)$, determined theoretically by Le Grand (1939). The results of these calculations are given in Table 2.11 (column 3).

In the worldwide subject literature we can now find several tens of empirical light absorption spectra of structurally pure liquid water (redistilled

TABLE 2.11. Light attenuation coefficient c_w and light absorption coefficients a_w in distilled water; light absorption coefficients a in pure natural waters.

Type of water	Distilled	Distilled	Distilled	Distilled	Distilled	Pure natural waters	Sargasso sea	Central pacific
Source	Clarke, James (1939)	Clarke, James (1939), Le Grand (1939)[a]	Shuleykin (1959)	Hale, Querry (1973)	Morel, Prieur (1977)	Smolu-chowski (1908)	Ivanov (1975)	Pelevin, Rostovtseva (2001)
	Measured Parameter							
λ [nm]	c_w [m^{-1}]	a_w [m^{-1}]	a_w [m^{-1}]	a_w [m^{-1}]	a_w [m^{-1}]	a [m^{-1}]	a [m^{-1}]	a [m^{-1}]
-1-	*-2-*	*-3-*	*-4-*	*-5-*	*-6-*	*-7-*	*-8-*	*-9-*
375	0.045	0.0383	—	0.117	—	—	—	
390	—	—	—	—	0.020	0.038	0.041	0.012
400	0.043	0.0379	—	0.056	0.018	—	—	0.012
410	—	—	—	—	0.018	0.037	0.034	0.011
425	0.033	0.0291	—	0.038	—	—	—	—
430	—	—	—	—	0.015	0.036	0.025	0.011
450	0.019	0.0159	—	0.028	0.015	0.037	0.016	0.012
470	—	—	—	—	0.016	0.039	0.014	0.012
475	0.018	0.0155	—	0.025	—	—	—	—
490	—	—	—	—	0.020	0.042	0.018	0.017
494	—	—	0.002	—	—	—	—	—
500	0.036	0.0340	—	0.025	0.026	—	—	0.019
510	—	—	—	—	0.036	0.054	0.023	0.026
522	—	—	0.002	—	—	—	—	—
525	0.041	0.0394	—	0.032	—	—	—	—
530	—	—	—	—	0.051	0.062	0.039	0.038
550	0.069	0.0676	—	0.045	0.064	0.074	0.05	0.052
558	—	—	0.038	—	—	—	—	—
570	—	—	—	—	0.080	0.094	0.067	0.075
575	0.091	0.0898	—	0.079	—	—	—	—
590	—	—	0.089	—	0.157	0.160	0.140	0.121
600	0.186	0.1850	—	0.230	0.244	—	—	0.162
602	—	—	0.173	—	—	—	—	—
607	—	—	0.200	—	—	—	—	—
610	—	—	—	—	0.289	0.260	0.240	—
612	—	—	0.233	—	—	—	—	—
617	—	—	0.234	—	—	—	—	—
622	—	—	0.239	—	—	—	—	—
625	0.228	0.2272	—	0.280	—	—	—	—
643	—	—	0.291	—	—	—	—	—
650	0.288	0.2873	—	0.320	0.3492	0.380	0.330	—
658	—	—	0.320	—	—	—	—	—
675	0.367	0.3664	—	0.415	—	—	—	—
690	—	—	—	—	0.5	0.540	0.520	—
700	0.500	0.4995	—	0.600	0.6499	—	—	—

[a] Values calculated as the difference $a_w = c_w - b_w$, where c_w is the light attenuation coefficient of pure water (after Clarke and James (1939)); b_w is the light scattering coefficient of pure water (after Le Grand (1939)).

water) or of pure natural waters (i.e., practically devoid of admixtures) meticulously plotted for a variety of spectral ranges (see the introduction to Section 2.2). Some of these empirical spectra of $a_w(\lambda)$ are compared in Figure 2.12. The plots in Figure 2.12a show such spectra of $a_w(\lambda)$ for a wide

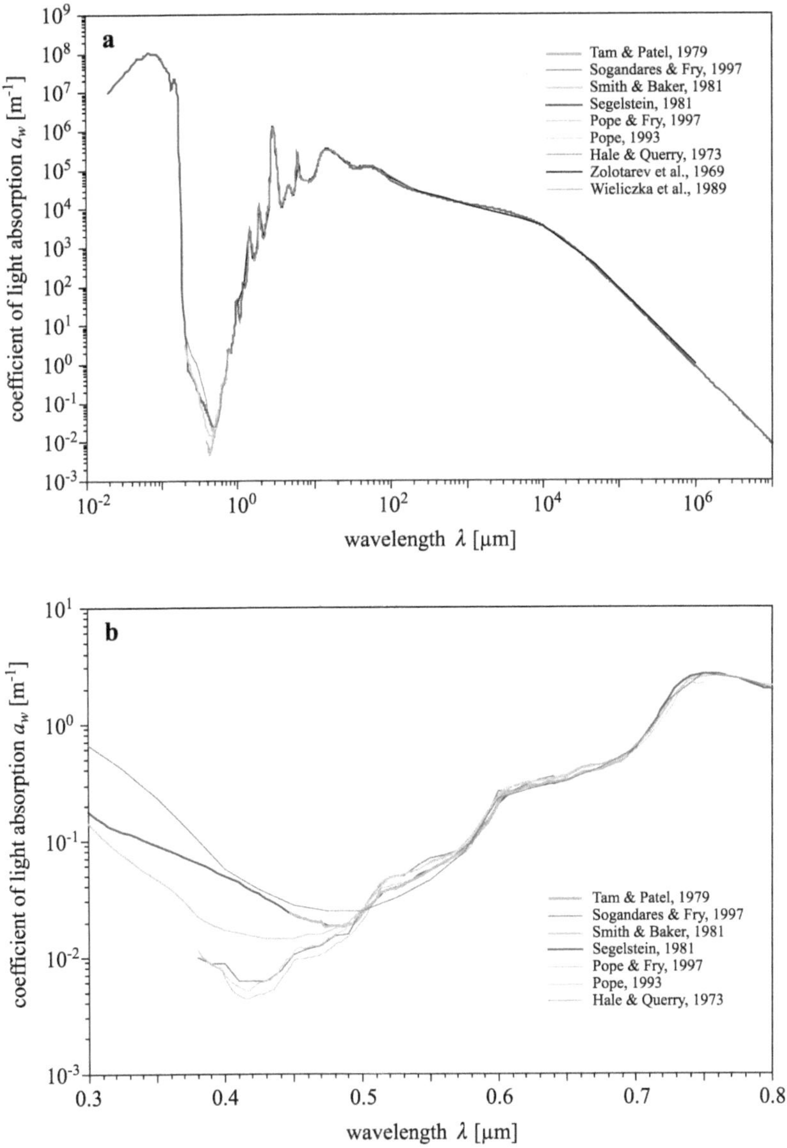

FIGURE 2.12. Empirical absorption spectra of liquid water, as determined by various authors: (a) in the wide spectral range from the ultraviolet to short radiowaves; (b) in the visible range and its nearest neighborhood. (See Colour Plate 2)

spectral range (from UV light to microwave and short radiowave radiation), and those in Figure 2.12b illustrate the spectra in the VIS region and its nearest neighborhood. Table 2.11, moreover, sets out some of these measurements in the visible part of the spectrum of $a_w(\lambda)$ in distilled water and in pure natural waters, whereas Table 2.12 gives examples (from Hale and Querry (1973)) of IR absorption coefficients $a_w(\lambda)$ of pure water.

We can see from the plots in Figure 2.12 and the data in Table 2.11 that there are discrepancies among the empirical absorption spectra of $a_w(\lambda)$ of liquid water obtained by various authors. With respect to those regions of the spectrum with high absorption coefficients (i.e., UV, IR, and longer waves; see Figure 2.12a), these discrepancies are small and can be neglected. But as far as visible and near-UV light is concerned, they are very significant and

TABLE 2.12. IR absorption coefficients $a_w(\lambda)$ in pure liquid water.

λ [μm]	a [m^{-1}]	λ [μm]	a [10^3 m^{-1}]	λ [μm]	a [10^3 m^{-1}]
-1-	*-2-*	*-1-*	*-2-*	*-1-*	*-2-*
0.70	0.60	1.40	1.23	4.00	14
0.725	1.59	1.60	0.067	4.10	17
0.75	2.60	1.80	0.802	4.20	21
0.775	2.40	2.00	6.91	4.30	25
0.80	2.00	2.20	16.5	4.40	29
0.81	1.99	2.40	50.05	4.50	37
0.82	2.39	2.60	15.32	4.60	40
0.83	2.91	2.65	31.77	4.70	42
0.84	3.47	2.70	88.4	4.80	39
0.85	4.30	2.75	269.6	4.90	35
0.86	4.68	2.80	516	5.00	31
0.87	5.20	2.85	815	5.10	28
0.875	5.60	2.90	1161	5.20	24
0.88	5.60	2.95	1269	5.30	23
0.89	6.04	3.00	1139	5.40	24
0.90	6.80	3.05	988	5.50	27
0.91	7.29	3.10	778	5.60	32
0.92	10.93	3.15	538	5.70	45
0.93	17.30	3.20	363	5.80	71
0.94	26.74	3.25	236	5.90	132
0.95	39.0	3.30	140	6.00	224
0.96	42.0	3.35	98	6.10	270
0.97	45.0	3.40	72	6.20	178
0.98	43.0	3.45	48	6.30	114
0.99	41.0	3.50	34	6.40	88
1.00	36.0	3.60	18	6.50	60
1.10	17.0	3.70	12	6.60	68
1.20	104	3.80	11	6.70	63
1.30	111	3.90	12	6.80	60

Selected from data published by Hale and Querry (1973).

cannot be ignored (Figure 2.12b). This applies both to the absolute values and to the position of the minimum of coefficient $a_w(\lambda)$ in the spectrum. These discrepancies are due primarily to the difficulties, mentioned earlier, in obtaining chemically (and therefore spectrally) pure water, free of all admixtures, especially of suspended particulate matter.

Regardless of these differences in the spectral dependences of light absorption on wavelength, all the empirical spectra of $a_w(\lambda)$, determined empirically by various authors, confirm the existence of a deep minimum of absorption intensity in the blue light area, probably at c. $\lambda_{min} \approx 415$ nm. A much trickier problem, however, is to discover which of these empirical results are characterized by absolute values of $a_w(\lambda)$ most closely approaching the real values in the visible and near-UV regions. As yet there is no definitive solution. What we can state with a fair degree of certainty, though, is that they may very well be the figures quoted in Table 2.13 for various light wavelengths in the 200 nm $< \lambda <$ 700 nm range. The absorption coefficients $a_w(\lambda)$ given in this table for various spectral subranges were gleaned from a number of works (see the footnotes to this table; Smith and Baker (1981), Sogandares and Fry (1997), and Pope and Fry (1997)). We should add that among the dozen or so most reliable light absorption spectra of water for this spectral region that we analyzed, these values are intermediate. The spectra of $a_w(\lambda)$ quoted in Table 2.13 can therefore be applied with confidence in practical model calculations[14] (see, e.g., Woźniak et al. (2005b[15])).

Finally, it is worth pointing out that, as Table 2.11 shows, the waters of pure natural bodies (some mountain lakes, such as Crater Lake in the United States), and also saline but pure areas of the ocean (e.g., the Sargasso Sea), absorb visible light in practically the same way as distilled water does, except for some subtle differences in the UV band. Some authors, for example, Pelevin and Rostovtseva (2001), would go further: they consider that these absorption coefficients of pure natural water bodies, determined in situ, most closely approach the real absorption coefficients of this light in structurally pure water. Their argument in support of this claim is that the accuracy of in vitro measurements in pure redistilled water is worse than that of in situ measurements, because of the small optical pathways obtainable in laboratory spectrophotometers.

[14] As far as the values of the coefficients $a_w(\lambda)$ from the IR region are concerned, we can use the data of, for example, Hale and Querry (1973) given in Table 2.12 in such model calculations to good effect. This is because the values of these coefficients given by other authors are similar.

[15] All tables and graphs taken for this book from the journal *Oceanologia* are reproduced with the kind permission of the Polish Academy of Sciences Institute of Oceanology and the National Committee on Oceanic Research.

TABLE 2.13. Spectral light absorption coefficients of liquid water in the 200–700 nm region suggested in this monograph.[a]

Light wavelength [nm]	a_w [m^{-1}]	Light wavelength [nm]	a_w [m^{-1}]	Light wavelength [nm]	a_w [m^{-1}]
-1-	-2-	-1-	-2-	-1-	-2-
200	3.070	370	0.0114	540	0.0474
205	2.530	375	0.0114	545	0.0511
210	1.990	380	0.0114	550	0.0565
215	1.650	385	0.0094	555	0.0596
220	1.310	390	0.0085	560	0.0619
225	1.120	395	0.0081	565	0.0642
230	0.930	400	0.0066	570	0.0695
235	0.820	405	0.0053	575	0.0772
240	0.720	410	0.0047	580	0.0896
245	0.640	415	0.0044	585	0.1100
250	0.559	420	0.0045	590	0.135
255	0.508	425	0.0048	595	0.167
260	0.457	430	0.0050	600	0.222
265	0.415	435	0.0053	605	0.258
270	0.373	440	0.0064	610	0.264
275	0.331	445	0.0075	615	0.268
280	0.288	450	0.0092	620	0.276
285	0.252	455	0.0096	625	0.283
290	0.215	460	0.0098	630	0.292
295	0.178	465	0.0101	635	0.301
300	0.141	470	0.0106	640	0.311
305	0.123	475	0.0114	645	0.325
310	0.105	480	0.0127	650	0.340
315	0.0947	485	0.0136	655	0.371
320	0.0844	490	0.0150	660	0.410
325	0.0761	495	0.0173	665	0.429
330	0.0678	500	0.0204	670	0.439
335	0.0620	505	0.0256	675	0.448
340	0.0325	510	0.0325	680	0.465
345	0.0265	515	0.0396	685	0.486
350	0.0204	520	0.0409	690	0.516
355	0.0180	525	0.0417	695	0.559
360	0.0156	530	0.0434	700	0.624
365	0.0135	535	0.0452		

Data from various authors, after Woźniak et al. (2005b).
[a] Values assumed on the basis of data from various authors and linearly approximated: in the 200–335 nm range, after Smith and Baker (1981); 340–370 nm, after Sogandares and Fry (1997); 380–700 nm, after Pope and Fry (1997).

The Absorption of Electromagnetic Radiation by Water: Its Influence on the Biological Evolution of Life

To conclude this chapter, we draw the reader's attention to the perfect adaptation of life on our planet to, among other things, the optical properties of water and to the most important features of solar radiation. Let us therefore take a look at the plots in Figure 2.13. They illustrate the dependence on

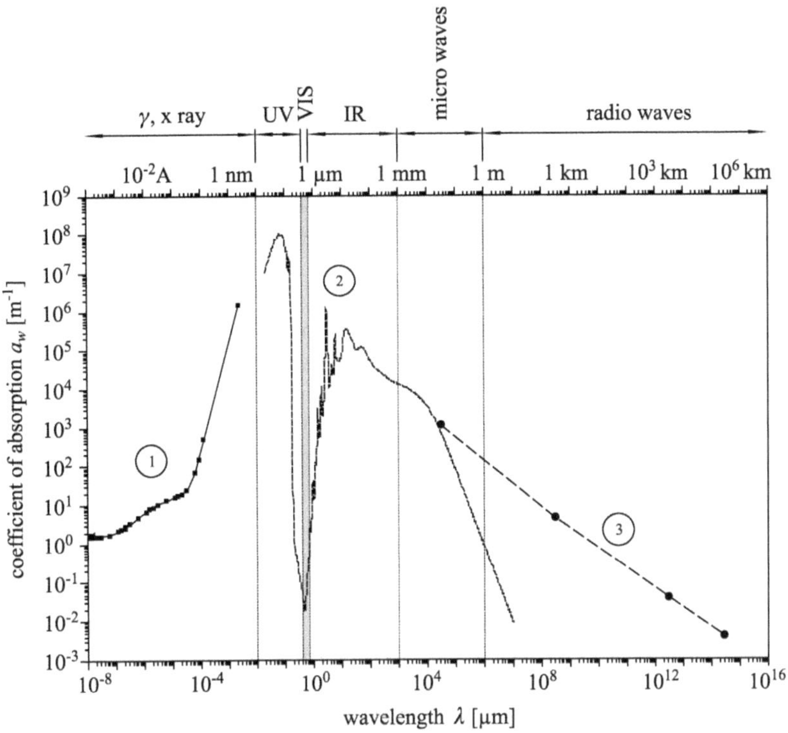

FIGURE 2.13. Electromagnetic waves absorption spectra of liquid water over the complete spectral range, from high-energy quanta γ ($\lambda \approx 10^{-4}$ Å) to long radio waves ($\lambda \approx 3 \times 10^5$ km). (1) Relationship among the radiation absorption coefficient of pure water and the wavelength for γ-radiation, X-radiation, and high-energy UV photons (based on data from Aglintzev (1961/1957), Jaeger (1962/1959), Ciborowski (1962), Adamczewski (1965), and others); (2) same relationship for pure water in the UV-short radiowave range (after Segelstein (1981)); (3) the same relationship for oceanic salt water in the microwave-longwave ($\lambda \approx 3 \times 10^5$ km) region determined approximately on the basis of data from Popov et al. (1979).

wavelength of the light absorption coefficients of water over their entire range, almost 11 orders of magnitude. The wavelengths we have taken into consideration here embrace the even greater range of more than 22 orders of magnitude, from short γ-radiation (c. 10^{-4} A) to extremely long radiowaves (c. 300,000 km), in other words, practically the entire range of electromagnetic radiation existing in Nature. First, the figure shows the spectrum of absorption in water of γ-, X-, and very-high-energy UV radiation. This absorption is due to interaction between the high-energy photons of this radiation and the inner electrons of the oxygen atom and the hydrogen and oxygen nuclei (plot 1 in Figure 2.13).

Next, plot 2 in this figure represents the spectrum of absorption in the UV-short-radiowave range, that is, absorption caused by standard electronic, vibrational (including librations and cluster vibrations), and rotational transitions, and also dipole-electric field interactions. Finally, plot 3 shows the spectrum of the absorption in saline oceanic water of radio waves (also very long). Such water absorbs these waves with greater intensity, an intensity that is all the more suited to the conditions prevailing on our planet inasmuch as the resources of salt water in the oceans are very much greater than the Earth's supplies of fresh water.

Figure 2.13 shows, therefore, that water—natural, omnipresent, and indispensable to life on Earth—is opaque to most electromagnetic radiation and is characterized by just the one deep spectral "window" for the transmission of radiation (of low absorption), in the visible light spectral range.[16] At the same time, we know that the maximum solar emission falls within this window, and that some 50% of the solar radiation reaching the Earth's surface lies within the VIS range. Is it surprising, then, that biological evolution on our planet, full of water as it is, has brought into existence plants that are green in color[17] (including marine phytoplankton), drawing their vital energy from the solar radiation flowing in through this window, and on the other hand, has caused animals to develop eyes with which they can perceive light, from red to violet, which readily penetrates the water? Mother Nature has undoubtedly made excellent use of her window. You, the reader, may wonder at this point: is this purely coincidence, or is it design?

2.2.3 The Absorption of Electromagnetic Radiation in Ice

As ice forms, all the water molecules (in the case of monocrystalline ice) or the vast majority of them (in the case of polycrystalline or amorphous ice) become attached to their neighbors, mostly by means of hydrogen bonds. As a result of these intermolecular interactions many possible geometrical structures of ice, both crystalline and amorphous, can come into and remain in existence. This depends on the thermodynamic conditions (temperature and pressure) prevailing during the formation of ice and while it lasts. The contemporary literature on the physics of ice describes some 20 possible

[16] Absorptions as low as those in the VIS region are only again attainable in the region of radiowaves thousands of kilometers long (in the case of sea water). Such low absorption is also conceivable with respect to shortwave radiation in the $\lambda < 10^{-4}$ A region, that is, for neutrino radiation.

[17] The green color of plants is, as we know, due to the absorption properties of chlorophyll and the other pigments essential to the photosynthesis of organic matter in plants. We return to these questions in Chapter 6.

geometrical structures of ice[18] (Hobbs (1974), Eisenberg and Kauzmann (1969), Chaplin (2006), and the papers cited therein). These structures differ not only in their geometry, but also in a number of physical properties, including such fundamental ones as density, the value of which can be from 0.92 (types Ih and Ic) to 2.51 (type X) and more (type XIII) times the density of liquid water under natural conditions.

They also differ considerably in their relative permittivity ε (dielectric constant; Table 2.1, item 8), one of the basic electromagnetic properties governing the interaction of a material with electromagnetic radiation. The relative permittivity ε for liquid water under normal conditions is c. 78.4; for the different types of ice it varies widely: from small values of c. 3.74 for type IX ice, through intermediate values of c. 97.5 for the common type Ih ice, up to the very high value of c. 193 for type VI ice (Chaplin 2006, Robinson et al. 1996).

In view of this, the optical properties of these different structural forms of ice, and their absorption capacities in particular, will vary widely as well (Warren (1984) and the papers cited therein). We do not deal with these problems here, because with the exception of type Ih ice, which occurs naturally all over the Earth, most of the other structures of ice do not occur in the sea or the atmosphere. They form only at the extremely high pressures produced in laboratories for experimental purposes.

An exception is type Ic ice, which forms under natural conditions in tropical cirrus clouds (Warren 1984); its absorption properties approach those of type Ih. That is why we limit our description of the absorption properties of ice to the monocrystals of type Ih ice, and primarily its polycrystalline forms (more easily obtainable), which consist of single crystals c. 0.1 cm in diameter. Studies have shown that the absorption properties of this polycrystalline ice are practically identical to those of an ice monocrystal (Hobbs 1974). The pure form of this ice has a density of c. 920 kg m^{-3} and comes into existence under natural conditions when distilled water, free of air bubbles and other particulate admixtures, freezes. In these conditions, water crystallizes into the hexagonal arrangement of tetrahedral unit cells characteristic of type Ih ice. Such an ice monocrystal is birefringent and uniaxial, with the optical axis coinciding with the crystallographic *c*-axis. The reader can find more detailed information on the structures of hexagonal type Ih ice and its fundamental properties in Eisenberg and Kauzmann

[18] Among them are the following crystalline and noncrystalline structures: hexagonal with density (expressed in g cm^{-3}) equal to $\rho = 0.92$, type Ih; cubic with $\rho = 0.92$, type Ic; noncrystalline with $\rho = 0.94$, LDA-type (low-density amorphous ice); noncrystalline with $\rho = 1.17$, HDA-type (high-density amorphous ice); noncrystalline with $\rho = 1.25$, VHDA-type (very high-density amorphous ice); rhombohedral with $\rho = 1.17$, type II; tetragonal with $\rho = 1.14$, type III; rhombohedral with $\rho = 1.27$, type IV; monoclinic with $\rho = 1.23$, type V; tetragonal with $\rho = 1.46$, type VIII; tetragonal with $\rho = 1.16$, type IX; cubic with $\rho = 2.51$, type X; orthorhombic with $\rho = 0.92$, type XI; tetragonal with $\rho = 1.29$, type XII; and hexagonal with $\rho > 0.92$, type XIII.

(1969), Hobbs (1974), Doronin and Chieysin (1975), Soper (2000), and Robinson et al. (1996).

In the first part of this section we characterize the principal spectral features of radiation absorption in pure type Ih ice. We also review the empirical absorption spectra of ice determined by various authors. The reader should bear in mind that natural ice forming in seas and in freshwater basins often has optical properties that deviate from those of the well-researched pure type Ih ice, which is usually obtained in the laboratory. We discuss the reasons for these discrepancies in the optical properties of natural and pure ice in the last part of this section.

The Absorption of Radiation in Pure Ice: Principal Spectral Features

The absorption spectrum of pure crystalline ice $a_{w,ice}(\lambda)$ in the near-UV-short radiowave range is illustrated in Figure 2.14; for comparison, this also

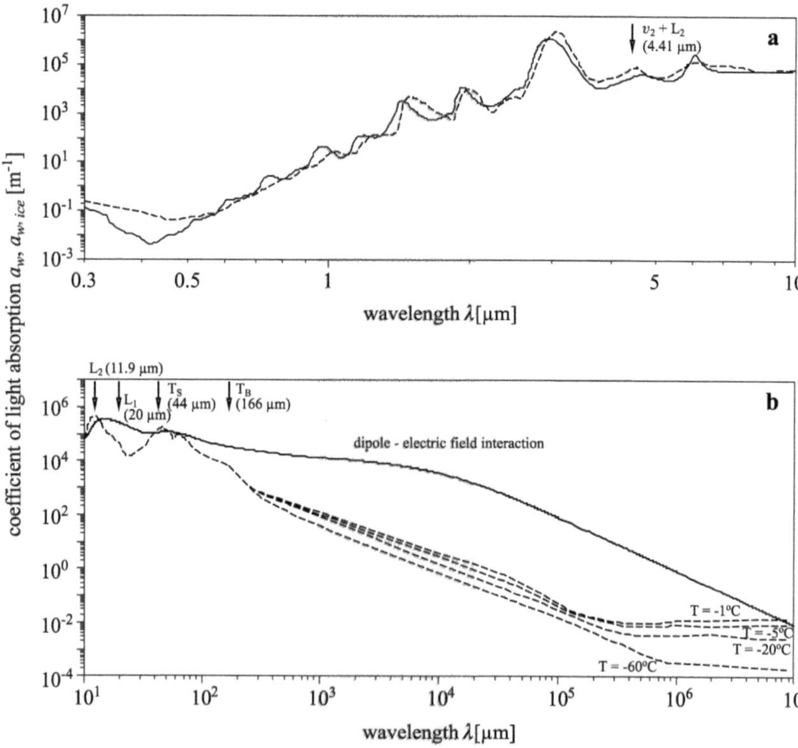

FIGURE 2.14. Visible light – short radio wave absorption spectra of liquid water and ice. The positions of the absorption band peaks due to librations and transitional vibrations are indicated, as is the area of "dipole-electric field interaction" in the ice crystal (denotations as in Figure 2.11). Dashed lines – ice (Warren 1984), Continuous lines – liquid water : 0.3 – 0.7 µm (Woźniak et al. 2005b), 0.7 – 10^7 µm (Segelstein 1981).

shows the corresponding absorption spectrum of pure water $a_w(\lambda)$. Because of intermolecular interactions, the spectra of both $a_{w,ice}(\lambda)$ and $a_w(\lambda)$ are continuous across the whole spectral range in question, which is not the case with the spectra of free molecules or water vapor. The absorptions of radiation in ice and liquid water have numerous spectral features in common, but there are also a number of features differentiating the absorption spectra in these two media. We discuss these features with respect to the following three spectral ranges: (1) UV, VIS, and near-IR light from $\lambda < 2.5$ µm, (2) intermediate IR in the $2.5 < \lambda < 8$ µm range, and (3) longwave radiation from $\lambda > 8$ µm.

In range (1) (UV–VIS–near-IR as far as λ = c. 2.5 µm) of the absorption spectrum of $a_{w,ice}(\lambda)$, as in the case of liquid water, there is a whole series of broadened and overlapping absorption bands, decreasing in intensity with shortening wavelength (see the left-hand side of Figure 2.14a; also Table 2.9, items 11–18). In the blue part of the VIS region there is an absolute minimum of absorption $a_{w,ice}(\lambda)$ in the spectrum of ice, as in the corresponding spectrum of liquid water. In comparison with the latter, however, the ice minimum is shifted slightly towards the longer waves and is located at $\lambda \approx 475$ nm. (The position of the liquid water minimum is at $\lambda \approx 415$ nm.) The increase in absorption coefficients on the shortwave side of these minima in the spectra of both $a_{w,ice}(\lambda)$ and $a_w(\lambda)$ is caused by absorption due to electronic transitions, whereas all the absorption bands on the longwave side are vibrational-rotational. The UV absorption spectrum of ice resembles that of liquid water (part of the UV region is visible in Figure 2.14). Should the reader wish for a more detailed account of this absorption, we refer him or her to the relevant literature: Daniels (1971), Browell and Anderson (1975), Seki et al. (1981), and Warren (1984).

The absorption bands on the longwave side of the absorption minimum in the spectra of $a_{w,ice}(\lambda)$ and $a_w(\lambda)$ in the VIS and near-IR regions correspond to vibrational-rotational transitions, as in the case of free molecules. In this spectral region they are combination and higher harmonic transitions from the ground state to excited state, exactly the same as for free molecules (Figure 2.3a). In comparison with the liquid water bands, the ice bands are shifted towards the longer wavelengths, in the same way as the absorption minimum, mentioned above. Moreover, the outline of these bands in the absorption spectrum of ice, and also all their irregularities, especially in the shortwave part of the spectral region in question, are expressed less distinctly than in the spectrum of liquid water. As a result, these three weak although discernible absorption bands, characteristic of liquid water in the wavelength ranges c. $510 < \lambda < 560$ nm, $600 < \lambda < 680$ nm, and $700 < \lambda < 800$ nm, are scarcely perceptible in the spectrum of ice.

Notice also that the absolute absorption coefficients in the longwave part of the spectral range $\lambda < 2.5$ µm are similar for ice and liquid water, if we ignore the shifts of the separate absorption bands. Only in the shortwave VIS range ($\lambda < 570$ nm) are these coefficients markedly larger than for liquid water. In the 350–450 nm range, the borderland between VIS and UV light, this difference is about one order of magnitude. This means that in this spectral range pure ice is much less transparent than pure liquid water.

Range (2) (intermediate IR; $2.5 < \lambda < 8$ μm) is characterized by the presence of "pure" fundamental vibrational-rotational absorption bands, or their combination with librations (see the right-hand side of Figure 2.14a and Table 2.9, items 6–10) against a background of an absorption continuum, as in the case of liquid water. These bands peak at $\lambda = 3.08$ μm, 4.41 μm, and 6.06 μm. The first of these, the absorption band in the vicinity of $\lambda = 3.08$ μm, is the strongest absorption band in the range of wavelengths longer than those of VIS light. It corresponds to band X of water vapor and is formed jointly by fundamental mode I and mode III transitions. The second band, with a maximum at c. $\lambda = 4.41$ μm, is a combination band of bending (υ_2) and librations (mode L_2), and the third one, peaking at c. $\lambda = 6.06$ μm, is due to frequent bending (υ_2), that is, mode II transitions.

Figure 2.14a shows that these principal absorption bands of ice closely resemble the analogous absorption bands of liquid water in that their peak positions, intensities, and breadths are all similar. The last-mentioned band (peaking at c. $\lambda = 6.06$ μm), however, is an exception: although somewhat less clearly expressed than its counterpart in the liquid water spectrum (peaking at $\lambda = 6.08$ μm), it is much broader. Nonetheless, the absorption spectra of liquid water $a_w(\lambda)$ and of ice $a_{w,ice}(\lambda)$ coincide most closely in this wavelength range (2.5–8 μm), certainly more so than in other spectral ranges.

Finally, range (3) (far-IR, microwaves, and radio waves; $\lambda > 8$ μm) exhibits the greatest differences between the spectra of water $a_w(\lambda)$ and of ice $a_{w,ice}(\lambda)$ (Figure 2.14b; Table 2.9 items 1–5). In comparison with the absorption coefficients of liquid water, those of ice in this range are usually from one to three orders of magnitude smaller. Exceptional in this respect are two narrower intervals in the IR region, where these coefficients for ice and water are similar (see the left-hand side of the plot in Figure 2.14b). As in the case of liquid water, we find here the L_1(20 μm) and L_2(11.9 μm) libration bands, as well as the T_S(44 μm) and T_B(166 μm) cluster vibration bands, characteristic of both condensed phases. But in the case of ice these bands are expressed much more distinctly than in the case of liquid water, and they are also shifted a little towards the shorter wavelengths. This is in part due to the fact that in ice, these bands overlap absorption bands due to molecular vibrations in the crystal lattice, which, of course, are characteristic only of the solid state (see Section 2.2.1).

In the UV, VIS, and IR regions, the absorption coefficient of ice varies only slightly with temperature changes between c. 0 and –30°C. Only when the temperature drops below –40°C does it begin to have a marked effect on the absorption coefficient of ice. On the other hand, where longer wavelengths are concerned (microwaves and radiowaves; $\lambda > 250$ μm), the absorption coefficient is significantly affected at all subzero temperatures (see the middle and right-hand parts of the plot in Figure 2.14b) in that the absorption intensity falls sharply as the temperature of the ice drops (Jiracek 1967, Johari et al. 1975, Warren 1984). Thus, in the radiowave range $\lambda = 10^6–10^7$ μm, the difference between the absorption coefficients of ice at around freezing point (0°C) and at –60°C is almost two orders of magnitude.

Empirical Light Absorption Spectra of Ice: A Review

The light absorption spectra of ice, like those of water, have been determined empirically by several dozen authors (see the introduction to Section 2.2), but in most cases within strictly limited spectral ranges. In a synthesis of all this work, Warren (1984) presented spectra of the real and imaginary part of refractive indices over a wide spectral range from UV (44.3 nm) to radiowaves (8.6 m) and at different temperatures from –1°C to –60°C. The spectrum of the absorption coefficient of ice[19] $a_{w,ice}(\lambda)$ plotted from these data is illustrated in Figure 2.15 (selected values of $a_{w,ice}$ for the spectral range from 195 nm to 6.75 μm are given in Table 2.14). The figure also shows, for comparison, three such empirical spectra postulated by other authors for somewhat narrower spectral ranges.

Hence, there are discrepancies also in the empirical spectra of $a_{w,ice}(\lambda)$ determined by various authors: they are marked in the VIS region and its nearest neighborhood, that is, in the range where absorption coefficients are very low. The reasons for this are the same as in the case of liquid water, namely, the difficulties of obtaining sufficiently pure ice, and technical imperfections on the other hand. It is thus hard to state definitively which of the absorption coefficients $a_{w,ice}(\lambda)$ in the spectra plotted in Figure 2.15 are closest to the real values in and around the VIS region. It does seem, however, that these coefficients in other spectral regions are encumbered with smaller errors.

Reasons for Discrepancies Between the Optical Properties of Natural and Pure Ice

The natural ice that forms on the surface of fresh- and salt-water bodies is a multicomponent substance, the properties of which differ from those of ice monocrystals (type Ih), or the similar polycrystalline ice produced artificially

[19] If the relevant coefficients of the imaginary part of refractive indices are known, light absorption coefficients can be calculated from the relationship $a = 4\pi n'/\lambda$, where n' is the coefficient of the imaginary part of the refractive index of light, called for short the imaginary refractive index (see Chapter 5).

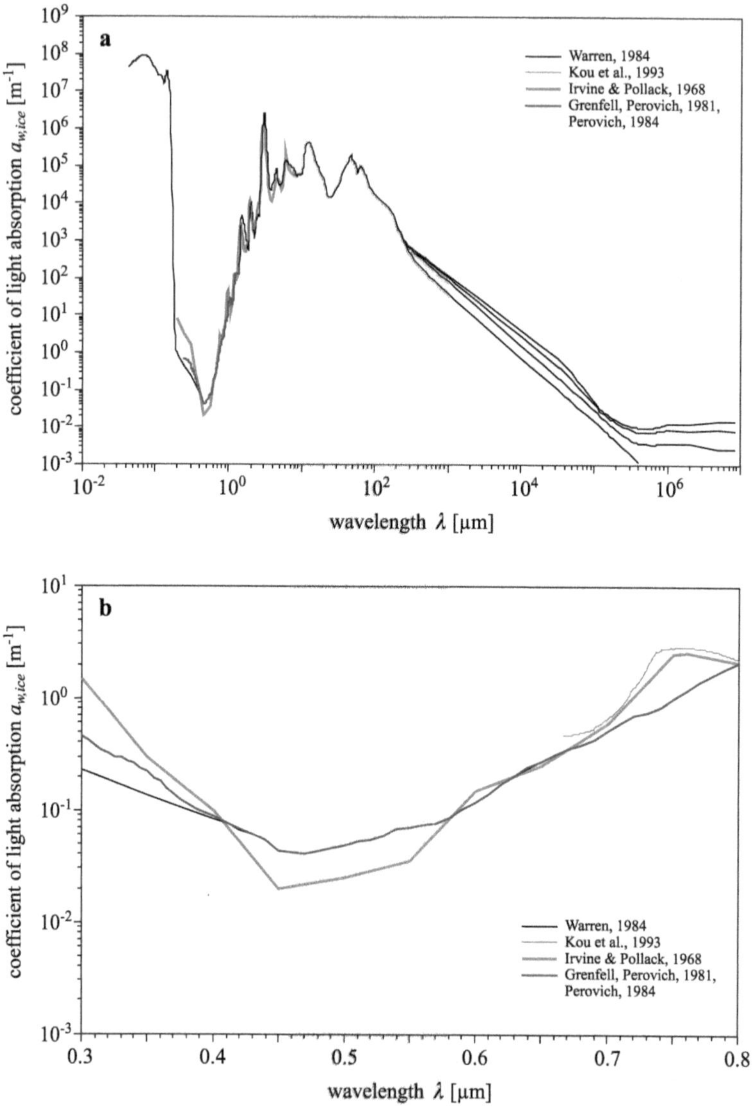

FIGURE 2.15. Empirical radiation absorption spectra of ice, as determined by various authors: (a) in the wide spectral range from the ultraviolet to short radio waves; (b) in the visible range and its nearest neighborhood. (See Colour Plate 3)

from distilled water or pure optically natural water and consisting predominantly of crystals c. 1 mm across.

The size and forms of the crystals constituting natural ice on bodies of open water depend on the degree of supercooling of the water, the rate of heat loss during crystallization, the rate of crystal nucleation, the salinity of the water,

TABLE 2.14. Light absorption coefficients $a_{w,ice}$ of type Ih ice; spectral range from 195 nm to 6.75 μm.

λ [μm]	a [m⁻¹]	λ [μm]	a [m⁻¹]	λ [μm]	a [m⁻¹]
-1-	-2-	-1-	-2-	-1-	-2-
0.195	1.025	0.660	0.316	2.350	2.93×10^3
0.210	0.793	0.670	0.354	2.500	4.65×10^3
0.250	0.433	0.680	0.386	2.565	4.26×10^3
0.300	0.231	0.690	0.437	2.600	4.88×10^3
0.350	0.135	0.700	0.521	2.817	1.70×10^5
0.400	0.085	0.710	0.609	3.003	1.83×10^6
0.410	0.077	0.720	0.703	3.077	2.55×10^6
0.420	0.068	0.730	0.740	3.115	2.17×10^6
0.430	0.061	0.740	0.835	3.155	1.74×10^6
0.440	0.055	0.750	0.984	3.300	3.92×10^5
0.450	0.043	0.760	1.171	3.484	6.71×10^4
0.460	0.042	0.770	1.400	3.559	3.74×10^4
0.470	0.041	0.780	1.643	3.775	2.20×10^4
0.480	0.043	0.790	1.877	4.099	3.40×10^4
0.490	0.046	0.800	2.105	4.239	4.30×10^4
0.500	0.048	0.820	2.191	4.444	7.35×10^4
0.510	0.053	0.850	2.705	4.56	8.27×10^4
0.520	0.055	0.910	6.131	4.904	3.31×10^4
0.530	0.060	0.970	1.20×10^1	5.000	3.02×10^4
0.540	0.068	1.030	2.84×10^1	5.100	3.08×10^4
0.550	0.071	1.100	1.94×10^1	5.263	3.34×10^4
0.560	0.074	1.180	5.08×10^1	5.556	5.43×10^4
0.570	0.078	1.270	1.34×10^2	5.747	8.31×10^4
0.580	0.088	1.310	1.26×10^2	5.848	1.12×10^5
0.590	0.104	1.40	1.78×10^2	6.061	1.43×10^5
0.600	0.12	1.504	4.93×10^3	6.135	1.43×10^5
0.610	0.142	1.587	2.93×10^3	6.250	1.35×10^5
0.620	0.174	1.850	5.45×10^2	6.369	1.22×10^5
0.630	0.207	2.000	9.98×10^3	6.452	1.11×10^5
0.640	0.24	2.105	5.02×10^3	6.579	1.05×10^5
0.650	0.276	2.245	1.13×10	6.757	1.08×10^5

Data selected from Warren (1984).

and also on the degree of wind-mixing of the water. This last factor applies to the surface layer of ice (Doronin and Chieysin 1975, Hobbs 1974). Strong supercooling and a slow rate of heat loss favor the formation of skeletal crystals, whereas the opposite situation—weak super cooling and a rapid loss of heat—promotes the formation of compact crystals. The subject literature distinguishes black and white ice. The former arises directly from water as this freezes, the latter when ice remelts together with rainwater, with water from melting ice, or water flowing over the surface of ice when this is melting, for example, under the weight of snow. Repeated melting and freezing of the surface layer of an ice covering causes this layer to become grainy and inhomogeneous, regardless of its original structure (Hobbs 1974, Grenfell 1991). The physical properties of such white ice therefore differ from those of black ice.

Furthermore, natural sea ice, because of the dynamically changing conditions of crystallization, is inhomogeneous and consists of alternate layers or patches of white and black ice. Apart from crystals of pure ice, sea ice also incorporates salt, gas bubbles, and other admixtures present in sea water (Grenfell and Maykut 1977, Zakrzewski 1982, Light et al. 2004).

These factors are the reason why the absorption of electromagnetic radiation in natural sea ice is a complex phenomenon. The radiation penetrating the surface of such ice is absorbed by both the ice crystals and by the admixtures contained in the ice; not only that, it is also scattered, and this scattering is generally stronger than the absorption (Perovich 1990, Grenfell 1991). The intensity of scattering in sea ice depends principally on its content of brine cells and capillaries (water that has become strongly saline as a result of the precipitation of salts during freezing) and of micropores remaining after the brine and the solid admixtures have trickled out (Grenfell and Hedrick 1983).

The light scattering at these constituents of the ice can be treated as the reflection and refraction of light at the successive interfaces of an inhomogeneous medium. As a result, the passage of light through a layer of ice is a process specific to ice, in which the resultant attenuation of the light by the ice depends on the number and orientation of the refracting surfaces, and also the refractive indices and absorption properties of the constituents of the ice (ice crystals, brine, air, solid admixtures). This is why it is difficult to define experimentally the spectral absorption coefficients of natural ice. In descriptions of the transmission of electromagnetic waves through a layer of natural ice, therefore, it is usual to employ the coefficient of irradiance transmission or the albedo of the ice. The reader can find detailed data on the value of these coefficients for sea ice from various parts of the Arctic and Antarctic in Grenfell and Maykut (1977), Grenfell and Perovich (1984), Perovich et al. (1986), Perovich (1994), and Hanesiak et al. (2001). The magnitudes of these coefficients, and also the influence of marine environmental parameters and the structural properties of sea ice on its optical properties and on radiation transfer in sea ice have recently been the subject of mathematical modeling (see, e.g., Mobley et al. (1998), Light et al. (2003, 2004), and the papers cited therein).

Snow is far less transparent to visible light than ice. This effect is due primarily to the structure of snow, which consists of ice grains interspersed with air. This causes light to be strongly scattered as a result of multiple reflections, whereby the degree of scattering is so large that the transport of radiant energy through snow can be treated as a diffusion process. As a consequence of this property, a 3 cm thick covering of snow on a surface of ice is enough to practically prevent any transmission of light through that ice. The optical properties of snow are described by Grenfell and Maykut (1977), Grenfell and Perovich (1986), Grenfell (1991), Perovich et al. (1986), and Warren (1984), among others.

2.3 Light Absorption by Atoms, Sea-Salt Ions, and Other Inorganic Substances Dissolved in Sea Water

Practically all natural elements are present in the World Ocean, although most of them in very low concentrations (e.g., Dera (2003)). They occur mostly as dissolved forms of inorganic and organic matter. The inorganic forms are represented not only by atoms and simple ions, but also by molecules and complex ions. Changes in the electronic states of the atoms and ions of the various elements (see Table 2.15) give rise to a narrow absorption band at the shortwave end of the electromagnetic spectrum, beginning with the visible region (and sometimes with the near-IR).

Radiation is absorbed by the atoms and ions of these elements as a result of the excitation of the weakly bound electrons in their unfilled *d* orbitals (e.g., Gołębiewski (1982), McQuarrie (1983), and Love and Peterson (2005)). The atoms and ions of elements with other configurations usually need higher excitation energies for absorption to occur—their absorption bands are located in the far ultraviolet—for instance, the chloride ion Cl^- absorbs radiation with a wavelength of 181 nm. These features of the absorption of radiation by inorganic substances, that is, the narrow absorption bands for the individual elements, although present in sea water, are scarcely detectable against the background of the total absorption of all the dissolved substances.[20]

The first empirical studies on the influence of all the inorganic substances dissolved in sea water on its optical properties (Hulburt 1928) showed that they strongly attenuated and absorbed UV and VIS radiation.

TABLE 2.15. The area of spectrally active elements in the periodic table.[a]

H																	He
Li	Be											B	C	N	O	F	Ne
Na	Mg											Al	Si	P	S	Cl	Ar
K	Ca	Sc	**Ti**	**V**	**Cr**	**Mn**	**Fe**	**Co**	**Ni**	**Cu**	Zn	Ga	Ge	As	Se	Br	Kr
Rb	Sr	Y	**Zr**	**Nb**	**Mo**	**Tc**	**Ru**	**Rh**	**Pd**	Ag	Cd	In	Sn	Sb	Te	I	Xe
Ca	Ba	La	Hf	**Ta**	**W**	**Re**	**Os**	**Ir**	**Pt**	**Au**	Hg	Ti	Pb	Bi	Po	At	Rn
Fr	Ra	Ac															

[a] The lanthanides and actinides have been omitted. The elements whose atoms and ions absorb visible and near-UV radiation (>300 nm) are highlighted in bold; the other elements absorb radiation in the far-UV region.

[20] Exceptional are the fairly narrow bands due to the absorption of radiation by metal ion complexes (Kondratev and Pozdniakov 1988), which can occur in the absorption spectra of waters mineralized to a high degree, for example, certain lakes; see the section 2.3.3 on inorganic complexes below.

These results were, however, overestimated as a result of methodological errors (contaminated water samples). Later work, for example, that by Clarke and James (1939) and by Sullivan (1963), indicated that the effect of inorganic substances on visible light absorption was so small as to be practically negligible. It is important to note, however, weak bands due to the absorption by inorganic substances of IR radiation were found (e.g., Visser (1967)), and the absorption of UV radiation by these substances in sea water was quite considerable in comparison to its absorption in pure water (e.g., Lenoble (1956) and Armstrong and Boalch (1961)). Nevertheless, these results were of a qualitative nature. In the light of subsequent research (see below) it has become possible to perform quantitative analyses, and in some cases to distinguish the elements and inorganic compounds responsible for certain spectral properties of sea water (Table 2.15 and Figure 2.16). Our

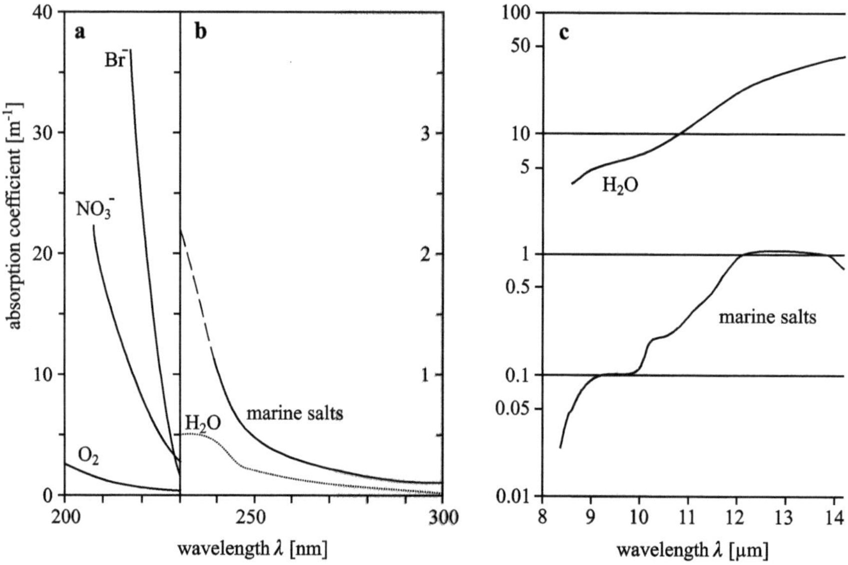

FIGURE 2.16. Comparison of empirical absorption spectra of the principal inorganic absorbers contained in sea water with the absorption spectra of pure water in the UV (a) and (b) and IR regions (c): (a) absorption by aqueous solutions containing: bromide ions Br^- of concentration 65 mg·dm^{-3}, nitrate ions NO_3^- of concentration 10 μmol·dm^{-3} and dissolved oxygen O_2 of concentration 5 mg·dm^{-3}; for the spectral region $\lambda < 230$ nm (after Ivanoff (1978)). (b) Absorption by dissolved salts from the Mediterranean Sea (salinity 37.8 PSU) (after Copin-Montegut et al. (1971)) and absorption by pure water (after Kopelevitch (1983)). Note that empirical data for UV absorption by water are rather inexact (Lenoble and Saint Guily 1955, Morel 1974, Smith and Baker 1981). (c) Absorption of all sea salts in model sea water (salinity 34.8 PSU) and the absorption of pure water. (After empirical data from numerous authors, gathered by Popov et al. (1979).)

estimates of the mass specific coefficients of UV and IR absorption by the major inorganic absorbers in sea water are given in Table 2.16. Any inaccuracies in this context are due to the broad scatter of the empirical data given in the literature, a situation which, in turn, results from the serious technical difficulties involved in the measurement of these coefficients (e.g., the problem of preparing suitably pure samples for spectral analysis, measurements of optical coefficients very low in value, around the detection limit of the spectrometer, etc.).

As can be seen from Table 2.16 and Figure 2.16, two groups of absorbents stand out among the dissolved inorganic substances—dissolved gases and sea salts—which actually or potentially affect the absorption properties of sea water. Moreover, in some cases (strongly mineralized waters), the absorption spectra can also be modified by the absorption of radiation by complexes of transition metal ions. We now discuss in turn the absorption properties of these three groups of absorbers.

2.3.1 *Dissolved Gases*

During the interaction between the sea and the air, atmospheric gases (oxygen O_2, nitrogen N_2, carbon dioxide CO_2, etc.) and a variety of chemical compounds contained in the aerosol enter the upper layers of the sea, where some of them are dissolved, and the remainder are present in the form of bubbles or suspensions (e.g., Dera (2003)). The small quantities of gases dissolved in sea water are of prime significance for the organisms inhabiting the sea and control of the rates of numerous geophysical processes (Vinogradov 1968, Hansell et al. 2002). But they have little if any effect on the optical properties of ocean waters, unless we take into account the wind-roughened, foam-covered surface of the sea. An exception to this rule is dissolved oxygen O_2: absorption by this is detectable in the UV region below 260 nm (see the O_2 curve in Figure 2.16).

The effect of this band on the overall absorption of radiation by dissolved inorganic substances is usually small; it is apparent mostly in the surface layers of well-aerated waters (Ivanoff 1978, Copin-Montegut et al. 1971).

2.3.2 *Salts*

In contrast to dissolved gases, it is the soluble salts of certain elements (see Table 2.15) and insoluble H_3BO_3 that make the principal contribution of the inorganic substances dissolved in sea water to the absorption of electromagnetic radiation (Jerlov 1976). The main components of these salts are NaCl, KCl, MgCl, $MgSO_4$, and $CaSO_4$. In open oceanic waters, the concentrations of these compounds (total average concentration = 34.7 g · dm^{-3}) and their relative proportions are temporally and spatially extremely stable (Tables 2.17 and 2.18; Dera (2003)).

TABLE 2.16. Spectral mass-specific light absorption coefficients of the major inorganic absorbers in sea water.[a]

A. UV Region

1	Wavelength λ [nm]	200	210	220	230	240	250
-1-	*-2-*	*-3-*	*-4-*	*-5-*	*-6-*	*-7-*	*-8-*
2	Dissolved oxygen O_2 $a^*_{O_2}$ [m^2g^{-1}]	0.46	0.18	0.10	0.07	~ 0	—
3	Bromide ions Br^- $a^*_{Br^-}$ [m^2g^{-1}]	—	~1.0	0.40	0.014	~ 0	—
4	Nitrates NO^-_3 $a^*_{NO_3}$ [m^2g^{-1}]	~40	29,8	13.0	3.2	~ 0	—
5	All salts a^*_S [m^2g^{-1}]	1.8×10^{-4}	1.1×10^{-4}	8.0×10^{-5}	5.5×10^{-5}	2.8×10^{-5}	1.2×10^{-5}
6	Percentage[b] [%]		> 80			~ 70	

1	260	270	280	290	300	400
-1-	*-9-*	*-10-*	*-11-*	*-12-*	*-13-*	*-14-*
5	8.1×10^{-6}	5.7×10^{-6}	4.0×10^{-6}	2.9×10^{-6}	2.6×10^{-6}	$< 1\times10^{-7}$
6			~ 70			< 10

B. IR Region

1	Wavelength λ [µm]	1.5–9.0	9.0	9.4	9.8	10.2	10.6	11.0
-1-	*-2-*	*-3-*	*-4-*	*-5-*	*-6-*	*-7-*	*-8-*	*-9-*
2	All salts a^*_S [m^2g^{-1}]	~ 0	2.9×10^{-6}	2.9×10^{-6}	2.9×10^{-6}	5.7×10^{-6}	5.7×10^{-6}	8.6×10^{-6}
3	Percentage[b] [%]	~ 0	2	1.9	1.7	2.9	2.4	2.8

1	11.8	12.3	12.6	13.0	13.4	13.8	14.2	14.6	15.5
-1-	*-10-*	*-11-*	*-12-*	*-13-*	*-14-*	*-15-*	*-16-*	*-17-*	*-18-*
2	1.1×10^{-5}	2.0×10^{-5}	2.9×10^{-5}	2.9×10^{-5}	2.9×10^{-5}	2.6×10^{-5}	2.4×10^{-5}	2.0×10^{-5}	1.7×10^{-5}
3	2.8	3.8	4.3	3.1	2.8	2.4	2.0	1.7	1.4

[a] Estimated on the basis of empirical data obtained from the following papers: Popov et al. (1979), Ivanoff (1978), Copin-Montegut et al. (1971), Armstrong and Boalch (1961), Kopelevich and Shifrin (1981), and Kopelevitch (1983).
[b] In Tables A (line 6) and B (line 3) we give the approximate percentage of absorption by the salt in the total absorption of light by model artificial sea water (an aqueous solution of all the main sea-salt components) with a salinity of c. 35 PSU, that is, typical of open ocean water.

At the same time, the salts listed in Table 2.18 are by weight the second largest group of sea water components after the water itself. In addition, nutrient salts, nitrates in particular (also listed in Table 2.18), although present in lower concentrations than the salts mentioned earlier, are also to some extent optically significant. Their concentrations fluctuate considerably in time, however, and are spatially extremely variable (e.g., Popov et al. (1979)).

When interacting with the strongly polar molecules of water, these dissolved mineral salts dissociate into positive metal ions and negative acid residues. The nondissociating H_3BO_3 is exceptional in this respect. As a result

TABLE 2.17. Occurrence and functions of the principal constituents of sea water in the World Ocean.

Constituent	Source literature	Total content in the world ocean (tons)	% of Total mass	Mean concentration [$\mu g \cdot dm^{-3}$]	Principal biological and optical functions
-1-	*-2-*	*-3-*	*-4-*	*-5-*	*-6-*
Water	Ivanenkov 1979a	1.36×10^{18}	96.6	9.92×10^{5}	The environment of life; sources of tissue fluids, Hydrogen and Oxygen; light absorption and scattering.
Total admixtures	Ivanenkov 1979a	4.78×10^{16}	3.40	3.49×10^{4}	
Ions of the principal salts					
Cl^-, Na^+, Mg^{2+}, SO_4^{2-}, Ca^{2+}, K^+, Br^-, HCO_3^-, Sr^{2+}, F^{2-}, H_3BO_3	Ivanenkov 1979a	4.78×10^{16}	3.40	3.49×10^{4}	Source of catalytic and some structural nutrients (low biological activity); optically practically inactive.
Trace elements					
Li, Rb, P, I, Ba, Mo, Fe, Zn, As, V, Cu, Al, Ti	Ivanenkov 1979a	3.00×10^{11}	2.13×10^{-5}	2.19×10^{-1}	
Other trace elements					
(c. 50 elements)	Ivanenkov 1979a	3.00×10^{10}	2.13×10^{-6}	2.308×10^{-2}	
Biologically active					
Dissolved gases					
N_2, O_2, CO_2, Ar	Ivanenkov 1979a	3.24×10^{13}	2.30×10^{-3}	2.36×10^{1}	O_2, CO_2: sources of oxygen and carbon; optically inactive.
Inorganic nutrients					
C	Alekin 1966	1.15×10^{14}	8.17×10^{-3}	8.27×10^{1}	Active sources of nutrients; optically inactive.
Si	Vinogradov 1967	3.60×10^{13}	2.55×10^{-3}	2.59×10^{1}	
N	Ivanenkov 1979b	2.90×10^{12}	2.06×10^{-4}	2.08	
P	Vaccaro 1965	5.80×10^{11}	4.12×10^{-5}	4.17×10^{-1}	
	Ivanenkov 1979b	9.90×10^{10}	7.04×10^{-6}	7.12×10^{-2}	

(*Continued*)

TABLE 2.17. Occurrence and functions of the principal constituents of sea water in the World Ocean.— Cont'd.

Constituent	Source literature	Total content in the world ocean (tons)	% of Total mass	Mean concentration [$\mu g \cdot dm^{-3}$]	Principal biological and optical functions
-1-	-2-	-3-	-4-	-5-	-6-
Optically active					
Organic substances in organisms					
Total[a]	Alekin 1966	6.00×10^{10}	4.26×10^{-6}	4.32×10^{-2}	Fountain of life; potential sources of nutrients; absorption and scattering.
	Bogorov 1968	3.30×10^{9}	2.36×10^{-7}	2.37×10^{-3}	
Chlorophyll	Whittaker & Likens 1975	1.8×10^{7}	1.28×10^{-9}	1.29×10^{-5}	
Organic substances in the environment					Potential or active sources of nutrients; light absorption and scattering.
(dissolved + suspended)[a]	Alekin 1966	2.00×10^{13}	1.42×10^{-3}	1.43×10^{1}	
	Skopintsev 1971	4.00×10^{12}	2.84×10^{-4}	2.88	

[a] The resources of inorganic carbon and the diverse forms of organic matter have been variously assessed by different authors. The relevant extreme values are given in the tables.

Table 2.18. Approximate average concentrations in ocean water of the principal salt anions and cations, and also nitrates.

Ion or Molecule	Average Concentration in ‰ by Weight[a]
	Principal Salts
Cl^-	18.98
SO_4^{2-}	2.65
HCO_3^-	0.139
Br^-	0.065
F^-	0.001
Na^+	10.56
Mg^{2+}	1.27
Ca^{2+}	0.40
K^+	0.38
Sr^{2+}	0.013
H_3BO_3	0.005
	Nitrates
NO_3^-	0.0015

[a] Based on data given by Popov et al. (1979), Vaccaro (1965), Ivanov (1975), and Brown et al. (1989).

of the change in their electronic, oscillational, and rotational states, neutral molecules and inorganic ions absorb radiation in various regions of the electromagnetic spectrum. Lenoble (1956) and Copin-Montegut et al. (1971) demonstrated empirically that the absorption of radiation by aqueous solutions of most of these fundamental mineral components of sea water takes place mainly in the ultraviolet region of the spectrum (230–300 nm; see Figure 2.16). The absorbed quanta of radiation induce a change in the electronic oscillational-rotational state of the ion or molecule, and in doing so, give rise to an absorption band of varying breadth and intensity. The intensity of these absorption bands in the UV region rises monotonically with the increase in the energy of the interacting quanta, that is, with a fall in wavelength (Armstrong and Boalch 1961). The further rise in the absorption coefficients in the $\lambda \in$ 230–300 nm region is due largely to the bromide ion Br^- (Ogura and Hanya 1967) (the Br^- curve in Figure 2.16). Nitrate ions NO_3^- can also absorb strongly in this region (the NO_3^- curve in Figure 2.16). Below 200 nm, the empirical spectra of radiation absorption by sea water components have not been subjected to detailed scrutiny. It should be noted further that the coefficients of UV light absorption by sea salts are larger in value than the corresponding coefficients for water molecules. (Figure 2.16b). So in this region, the percentage share of salts in the total absorption of radiation in model (artificial) sea water of salinity 35 PSU is from 70 to >80% (see line 6 in Table 2.16). In less saline waters this value will, of course, be proportionally less.

The IR absorption properties of sea salts have been characterized by Friedman (1969), Pinkley and Williams (1976), and Pinkley et al. (1977), among others. These authors have shown that mineral salts and their ions absorb IR radiation mainly in the wavelength interval $9 < \lambda < 15$ μm (Figure 2.16c), forming oscillational-rotational absorption bands consisting of numerous overlapping spectral lines. The intensity of these bands, however, is weak, far more so than that of the corresponding bands for the water itself. As we can see from line 3 in Table 2.16B, the contribution of salts to the total absorption of IR radiation in sea water of salinity 35 PSU is no more than a few percent. As a result, these bands are practically invisible against the background of the IR absorption band of water. That is why the empirical spectrum of IR absorption by sea water is practically identical in all seas, that is, is independent of salinity.

As far as visible radiation is concerned, solutions of mineral salts behave as if they were optically transparent; that is, they transmit the entire range of the VIS spectrum, with hardly any being lost through absorption. For one thing, this is because none of the metals in the principal components of sea salt are spectrally active, either in the visible or in the near ultraviolet (see Table 2.15). What is significant, however, is that salts exert an important influence on the molecular scattering of light in sea water: in oceanic waters the components of sea salt are responsible for c. 30% of the scattering due to all the molecules present in sea water (Morel 1974, Kopelevitch 1983).

2.3.3 Inorganic Complex Ions

Mineral salts and dissolved gases have practically no effect on the absorption properties of sea water with respect to visible radiation: this is a characteristic feature of the vast majority of sea and oceanic waters. Nevertheless, Kondratev and Pozdniakov (1988) have shown that such an effect can be detected in highly mineralized waters, such as saline lakes with high concentrations of transition metals,[21] mostly manganese and iron (Alekin 1970, Kondratev et al. 1990). Ions of these metals absorb visible light (Table 2.15). Moreover, under appropriate environmental conditions (suitable pH and temperature), these ions can form complexes with water molecules and other inorganic compounds that function as ligands. Complex formation between metals and ligands can potentiate absorption capabilities of the former. The basic theory of the absorption spectra of such complexes can be found, for example, in Gołębiewski (1982), Babko and Pilipienko (1972), and Barrow (1969). Figure 2.17 illustrates visible extinction spectra $\varepsilon(\lambda)$[22] of aqua-complexes of manganese and nickel.

---

[21] For example, the average mineralization of Lake Ladoga is 56 $mg \cdot dm^{-3}$ (after Raspletina et al. (1967)), whereas the metal content in Lake Balkhash near the mouth of the river Ili may be several times greater (Alekin 1970).

[22] See the definition of ε in the footnote 2 to Chapter 3.

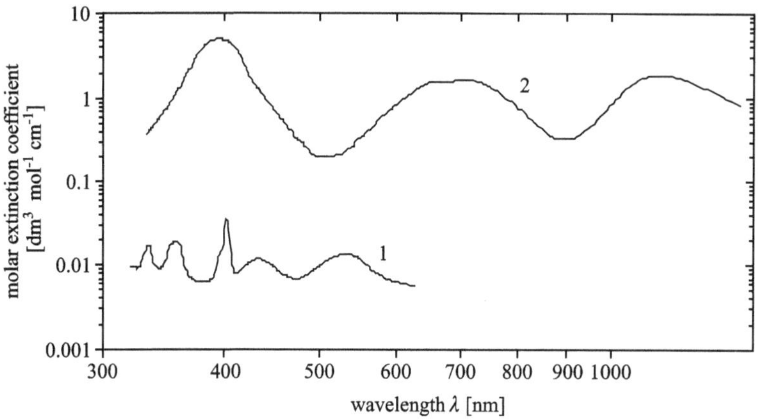

FIGURE 2.17. Spectra of molar extinction coefficients ε for aqua-complexes of manganese ($[Mn(H_2O)_6]^{2+}$) (1) and nickel ($[Ni(H_2O)_6]^{2+}$) (2). (Data from Barrow (1969).)

TABLE 2.19. Absorption band peaks (λ_{max}) and colors of selected complex ions.[a]

Complex	λ_{max} [mn]	Color
-1-	*-2-*	*-4-*
$[Co(CN)_6]^{3-}$	300	Colorless
$[Co(NH_3)_6]^{3+}$	330, 470	Yellow red
$[Cr(H_2O)_6]^{3+}$	405, 580	Purple
$[Ti(H_2O)_6]^{3+}$	545	Purple red
$[Co(H_2O)_6]^{2+}$	500, 900	Pink
$[CoCl_4]^{2-}$	665	Blue
$[Cu(NH_3)_4]^{2+}$	640	Blue
$[Cu(H_2O)_6]^{2+}$	790	Pale blue
$[Ni(H_2O)_6]^{2+}$	500, 705, 880	Green
$[Mn(H_2O)_6]^{2+}$	305, 350, 400, 465, 555	Very pale pink

[a] Data taken from Phillips and Williams (1966); see also Campbell and Dwek (1984).

Table 2.19 gives the rough positions of absorption band peaks and the colors of selected complexes of transition metals with inorganic ligands. This absorption can cause a change in the color of natural waters, especially of inland waters, whose optical properties are far more complicated and much less well understood that those of sea waters. Nonetheless, the effects just described may well occur in certain regions of the sea, although on a smaller scale. Unfortunately, however, we are not aware of any papers that can provide confirmation of this.

3 The Interaction of Light with Organic Molecules Present in Sea Water: Physical Principles

In sea water, light is absorbed primarily by the molecules of H_2O and of complex organic compounds. The latter may be dissolved in the water, or contained in living organisms (mainly plankton) and organic detritus. By "light" we are thinking here in particular of the ultraviolet (UV) and visible (VIS) regions, where the energies of the photons are sufficiently high for transitions between the electron levels of the molecules to occur. The reactions between photons and the electron states of organic molecules are thus of fundamental importance to a whole range of biological processes taking place in living organisms, such as photosynthesis; they also determine the course of photochemical processes occurring in conjunction with nonliving organic matter. In this chapter we briefly discuss the mechanisms by which the electron spectra of the organic molecules ubiquitous in seas and oceans are formed.

3.1 The Characteristic Absorption Properties of Simple Chromophores in Organic Molecules

The energies of photons (corresponding to wavelengths) absorbed or emitted as a result of discrete changes in the energy states of molecules are fixed, and are further limited by quantum mechanical rules of selection. In the case of small molecules, for example, diatomic ones, and some linear or regular molecules made up of quite a large number of atoms, the quantitative definition of these states and selection rules with the aid of quantum mechanical equations is a tedious process, but nonetheless achievable with almost 100% precision. However, the matter becomes very much more complicated when we wish to determine these states in the nonlinear case for polyatomic molecules, for example, organic substances in the sea, or the pigments in the cells of marine phytoplankton. Despite the tremendous advances in the methodology of quantum-mechanical calculations brought about by computerization, the mathematical description of such complex molecules is extremely arduous and time-consuming. Based as it is on highly complicated theoretical and semi-empirical models, this description and its discussion go beyond the scope of this book: readers interested in this question are referred to monographs on quantum mechanics, quantum chemistry, and molecular spectroscopy, for example: Brehm and Mullin (1989), McQuarrie (1983), Gołębiewski (1982, 1973, 1969), Rao (1981), Cohen-Tannoudji et al. (1977), Bergs (1975), Kęcki (1992), Herzberg (1950, 1963–1967, 1992), and also to the hundreds of papers published in specialist journals.[1] Here we only explain briefly and very simply the most important concepts and facts enabling the reader to gain an understanding of the optical properties of the complex polyatomic molecules present in sea water.

There can be no doubt that the molecular spectra of the absorption or emission of radiation are determined by the structure of the whole molecule. Nevertheless, the interpretation of such spectra of polyatomic molecules becomes much easier if we assume that the various components of these spectra are due not to the molecule in its entirety, but to larger or smaller fragments of it known as chromophoric groups, or simply chromophores. Originally, the concept of a chromophore was linked to the visible color of chemical compounds. As we know, color is produced when light components of some of the wavelengths of the white light passing through a substance are absorbed by it and the emergent light is of a complementary color. That component of a molecule imparting color to the substance is called a chromophore. Substances containing the "blue-violet" chromophore are very common in the sea, and, absorbing blue and violet light as they do, these substances have a characteristic yellow color. The concept of a chromophore was later extended to include UV light, which is invisible to the eye (e.g., Kęcki (1992)). It is now commonly used in oceanography (e.g., Vodacek et al. (1997, 1995) and Matsunaga et al. (1993)). In this book we take the concept of a chromophore to mean a group of atoms and chemical bonds with characteristic spectral features in the UV–VIS range that depend only marginally on the other properties of the other parts of the molecule.

The electron states of linear molecules at different levels of excitation n are described by the electron quantum number Λ, which signifies the absolute value of the projection of the total angular momentum on its axis. As applied to the simple molecules described in Chapter 2, such a description is possible. In nonlinear molecules, however, the magnitude of this quantum number loses its physical sense because the internuclear axes are multidirectional. In this case, the electron energy states can be defined in relation to various components of the molecule's symmetry with the aid of group theory symbolism (e.g., Barrow (1969)).

With respect to organic molecules occurring in the sea, such a procedure is complex in the extreme because, for one thing, most of these molecules

[1] For example: *Theor. Chem. Phys.*, *J. Chem. Phys.*, *Intern. J. Quantum Chem.*, *J. Am. Chem. Soc.*, *Mol. Phys.*, *Naturforsch*, *Adv. Quantum Chem.*, *Chem. Phys. Lett.*, *Photochem. Photobiol.*, and *Acta Phys. Polon.*

contain a very large number of diverse, independent chromophore groups, each one of which merits a separate description. In this case, a simplification often applied in the spectroscopy of organic compounds is to analyze the energy states not of the whole molecule or chromophore, but its separate, single valence electrons, in other words, the electrons potentially responsible for the formation of electron spectra in the UV–VIS spectral interval. For a quantitative description of these states we use the so-called "single-electron" approximation (e.g., Gołębiewski (1982) and Haken and Wolf (1995 [1998])). In accordance with this approximation we consider the motion of any electron in an atom or molecule in relation to the total resultant of the electric field of the atom's nucleus (or molecule's nuclei) and the remaining electrons. We further assume that a change in the energy level of this electron has no effect on the other electrons. By making this assumption we can obtain descriptions of the ground and excited energy states of the valence electrons. Although not always wholly correct quantitatively, such a description does provide a clear picture of the spectrum of light absorption by complex molecules, which makes interpreting the absorption process that much easier.

As we already mentioned in Section 2.1, the quantum mechanical magnitude characterizing the state of an electron in an atom or molecule is the single-electron wave function $\varphi_e(x,y,z,s)$, known as the spin orbital (see Equation (2.4)). Atomic orbitals and molecular orbitals, related to the spin orbital, describe the spatial distribution of the electron charges and are characterized by various shapes, sizes, and localizations within the atom or molecule. These depend on the type of atoms or molecular bonds and on the electronic configuration. Examples of valence electron orbitals (outermost shells), in particular, atoms and molecules of carbons and hydrocarbons, are shown in Figures 3.1 and 3.2.

Responsible as they are for light absorption in the UV–VIS range, valence electrons can thus be involved in chemical bonds to a greater or lesser extent. In molecules of organic compounds we usually come across three kinds of such electrons and their corresponding molecular orbitals: σ, π, n. The principal differences between them are in their positions, orbital shapes, and energy levels (see Figures 3.2 and 3.3). These features depend not only on the type of chemical bond, but also on the kind of atoms present in the chromophore groups of the molecule.

Characteristic of chromophores with single bonds are σ electrons. A pair of these electrons forms a σ orbital, whose axis of symmetry is the axis of the chemical bond (see, e.g., the σ_{CH} orbital in the formaldehyde molecule in Figure 3.2). These σ orbitals are typical of saturated compounds (i.e., with single bonds), such as aliphatic hydrocarbons.

In the case of chromophores containing unsaturated bonds, for example, double bonds, two-electron σ bonds also occur (see σ_{CO} in Figure 3.2); in addition, the other two electrons are of the π type and form a two-electron π orbital (see π_{CO} in Figure 3.2). These orbitals, which typically transfer charge beyond the bond axis, form electron clouds whose axis of symmetry

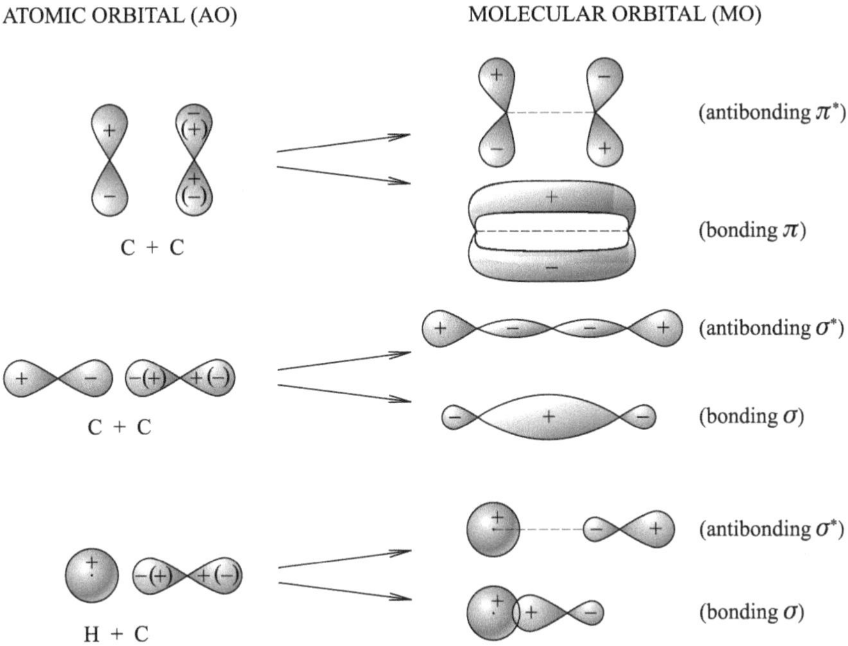

FIGURE 3.1. Ways in which atomic orbitals (AO) combine to form molecular orbitals (MO).

is perpendicular to the bond axis. The same applies to triple bonds: in addition to the two-electron σ bond, there are two 2-electron π orbitals, with two different energies. The simultaneous occurrence of σ and π electrons is, among other things, characteristic of unsaturated hydrocarbons and aromatic compounds.

Because of the position of the electron charge, both σ and π orbitals are said to be "delocalized", which means that the electron cloud has shifted from a single atom onto the entire bond. Such delocalized orbitals can be bonding or antibonding (see σ_{CH}, σ_{CO}, σ_{CC}, π_{CO}, π_{CC}, and σ^*_{CH}, σ^*_{CO}, σ^*_{CC}, π^*_{CO}, π^*_{CC} in Figures 3.1 and 3.2). In the first case the charge clouds coalesce and are characteristic of molecules in the ground state. But after excitation, the electron clouds separate to a greater or lesser extent and the orbitals become antibonding. Note that excited antibonding orbitals are similar to atomic orbitals.

Unlike the σ and π bonding electrons we have just been discussing, the third kind of electron characteristic of organic molecules, the n electrons, hardly participate at all in chemical bonds and are referred to as nonbonding. They are typical of molecules containing atoms of such elements as

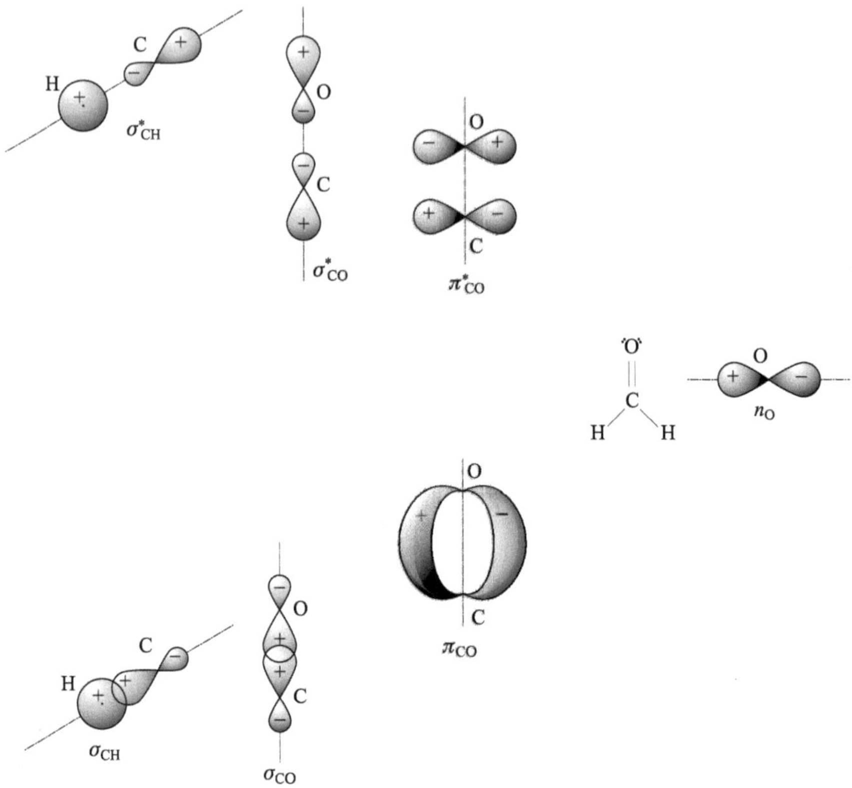

FIGURE 3.2. Orbitals of the H_2CO molecule: there is one lone pair of electrons in the nonbonding orbital n_O. In fact, however, there are two such orbitals with different energies, because two of the six valence electrons in the oxygen atom are used to form a double bond σ_{CO} and π_{CO} (explanation in the text). (Reproduced from Z. Kęcki (1992), *Fundamentals of Molecular Spectroscopy,* third edition, amended and expanded (in Polish), with the kind permission of PWN Publishers, Warszawa.)

oxygen, nitrogen, and sulfur, and appear in chromophore regions such as C = O, C = S, and – N = N –. They are therefore valence shell electrons not used in chemical bonds. Their charges spread out in the neighborhood of the "maternal" atom and, unlike the delocalized σ and π electrons, n electrons correspond to the so-called localized n molecular orbitals, with shapes approaching those of atomic orbitals. Such orbitals are occupied, for instance, by the two lone pairs of electrons of the oxygen atom in the formaldehyde molecule (see the n_O orbital in Figure 3.2).

Recapitulating then, we can distinguish simple chromophores, which in a molecule can take the form of single, double, or triple bonds as well as lone electron pairs. Depending on where these structural elements occur, an organic molecule can have a larger or smaller number of valence electrons of different kinds occurring in a diversity of mutual proportions:

Type of chromophore:	Number and type of bonding electrons	Type of compound
Single bonds	2σ	Saturated aliphatic hydrocarbons
Double bonds	$2\sigma + 2\pi$	Unsaturated and aromatic hydrocarbons
Triple bonds	$2\sigma + 2\pi_1 + 2\pi_2$	Unsaturated hydrocarbons
A lone pair of electrons and:		Organic molecules containing, e.g., oxygen, sulfur, nitrogen, and halogens (not occurring in pure saturated hydrocarbons)
—a single bond	$2\sigma + 2n$	
—a double bond	$2\sigma + 2\pi + 2n$	
—a triple bond	$2\sigma + 2\pi_1 + 2\pi_2 + 2n$	

Quantitatively, we can approximately determine the quantized energy of electrons by assuming the single-electron approximation discussed above and solving the Schrödinger equation (Love and Peterson 2005). The results of these solutions, which present the relative mutual relations between the energies of various electrons in the ground state and in excited states as well as the possible energy transitions (which are accompanied by absorption or emission of quanta of radiation), are illustrated in Figure 3.3 by the example of a chromophore (molecule) containing all three kinds of electrons: σ, π, n.

Let us consider first of all a molecule in the ground state. Because the n electron does not participate in the chemical bond, its energy is by convention equal to zero. The formation of σ or π bonds involves the release of

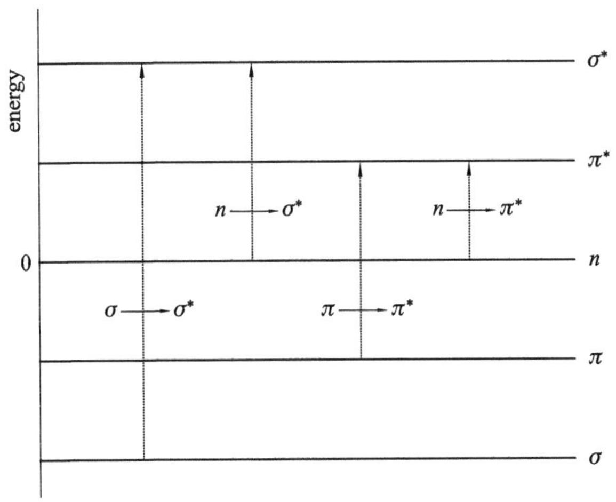

FIGURE 3.3. Energy levels and possible electronic transitions in a hypothetical molecule containing all three kinds of electrons: σ, π, n. The energy state of the nonbonding n electron is by convention put equal to zero. (Reproduced from Z. Kęcki (1992), *Fundamentals of Molecular Spectroscopy*, third edition, amended and expanded (in Polish), with the kind permission of PWN Publishers, Warszawa.)

energy, hence the energies of bonding electrons are by convention negative. A σ bond is more stable than bonds formed by the mobile π electrons, hence the energies of σ electrons are lower (larger negative values) than those of π electrons. Following the absorption of a quantum of energy by any one of the valence electrons (π and σ or n), its orbital becomes antibonding, correspondingly σ^* or π^*. The energies of antibonding states are higher than those of the nonbonding state n, so by convention the energies of σ^* and π^* electrons are positive. This implies repulsion between the atoms, and thus a weakening of the chemical bond. Figure 3.3 illustrates only the positions of the lowest of the many possible excited σ^* and π^* energy states, that is, those corresponding to the absorption of the least energetic (longest-wave) quanta. When the excitation of all the bonding electrons is sufficiently great, the chemical bond breaks.

Shown diagrammatically in Figure 3.3, the energy distances between the ground states of σ, π, and n electrons and the excited states σ^* and π^* in individual molecules can vary. Usually, however, the possible spectral transitions arrange themselves in the following series of decreasing transition energies,

$$\sigma \rightarrow \sigma^* > n \rightarrow \sigma^* > \pi \rightarrow \pi^* > n \rightarrow \pi^*,$$

which yields relations between the wavelengths of the quanta corresponding to these transitions:

$$\lambda_{\sigma \rightarrow \sigma^*} < \lambda_{n \rightarrow \sigma^*} < \lambda_{\pi \rightarrow \pi^*} < \lambda_{n \rightarrow \pi^*}.$$

This is confirmed by the data given in Table 3.1, which sets out selected optical characteristics (as decimal molar extinction coefficient ε and mass-specific absorption coefficients: a^*_{OC} and a^*_{OM})[2] of the simple chromophores most frequently occurring in organic molecules.

---

[2] In molecular spectroscopy the molar, decimal extinction coefficient ε [$dm^3\ mol^{-1}\cdot cm^{-1}$] was traditionally used to describe the absorption properties of a solution of a substance. The coefficient ε [$dm^3\ mol^{-1} cm^{-1}$] is given by the equation $\varepsilon = \frac{1}{M \cdot l} \log T^{-1}$, where M [$mol\ dm^{-3}$] is the concentration of the solution; l [cm] is the length of the optical cell containing the solution under investigation; and T is the transmittance of the solution in the cell, that is, the ratio of the light beam intensity I transmitted through the solution under investigation to the incident intensity I_0 (loss of light due to scattering is neglected as it is very small compared to its loss due to absorption). The coefficient ε defined in this way is related to the corresponding mass-specific absorption coefficients a^* of the substance in question as follows.

$$a^*_{OM}\left[m^2 g^{-1}\right] = \frac{(\ln 10)\,\varepsilon}{10 A_M} \approx 0,23 \frac{\varepsilon}{10 A_M} \text{ and } a^*_{OC}\left[m^2 g^{-1}\right] \approx 0,23 \frac{\varepsilon}{10 A_C},$$

where A_M is the mass number of its molecules (i.e., the mass [g] in 1 mol of the substance in the solution); and A_C is the mass number of all the atoms of carbon in the molecule (i.e., the mass of carbon molecules [g] in 1 mol of the substance in the solution). In oceanography absorption coefficients a and a^* are commonly applied.

TABLE 3.1. The principal absorption bands of the most important simple organic chromophores occurring in the sea.[a]

Chromophore	Molecule	λ_{max}[b] [nm]	ε_{max}[c] [dm^3 mol^{-1} cm^{-1}]	a^*_{OC}[d] [m^2g^{-1}]	a^*_{OM}[e] [m^2g^{-1}]
-1-	*-2-*	*-3-*	*-4-*	*-5-*	*-6-*
		$\sigma \rightarrow \sigma^*$ Transitions			
H–C	CH_4 (Gaseous phase)	120	15000	287	216
C–C	$H_3C–CH_3$ (Gaseous phase)	135	high	high	high
		$n \rightarrow \sigma^*$ Transitions			
C–Cl	$H_3C–Cl$	173	200	3.83	0.92
C–O	$H_3C–OH$	184	150	2.88	1.08
C–O–C	$H_3C–O–CH_3$	188	1900	18.2	9.5
C–S–C	$H_3CH_2C–S–CH_2CH_3$	194	5000	24.0	12.8
S–S	$H_3CH_2CS–S–CH_2CH_3$	194	5000	24.0	9.42
C–Br	$H_3C–Br$	204	200	3.83	4.84
>N–	CH_3NH_2	215	600	11.5	4.45
	$(CH_3)_3N$	227	900	5.75	3.51
C–I	$H_3C–I$	259	360	6.90	0.58
		$\pi \rightarrow \pi^*$ Transitions			
C=C	$H_2C=CH_2$	162.5	15800	151	130
	$RHC=CH_2$	175	12600		
	$R_2C=CH_2$	187	8000		
	$(H_3C)_2C=(CH_3)_2$	196.5	12600	40.3	34.5
C≡C	$HC≡CH_3$ (Gaseous phase)	173	6300	60.4	55.7
	RC≡CH	187	450		
	RC≡CR	191	850		
C=O	$H_2C=O$	175	20000	383	153
	$(CH_3)_2C=O$	188	2000	12.8	7.93
C=N	$(CH_3)_2C=NOH$	193	2000	12.8	6.30
C=C–C=O	$H_2C=CH–CH=O$	208	10000	63.9	41.1
C=C=C	$H_5C_2HC=C=CH_2$	225	500	2.0	1.7
		$n \rightarrow \pi^*$ Transitions			
C=O	$CH_3OHC=O$	197	63	0.60	0.21
	$H_2C=O$	280	20	0.38	0.15
	$(CH_3)_2C=O$	279	16	0.102	0.063
	$H_3CHC=O$	294	12	0.115	0.063
NO_2	CH_3NO_2	278	20	0.383	0.075
N=O	$(CH_3)_3CN=O$	300	100	0.479	0.264
	$(CH_3)_3CN=O$	665	20	0.096	0.053
C=C–C=O	$H_2C=CH–CH=O$	328	12	0.077	0.049
N=N	$H_3CN=NCH_3$	350	4.5	0.043	0.018
>C=S	H2CS	330			

[a] After various sources, for example, Scott (1964), Kęcki (1992), and Garaj et al. (1981).
[b] The wavelength defining the position of the main absorption peaks in the spectrum.
[c] Decimal molar extinction coefficient ε at the maximum.
[d] The specific light absorption coefficient at the absorption band maximum referred to the mass of carbon in the compound.
[e] The specific absorption coefficient at the absorption band maximum referred to the whole mass of the compound;
the symbol R in the table indicates a substituent, usually the CH_3 group.

Among other things, this table gives the wavelength λ_{max} at the absorption band maximum corresponding to the lowest energy transitions, as well as various optical parameters such as molar extinction coefficients and specific absorptions, which characterize the intensity of these bands. All the numerical values of these parameters given in Table 3.1 should be treated as approximate; in actual fact they are dependent on the environment, for example, on the type of solvent, the pH of the solution, or the temperature. We return to the explanation of the characteristic intensities of the absorption bands a little later.

In the meantime, let us turn our attention to the regularities relating to the wavelengths λ_{max} of photons for the various types of energy transition. As we can see, $\sigma \rightarrow \sigma^*$ transitions are characteristic of the gaseous phase and yield bands lying in the far, so-called vacuous ultraviolet. For example, the absorption band due to the H–C chromophore in methane (CH_4) has a maximum for the wavelength of c. 120 nm. Not much longer is λ_{max} for the C–C chromophore, which in ethane (C_2H_6) is c. 135 nm. However, this type of radiation does not occur in the sea, and so light absorption by all forms of "pure" saturated hydrocarbons (i.e., containing only H–C and C–C groups) can be disregarded as far as the marine environment is concerned. What is significant, however, is absorption by π-electron molecules: by this we mean all molecules containing π electrons (as well as other kinds of electrons). These organic compounds can be classified into two groups:

- "Pure" unsaturated and aromatic hydrocarbons, that is, containing σ and π electrons. In the sea these come from petroleum-derived pollutants, and are present in carotenoid pigments, to mention two examples.
- Compounds with admixtures forming lone electron pairs, that is, containing σ, π, and n electrons. A very common example of such compounds is molecules containing a carbonyl group (C=O) (e.g., formaldehyde), and also a whole range of complex natural organic substances occurring in the sea. These include molecules with atoms of nitrogen (N), sulfur (S), and others (e.g., pigments from the chlorophyll groups and phycobilin).

In the first group we have $\sigma \rightarrow \sigma^*$ and $\pi \rightarrow \pi^*$ transitions, in the second group, not only $\pi \rightarrow \pi^*$ but also $n \rightarrow \pi^*$ and $n \rightarrow \sigma^*$ transitions. As Table 3.1 shows, the light absorption bands of $n \rightarrow \sigma^*$ and $\pi \rightarrow \pi^*$ transitions are in the mid-UV region of the electromagnetic spectrum, and those of $n \rightarrow \pi^*$ transitions are in the near-UV and VIS regions. Because the $\pi \rightarrow \pi^*$ transitions produce the most intense absorption bands, the use of the term "π-electron molecules" to describe all organic molecules except saturated hydrocarbons is justified.

These regularities refer to the wavelengths of photons absorbed or emitted as a result of electronic transitions. It remains to explain the problem of the intensity of these absorption bands. Without involving ourselves in a detailed discussion of quantum mechanics, it is enough to state that on the basis of probability considerations, we can derive the following approximate formula for the molar extinction coefficient ε_{max} (i.e., ε at the maximum; e.g., Garaj et al. (1981)):

$$\varepsilon_{max}\ [\mathrm{dm^3(mol{\cdot}cm)^{-1}}] = 0.87 \times 10^{20} Pa, \tag{3.1}$$

where P is the probability of a transition, and a is the cross-section of a molecule, that is, the effective area of the chromophore absorbing radiation, expressed in cm^2. The magnitudes of these molecular cross-sections are of the order of 10^{-16} cm^2.

The value of the probability P depends on the degree of conformity of a given transition with a range of factors and selection rules. Here we give the most important ones.

1. For the probability of a transition to be other than zero, there must exist two such quantum states of a molecule, for which the energy difference ΔE between the electron states corresponds to the energy of an incident photon: $\Delta E = h\nu$. This is the principal rule of selection, to which there are no exceptions. The conformity of these energies is therefore a necessary condition, but not a sufficient one for an energy transition to take place.
2. According to quantum theories, only those electronic transitions are possible for which the initial and final states have the same multiplicity. By multiplicity we mean the magnitude $M = 2S + 1$, where $S = |\Sigma s_i|$ is the resultant spin of all the electrons s_i $(= \pm 1/2)$ in the molecule.[3] In the ground state a molecule is usually characterized by a spin of $S = 0$, which means that its multiplicity is $M = 1$, that is, a singlet state usually denoted by S_0 (Figure 3.4a). This is because the electrons in all the orbitals at lower energy levels and in the highest nonexcited orbital are paired and, in accordance

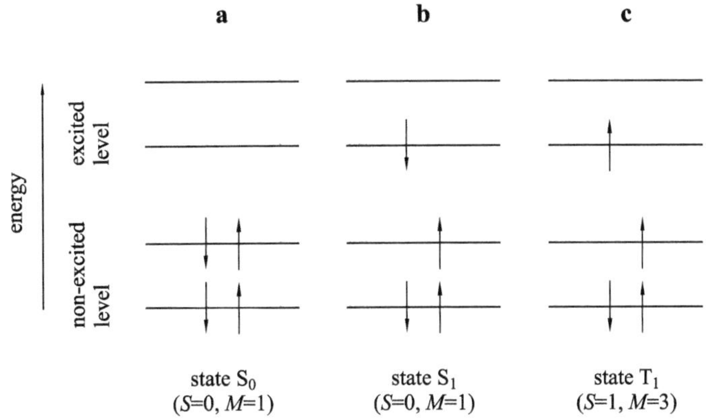

FIGURE 3.4. A simplified diagram of the multiplicity of molecular states: (a) singlet ground; (b) singlet excited; (c) triplet excited. The arrows denote the position and sense of electron spins: $s = -\frac{1}{2}\downarrow$ and $s = +\frac{1}{2}\uparrow$.

[3] When $S = 0$, the multiplicity $M = 1$, and the state is known as a singlet state; when $S = 1/2$, $M = 2$, such a state is a doublet state, and is typical of certain atoms or ions with an odd number of electrons; when $S = 1$, $M = 3$, we have a triplet state, and so on.

with the Pauli exclusion principle, have opposite spins. When one of the electrons becomes excited, the exclusion principle no longer applies, and two situations are possible: the first, when the electron has not changed its spin, that is to say, the resultant spin is still $S = 0$ and the excited state is also a singlet state denoted by S_1 (Figure 3.4b), and the second, when during a transition the direction of spin is reversed, so that the resultant spin is now equal to $S = 1$; that is, $M = 3$. This excited state is thus a triplet state denoted by the symbol T_1 (Figure 3.4c). In Figure 3.4 the states S_1 and T_1 are denoted as though the electrons in these states had the same energies. In fact this is an approximation, because the energies of triplet states are usually somewhat lower than those of singlet states.

The second of the above-mentioned rules, that is, the rule of conservation of multiplicity, can be written thus:

$$\Delta S = 0, \tag{3.2}$$

which means that the only permissible transitions are those between states of the same multiplicity, for example, singlet–singlet, doublet–doublet, or triplet–triplet. In principle, then, transitions between states of different multiplicities, for example, singlet–triplet, are forbidden. In reality, however, this selection rule, as with many other quantum-mechanical models, is an approximation. It does not take into account the more subtle mutual interactions between the various parts of the molecule's structure. The upshot is that permitted transitions ($\Delta S = 0$) are mostly characterized by a high probability approaching unity or of the order of tenth parts of unity. In addition to this, bands corresponding to forbidden transitions, that is, for $\Delta S \neq 0$, also appear in the spectra, although with much lower intensities. Table 3.2 illustrates this.

Most of the $\pi \rightarrow \pi^*$ transitions discussed above and practically all $\sigma \rightarrow \sigma^*$ transitions are permissible ones. Some $n \rightarrow \sigma^*$ transitions are also allowed. In contrast, nearly all $n \rightarrow \pi^*$ transitions take place with a change in multiplicity and are theoretically forbidden.

With the exception of selected $n \rightarrow \pi^*$ transitions, which produce the weakest absorption bands, the spectral features presented in Table 3.1 practically do not encroach into the visible region. On the other hand, we know that natural "associations" of organic compounds existing in Nature, such as organic substances in the sea, in the soil, or in animal and plant tissues, are

TABLE 3.2. Probability of transition and the molar extinction coefficient.

Kind of absorption band	Probability P	Molar extinction coefficient ε_{max} [$dm^3 mol^{-1} cm^{-1}$]
-1-	*-2-*	*-3-*
Strong bands (permitted)	0.1–1	10^4–10^5
Weak bands (forbidden)	<0.01	$<10^3$

frequently endowed with a more or less intense visible coloration, for example, the "green" chloroplasts, the parts of plant cells responsible for photosynthesis. This coloration cannot be explained exclusively on the basis of the spectral properties of the simple chromophores that we have been discussing so far. Nevertheless, it is the result of the same physical mechanisms and can be explained by the effects the chromophores have interacting on each other, both within the complex molecule itself, and between distinct molecules. This leads to the formation of complex chromophores. There are quite a number of these effects. The most important ones, responsible for the absorption of visible light, are:

1. Effects leading to the formation of conjugated chains of π-orbitals
2. Effects leading to the formation of various types of stable bonds and auxochromic complexes, as well as effects giving rise to "charge transfer"

The following discussion of these questions focuses mainly on the first of these effects, which determines the visible color of the associations of organic compounds occurring in seas and other natural bodies of water.

3.2 The Absorption Properties of Complex Organic Molecules with Conjugated π-Electrons

The great mobility of π-electrons and their fairly loose association with chromophore groups (small, negative bonding energies) means that π-orbitals belonging to the same molecule and localized close to each other coalesce. This effect is known as *conjugation* of π-orbitals.

3.2.1 Linear Polyenes

The most common example of π-orbitals in organic compounds are linear chain molecules in which C=C double bonds alternate with single bonds (Figure 3.5).

Known as linear polyenes, these compounds are biologically crucial: their derivatives include vitamins and organic pigments, among the latter, all the photosynthetic phytoplankton pigments of the carotenoid group.

The formation of a conjugated π-orbital in polyenes is illustrated in Figure 3.6. An "ordinary" π-orbital filled by two electrons is formed (with opposite spins) around the simple, starting C=C chromophore in the ethylene molecule.

However, in the butadiene molecule there are four π-electrons not connected with any particular bond, but that can move freely along the whole C=C–C=C chain, forming a doubly conjugated π-orbital. As a result of conjugation, each of the π and π^* energy levels is split into two in the butadiene molecule, into three in hexatriene, into four in octatetraene, and so

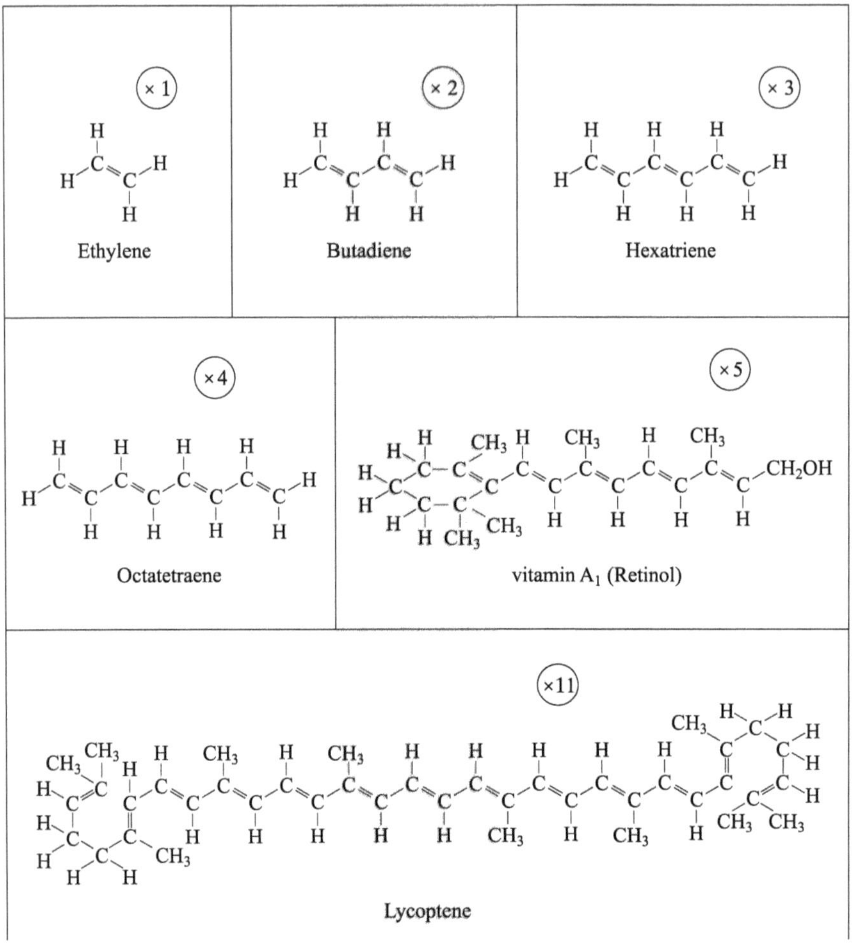

FIGURE 3.5. Conjugated C = C bond systems in the molecules of selected linear polyenes and their derivatives (the multiplicity of the conjugations is given in the circles).

on, as shown in Figure 3.7. This is, of course, a consequence of the Pauli exclusion principle, according to which each energy level can be filled with at most two electrons with opposite spins. As a result of this splitting, adjacent π and π^* energy levels approach each other; the closer, the more multiple bonds are conjugated with each other. As the multiplicity of the conjugation increases, so does the wavelength λ_{max}, the longest-wave absorption band corresponding to the $1 \rightarrow 1'$, or $2 \rightarrow 1'$, or $3 \rightarrow 1'$ transitions (see $h\nu_{min}$ in Figure 3.7), and also the probability of this transition, which is manifested by a rise in the molar extinction coefficients. This is confirmed by the data in Table 3.3A.

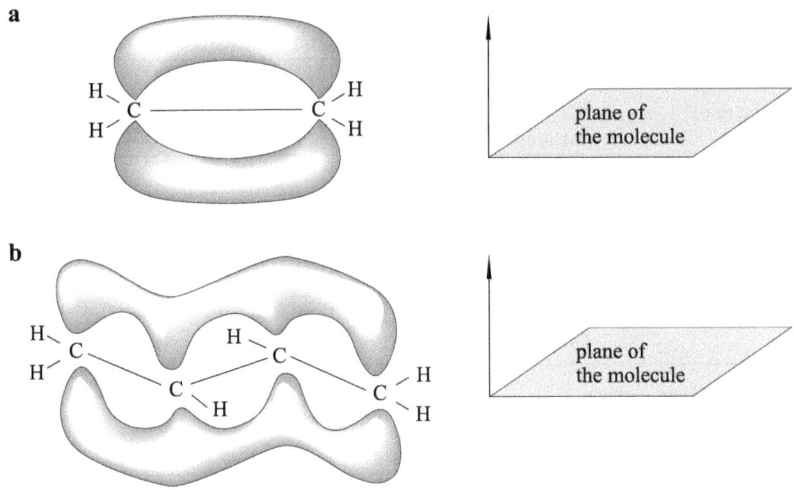

FIGURE 3.6. The approximate shape of simple (a) and conjugated (b) π-type bonding molecular orbitals: (a) in the ethylene molecule, (b) in the butadiene molecule. The figure shows the orbital cross-sections in the plane perpendicular to the plane of the molecule.

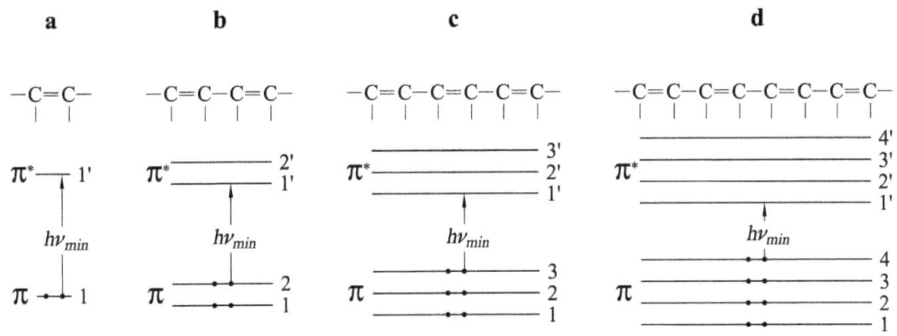

FIGURE 3.7. Diagram of π and π^* energy levels in polyenes. (a) ethylene; (b) butadiene; (c) hexatriene; (d) octatetraene. The dots indicate valence electrons in the ground state: $h\nu_{min}$ illustrates the minimum energy of absorbed light quanta corresponding to the $1 \rightarrow 1'$, $2 \rightarrow 1'$, $3 \rightarrow 1'$, $4 \rightarrow 1'$, . . . , transitions; the Arabic numerals 1,2,3,4 denote the successive nonexcited levels of valence electrons, and the numerals $1'$,$2'$, $3'$, $4'$ denote their corresponding excited levels.

As we can see, the last of these compounds, constructed on a foundation of an elevenfold conjugated C=C bond, called lycoptene (or γ-carotene in the nomenclature of photosynthetic pigments), is a very strong absorber of visible light, with a maximum for a wavelength of c. 470 nm. The enhanced multiplicity of the conjugation is also the reason for the enriched electronic spectral structures of the molecules. Apart from the longest-wave band

TABLE 3.3. The effect of chromophore conjugation on the absorption bands of organic compounds, on linear polyenes and aromatic hydrocarbons.

A. Linear Polyenes

Multiplicity of conjugation	Molecule	λ_{max} [nm]	ε_{max} [dm^3 mol^{-1} cm^{-1}]	a^*_{OC} [m^{2}g^{-1}]	a^*_{OM} [m^{2}g^{-1}]
-1-	*-2-*	*-3-*	*-4-*	*-5-*	*-6-*
Basic Chromophore	Ethylene	163	10000	95	82
2	Butadiene	217	21000	101	89
3	Hexatriene	252	35000	111	101
4	Octatetraene	304			
5	Vitamin A	328			
11	Lycoptene	470	128000	61	55

B. Aromatic Hydrocarbons

Molecule	λ_{max} [nm]	ε_{max} [dm^3 mol^{-1} cm^{-1}]	a^*_{OC} [m^{2}g^{-1}]	a^*_{OM} [m^{2}g^{-1}]
-1-	*-2-*	*-3-*	*-4-*	*-5-*
	180	65000	208	192
benzene	258	200	0.64	0.62
	215	71000	136	128
naphthalene	309	316	0.61	0.57
	250	175000	240	226
anthracene	378	5300	7.2	6.8
	275	310000	330	313
tetracene	475	10000	10.6	10.1

Columns 3 to 6: see the explanations to Table 3.1; for each compound Table B gives the characteristics of the principal absorption maximum (upper lines) and the characteristics of the longest wave of the distinct absorption maximum (lower lines).

associated with the 1→ 1′, or 2→ 1′, or 3→ 1′ transitions, there are also shorter-wave bands in these spectra corresponding to 2 → 2′, 3 → 3′, and so on transitions, and also less intense bands due to transitions involving a change in the electronic configuration, for example, 1 → 2′, 2 → 1′.

Qualitatively, the effect of the length of the conjugation chain on the color of these linear polyenes can be explained on the basis of fairly straightforward quantum-mechanical considerations using the single-electron approximation. Here we assume the so-called electron gas model (Kuhn 1949, 1951). According to this, all the π electrons belonging to the conjugation chain are in a one-dimensional energy potential cavity of length a, exactly or approximately equal to the length of the chain (see, e.g., Figure 3.8).

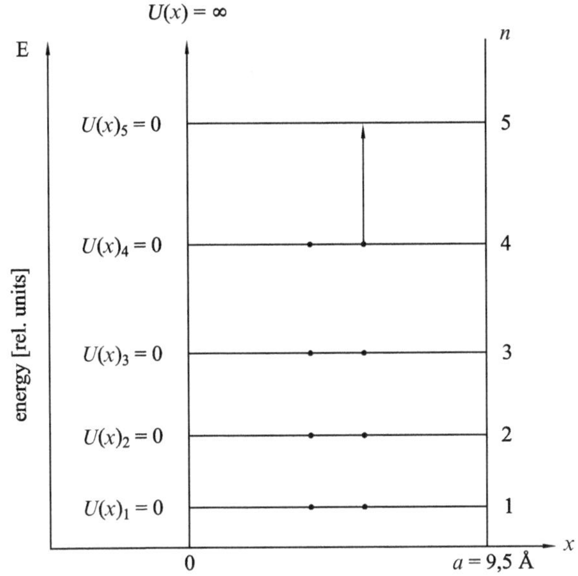

FIGURE 3.8. Energy diagram of the valence electrons of octatetraene in the ground state. The magnitude of n is the quantum number of each successive energy level. The arrow illustrates the transition corresponding to the appearance of the longest wavelength absorption band.

Within this cavity the potential is homogeneous and the potential energy of every discrete electron is $U(x) = \text{const} = 0$ for $0 < x < a$. In contrast, the potential outside the cavity tends towards infinity (i.e., $U(x < 0) \to \infty$ and $U(x > a) \to \infty$).

The timeless Schrödinger equation for this case,

$$-\frac{h^2}{8\pi^2 m}\frac{d^2\psi_e(x)}{dx^2} + U(x)\psi_e(x) = E\psi_e(x) \tag{3.3}$$

thus reduces to the form:

$$-\frac{h^2}{8\pi^2 m}\frac{d^2\psi_e(x)}{dx^2} = E\psi_e(x), \tag{3.4}$$

where E is the total energy of the electron, h is Planck's constant, m is the mass of the electron, and Ψ_e is the single-electron wave function

The solution to Equation (3.4) is the family of single-electron wave functions

$$\psi_e(x) = A \cdot \sin\frac{n\pi x}{a} \quad \textit{for } 0 < x < a, \tag{3.5}$$

where A is a constant coefficient, and $n = 1,2,3, \ldots$ is the quantum number denoting the successive energy levels of the electron. We obtain the total energy of the electron E_n at the various energy levels n corresponding to these equations by substituting Equation (3.5) in Equation (3.4) and transforming:

$$E_n = \frac{n^2 h^2}{8ma^2}. \tag{3.6}$$

These energies thus depend on the length of the conjugation chain a. Bearing in mind that this length is equal to the product $a = N\,2R_{CC}\sin(\alpha/2)$ (where N is the number of C=C–C links, R_{CC} is the average length of two bonds C=C and C–C, and α is the angle between the bonds), the expression for the energy reduces to the form:

$$E_n = \frac{n^2 h^2}{8ml^2 N^2}, \tag{3.7}$$

where $l = 2R_{CC}\sin\alpha/2$. Taking the Pauli exclusion principle into account, we can conclude that in the ground state the highest filled electron level in an N-times conjugated molecule is described by the quantum number $n \equiv n_1 = N$, whereas the lowest unfilled level corresponds to the quantum number $n = n_2 = N + 1$.[4] This means that the lowest excitation energy of such a molecule is equal to

$$\Delta E = E_{N+1} - E_N = \frac{h^2}{8ml^2}\frac{2N+1}{N^2}, \tag{3.8}$$

which yields the following equation for the wavelength of the first longwave absorption band.

$$\lambda_{max} = \frac{hc}{\Delta E} = K\frac{N^2}{2N+1}, \tag{3.9}$$

where K is a constant taking the value

$$K = \frac{8mcl^2}{h}. \tag{3.10}$$

The set of equations (3.9) and (3.10) is a good qualitative description of the regularity of formation of the color of polyenes. It predicts an increase in the wavelength of absorbed light λ_{max} together with the rise in the multiplicity of conjugation N in the molecule. For some polyenes, this reasoning, despite its very considerable simplification, can also be used to obtain approximate quantitative descriptions. This applies, for example, to

---

[4] For example, for octatetraene, a polyene with a four-times conjugated double bond ($N = 4$), the corresponding quantum numbers are $n_1 = 4$, $n_2 = 5$ (see Figure 3.8).

cyanopolyenes, which contain multiple conjugated double bonds. The geometrical aspects of such molecules (R_{CC} and α) have practically fixed values: $R_{CC} \approx 0.150$ nm, $\alpha \approx 120°$ (Gołębiewski 1982). On this basis the value of the constant K was calculated at ≈255 nm. Below we give the relevant calculated and empirical data, quoted from Babko and Pilipienko (1972), for various multiplicities of N conjugations for these cyanopolyenes, which are characterized by the general formula:

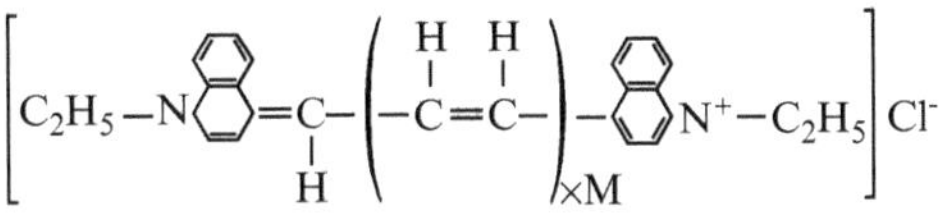

The number of links in the chain, beginning with C_2H_5–N (on the left) to N^+–C_2H_5 (on the right) for this molecule is $N = 5 + M$ (M is the number of multiplicity of conjugation bonds). Hence, the wavelengths of the first long-wave absorption bands can be calculated on the basis of Equation (3.9) and the values of the constant K (see Table 3.4).

As we can see from this example, the coincidence of observed and calculated wavelengths of the absorbed light is surprisingly good, although very probably accidental. For other polyenes such descriptions do not yield correct quantitative results, which is due to the simplifications applied to the method used and the very nature of the single electron approximation. It is often the case, however, that Equation (3.9) can be corrected by suitable semi-empirical modeling. For instance, for simple polyenes, such as those in Table 3.3A, we can obtain a good quantitative description by taking into account the dependence of parameter K on the multiplicity of conjugation N; in accordance with the formula (3.9) we have derived:

$$K\,[\text{nm}] = 447\ N^{-0.69}. \tag{3.11}$$

The differences between the theoretically calculated and empirically determined absorption band positions of the first dozen or so polyenes are on average 0.4% and vary within limits of ±5% in relation to the observed values of λ_{max}.

TABLE 3.4. Wavelengths of the first long-wave absorption bands, calculated and empirical, for various multiplicities of N conjugations of cyanopolyenes.

Number M	0	1	2	3
Number N	5	6	7	8
$\lambda_{calculated}$ [nm]	580	706	834	959
$\lambda_{empirical}$ [nm]	590	710	820	930

Quoted from Babko and Pilipienko (1972).

3.2.2 *Cyclic Polyenes*

Apart from the linear polyenes, the other important group of pure organic compounds consists of the cyclic polyenes, a characteristic property of which is the absorption of radiation due to the $\pi \rightarrow \pi^*$ transition of conjugated electrons. These compounds occur commonly in natural associations of humic substances in seas, and they are most numerously represented by 6-carbon aromatic ring compounds, that is, based on benzene (C_6H_6) and its various polyaromatic condensation products. Benzene itself absorbs light mainly in the UV region and is colorless. The polyaromatic compounds, however, are strong absorbers of visible light (see Figure 3.9 and Table 3.3B), as are the derivatives of aromatic compounds that form chemical bonds or complexes with various auxochrome groups and metals (see below). Of the 30 valence electrons in the benzene ring, six π electrons form three strongly delocalized molecular orbitals with a far more complicated structure than that of their counterparts in linear polyenes (Figure 3.9). The electron structure of aromatic compounds is thus highly complex, and we do not discuss it in detail. Nevertheless, we would like to draw the reader's attention to two points:

- Analysis of the various possible electronic configurations shows that three $\pi \rightarrow \pi^*$ transitions of conjugated electrons are typical of aromatic compounds, and their spectra each display three more or less complex, but distinct absorption bands (Figure 3.10). The structure of these absorption spectra is far more complex than the absorption spectra of linear polyenes. The spectral properties of aromatic rings are highly individual and are excellently suited to spectroscopic identification.
- As far as organic compounds are concerned, aromatic hydrocarbons and their derivatives are some of the strongest absorbers of radiation. This is evidenced by their high coefficients of molar extinction and of light absorption (Table 3.3B). That is why, even though they may be present in sea water only in small quantities, these small concentrations can "raise" quite con-

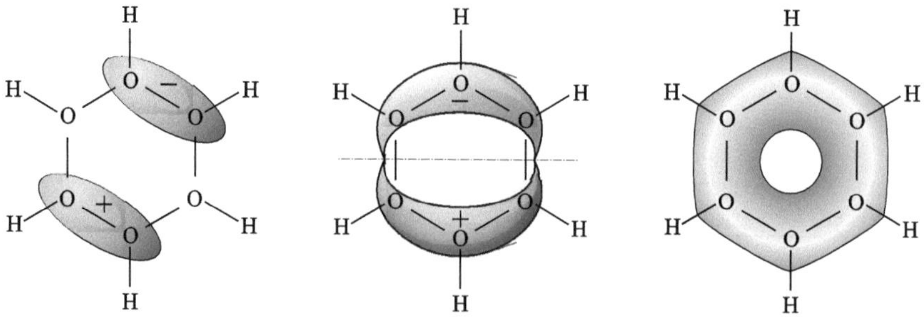

FIGURE 3.9. The approximate shapes of the conjugated bonding π.molecular orbitals in the benzene molecule (projection onto the molecular plane).

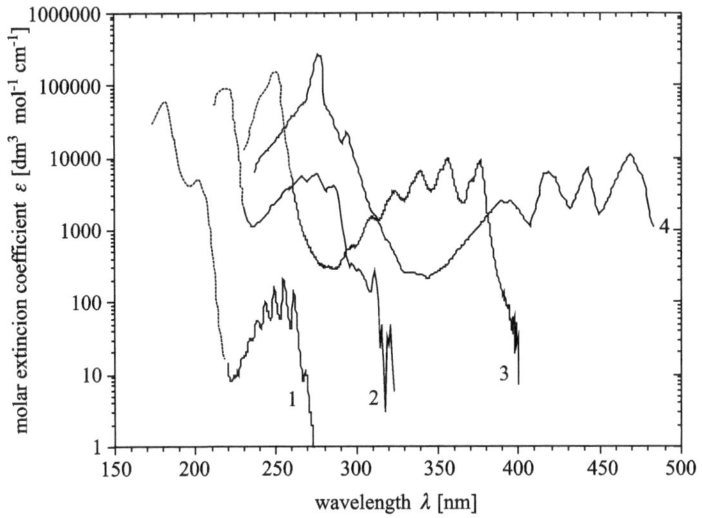

FIGURE 3.10. Spectra of the molar extinction coefficients of selected aromatic hydrocarbons: benzene (1); naphthalene (2); anthracene (3); tetracene (4). (Based on data from Du et al. (1998) and Berlman (1971).)

siderably the values of the specific coefficients of light absorption by natural associations of organic substances in the sea, for example, yellow substances in the littoral zone (for further details, see Chapter 4).

3.2.3 *Mixed Conjugations (π- and n-Electron) and Photosynthetic Pigments*

Conjugated π electron chromophores are characteristic not only of pure hydrocarbons, but also of compounds containing chains or rings with a hetero-atom, that is, organic compounds in which atoms other than hydrogen are attached to the carbons. Then, if the chain of conjugated π bonds contains a link with nonbonding n electrons from a hetero-atom, this lengthening of the conjugation chain also leads to an increase in the length of absorbed light waves due to the $n \rightarrow \pi^*$ transitions. This effect is typical of oxidized polyenes, such as chains terminating in a carboxyl group, which are commonly found in the sea and elsewhere in nature. Table 3.5 illustrates this (data taken from Banwell (1985)).

The same applies to the molecules of plant pigments of the chlorophyll group and phycobilin, for example, in phytoplankton. The chemical structures of these compounds are discussed further in Chapter 6 (see Figures 6.7 to 6.9). Among other things, they contain pyryl rings, in which

TABLE 3.5. Absorption bands due to the $\pi \rightarrow \pi^*$ and the $n \rightarrow \pi^*$ transitions.

	$\pi \rightarrow \pi^*$	$n \rightarrow \pi^*$
Chromophore	Absorption band λ [nm]	Weak absorption band λ [nm]
$-\overset{\vert}{C}=O$	166	280
$-\overset{\vert}{C}=\overset{\vert}{C}-\overset{\vert}{C}=O$	240	320
$-\overset{\vert}{C}=C=\overset{\vert}{C}-\overset{\vert}{C}=O$	270	350
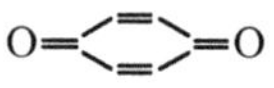	245	435

nitrogen is present. As a result of the considerable concentration of conjugated valence electrons in a small space (both π and n, but the former in particular), these compounds are strong absorbers of VIS and even IR radiation (e.g., bacteriochlorophylls). A theoretical description of the electron states in molecules such as these is much more difficult than in the case of linear polyenes or pure aromatic hydrocarbons. Such a description requires the use of complicated, approximate quantum-mechanical models, which we do not go into here. It is, moreover, not possible to ascribe any particular feature in the absorption spectra of these molecules unequivocally to this or that $\pi \rightarrow \pi^*$ or $n \rightarrow \pi^*$ transition. Furthermore, the spatial irregularities of the electron distributions make the electron structures of these pigment molecules (especially chlorophyll) and their absorption spectra exceedingly complex. This complexity is exemplified in Figures 3.11 to 3.13.

These figures show the results of theoretical quantum-mechanical analyses from the papers of Petke et al. (1979) and Shipman (1982). These authors analyzed the energy states of the four following compounds: chlorophyll *a*, bacteriochlorophyll *a*, pheophytin *a*, and bacteriopheophytin *a*. By way of example, Figure 3.11 shows maps of two extreme configurations of molecular orbitals of chlorophyll *a* in ethyl ether in the ground state MO_x and MO_y and in the lowest excited state $MO_x{}^*$ and $MO_y{}^*$ determined by the FSGO method.[5] The orbitals are shown in the form of spatial distributions of the wave function of the valence electrons ψ_e (where, as we remember, ψ_e is the probability of finding an electron) in the plane of the main tetrapyryl ring. The thus-determined distributions of the functions ψ_e for the ground and excited states of all four compounds served to determine their characteristic electron energy levels (Figure 3.12), as well as their positions in the spectra

[5] FSGO, floating spherical Gaussian orbitals.

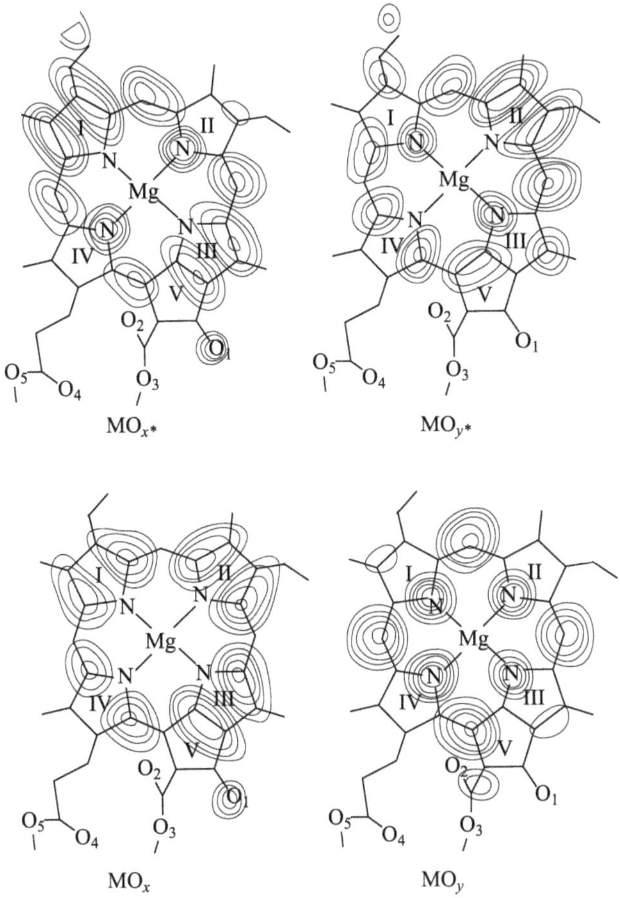

FIGURE 3.11. Electron density maps of the "four-orbital model" molecular orbital of ethyl chlorophyllide *a* is in the ground state MO_x and MO_y; is in the lowest excited state MO_x^* and MO_y^*. (Reproduced from J.D. Petke et al., *Photochemistry and Photobiology* (1979), 30, with the kind permission of the American Society for Photobiology.)

and the intensities of the absorption lines due to them. This is illustrated in Figure 3.13, where chlorophyll *a* is the example.

As can be seen from these figures, photosynthetic pigments of the chlorophyll group are characterized by a large number of possible excited singlet states, as a result of which they can absorb visible light in numerous electron absorption bands (from 7 to 9 bands). A further feature of these pigments is the large number of possible triplet states within the same energy range (c. 10), that is, corresponding to excitations by photons from the blue-green and green region of the visible light spectrum. This is of the utmost importance

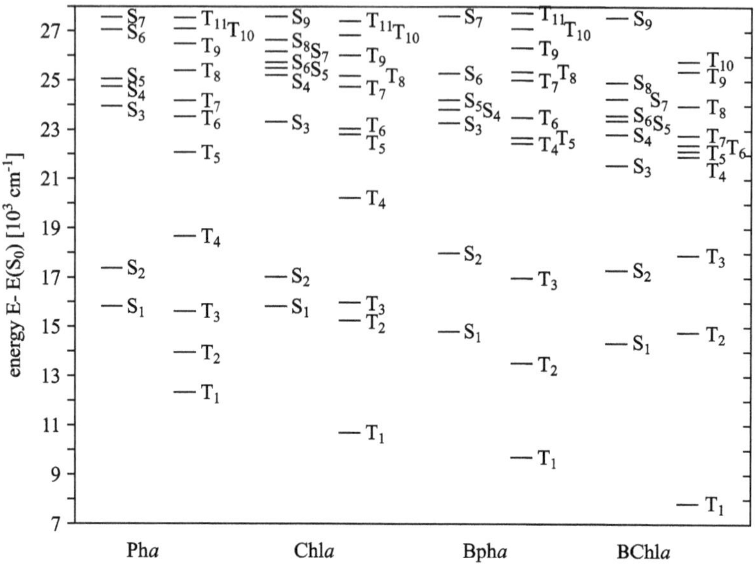

FIGURE 3.12. Calculated $S_0 \rightarrow S_n$ and $S_0 \rightarrow T_n$ transition energies for the low-lying singlet and triplet states of ethyl pheophorbide Pha, ethyl chlorophyllide Chl*a*, ethyl bacteriopheophorbide Bph*a*, and ethyl bacteriochlorophyllide BChl*a*. All the transition energies $E - E(S_0)$ are computed. Energies are expressed in wave number units ν^*. (Reprinted from Govindjee (Ed.), *Photosynthesis*, Vol.1, p. 283, Figure 5, by Shipman L.L., *Electronic Structure and Function of Chlorophylls and Their Pheophytins*, copyright 1982, with the kind permission of Elsevier Publishers.)

for the process of photosynthesis. Because this spectral region of photon energies is also characteristic of the excitations of other pigments involved in the photosynthetic apparatus of plants (carotenoids and phycobilin), it is highly likely that these excitations take place nonradiatively (Frąckowiak et al. 1997) by means of energy transferred from chlorophyll and pheophytin molecules in triplet states, for example:

$$\mathrm{Chl^*(T) + Car(S_0) \rightarrow Chl(S_0) + Car^*(S_1)},$$

where Chl is chlorophyll and Car is carotenoid.

Moreover, some triplet and singlet levels in chlorophylls are approximately the same; for example:

$$\mathrm{Chl^*(T_3) \approx Chl^*(S_1) \text{ and } Chl^*(S_3) \approx Chl^*(T_6) \approx Chl^*(T_7)}.$$

Both these aspects facilitate the migration of excitation energy at photosynthetic centers by means of internal conversion. We return to the problem of the absorption properties of plant pigments in Chapter 6.

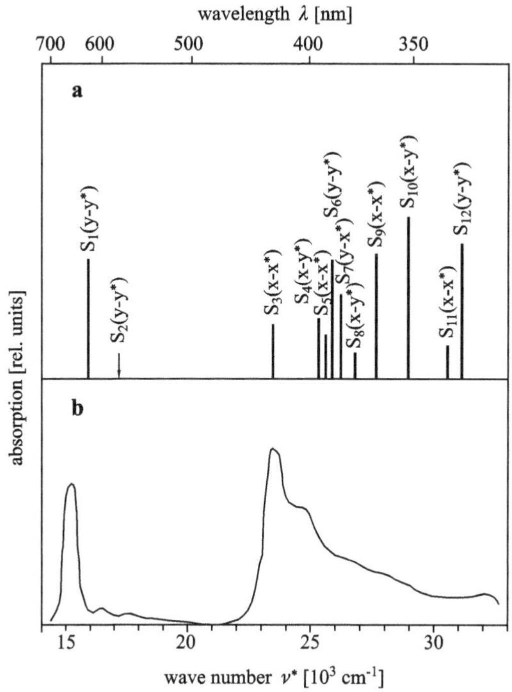

FIGURE 3.13. Light absorption spectra of chlorophyll *a*: (a) theoretical (the heights of the columns reflect the relations between the probabilities of the particular energy transitions); (b) experimental, in ethyl ether. Note: the indices S_1–S_{12} denote the multiplicity of the relevant excited singlet states for transitions from the ground state S_0, corresponding to oscillation numbers $v' = 0$ (in the ground state) and $v'' = 0$ (in excited states). In brackets, the configurations of the orbitals between which transitions occur. (Reproduced from J. D. Petke et al., *Photochemistry and Photobiology* (1979), 30, with the kind permission of the American Society for Photobiology.)

3.3 The Influence of Auxochromic Groups and Complexes on the Optical Properties of Organic Compounds in the Sea

So far we have been concentrating mainly on the spectral properties of "pure" organic substances. In Nature, these properties are subject to variation, for example, absorption bands shift towards the longer wavelength of the light spectrum as a result of intra- and intermolecular interactions.

The presence of water as well as a whole range of active inorganic compounds such as mineral salts and trace element ions, not to mention the reactions between diverse organic compounds, facilitate the addition of various functional groups to organic molecules or the formation of more or less stable complexes, particularly with the hydrated ions of metals. The spectral effects of these processes were identified qualitatively and classified empirically quite a long time ago (e.g., Kęcki (1992)). The following concepts were introduced to describe them.

- *Auxochrome*: When attached to a chromophore thisgroup alters the position and intensity of an absorption band, for example, OH, NH_2, Cl.
- *Bathochromic shift*: This is an increase in the wavelength of the absorbed radiation caused by the action of a substituent or the solvent.
- *Hypsochromic shift*: This is a reduction in the wavelength of the absorbed radiation caused by the action of a substituent or the solvent.
- *Hyperchromic effect*: This takes place when the substituent or solvent causes the intensity of absorption to rise.
- *Hypochromic effect*: This occurs when the substituent or solvent causes the intensity of absorption to fall.

3.3.1 Intramolecular Interactions

To begin with, let us analyze the effect of substituents on the spectral properties of organic molecules. Natural functional groups, especially those that undergo dissociation, are mostly hyperchromic auxochromes and give rise to a bathochromic shift. This means that the absorption band shifts towards the long-wavelength region and that its intensity is much enhanced. Evidence in support of this is given in Table 3.6 (see also Figure 3.14), which sets out the positions and molar extinction coefficients of the peaks of the three principal absorption bands of benzene and its derivatives. The effect of substituents on the position of λ_{max} and the intensity of absorption of the other organic molecules is similar.

As we can see, this effect is varied. Atoms of Cl, Br, I, and some groups as $-CH_3$, $-C(CH_3)_3$ and SO_3H (this last is not given in the table), as well as many other groups with stable electronic charges, exert hardly any effect at all. On the other hand, strong effects (shifts of the order of several tens of nm) are elicited by groups containing unshared electron pairs, for example, $-NH_2$ ($-NR_2$), –OH, –SH, likewise electrophilic groups or electron acceptors such as $-NO_2$, $>C{=}O$, and numerous substituents with multiple conjugated bonds. In a few rare cases, even larger bathochromic shifts (up to 150 nm) are possible: they occur in the presence of two auxochromes of different types. This is illustrated by the positions of the principal absorption band of the following compounds (after Babko and Pilipienko (1972)).

TABLE 3.6. The parameters of the three principal electron absorption bands of benzene and its selected derivatives.

Substituent in the benzene ring	Absorption band parameters					
	Wave-length λ_1 [nm]	Molar extinction ε_1 [$10^3 dm^3$ $mol^{-1}cm^{-1}$]	Wave-length λ_2 [nm]	Molar extinction ε_2 [$10^3 dm^3$ $mol^{-1}cm^{-1}$]	Wave-length λ_3 [nm]	Molar extinction ε_3 [$10^3 dm^3$ $mol^{-1}cm^{-1}$]
-1-	*-2-*	*-3-*	*-4-*	*-5-*	*-6-*	*-7-*
Pure benzene(-H)	255	0.2	202	7.3	182	65
–Cl	265	0.3	215	8	—	—
–Br	265	0.3	215	9	—	—
–I	257	0.7	231	12	—	—
$-CH_3$	262	0.3	205	8.1	187	55
$-CCl_3$	268	0.6	224	7	—	—
–OH	270	1.8	213	6	—	—
$-NH_2$	285	1.7	233	8	200	20
$-NH(CH_3)$	288	1.8	238	14	200	21
–CHO	280	1.5	242	14	—	—
$-NO_2$	280	1.5	252	9.6	—	—
$-CH{=}CH_2$	282	0.8	245	13	203	22
$-CH{=}CH{-}NO_2$	—	—	300	17	229	10

Data from Du et al. (1998), Berlman (1971), Minczewski and Marczenko (1985), and Kęcki (1992).

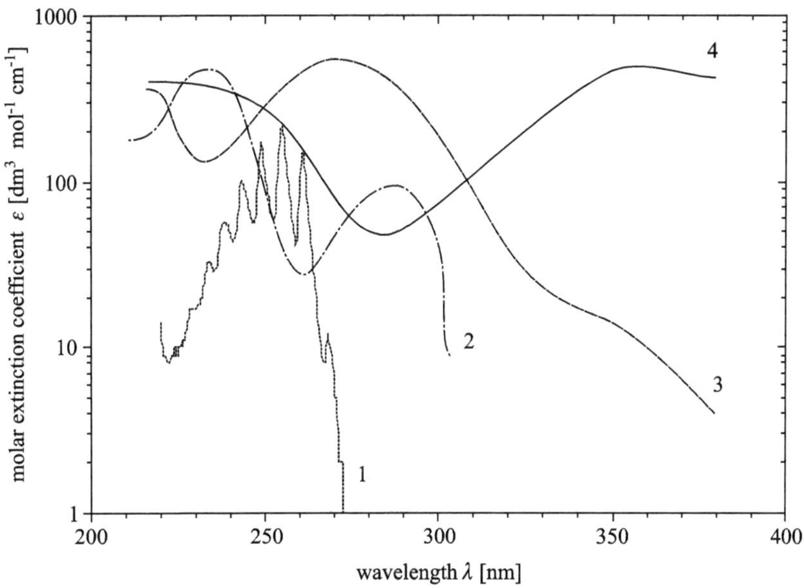

FIGURE 3.14. Absorption spectra of benzene and its derivatives: (1) pure benzene (based on data from Du et al. (1998) and Berlman (1971)); (2) aniline (from Minczewski and Marczenko (1985)); (3) p-nitrophenylacetic acid (from Minczewski and Marczenko (1985)); (4) trinitrophenol (from Minczewski and Marczenko (1985).)

Benzene (λ = 255 nm)

One Auxochrome:

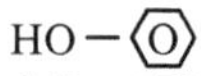

Phenol (λ = 275 nm)

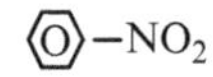

Nitrobenzene (λ = 268 nm)

Two Auxochromes:

p-nitrophenol (λ = 315 nm

p-nitrophenol anion (λ = 400 n

Not only intramolecular auxochromic interactions, but also the following intermolecular interactions strongly affect the parameters of absorption bands.

1. Interactions between organic molecules and the solvent
2. The formation of stable organometallic complexes
3. "Charge–transfer" interactions
4. The formation of organic aggregations

The last-mentioned of these interactions, that is, various aggregations of organic molecules, is especially significant for the formation of so-called native forms of pigments (see Section 6.2). Here we briefly describe the first three effects.

3.3.2 The Solvent Effect

The effect of the solvent on the position of absorption bands depends on its polarity. In general, interactions between nonpolar solvents and chromophores are weak. Polar solvents, on the other hand, can affect absorption properties almost to the same extent as substituents. The almost universally applicable rule here is that all the $n \rightarrow \pi^*$ bands that have been investigated undergo a hypsochromic shift in solvents such as saturated hydrocarbons, alcohols, and, to a lesser degree, water. In contrast, $\pi \rightarrow \pi^*$ transitions in these same solvents are almost always subject to a bathochromic shift. It is often the case, then, that when in doubt, the solvent effect can be used to distinguish these two types of transition ($n \rightarrow \pi^*$ or $\pi \rightarrow \pi^*$).

A further point to make is that the spectra of organic molecules containing dissociating groups are often strongly dependent on the pH of the solution. This emerges from the equilibrium between protonated and nonprotonated forms. Any change in pH causes a shift in the equilibrium, as a result of which opposite changes in the absorption band intensity of these two forms occur.

3.3.3 Organometallic Complexes

The atoms and ions of a great many naturally occurring metals, especially the transition metals, possess chromophoric properties; that is, they absorb visible light and near-ultraviolet radiation (see Section 2.3 and Table 2.15). These metals include Ti, V, Cr, Mn, Fe, Co, Ni, Cu, Nb, Mo, Tc, Ru, Rh, Pd, Ta, W, Re, Os, Ir, Pt, and Au. As we have already mentioned, many of them are present in sea water as trace elements (see, e.g., Wells (2002)), in the form of free cations or hydrated cations, for example, $[Mn(H_2O)_6]^{++}$, $[Cu(H_2O)_6]^{++}$, $[Ni(H_20)_6]^{++}$, $[Co(H_20)_6]^{++}$, $[Ti(H_2O)_6]^{+++}$, and $[Cr(H_2O)_6]^{+++}$. Moreover, they can combine to form colored ionic complexes with organic molecules.

Clearly, then, these complexes absorb VIS–Near UV radiation in bands very close to the absorption bands of the metals themselves. Regardless of this, particularly when the metal ion is situated on the extension to a conjugated polyene chain (although not only), additional bathochromic shifts of as much as 100 nm are observed in the absorption bands of this chain. Interestingly, this happens in the case of metals that by themselves are not chromophores, such as Al, Ge, Ti, Zr, and Th. The formation of such a complex with a metal cation is thus analogous to the appearance of an auxochrome in the conjugated chain (Tcherkasov 1957). We may assume that in attaching itself to the electronegative end of a chain of conjugation, a metal ion attracts electrons, thereby elongating the conjugation chain by one link but without increasing the number of conjugated electrons n. The length of such a link depends on the ionic radius of the metal. Setting out from this assumption, Sano (1960) considered the question of the effect of particular metals on the magnitude of the bathochromic shift using the electron gas model that we discussed earlier in connection with linear polyenes (Section 3.2.1). The relationship we can obtain on this basis, analogous to Equation (3.9), thus takes the form:

$$\frac{1}{\lambda_{max}} = \frac{V_0}{hc}\left(1 - \frac{1}{2N}\right) + \frac{h}{8mcl^2}\frac{2N+1}{N^2}, \tag{3.12}$$

where V_0 is the amplitude of the potential along the chain of conjugation; the other denotations are as for Equation (3.9).

The values of V_0 are determined by the properties of the metal. The semi-empirical method of Sano (1960) showed that V_0 depends on the degree of ionicity of the metallic bond with the organic anion and on the ionic radius of the metal. The larger this radius, the greater the shift in λ_{max} of the complex in comparison with the undissociated compound, and the closer λ_{max} approaches the wavelength of the light absorbed by a free anion of this compound. This last wavelength is the limit of the bathochromic shift. To illustrate this we give below the results of the relevant calculations (consistent with observations) obtained by Sano (1960) for phenylfluorene complexes with tetravalent metals (denoted below as $Metal^{4+}$):

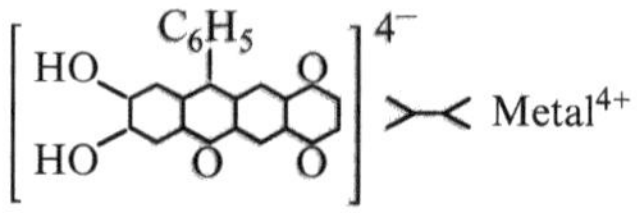

Ion	H^+	Ge^{4+}	Ti^{4+}	Zr^{4+}	Free Anion
Ion radius of the metal [nm]	—	0.50	0.65	0.83	—
λ_{max} of the absorption band [nm]	468	508	525	540	560

Despite these early advances, the question of the effect of metal ions on the optical properties of organometallic complexes in sea water has not yet been finally elucidated. In particular, the relations between the structure of the complexes and the intensity of absorption remain unexplained. A range of additional information on these problems can be found in such works as Babko and Pilipienko (1972), Gołębiewski (1982), and Campbell and Dwek (1984).

3.3.4 Charge–Transfer Complexes

The stability of the organometallic complexes we discussed in the previous section is variable, but practically independent of interactions with light. A separate group of electronic transitions are donor–acceptor transitions, otherwise known as charge–transfer transitions, which are characteristic of less stable complexes; their mechanism was described by Mulliken and Person (1962) and by Prochorow and Siegoczyński (1969). They occur in two-component systems including atoms or molecules of different elements or chemical components (e.g., organic molecules or mixed organic–inorganic molecules) that, prior to the absorption of a photon or after its emission, do not have to be linked in any way (Siegoczyński et al. 1975). A charge–transfer complex can, but need not, exist only when a system is in an excited state. Once a photon has been absorbed by a π or n electron of one compound—the donor—which has a low ionization potential, this electron transfers to an unoccupied orbital in the second compound—the acceptor—which has a high electron affinity. Three examples of such systems are phenol and 1,3,5-trinitrobenzene, hexamethylbenzene and tetracyanoethylene, and iodine and methylpyridine.

transfer $\pi \longrightarrow \pi^*$

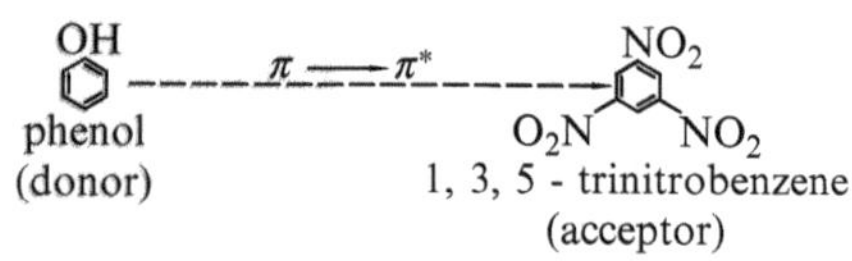

transfer $n \longrightarrow \pi^*$

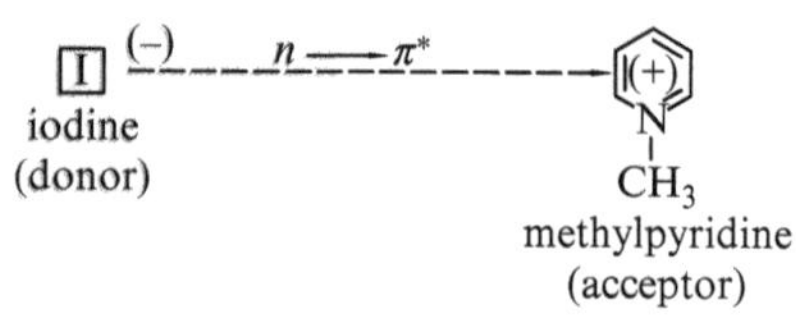

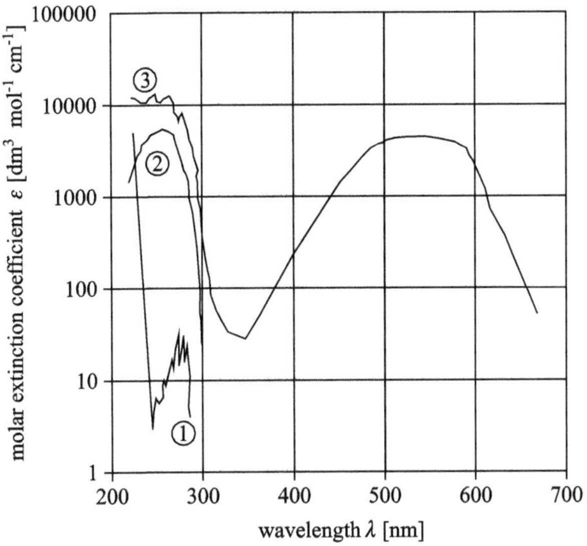

FIGURE 3.15. Absorption spectra of (1) hexamethylbenzene, (2) tetracyanoethylene, (3) a mixture of tetracyanoethylene (acceptor) and hexamethylbenzene (donor). The broad absorption band in the visible region ($\lambda_{max} \approx 540$ nm) in spectrum (3) is the effect of charge transfer. (Based on data from Prochorow (1983).)

The various types of donor–acceptor transitions can be illustrated as follows (see Garaj et al. (1981)).

A feature of charge–transfer transitions is the appearance of a new absorption band independent of the absorption bands of the separate components still present in the spectrum of the mixture (Figure 3.15). As we can see, this new band lies at some distance from the component bands, usually in the visible or the near ultraviolet regions. Moreover, charge–transfer bands are broad and intense, and have no vibrational structure. More detailed information on charge–transfer complexes can be found in, for example, Murrel (1963), Mulliken and Person (1962), Briegleb (1961), Ejchart et al. (1978), Banwell (1985), and Siegoczyński and Ejchart (2004). We should emphasize that under natural conditions, charge–transfer complexes do not greatly affect the optical properties of the majority of organic substances dissolved in sea water. Be that as it may, these interactions are highly significant in biological systems, for instance, in the functioning of the photosynthetic apparatus in marine phytoplankton cells. We hope to describe this process in a separate volume dealing with the secondary processes involved in the interaction of light with sea water.

4
Light Absorption by Dissolved Organic Matter (DOM) in Sea Water

In Chapter 3 we described the physical principles underlying the formation and interpretation of the electronic radiation absorption spectra of organic molecules. In the present chapter we apply that information to characterize and analyze the processes by which a vast number of natural organic substances dissolved and suspended in sea water absorb light (Hedges 2002).

The ultraviolet (UV) and visible (VIS) absorption coefficients vary in different seas by as much as two, three, and more orders of magnitude. However, because the influence of sea salt and other inorganic substances is minimal (see Section 2.3), it is the organic components of sea water that are almost entirely responsible for this variation (Figure 4.1). As we have already stated, they absorb light in the UV–VIS spectral region as a result of photons interacting with the valence electrons of the molecules. This leads to the excitation of these electrons, and in some cases can give rise to different types of photoreactions (e.g., predissociation, bond breaking, and synthesis of new compounds). These processes have been analyzed in detail, for example, by Simson (1976).

Organic substances also absorb infrared (IR) radiation (usually in the spectral range from c. 2 μm to c. 30 μm), as a result of which the vibrational energies of the molecules change (Alpert et al. 1974, Herzberg 1992). These changes are due to the effect of all the atomic groups in a molecule that are capable of vibrating. Figure 4.2 shows the ranges of IR absorption by the more important functional groups in organic molecules. Owing to the complex chemical structure of organic substances in the sea and their many possible types of vibration, their IR absorption spectra are far more complex than their UV–VIS counterparts. At the same time, however, they are much less ambiguous, displaying as they do numerous peaks characteristic of particular atomic groups. That is why IR spectroscopy is widely applied in chemical analysis as one of a battery of methods for identifying chemical structures and determining the contents of various compounds, especially when suitably prepared and concentrated. (e.g., Banwell (1985), Williams and Fleming (1980), and Sharp (2002)). With respect to organic substances in the sea, this particular problem has been investigated and described in detail by, for example, Pempkowiak (1989).

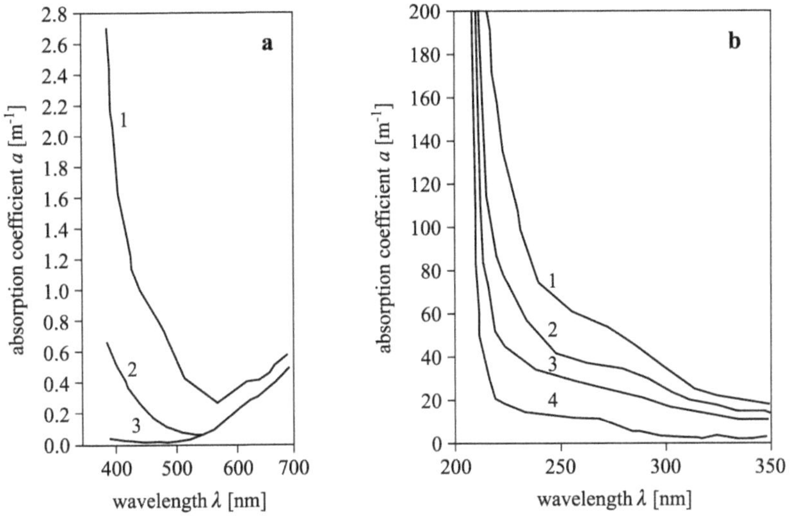

FIGURE 4.1. Light absorption spectra of sea waters containing different amounts of organic substances. (a) In the VIS region: (1) Baltic Sea, Gulf of Riga; (2) Baltic, Gotland Deep; (3) Sargasso Sea. (b) In the UV region (1) Vistula River mouth; (2–4) Baltic, Gulf of Gdańsk. (Reproduced from J. Dera, *Marine Physics*, second edition, updated and supplemented (2003) (in Polish), with the kind permission of PWN publishers, Warsaw.)

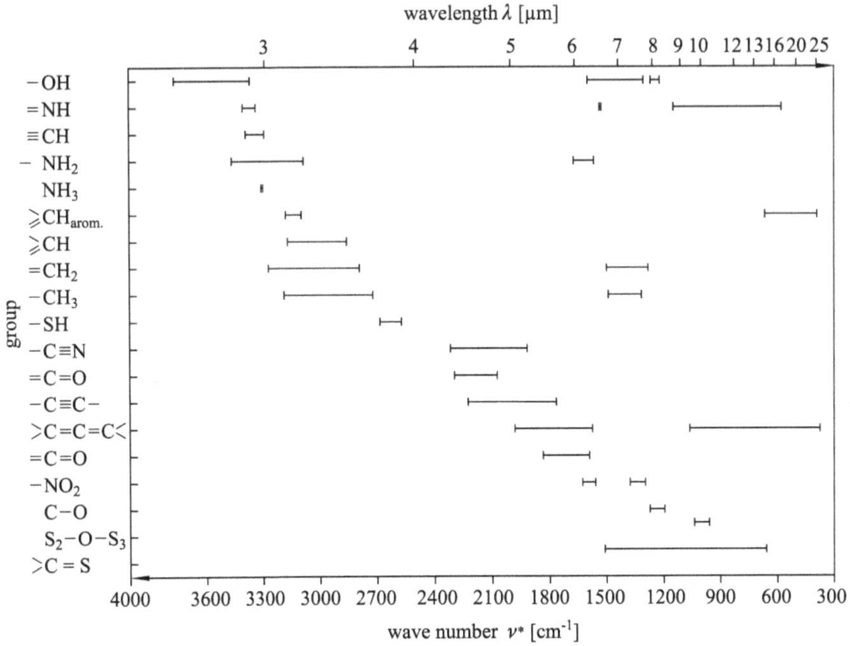

FIGURE 4.2. IR spectral absorption ranges of some organic functional groups occurring in the sea. (Based on data from various sources, e.g., Alpert et al. (1974), Minczewski and Marczenko (1985), and Pempkowiak (1989).)

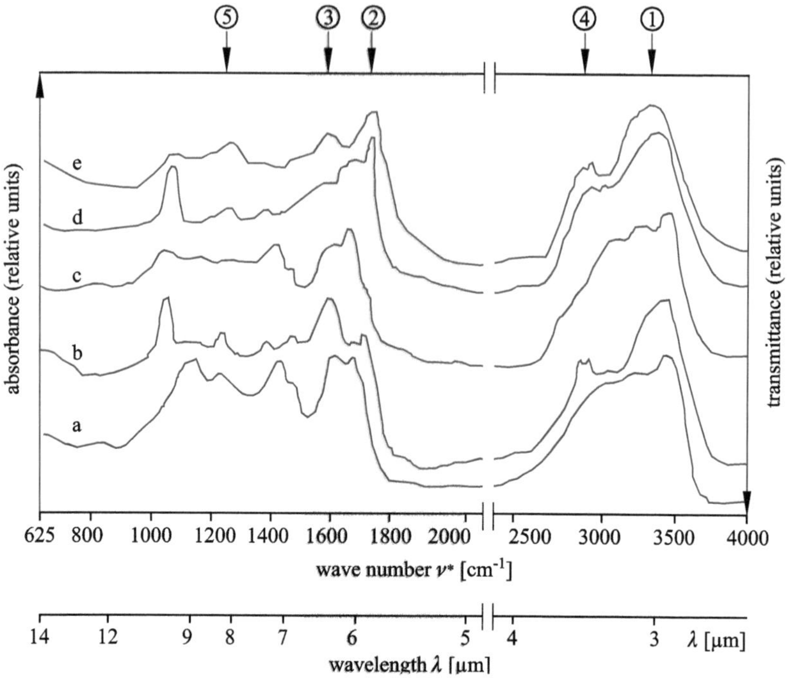

FIGURE 4.3. IR absorption spectra of selected classes of organic compounds extracted from natural waters: (a) dissolved humic acids in Vistula River water; (b) suspended humic acids in Vistula River water; (c) dissolved humic acids in Baltic Sea water; (d) and (e) humic acids from surface (d) and subsurface (e) bottom sediments in the Baltic. These absorption spectra enabled the following classes of compounds to be detected: (1) hydroxyl and amine groups (~3300 cm^{-1}); (2) carbonyl groups in aldehydes, ketones, and carboxylic acids, (1720 cm^{-1}); (3) amide bonds (c. 1500–1600 cm^{-1}); (4),(5) methyl and methylene groups (~2950 cm^{-1} and ~1370 cm^{-1}). (Adapted from Pempkowiak (1989).)

Figure 4.3 illustrates the IR absorption spectra of various classes of substances isolated from natural waters. It also shows the positions of absorption bands associated unequivocally with the vibrational transitions of various atomic groups. Nevertheless, it should be borne in mind that against the background of the overall absorption spectra of sea water the natural IR absorption bands of organic substances in the sea are undetectably small. Owing to the fact that the concentrations of these compounds are relatively low, their IR absorption is also much weaker than that of inorganic salts. For this reason, we now restrict our further discussion to the analysis of UV–VIS absorption by organic substances in the sea, that is, to electronic absorption. The nature of this absorption depends, among other things, on the physical forms and the chemical properties of the principal organic absorbers in the sea.

4.1 Classification, Origin, and General Characteristics of Light Absorption by the Principal Groups of Organic Absorbers in Sea Water

4.1.1 Occurrence and Origin of Organic Matter in the Ocean

Organic substances in the sea occur mainly as:

1. Suspended organic matter, SOM (or particulate organic matter POM), which covers a variety of organisms and organic detritus, or
2. Dissolved organic matter DOM[1]

The approximate total resources in the World Ocean, the mean concentrations, and annual production of POM and DOM are given in Table 4.1. Clearly, DOM is predominant, as it probably makes up over 98% of the mass of all organic substances.[2] These do not, however, form a molecular solution in the literal sense of the word. In marine chemistry, the concept of "dissolved organic matter" covers only those substances that can pass through a 0.45 μm pore filter (Romankevich 1977, Riley and Skirrow 1965). It emerges from this definition that by DOM we should understand a molecular solution that may also contain very fine undissolved organic particles (Yentsch 1962). The division of organic matter into DOM and POM is thus a matter of convention. In actual fact the dimensions of DOM and POM overlap to make up an "organic carbon continuum" (Thurman 1985). This is illustrated by Figure 4.4, which shows to a good approximation the typical ranges of dimensions and molecular masses of the various compounds and diverse forms of marine organic matter. Figure 4.5 gives these proportions of the DOM molecular fractions (i.e., of different molecular masses) as percentages for Pacific Ocean water. Compounds with small and medium-sized molecules (molecular mass <1000 Daltons) predominate. In the central Pacific, such molecules make up as much as 70% of the total DOM; the figure is lower in littoral zones, but is usually in excess of 50% (Romankevich 1977). Only in basins seriously polluted by the inflow of terrigenous organic substances, for example, in the vicinity of river mouths, is the proportion of these small and medium-sized organic molecules smaller still (Pempkowiak 1989). This composition of the molecular fractions strongly affects the overall absorption properties of DOM in the different basins, in

---

[1] In oceanographic practice the following abbreviations are used to describe the physical forms of organic matter or organic carbon and their concentrations in sea water. SOM or POM: suspended or particulate organic matter; SOC or POC: suspended or particulate organic carbon; DOM: dissolved organic matter; DOC: dissolved organic carbon.

[2] This applies to the total volume of water in the World Ocean. In surface layers the content of organic suspended matter is more than 2%, and in some cases (in eutrophic littoral zones) can exceed 20%. We return to this problem in Chapter 5.

TABLE 4.1. The occurrence and origin of organic forms in the World Ocean.

Form of Occurrence			Organic mass (OM) 10^9t	Total OM (%)	Mean concentration [μg·dm^{-3}]	Annual production: Processes	Flux 10^9t year^{-1}
-1-	-2-	-3-	-4-	-5-	-6-	-7-	-8-
		Producers:					
		Phytoplankton	0.80[6]	1.97×10^{-2}	0.584	Primary production	60[2]
		Phytobenthos	0.02[1]	4.92×10^{-4}	0.015	Primary production	0.224[4]
POM (except phytobentos and zoobentos)		*Consumers:*					
	Bio-sphere	Zooplankton	1.48[6]	3.64×10^{-2}	1.08	Secondary production	1.5[3]
		Zoobenthos	1.00[1]	2.46×10^{-2}	0.730	Secondary production	0.3[1]
		Nekton	0.10[1]	2.46×10^{-3}	0.073	Secondary production	0.021[1]
		Decomposers:					
		Bacteria	0.007[1]	1.7×10^{-4}	5.1×10^{-3}	Bacterial production	7.0[1]
	Detritus		60[4]	1.48	43.8	Influx from the biosphere:	
						1. Water	40[2]
						2. Sea-bed	0.224[4]
DOM	Dissolved organic matter		4000[5]	98.44	2.92×10^3	Influx from external sources:	
						1. Rivers	1.2[4]
						2. Others	1.0[4]

[1]Bogorov (1969, 1974), [2]Koblentz-Mishke (1977), [3]Moizeev 1969, [4]Romankevich (1977), [5]Skopintsev (1971), [6]Vinogradov and Shushkina (1987).

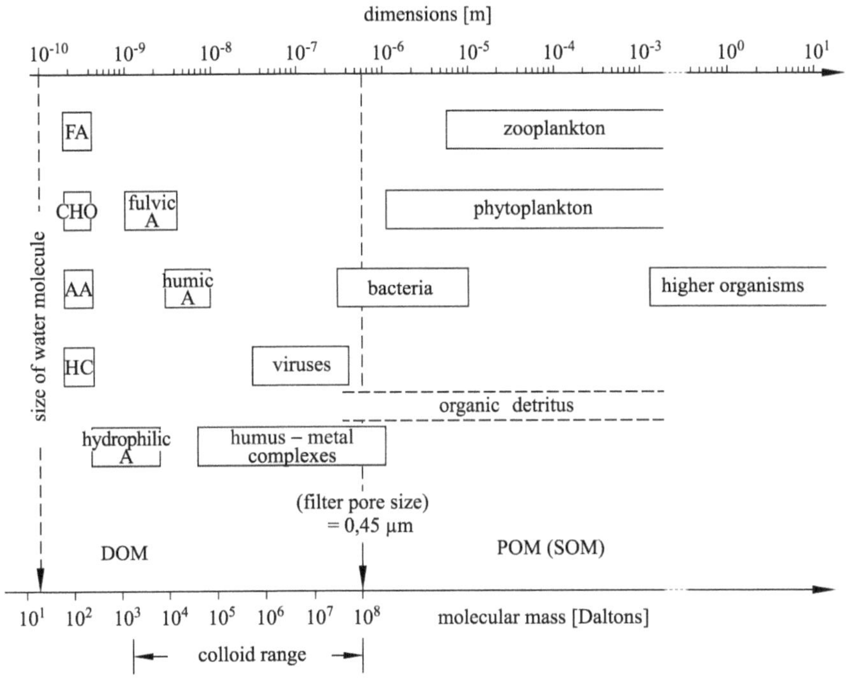

FIGURE 4.4. Typical dimensions of various forms of organic matter in the sea. FA: fatty acids, CHO: carbohydrates, AA: amino acids and other molecules containing nitrogen, HC: hydrocarbons, fulvic A: fulvic acids, humic A: humic acids, hydrophilic A (hydrophilic acids): various medium-sized hydrophilic compounds, that is, dissolved in water, or undissolved, which for reasons of size are classified as DOM. (Adapted from Thurman (1985).)

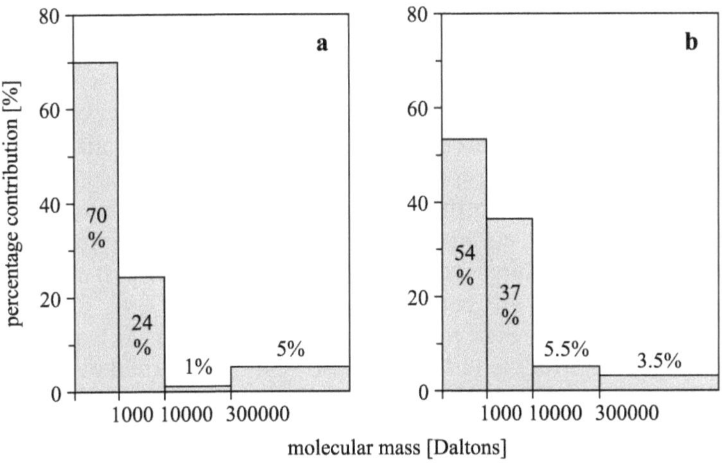

FIGURE 4.5. Percentage contributions of the different organic fractions to the total DOM concentration in selected basins of the Pacific Ocean: (a) central Pacific ($DOM \approx 2$ mg · dm^{-3}); (b) littoral zones ($DOM \approx 4.4$ mg · dm^{-3}). (Data from Romankevich (1977).)

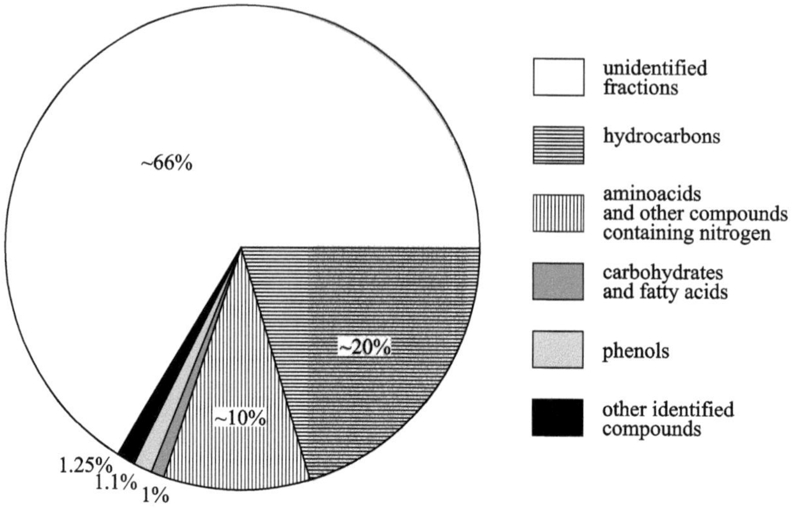

FIGURE 4.6. The probable proportions of compounds in the total DOM content in the World Ocean. (Adapted from Ittekkot (1981).)

particular the total mass specific absorption coefficient, of which more in Section 4.2.

As Table 4.1 shows, the mean concentration of DOM in the World Ocean is relatively low at ≈3 mg dm^{-3} (converted to DOC this gives $C_{org} \approx 1.5$ mg dm^{-3}). Under real conditions, for example, in the surface layers of oceans (0–100 m), DOC concentrations also vary within a fairly narrow range, from c. 1.3 mg C_{org} dm^{-3} in oligotrophic areas to c. 4 mg C_{org} dm^{-3} in areas of high productivity (Skopintsev 1971, Romankevich 1977). Only in coastal zones, particularly those of shallow enclosed seas, can this figure rise to over a dozen mg C_{org} dm^{-3} and more. These are concentrations around one thousand times smaller than those of mineral salts dissolved in sea water. The question thus arises as to why these very small amounts of organic substances dissolved in sea water can have such a strong influence on light absorption and can so clearly differentiate sea waters with respect to their optical properties.

In general, this is due to the molecular structure of these organic compounds, a topic that we covered in the previous chapter. Despite very intensive research, this structure is still poorly understood as far as natural organic substances in the sea are concerned (Pempkowiak (1989), Romankevich (1977), Ittekkot (1981), Benner (2002), and many others). As can be gleaned from the work of Ittekkot (1981) (see also Spitzy and Ittekkot (1986)), the chemical structures of only c. 30–40% of the total number of organic compounds dissolved in the World Ocean have so far been established precisely. They are primarily small and medium-sized molecules of compounds secreted from organisms or that are direct products of their decay, among them, carbohydrates, amino acids, hydrocarbons, fatty acids, and phenols.

Figure 4.6 gives their approximate mean proportions. In actual fact, the contents of these compounds vary greatly in time and space. A detailed list of organic substances identified in sea water can be found, for example, in Popov et al. (1979) and Hansell and Carlson (2002).

The other 60–70% of DOM is much less well understood. They include mostly medium-sized and large molecules such as proteins, lipids, or combinations of them, as well as very large aggregates containing fragments of phenols and other organic groups, and their metal complexes. Some of these compounds, usually the most stable of them, are the humic substances (acids), referred to by specialists in marine optics as yellow substances (Kalle 1966, Jerlov 1976, Blough and Del Vecchio 2002). Some of their properties are discussed later in this chapter.

4.1.2 Principal Organic Absorbers of Light in the Ocean

Different organic compounds affect light absorption in sea water to different extents. Because of their vast number and our poor understanding of the chemical structure of many of them, it is impossible to discuss separately the optical properties of every single one of these compounds. We therefore cover only those compounds or groups of compounds known to have a significant effect on the overall UV–VIS absorption by organic matter in the sea.

There are four types of absorber. Most of the small and medium-sized molecules are compounds containing nitrogen:

a. Amino acids and their derivatives (peptides and protein fragments) absorbing mainly UV radiation
b. Purine and pyridine compounds, and nucleic acids, also absorbing UV

The other compounds in this fraction of size (small and medium-size) to all intents and purposes do not absorb light. Numerous authors have shown (see the data collected in Popov et al. (1979) and in Herzberg (1992)) that most of the simple carbohydrates, hydrocarbons, and fatty acids contained in sea water are compounds that are to a large extent saturated, with only a small number of *n*-electron multiple bonds. As we showed in Chapter 3, their absorption capabilities in the $\lambda > 200$ nm range are relatively weak. Likewise, free phenols, strong absorbers though they are, hardly affect the overall absorption of light in the sea owing to their tiny concentrations in the water. Far more often it is the case that phenols occur as components of various amino acids, humic acids, or lignin: these are the principal absorbers and are accorded separate treatment.

The other two groups of absorbers are polyatomic compounds with large single molecules:

c. Humic acids, which absorb UV–VIS radiation
d. Lignins, which are chiefly UV absorbers

The first three groups of these absorbers (a, b, and c) are present in various concentrations in all seas (see Table 4.2). Because of the broad scatter

TABLE 4.2. Typical concentrations of the principal organic absorbers dissolved in surface (euphotic) waters of seas and oceans.

A. Amino acids and their derivatives (Peptides + Proteins)

Region	Concentration [μg dm^{-3}]	% DOM	Taken from, or estimated estimated on the basis of
World Ocean			
Range of variation	50–300	3–8	Romankevich 1977
Incl. aromatic and bound aminoacids		(2–7)	
Boundary values:			
East-central Pacific	<100	(1)	Bordovskiy 1985, Zlobin et al. 1975
Northeastern Atlantic	7990	57	Morris and Foster 1971

B. Purine and pyridine compounds (Nucleic Acids)

Region	Concentration [μg dm^{-3}]	% DOM	Taken from, or estimated on the basis of
World Ocean			Takahashi et al. 1974,
Range of variation	6.4–59	(0.3–2)	Popov et al. 1979

C. Lignins

Region	Concentration [mg dm^{-3}]	% DOM	Taken from, or estimated on the basis of
Baltic			
Range of variation	0.18–0.35		Pempkowiak 1989
Baltic		18–27	Nyquist 1979

D. Humic substances

Region	Concentration [mg dm^{-3}]	% DOM (Degree of humification)	Comments
Open ocean		4.9–5.1	From various sources collected
Coastal oceanic waters		3.7–4.5	by Popov et al. 1979
Coastal oceanic waters		6–30	
Atlantic Ocean	0.1–0.2	10–17	From various sources gleaned
Sargasso Sea	0.2	23	from Pempkowiak 1989
Narragansett Bay	1.6	55	
Narragansett Bay	0.4–1.0	25–38	
World Ocean	0.12	15	
Open Baltic	2.30	30	
Gulf of Gdańsk and the area around the Vistula River mouth	3.9–7.3	44–56	

The data given in parenthesis are less accurate.

of the empirical data taken from different sources, the figures quoted in this table should be treated as approximate. One of the reasons for the discrepancies in these data is the use of different analytical methods for identifying DOM (for reviews of these methods, see, for example, Romankevich (1977) and Sharp (2002)). On the other hand, the lignin sulfates are a group of terrigenous compounds found mostly in the littoral zones of oceans and in enclosed seas, especially shallow ones such as the Baltic.

Table 4.3 gives the spectra of the mass-specific absorption coefficients of the strongest absorbers among the above-named compounds. Later we discuss the origin and absorption capabilities of the main groups of absorbers specified above.

Apart from these four types of absorbers, oil pollutants can also have a significant effect on absorption in some maritime regions (see, e.g., Otremba and Król (2002) and Otremba and Piskozub (2001); see also Section 5.2.4).

4.2 Analysis of the Conditions Governing UV–VIS Absorption by the Principal Organic Absorbers in the Sea

4.2.1 Amino Acids and Their Derivatives

Containing nitrogen and being strong UV absorbers, certain free amino acids and their diverse associations and compounds with other substances are some of the most common groups of organic compounds in sea water. They probably make up c. 3–8% of the total DOM in the World Ocean (Table 4.2A). The optical properties of the individual components of this group are determined by the chemical structures of their molecules. The most active absorbers among them are the compounds containing aromatic rings. Because they are small (the molecular mass is usually smaller or much smaller than 400 Daltons), discrete amino acid molecules enter sea water from organisms directly via their cell membranes. They can also come into existence as the decomposition products of proteins (of which they are component parts) from excreta or dead organisms. In the sea these compounds are quite active and readily combine with other substances to form medium-sized molecules with masses from 400 to 1000 Daltons (Degens 1970). Thus, the amounts of bound amino acids in sea water are 3–10 times greater than those of free amino acids (Romankevich 1979). Amino acid molecules consist of two characteristic functional groups, that is, an amino group H_2N– and a carboxyl group –CO–OH

TABLE 4.3. Spectral mass-specific UV–VIS absorption coefficients [m^2g^{-1}] of the strongest organic absorbers in DOM.[a]

	Aromatic Amino Acids			Amino Acids	Purines and Pyridine			Lignins	Humic Substances		
λ[nm]	TRP	TYR	PHE	MAAs	A	G	C	LIG-S	HUMUS	HA	FA
-1-	-2-	-3-	-4-	-5-	-6-	-7-	-8-	-9-	-10-	-11-	-12-
188	26	—	84	—	—	—	—	—	—	—	—
193	33	60	18.3	—	—	—	—	—	—	—	—
195	31	35	14.7	—	—	—	—	—	—	—	—
200	29	8.9	12.7	—	—	—	17.3	21.2	—	—	—
205	31	5.8	12.8	—	—	16.1	24.1	—	—	—	—
210	41	6.6	6.6	—	—	14.4	24.5	15.2	8.0	—	—
215	48	8.1	1.4	—	16.4	10.3	23.5	—	—	—	—
220	53	10.6	0.14	—	6.6	7.6	21.9	9.0	6.1	—	—
225	20	9.6	0.054	—	4.1	6.8	12.6	—	—	—	—
230	5.1	2.8	0.058	—	4.4	8.2	6.3	7.9	5.1	—	—
235	2.0	0.62	0.08	—	6.6	11.9	4.1	—	—	—	—
240	1.8	0.20	0.13	—	10.2	15.3	3.2	5.3	4.6	1.26	0.31
245	2.1	0.21	0.16	—	13.1	16.1	4.1	—	—	—	—
250	3.3	0.32	0.23	—	16.4	15.0	6.3	2.8	3.9	1.20	0.25
255	3.5	0.58	0.28	—	21.3	12.9	10.0	—	—	—	—
260	4.4	0.89	0.20	—	22.3	11.2	12.9	2.2	3.6	1.10	0.21
270	5.8	1.7	0.01	—	16.3	12.0	23.2	2.6	3.3	0.99	0.17
275	6.0	1.7	—	—	6.6	12.6	26.6	—	—	—	—
280	6.3	0.80	—	—	2.3	11.7	27.3	2.8	2.9	0.88	0.14
285	4.4	0.22	—	—	0.5	9.6	25.1	—	—	—	—
290	2.0	0.04	—	—	0.07	5.9	21.3	2.3	2.6	0.80	0.10
295	0.78	0.01	—	—	—	2.9	15.7	—	—	—	—
300	0.21	—	—	277	—	1.5	4.7	1.5	2.2	0.68	0.080
305	0.06	—	—	244	—	0.8	0.8	—	—	—	—
310	0.01	—	—	233	—	—	—	1.2	1.9	0.62	0.063
320	—	—	—	259	—	—	—	0.92	1.6	0.52	0.051

330	—	—	—	348	—	—	—	0.30	1.46	0.47	0.042
340	—	—	—	358	—	—	—	0.64	1.2	0.42	0.036
350	—	—	—	292	—	—	—	0.53	1.05	0.38	0.031
360	—	—	—	147	—	—	—	0.35	0.90	0.34	0.025
370	—	—	—	—	—	—	—	0.21	0.73	0.31	0.021
380	—	—	—	—	—	—	—	0.11	0.62	0.28	0.018
390	—	—	—	—	—	—	—	0.066	0.53	0.26	0.015
400	—	—	—	—	—	—	—	0.055	0.46	0.23	0.013
450	—	—	—	—	—	—	—	—	0.21	0.13	0.0058
500	—	—	—	—	—	—	—	—	0.12	0.072	0.0026
550	—	—	—	—	—	—	—	—	0.066	0.041	0.0013
600	—	—	—	—	—	—	—	—	0.035	0.025	0.00058

Data from the literature as cited, or our own estimates.
TRP: tryptophan, TYR: tyrosine, PHE: phenylalanine, MAAs: mycosporinelike amino acids, A: adenine, G: guanine, C: cytosine, LIG-S: lignin from the Baltic Sea, HUMUS: humic substances from the Baltic Sea, HA and FA: humic acids and fulvic acids, respectively, from the Gulf of Mexico.
[a] Data from numerous works, including: Campbell and Dwek (1984), Filipowicz and Więckowski (1983), Nyquist (1979), Carder et al. (1989), Moisan and Mitchell (2001), and Woźniak et al. (2005b).

as well as an amino acid residue R, all of which are combined in the following structure:

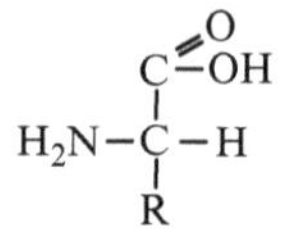

In all there are more than 20 amino acids of biological origin, differing in the structure of the residue R (Filipowicz and Więckowski 1983). Figure 4.7 shows a number of the amino acids most frequently found in sea water. All free amino acids absorb light in the far and medium UV, although with an intensity that varies depending on the structure of the amino acid residue. In general, there are two principal types of free amino acids, which we can name

- "saturated" amino acids (weak absorbers)
- "aromatic" amino acids (strong absorbers)

The first group encompasses most amino acid molecules, for example, those of Figure 4.7a, whose residues R are saturated. In this case, as long as the residue does not contain any heteroatoms with *n* electrons, absorption of radiation is due entirely to the electronic transitions characteristic of carboxyl and amino groups (Table 4.4A, rows 1–3). However, the intensity of these transitions is very low (apart from $\pi \rightarrow \pi^*$ transitions in the far UV) in comparison to that of other organic absorbers contained in DOM.

The participation of "saturated" amino acids in absorption by all DOM is thus rather small. We can also disregard the effect of the more complicated "saturated" amino acids (or weakly "unsaturated" amino acids), which contain, for example, two carboxyl or amino groups[3] such as, for example, in Figure 4.7b. In some cases, when two double bonds are fairly close to each other, weak π-electron conjugation may occur. That is why some amino acids have an absorption band around $\lambda = 210$ nm due to a conjugated $\pi \rightarrow \pi^*$ transition. Nevertheless, it is relatively weak, achieving its greatest intensity in pure histidine (Table 4.4B, row 1). Finally, in cysteine (Table 4.4B, row 2) there is one more feature, due to the $n \rightarrow \sigma^*$ transition of the disulphide group (–S–S–), but which also gives rise to a very weak absorption band.

In contrast to these "saturated" amino acids, natural "aromatic" amino acids can make a very significant contribution to the absorption of UV radiation, particularly in the 200–300 nm range. As the data in Popov et al. (1979) indicate, these "aromatic" amino acids make up only c. 4% of the total free amino acid content in sea water, but their absorption capabilities are two orders of magnitude greater than those of the "saturated" amino acids (see Table 4.4B). These considerable absorption capabilities are due to the fact that the amino acid

[3] Note that in the case of these molecules the molar extinction coefficient ε doubles, but that at the same time the molecular mass A_M also increases by a factor of nearly two, so that the specific absorption ($a^*_{OM} = 2.3\varepsilon/10A_M$) hardly changes.

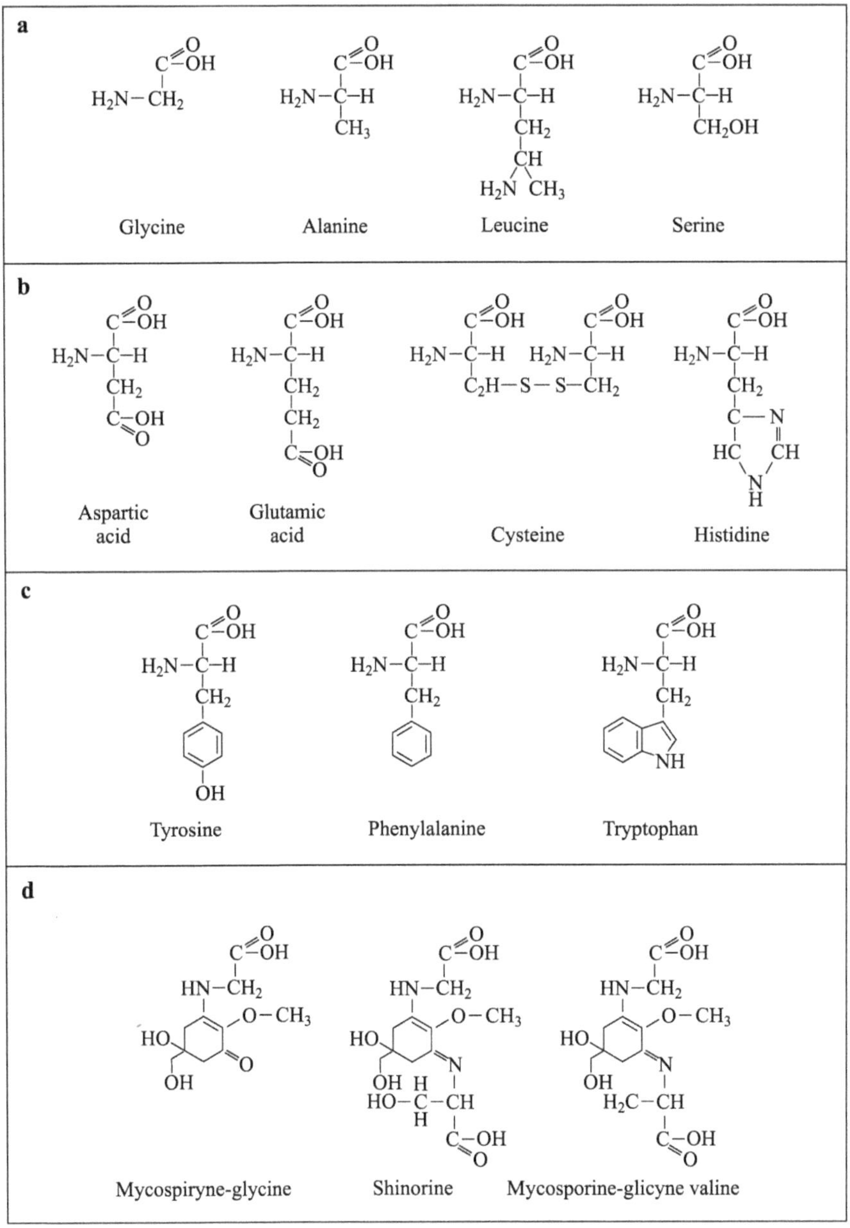

FIGURE 4.7. The chemical structures of the amino acids most commonly occurring in sea water: (a) with one double bond (one carboxyl group); (b) with two or more double bonds (two carboxyl or other groups); (c) with an aromatic ring; (d) mycosporine-like amino acids (MAAs).

TABLE 4.4. The principal light absorption properties of amino acids and peptides.

A. Characteristic Functional Groups

No.	Functional Group	Transition	Spectral Range λ_{max} [nm]	Ranges of ε [$dm^3\ mol^{-1}\ cm^{-1}$]	a^*_{OM} [$m^2\ g^{-1}$]
-1-	*-2-*	*-3-*	*-4-*	*-5-*	*-6-*
	Aminoacids, Peptides, and Proteins				
1	Groups: amino (NH_2) imino (NH) ammonia (NH_3)	$n \rightarrow \sigma^*$	210–230	300–900	< 0.3–2
2	Carboxyl group: (C=O bond)	$\pi \rightarrow \pi^*$	170–190	~ 2000	< 0.6–5
3	(–C–OH bond)	$n \rightarrow \pi^*$	190–200	~ 100	< 0.1
	Peptides and Proteins				
4	Peptide bonds: (–C–NH–)	$\pi \rightarrow \pi^*$	~ 190	~ 7000	~ 8
5	as above	$n \rightarrow \pi^*$	~ 210–220	~ 100	~ 0.1

B. Selected Amino Acids

No.	Compound	Transition	λ_{max} [nm]	ε [$dm^3 mol^{-1} cm^{-1}$]	a^*_{OM} [$m^2 g^{-1}$]
-1-	*-2-*	*-3-*	*-4-*	*-5-*	*-6-*
1	Histidine	$\pi \rightarrow \pi^*$	211	5900	8.7
2	Cysteine (–S–S– bonds)	$n \rightarrow \sigma^*$	250	300	0.3
	Aromatic Amino Acids				
3	Tryptophan	$(\pi \rightarrow \pi^*)_{arom}$	192	75000	33
4			219	47000	53
5			280	5600	6.1
6	Tyrosine	$(\pi \rightarrow \pi^*)_{arom}$	193	48000	61
7			222	8000	10
8			274	1400	1.8
9	Phenylalanine	$(\pi \rightarrow \pi^*)_{arom}$	188	60000	84
10			206	9300	13
11[a]			257	300	0.28

Data taken from or estimated on the basis of various sources, for example, Campbell and Dwek (1984).
[a] Spin-forbidden transition.

residues in these compounds contain aromatic rings (Figure 4.7c). In nature, three such compounds are biosynthesized: tyrosine, which contains phenol; phenylalanine, which contains benzene; and tryptophan, where the residue contains indole. These compounds are thus auxochromic derivatives of benzene, and elicit the hyperchromic effect (we dealt with this question in detail in Section 3.3). It is for this reason that "aromatic" amino acids are among the strongest absorbers of light of all the organic substances present in sea water, except for certain pigments and polyaromatic hydrocarbons and also so-called mycosporinelike amino acids (see the end of this section). Hence their very high specific absorption coefficients, the spectra of which are shown in Figure 4.8a (see also Table 4.3).

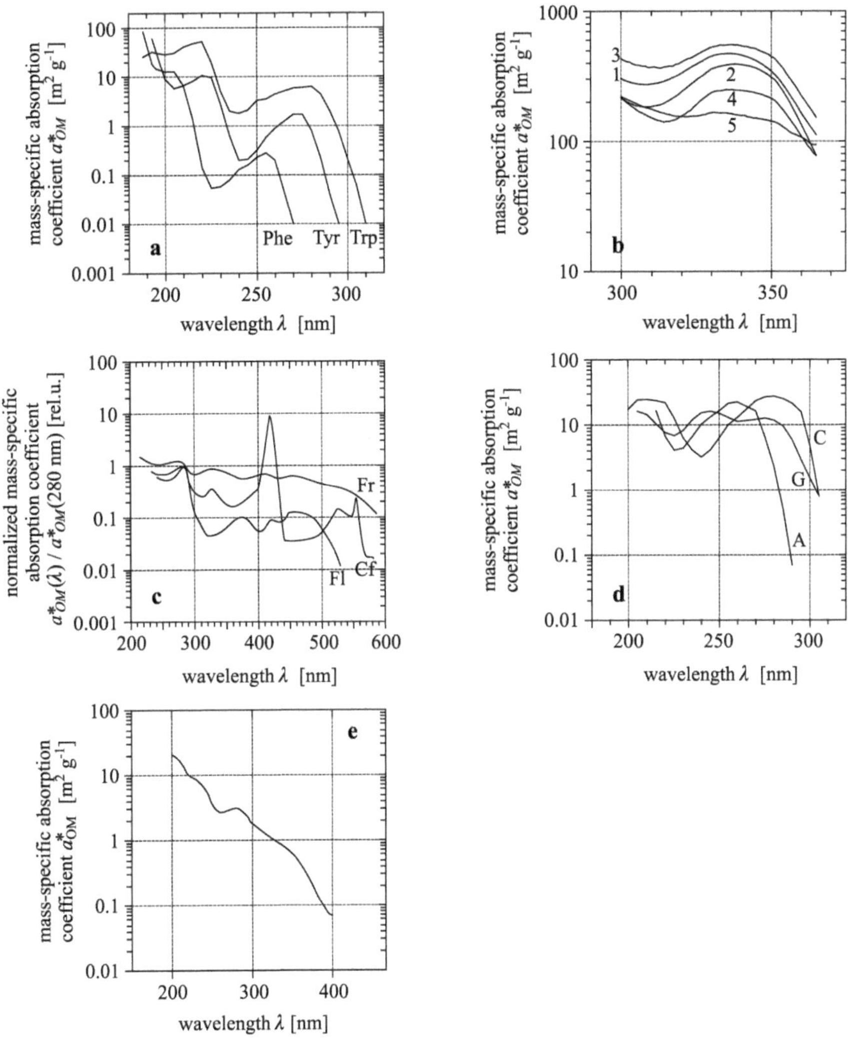

FIGURE 4.8. The spectra of the mass-specific absorption coefficients of the strongest identified absorbers of UV and/or VIS light in the sea and some of their derivatives: (a) "aromatic" amino acids: tryptophan (Trp), tyrosine (Tyr), and phenylalanine (Phe) (after data taken from Campbell and Dwek (1984)); (b) mycosporinelike amino acids including mycosporine-glycine, shinorine, and mycosporine-glycine valine in proportions characteristic of mixtures of these compounds occurring naturally in *Phaeocystis antarctica*, incubated under various irradiance conditions (under PAR irradiance levels of 542, 400, 259, 151, and 85 μEin m^{-2} s^{-1}, curves 1 to 5, respectively; these spectra were estimated from empirical data taken from Moisan and Mitchell (2001) and on the basis of private communications from Dr. T. A. Moisan); (c) various conjugated proteins of plant origin: cytochrome (containing haem(b) groups), Cf; ferredoxin (with iron not bonded via haem groups), Fr; and flavoproteid (reductose – ferredoxin – NADP) – Fl (after different authors; data taken from Grodziński (1978)); (d) purine and pyrimidine derivatives: the main constituents of nucleic acids: adenine, A; guanine, G; and cytosine, C (after data taken from Filipowicz and Więckowski (1983)); (e) terrigenous lignins in the Baltic (spectrum drawn on the basis of approximate empirical data taken from Nyquist (1979)).

In analytical chemistry we often make use of the important property of the absorption spectra of "aromatic" amino acids, namely, the occurrence of a strong absorption band around the wavelength $\lambda = 280$ nm (with the exception of phenylalanine, whose fairly weak band is shifted towards the shorter wave region, because the relevant $\pi \rightarrow \pi^*$ transition is spin-forbidden). The presence of "aromatic" amino acids in the vast majority of proteins enables approximate quantitative determinations of proteins in solutions using spectral measurements in the above wavelength range (see, e.g., Campbell and Dwek (1984), Williams and Fleming (1980), and Scott (1964)).

The most numerous group of amino acids dissolved in sea water includes, as we have already said, bound amino acids (Benner 2002, Burdige 2002). They occur mainly in the form of various phenol–quinone complexes[4] and, according to the hypotheses of some authors (e.g., Duursma (1965)), may be precursors of humus substances, and according to other authors (e.g., Romankevich (1979)) are one of the components of these substances. Owing to the difficulty of isolating these substances from DOM without damaging their structure, it is not possible to define precisely their individual absorption properties in the natural state. However, like free "aromatic" amino acids, they are also auxochromic compounds of benzene. We would therefore be justified in thinking that the absorption properties (spectral ranges and absorption intensity) of both these types of compound are similar, or that some bound amino acids will also absorb longer wavelength radiation (as with lignins; see below). This last may be the case if, for example, these substances contain multiple condensed rings (see Table 3.3B in Section 3.2) or two opposed auxochromes, that is, an electron donor and an electron acceptor (see the discussion on the auxochromic compounds of benzene in Section 3.3).

The above-mentioned amino acids (present in DOM) are biosynthesized on a large scale by marine organisms. Under certain conditions, these organisms may produce a further group of amino acids with somewhat different properties. They are the mycosporinelike amino acids (MAAs); see the examples in Figure 4.7d. Their structure and chemistry have been described only quite recently (e.g., Favre-Bonvin et al. (1976), Bandaranayke (1998), and Whitehead and Hedges (2003)). MAAs are synthesized by a variety of organisms, but principally by a species of algae from polar regions, which is exposed to high doses of UV radiation (Carreto et al. 1990a,b, Karentz et al. 1991, Miller and Carpentier 1991).

Several studies (e.g., Vernet and Whitehead (1996) and Garcia-Pichel (1994)) have suggested that MAAs act as natural sunscreens to UV radiation and as such should be regarded as photoprotecting; there are numerous empirical indications of this. Indirect confirmation of this hypothesis was also obtained from the detailed laboratory studies carried out by, for example, Moisan and

---

[4] Quinone is an oxidized form of benzene: .

Mitchell (2001). These researchers demonstrated that the MAAs in *Phaeocystis antarctica* (in particular, mycosporine-glycine, shinorine, and mycosporine-glycine valine) act in a very similar way to photoprotecting carotenoids.

Figure 4.7d shows that the molecules of these MAAs contain not the usual amino functional group, but an imino group (HN) along with a substituent usually containing a highly saturated carbon ring with one double bond which, functioning as an auxochrome, gives rise to a hyperchromic effect and a bathochromic shift of the MAA absorption bands with respect to other amino acids. This is illustrated by the plots in Figure 4.8b and the data in Table 4.3 (column 5), which present spectra of the mass-specific light absorption coefficients of these substances; we ourselves compiled these spectra (Woźniak et al. 2005a) on the basis of empirical data kindly supplied by Dr. T. Moisan. The bathochromic shift causes the absorption band peaks of these MAAs to lie in various segments of the 310–370 nm spectral interval, and as a result of the hyperchromic effect, their mass-specific absorption coefficients are far in excess of the corresponding coefficients of the other amino acids, even the aromatic ones (Figure 4.8a and Table 4.3, columns 2–4). Generally speaking, even though MAAs are the strongest absorbers of radiation in the near-UV region, adjacent to the VIS, their participation in the absorption of light by DOM in oceans on a global scale is only very sporadic and can thus be ignored.

4.2.2 Peptides and Proteins

In addition to the small and medium-sized molecules characteristic of amino acids (free and bound), fractions with a larger molecular mass containing peptides and whole protein molecules or fragments of them make up a further sizeable proportion of the total DOM concentration. Biosynthesized in organisms from amino acids, peptides and proteins play a variety of essential parts in the maintenance of life. We refer readers wishing to acquire a more detailed knowledge of this complex question to the extensive monographs on the biochemistry of organisms and metabolic processes, (e.g., Baryła (1983), Barber (1976, 1977), and others). Here we merely mention the most important physical and chemical properties of these compounds, which explains why they are present in DOM and why they are able to absorb light. They may be linear or cyclic, with complex spatial structures comprising chains of amino acid molecules linked by so-called peptide bonds –(C=O)–(NH)– as the diagram below shows.

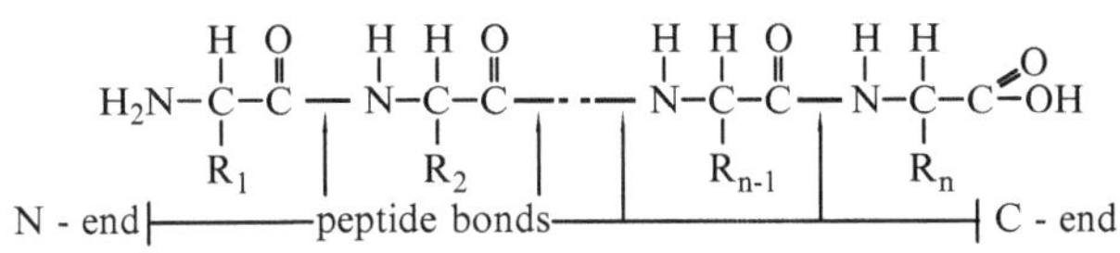

Here R_1, . . . , R_n denote the amino acid residues of the successive amino acid molecules, and the thick lines indicate the peptide bonds. These bonds constitute a new chromophoric group, the absorption properties of which are given in Table 4.4A, rows 4 and 5.

The reaction giving rise to peptide bonds is endothermic and requires the –C–O and N–H bonds to be broken, and the formation in their place of –C–H and HO–H bonds; one molecule of H_2O is thereby released. Peptide bonds are quite stable (although not as stable as C–C bonds) and are arranged in chains with an amino group at one end and a carboxyl group at the other (the C-end; see the above diagram). These groups need not be present if the chains are closed. Because of the number and variety of amino acids that can make up such a chain, the number of such possible molecules is vast. On the basis of the number of links in the chain, or the molecular mass, these compounds are conventionally divided into oligopeptides, polypeptides, and simple proteins (in contrast to complex proteins; see below).

	Number of amino acid molecules	Molecular mass [daltons]
Oligopeptides	<10	<1000
Polypeptides	10–100	1000–10000
Proteins	>100	>10000

The molecular mass of some proteins, especially complex ones, can be as high as several hundred thousand Daltons (e.g., haemocyamine, the blood pigment in invertebrates, has a molecular mass of c. 760,000 Daltons).

The criterion for dividing amino acid chains into oligo- and polypeptides and proteins is not just a matter of convention. Owing to their large molecular mass and the fairly large diameters of protein molecules, these compounds do not pass through dialyzing membranes or cell membranes. Their presence in sea water is due entirely to excreta and the decomposition of dead organisms. However, peptides, like amino acids, can pass directly into water from a living organism. Moreover, living organisms produce proteins with a complex structure, the so-called proteids. Unlike proteins consisting exclusively of amino acids linked by peptide bonds, proteids contain an additional nonprotein group known as a prosthetic group. Such a group is chemically bonded to or enters into various complexes with proteins.

A variety of organic compounds,[5] including those containing metals, can function as prosthetic groups. Diverse proteids also occur in sea water. Unfortunately, the proportions between the proteins and proteids dissolved in

[5] Because of the different types of compound making up prosthetic groups, numerous groups and subgroups of proteids are distinguished, for example, lipoproteids (containing lipids), glycoproteids (containing carbohydrates), nucleoproteids (containing nucleic acids and regarded as the most important proteins, constituting in a way the simplest form of life), and chromoproteids (containing pigments), for example, with the subgroups of haemoproteids and metal proteids (proteins bonding with metals), and many others (see, for example, Baryła (1983) and Filipowicz and Więckowski (1983)).

sea water are not known precisely. This is because of the methods used to analyze the composition of DOM, which usually permit the determination only of the overall concentration of proteins and the protein fragments of proteids (in practice it is the component amino acids that are usually determined). Our ignorance of the proportions between proteins and proteids dissolved in water makes it very hard to interpret the origin of the overall absorption properties of DOM in the sea, because the absorption capabilities of peptides and proteins and those of proteids on the other hand are very different.

The absorption capabilities of peptides and proteins depend to a large extent on the absorption properties of their component amino acids, and to a first approximation are their resultant. The differences between the chemical structures of these compounds and the structure of a mixture of free amino acids are not great enough to differentiate the electronic structures of these two groups. Peptide chains do not normally contain carboxyl groups (except at the C-end); they are replaced by carbonyl groups (C=O) situated in the vicinity of the peptide bond, contiguous with the nitrogen atom. This leads to a partial shift of the delocalized orbitals of the π- and π^*-electrons of the C=O double bond on to the N atom. The upshot is that the absorption bands of the $\pi \rightarrow \pi^*$ and $n \rightarrow \pi^*$ transitions in the peptide bond are slightly stronger and shifted towards longer wavelengths than would be the case for the carboxyl group (see Table 4.4A, rows 2, 3 and 4, 5).

However, these features are not very intensive and in practice do not affect the resultant spectra of peptides and proteins. This is mainly due to the high proportion of links containing "aromatic" amino acids in natural peptide chains, which, as we have already shown, are especially strong absorbers of radiation, mainly in the 200–300 nm range. From 3 to 5% of most proteins consists of such links containing "aromatic" amino acids, although in some proteins this percentage may be very much higher (20–50%; Filipowicz and Więckowski (1983)). The shapes of the light absorption spectra in the 230–300 nm region of natural peptides and proteins in sea water are therefore similar to those of the absorption spectra of "aromatic" amino acids. The absolute magnitudes of their specific absorption coefficients are smaller, differing by the factor k from the specific absorption coefficients of "aromatic" amino acids, where k expresses the mean ratio of the mass of all aromatic molecules in the peptide chain to the total mass of the chain.

Taking into consideration the above percentages of "aromatic" amino acids and also the molecular masses of the various amino acid components in the chain, we reckon that we can take $k \approx 0.05$–0.1 (mean $k \approx 0.075$). In the $\lambda < 230$ nm spectral range, this absorption is slightly higher because here we have a series of absorption bands due to the $n \rightarrow \sigma^*$ transition of numerous amino NH groups in the peptide chain (see Table 4.4A, row 1).

For practical estimates, we can write down an approximate form of the formula for the mean spectrum of the specific absorption coefficient in the 200–300 nm range for all peptides and proteins dissolved in sea water:

$$a^*_{OM,pep}(\lambda) \approx a^*_N(\lambda) + a^*_{OM,arom}(\lambda)\,k, \tag{4.1}$$

where $a^*_N(\lambda)$ is the mean specific absorption coefficient of ammonium, amino, and imino groups given in [m^2g^{-1}] (Table 4.4A, row 1), which affects the overall absorption in the 200–300 nm range in sea water only weakly; $a^*_{OM,arom}$ is the mean specific absorption coefficient of "aromatic" amino acids (Table 4.4B or Figure 4.8a), which strongly affects the overall absorption over the whole range of wavelengths 200–300 nm; $k \approx 0.075$ is the mean, large-scale participation of aromatic hydrocarbons in peptide chains (mentioned above).

We can also give similar forms of the spectra of the specific absorption coefficients for amino acids:
free

$$a^*_{OM,af}(\lambda) \approx a^*_N(\lambda) + a^*_{OM,arom}(\lambda)\, k' \qquad (4.2a)$$

and bound

$$a^*_{OM,ab}(\lambda) \approx a^*_{OM,arom}(\lambda), \qquad (4.2b)$$

where $k' \approx 0.04$ is the mean, large-scale participation of "aromatic" amino acids in the total concentration of free amino acids dissolved in sea water, calculated on the basis of data taken from Popov et al. (1979).

In contrast, the absorption properties of proteids are far more complex and harder to describe quantitatively. Their capabilities of absorbing radiation are highly diverse and differ markedly from the absorption of light by "pure proteins" (see, e.g., Figure 4.8c, which shows light absorption spectra of proteids present in phytoplankton, and therefore also in sea water). This is due to the occurrence of various conjugations and auxochromic interactions between protein proper and prosthetic groups. They also depend on the optical properties of these latter groups. A feature common to the absorption spectra of most proteids is undoubtedly the presence, as in proteins, of a distinct absorption band in the 270–280 nm range, due to "aromatic" amino acids.

The intensity of this band varies, however, and usually diminishes as the mass of the nonprotein part increases, especially when the prosthetic group is to a large extent saturated, as is the case with many glycoproteids and lipoproteids. The absorption spectra of many proteids have a very complex and intricate structure, possibly extending into the visible region and even into the near-IR. Just a few examples of these compounds are the various chromoproteids, such as the protein complexes with photosynthetic pigments present in phytoplankton, (e.g., the phycobiliproteids, described by Grabowski (1984), among others), the protein–chlorophyll complexes discussed by Shlyk (1974), as well as the haemoproteids (see plot Cf in Figure 4.8c) and metal proteids (plots Fr and Fl in Figure 4.8c) that function as cell enzymes. In view of this complexity, the mass-specific absorption spectrum for all proteins in the sea can be estimated only very roughly, using, for instance, the following expression that we derived

$$a^*_{prot}(\lambda) = k a^*_{OM,arom}(\lambda) + 0.93e^{-0.00768(\lambda - 220)}, \quad (4.3)$$

where λ is expressed in [nm]; a^*_{prot} and $a^*_{OM,arom}$ are expressed in [m^2g^{-1}].

When deriving this expression, we assumed that this spectrum is to a first approximation the resultant effect of light absorption by amino acids and possibly by other admixed proteid submolecules. From the optical point of view, all these compounds can, for the sake of simplicity, be divided into two groups: aromatic amino acids (mentioned above) and other compounds. The former are strong UV absorbers. The latter, mainly saturated amino acids (ammonium, amino, or imino) and others, absorb light somewhat less strongly, but do so in different regions of the UV and VIS. Accordingly, the bulk spectral coefficients a^*_{prot} for proteins are described as the sum of two such coefficients derived from the above two groups of absorbers. One is the spectrum of $a^*_{OM,arom}(\lambda)$ for aromatic amino acids. This enters the expression for the total absorption with a weight k ($k \approx 0.075$) representing the mean ratio of the mass of all aromatic molecules in the peptide chain to the total mass of the chain. The second component of this coefficient, responsible for light absorption by the other constituent amino acids of various proteins and the other constituents of proteids, we approximated by means of a curve diminishing exponentially with increasing light wavelength λ. The parameters of this approximating curve were established by statistical analysis of over a dozen empirical absorption spectra of various complex proteins (see the examples in Figure 4.8c).

It should also be noted that as far as optical criteria are concerned, numerous proteids are classified among the yellow substances, even though they are not humus substances. Unfortunately, given the present state of research, it is not possible to separate and quantitatively characterize with any precision the contributions to the total absorption of light of humus substances on the one hand and of various amino acids and other protein fragments. We return to this question in Section 4.3.1.

4.2.3 Purines, Pyrimidines, and Nucleic Acids

A second group of organic compounds containing nitrogen that are listed among the principal absorbers of UV radiation includes the nucleic acids and some of their components. As is well known, two types of nucleic acid occur in living organisms[6] (see, e.g., Baryła (1983), Filipowicz and Więckowski (1979, 1983), and Barber (1976)). They are:

---

[6] They are components of chromosomes and determine the quality of genes. They are also capable of autocatalytic reproduction and of transmitting inherited characteristics in the genetic code. They are thus the most fundamental compounds as regards the existence of life, and are probably its most profound, knowable essence. If we were to analyze this essence in any greater depth, we would be entering the realms of metaphysics (God?).

Deoxyribonucleic acids (DNA), the components of which are

1. Nitrogenous bases (purines: adenine and cytosine; pyrimidines: guanine and thymine)
2. Sugar: deoxyribose
3. A phosphoric acid residue

Ribonucleic acids (RNA), consisting of

1. Nitrogenous bases (purines: adenine and cytosine; pyrimidines: guanine and uracil)
2. Sugar: ribose
3. A phosphoric acid residue

The chemical structures of some of these compounds are illustrated in Figure 4.9. In sea water they are present as a result of excretion or the death of organisms. The proportions of purines, pyrimidines, and nucleic acids in the overall DOM concentration are not known. According to Takahashi et al. (1974), their concentrations are c. four to six times lower than the levels of amino acids and proteins. Hence, they probably make up around 2% or less of DOM (see Table 4.2B in Section 4.1). Nevertheless, in certain UV bands, nucleic acids absorb light more strongly than amino acids.

Among the radiation-absorbing purine and pyrimidine derivatives that are products of nucleic acid decay, the following are most likely to be found in sea water.

1. The five purine and pyrimidine bases mentioned above; see Figure 4.9b.
2. Nucleosides, that is, combinations of these bases with ribose or deoxyribose (names outside parentheses); see Figure 4.9c.
3. Nucleotides, that is, the esters of orthophosphoric acids and nucleosides (names within parentheses): see Figure 4.9c. These compounds are commonly referred to as acids; for example, adenosine-5′-phosphate, is also called adenylic acid. Some nucleotides can have one or two additional phosphoric acid molecules attached to them.[7]
4. "Twisted" nucleic acid fragments of various sizes (Figure 4.9d), for example, component sugar molecules (ribose and deoxyribose). But because these are saturated carbohydrates, they are optically almost transparent.

[7] Four such compounds play a particularly significant part in sunlight-stimulated processes in the sea, apart from the mere absorption of light. They are adenosine-5′-phosphate, commonly known as ADP, adenosine-5′-triphosphate (ATP), plus two, more complex nucleotides, namely, nicotinamide adenine dinucleotide (NAD) and its phosphate NADP. Their electronic structures enable these compounds to participate in the transport of electrons and in the conversion of radiant energy into the chemical energy of the biomass, which takes place during the photochemical phase of photosynthesis (Govindjee 1975).

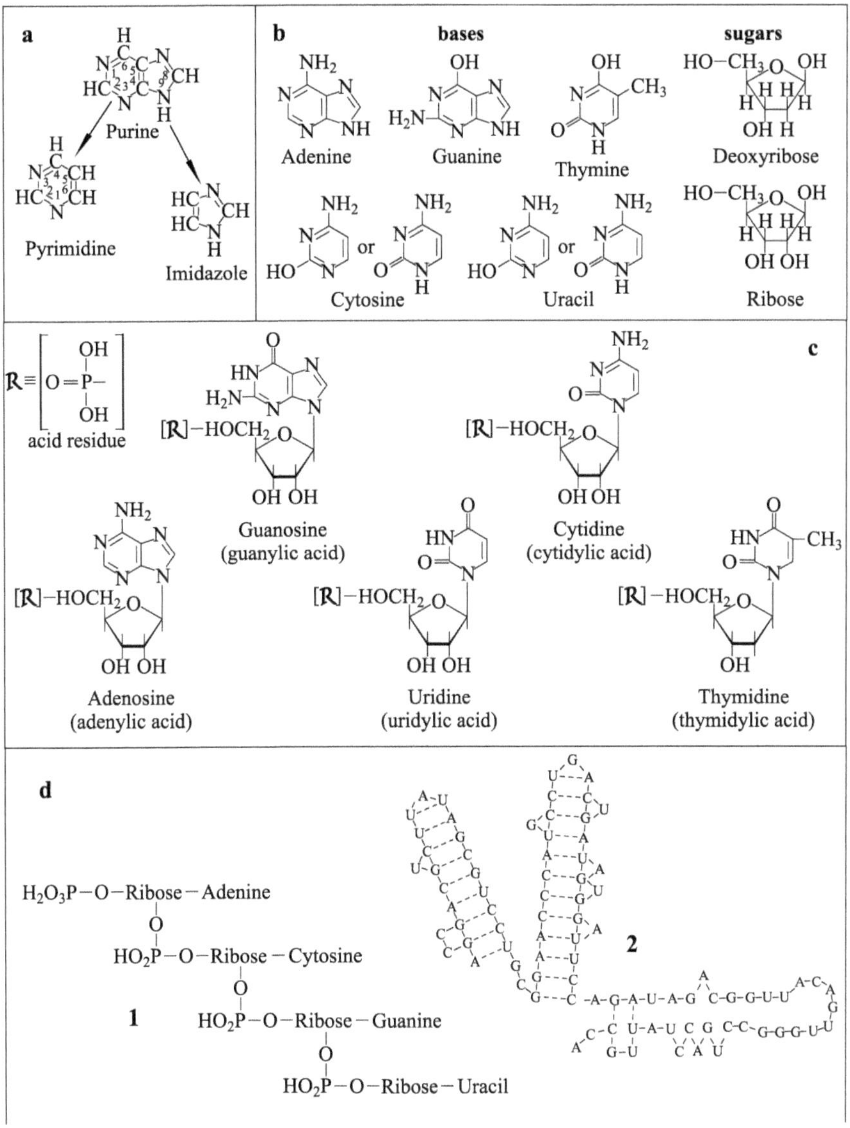

FIGURE 4.9. The chemical structures of nucleic acids and their components (a) initial compounds; (b) the bases and sugars in DNA and RNA; (c) examples of nucleosides (names outside parentheses) and nucleotides (names within parentheses); in the nucleotides, the H atom in the $HOCH_2$ group is replaced by an acid residue R; (d) the structure of nucleic acids (RNA): 1, simplified structural element (tetranucleotide); 2. example of a repeating unit in double-stranded RNA (A, adenine; G, guanine; C, cytosine; U, uracil). (Data adapted from Filipowicz and Więckowski (1983).)

As Figure 4.9 shows, it is the conjugated double bonds of the single and double nitrogen-containing purine and pyrimidine rings that are responsible for the absorption properties of nucleic acids. As their electronic structure is quite complex, we analyze just the most intensive absorption band, due to the $\pi \rightarrow \pi^*$ transition of conjugated electrons. Williams and Fleming (1980) and others found that the absorption peak for the wavelength $\lambda \approx 255$ nm is due to the chromophore groups with a pyrimidine structure, whereas that at $\lambda \approx 265$ nm is due to the purine ring. The attachment of different substituents to the rings in purines and pyrimidines produces only a marginal auxochromic effect; that is, the peak positions and molar extinction intensities change only slightly. These features are displayed in Table 4.5: clearly, the properties of the various purine and pyrimidine derivatives are very similar.

The attachment of sugars and phosphate groups to these molecules causes the molar extinction coefficients of the nucleosides and nucleotides to rise slightly in comparison with the bases. At the same time, a substantial increase in the molecular masses takes place; hence, the largest specific absorptions are characteristic of the nitrogen bases, and they are the strongest absorbers in the c. 240–300 nm range of all the substances naturally occurring in the sea. Their specific absorption coefficient spectra are illustrated in Figure 4.8d (see also Table 4.3). There we see that the absorptions in this spectral range exceed the similar absorptions of amino acids by at least one order of magnitude.

Determining the molar extinction coefficients of nucleic acids is rather difficult owing to their large and undefined molecular mass. For this reason, these coefficients are defined with respect to the number of moles of phosphorus (because one purine or pyrimidine unit corresponds to each atom of phosphorus) and are denoted $\varepsilon(P)$. Theoretically, the molar coefficients $\varepsilon(P)$ should be the resultants of the component extinction coefficients of the nucleotides. According to Table 4.5, however, $\varepsilon(P)$ for DNA and RNA are much lower (by c. 40%) than the values predicted from the molar extinction coefficients of the nucleotide mixtures that are classified as nucleic acids. This effect is known as hypochromism (e.g., Filipowicz and Więckowski (1983)); no complete theoretical explanation for it has yet been put forward.

This reduction in absorption capabilities is probably due to the competitive interaction of the electron states of adjacent nitrogen bases (Figure 4.9d). The effect is thus the opposite of that produced by the conjugation of π-orbitals. This is why hypochromism is the most pronounced in the coiled chains of DNA. On the other hand, the transition to the random state of these chains causes the hypochromism to diminish and absorption to increase. This probably applies to DNA and RNA fragments dissolved in sea water. In order to estimate the approximate total absorption due to nucleic acids in the sea, it seems justified to assume that their specific absorption coefficients are higher by a factor of around 1.7 than those given in Table 4.5, that is, the values characteristic of individual nucleotides.

TABLE 4.5. Principal spectral features of purine and pyrimidine derivatives.[a]

Compound	λ_{max} [nm]	ε [dm^{3}mol^{-1}cm^{-1}]	a^*_{OM} [m^{2}g^{-1}]
-1-	*-2-*	*-3-*	*-4-*
	Nitrogen Bases (Purines and Pyrimidines)		
Adenine	260–264	12300–13400	22.8
Guanine	273–275	7350–8100	12.3
Cytosine	267–274	6100–10200	12.6
Uracil	259	8200	16.8
Thymine	264	7900	14.4
	Nucleosides (Bases + Sugars)		
Adenosine	257–260	14600–14900	12.9
Deoxyadenosine	258	14100	13.2
Guanosine	256–276	9000–12200	8.7
Deoxyguanosine	255	12300	10.8
Cytidine	271–280	9100–13400	10.8
Deoxycytidine	280	13200	13.7
Uridine	261–262	10100	9.7
Thymidine	267	9700	8.7
	Nucleotides (Nucleosides + Phosphoric Acid Residue)		
Adenylic acid	257–259	15100–15400	10.2
ATP	257–259	14700–15400	6.9
NADP	259	14400	4.0
	340	6230	1.7
NAD^+	260	18000	5.6
Deoxyadenylic acid	258	14300	10.1
Guanylic acid	257	12200	7.8
Deoxyguanylic acid	255	11800	8.0
Cytiylic acid	281	13600	9.8
Deoxycytidylic acid	280	12300	9.4
Uridylic acid	261	9700	7.0
Thymidylic acid	267	9600	6.6
	Nucleic Acids		
DNA	258	6100–6900	4.4
RNA	258	7400	4.5

[a] Data taken from or estimated on the basis of various sources, for example, Campbell and Dwek (1984) and Filipowicz and Więckowski (1983).
Column 3: in some cases the range of variability of the molar extinction coefficients in acidic and basic media (different pH) is given; it corresponds to the range of λ_{max} (column 2). Column 4 gives specific absorptions in nearly neutral solutions (pH ≈ 7).

To conclude this brief discussion of the absorption of radiation by nucleic acids and their components, we remind the reader that UV radiation destroys or brings about mutations in marine microorganisms and phytoplankton. One of the reasons for this is that photochemical changes take place in the nucleic acids (especially DNA), that is, in the chromosomes governing life processes.

4.2.4 *Lignins*

The third group of organic UV absorbers in sea water comprises the lignins (Nyquist 1979, Anderson 2002). Unlike the previous two groups of absorbers, which are autogenous components of the sea, lignin sulfates are not produced in the marine environment. Synthesized in higher terrestrial plants, they are carried to the sea in riverine waters. This is why they occur only in shelf waters, and especially in enclosed seas, where their concentrations can reach high levels (see Table 4.2C). One of the reasons for this is that in a reducing environment, these compounds are particularly stable and resistant to biochemical decomposition. By contrast, lignins are absent from open oceanic waters. Waters containing lignins are thus classified as Case 2 sea waters (see Chapter 1).

Chemically, lignins are the complex products of the biosynthesis of carbohydrates and aromatic amino acids (phenylalanine and tyrosine) and form rigid polymeric structures that cement the cellulose fibers of higher plants (Meyers-Schulte and Hedges 1986, Pempkowiak 1989). This is one of the reasons for the amazingly diverse shapes and sizes of trees. In the sea we come across fragments of these structures, usually in the form of lignin sulfates. The depolymerization of lignins by various methods causes their component monomers to be released; Figure 4.10 gives examples. All chromophores, they are auxochromic compounds of phenol, and it is this that determines their optical properties.

The approximate spectrum of the specific absorption of light by the lignins typical of Baltic Sea waters, as measured by Nyquist (1979), is given by the plot in Figure 4.8e (see also Table 4.3). Because of the great similarity between the

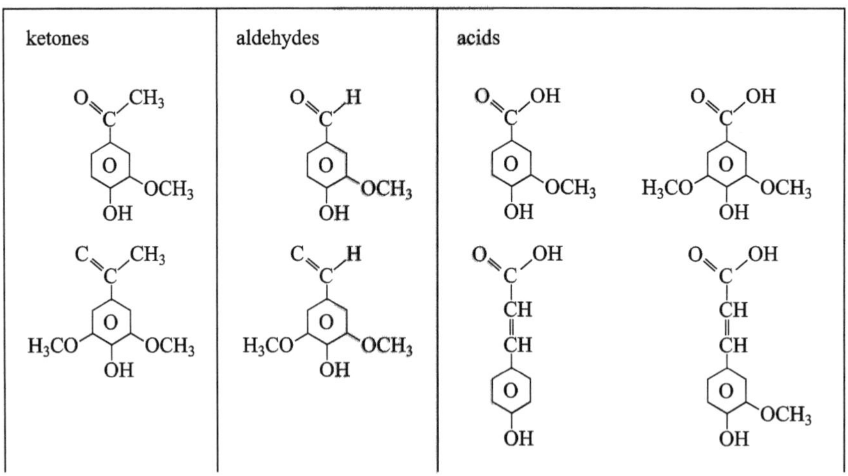

FIGURE 4.10. Selected monomers released from lignins; they are chromophoric groups typical of lignins. (After Pempkowiak (1989).)

chemical structures of lignins and "aromatic" amino acids, likewise the absorption characteristics in the 200–300 nm range of both types of compounds are similar. There is thus a distinct absorption peak at wavelengths $\lambda \approx 280$ nm and an abrupt increase in absorption at $\lambda < 250$ nm. Moreover, unlike the "aromatic" amino acids, lignins also quite strongly absorb wavelengths longer than 300 nm. This is due to the greater variety of auxochromic substituents in the benzene ring, and also to the possible conjugations between the individual aromatic rings in the polymer structure.

It is also worth mentioning that lignins are the precursors of humus substances (HS) forming in the soil (Hedges 1980). Hence, their presence in some HS extracted from sea water can be used to determine the proportion of terrigenous HS in the overall mass of HS in the sea (e.g., Pempkowiak (1984) and Pempkowiak and Pocklington (1983), Staniszewski et al. (2001), Grzybowski and Pempkowiak (2003)).

4.2.5 Colored Dissolved Organic Matter (CDOM)

The organic absorbers we have been discussing affect the total absorption of radiation in the sea by all DOM in principal only in the UV region. On the other hand, the wide diversity of more or less complex substances (chemically unidentified) making up the majority of DOM is responsible for the absorption of light in the entire spectral region characteristic of electronic transitions, that is, the UV–VIS region, and to some extent the near-IR (see Figure 4.3). These compounds impart a yellowish tinge to sea water, and in marine optics they are covered by the general appellation of *yellow substances* or *Gelbstoff* (from the German), first introduced by Kalle (1961, 1962, 1966).

Quite early chemical studies (e.g., Oden (1919), Shapiro (1957), and Skopintsev (1950)) showed that they are similar in nature and in some of their chemical properties, although not identical, to the products of the advanced decomposition of organic matter in the soil. Thus, to describe them in marine chemistry we use terms borrowed from soil chemistry: humus substances or humus acids, or quite simply, aquatic humus.[8] It is not possible, however, to define these terms unequivocally, that is, to find criteria for establishing which of the natural organic compounds occurring in the sea, produced in a chain of diverse transformations, can or cannot yet be classified among the HS. Some authors (e.g., Romankevich (1977, 1979), Spitzy and Ittekkot (1986), and Hansell and Carlson (2002)) lump practically all unidentified organic substances under this label, that is, around two thirds or more of the DOM in the sea. More often, however (e.g., Aiken et al. (1985), Rashid (1985), Harvey et al. (1983), and Pempkowiak (1989)), and in the present book too, the term "humus

[8] Other terms include *aquatic fulvic acid* (Oden 1919), *yellow organic acid* (Shapiro 1957), *gilvin* (Kirk 1976), and *unknown photoreactive chromophores, UPC* (Blough and Zafiron,1985), and so on.

acids" is used to define only the most stable end products of the natural transformations of organic matter in the environment. These include the so-called fulvic acids (FA) and humic acids (HA).

These make up from 10 to 50% of DOM, depending on the part of the world and the sea's depth (Sieburth and Jensen (1968), Harvey et al. (1983), Pempkowiak (1989), Blough and Del Vecchio (2002), Nelson and Siegel (2002), and Cauwet (2002); see also Table 4.2D). The reason for distinguishing only these groups of compounds as yellow substances or HS is because they are primarily responsible for the absorption of visible light in the sea by DOM (Stuermer 1975, Nyquist 1979); moreover, it is possible to separate them—by absorption on solid absorbers—from all the other DOM. By means of acidification ($pH \approx 1$ to 3), HS can be further separated into fulvic acids and humic acids. FA are soluble in such weak aqueous solutions, but the insoluble HA are precipitated. These procedures and their modifications are described in detail in, for example, Stuermer and Harvey (1974), Aiken et al. (1985), and Carder et al. (1989).

The optical properties of the various HS are very diverse and depend on their type (fulvic or humic) and origin (auto- or allogenous): the diverse chemical structures of HS are governed by the mechanisms of their formation. In all environments, but particularly in sea water, these are poorly understood (e.g., Carlson (2002) and Bronk (2002)). Based on a recent model (Rashid 1985), Figure 4.11 is a simplified diagram of the successive transformations of organic compounds secreted by organisms into the environment.

In addition to the gradual decomposition of organic compounds into inorganic forms, some of the former at a certain stage in their conversion undergo a variety of multistage condensation and polymerization processes, the end result of which is the formation of HS. It is not yet known which factors facilitate these processes and to what extent, neither has the full chemical picture of the reactions accompanying them been established. The plethora of hypotheses advanced to explain these processes have only partially achieved their objective. Nevertheless, we do not discuss these attempted explanations: we refer interested readers to the literature, for example, Schnitzer and Khan (1972), Stevenson (1982), Duursma (1965), Duursma and Dawson (1981), Pempkowiak (1989), and Hansell and Carlson (2002).

Despite the complexity of the mechanisms of humus acid formation in natural environments, current research has been able to establish quite a number of their physical and chemical properties. Up to a point this has made it easier to interpret their absorption spectra. The following groups of compounds with different physical and chemical properties are worth examining.

1. *Compounds with complex molecules, generally of large molecular mass.* With respect to HS forming in the soil and in sea water, this mass may range from 500 to 10,000 Daltons (Stuermer and Harvey 1974, Thurman 1985). The lighter fractions among them tend to be typical of FA, and the heavier ones of HA. In rare cases, such as in bottom sediments and in soil, the molecular

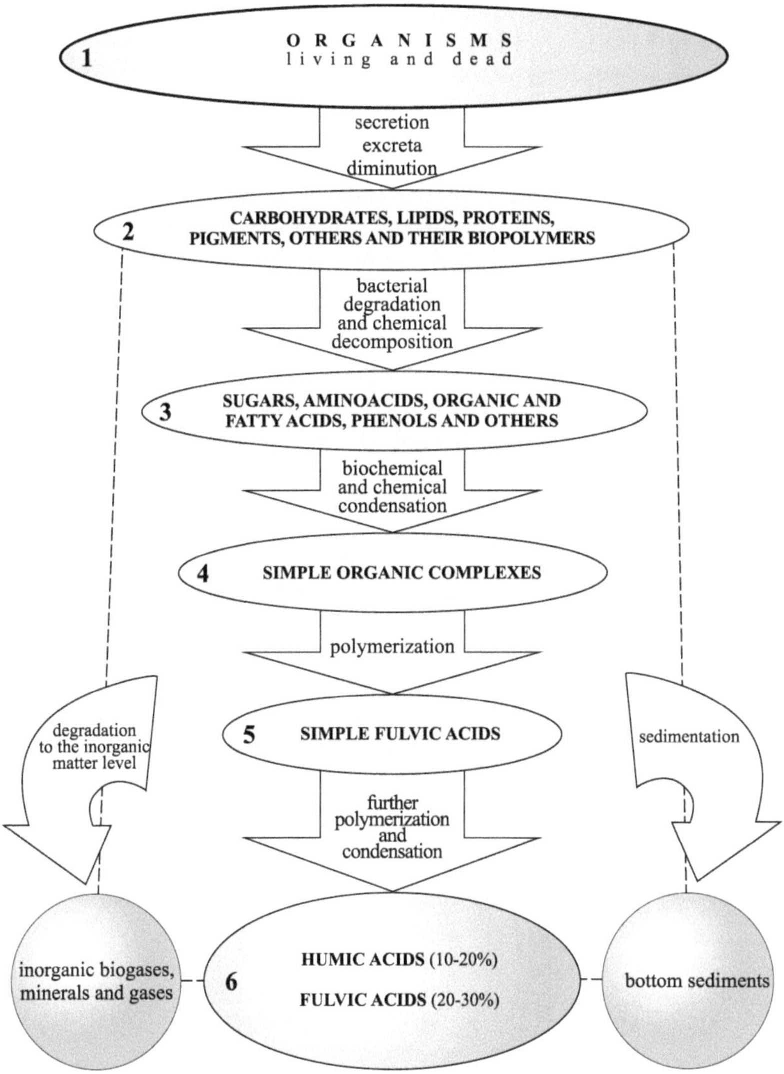

FIGURE 4.11. A simplified diagram of the formation of humus substances (HS) in the marine environment. The percentages given in block 6 are characteristic of allogenous humus substances in the sea. (Based on an idea of Rashid (1985); some data taken from Spitzy and Ittekkot (1986).)

masses of humus acids can be as high as several hundred thousand Daltons (Hayase and Tsubota 1985, Zepp and Schlotzhauer 1981).

2. *Humus acids*. This term covers a wide range of relatively simple organic compounds produced by condensation reactions. Numerous authors have suggested hypothetical structures for these compounds (Figure 4.12).

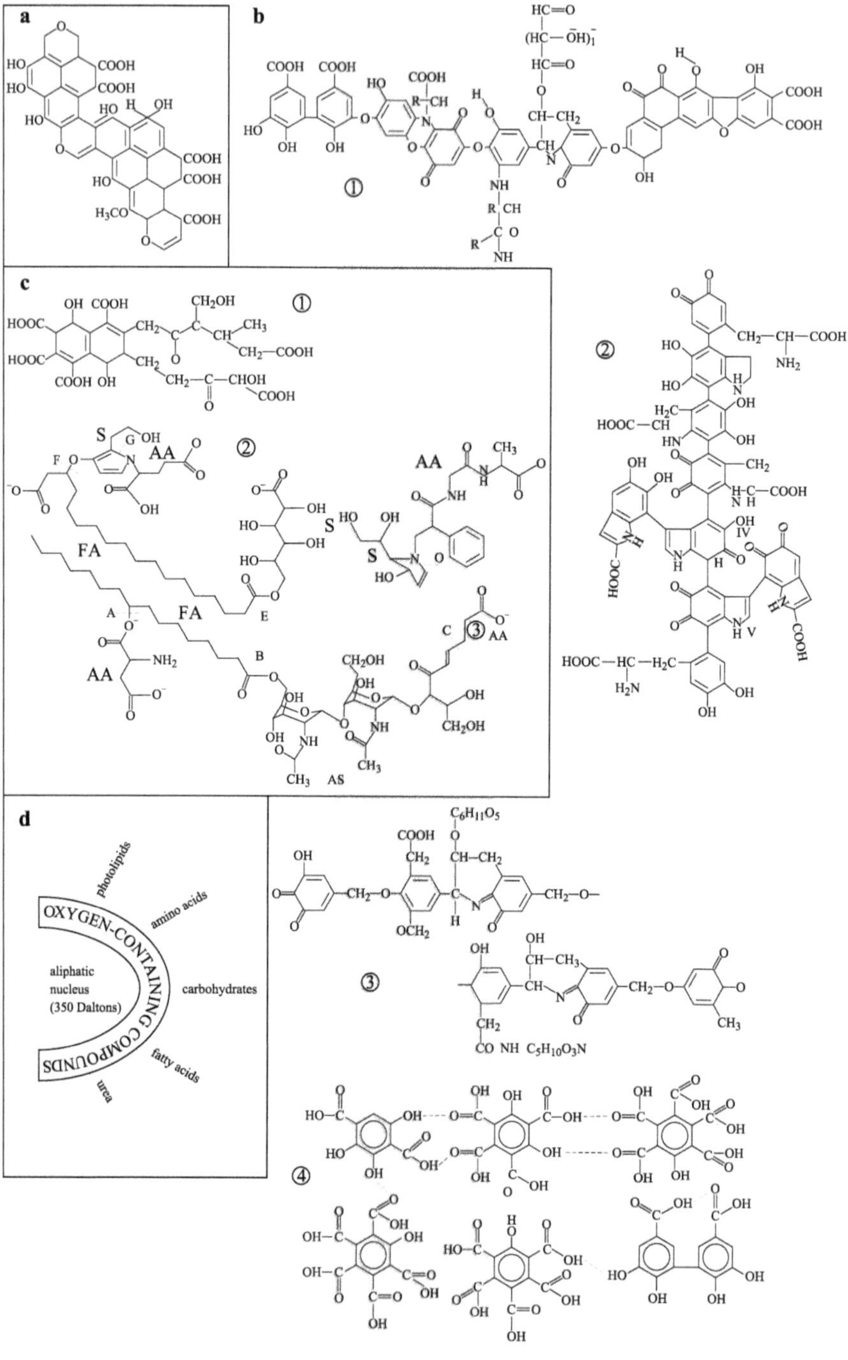

FIGURE 4.12. Hypothetical structures of humus substances: (a) humus substances extracted from the soil; (b, 1-4) humic acids extracted from the soil; (c) fulvic acids: 1. extracted from the soil; 2-4. extracted from sea water; (d) probable structure of autogenous humus substances in sea water forming around an aliphatic nucleus of molecular mass 300–350 Daltons. (After various authors; based on data taken from Spitzy and Ittekkot (1986) and Pempkowiak (1989).)

They can consist of fragments of amino acids, sugars, fatty acids, phenols, and hydrocarbon chains, in various proportions to one another, depending on the type of environment in which they are formed. HA are usually highly aromatic, whereas FA are mostly aliphatic compounds (Figures 4.12b,c). Moreover, humus compounds forming in sea water generally contain far fewer phenolic components and are saturated to a much greater extent than their counterparts forming in the soil (see compounds 1 and 2 in Figure 4.12c).

Numerous authors, chemists in particular, have investigated the light absorption spectra of various HS extracted from the sea, for example, Ishiwatari (1973), Gorshkova (1972), Flaig et al. (1975), Kalle (1961), Pempkowiak (1977, 1989), and Pempkowiak and Kupryszewski (1980). Although they treated the light absorption measurements as one of the analytical methods for identifying particular chemical structures, they did not define absolute specific absorption coefficients.

Also of not much use as regards the definition of these coefficients are papers in hydrooptics (cited later), which, although containing valuable data on the absorption properties of yellow substances in the sea, are devoid of additional information on their concentrations. In view of this, there are only a few literature sources providing absolute values of the spectra of the specific absorption coefficients of humus substances, or data from which these spectra can be calculated. These few sources, which came about as a result of research into remote-sensing methodology, include the work done by teams led by Stuermer, Harvey, Carder, and others (later quoted in detail), a little-known dissertation by Nyquist (1979), and the papers by Zepp and Schlotzhauer (1981) and Hayase and Tsubota (1985). Unfortunately, they refer only to certain regions of the sea or to given situations in the sea. We discuss their results in this and the following section.

Among the most meticulously performed research into the absorption properties of marine humus acids—HA and FA are given separate treatment—is that by Carder et al. (1989) dealing with humus acids extracted from Gulf of Mexico waters. The spectra of the specific absorption coefficients in the λ = 240–675 nm range are shown in Figure 4.13a (see also Table 4.3). For comparison, Figure 4.13b presents examples of the ranges of specific absorption coefficients typical of these compounds extracted from other environments. It is clear from Figure 4.13a that a monotonic decrease in absorption with increasing wavelength is characteristic of all these spectra. There are hardly any local maxima.[9] Furthermore, because these spectra

[9] This is a property distinguishing humus acids of marine origin from those occurring in other environments. For example, for HA extracted from bottom sediments, and also from organic detritus, we find bulges or inflexion points (e.g., in the 440, 460–470, and 660–665 nm wavelength ranges) due to the plant pigment fragments present in these compounds (see, e.g., Ishiwatari (1973), Gorshkova (1972), and Pempkowiak (1989)).

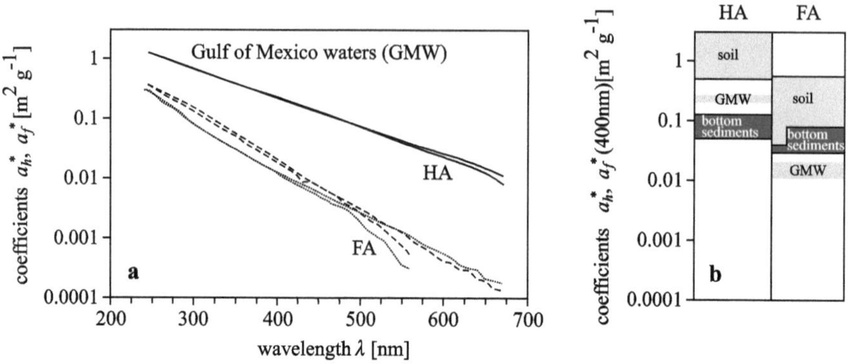

FIGURE 4.13. Characteristics of the mass-specific light absorption coefficients of humic and fulvic acids extracted from various sea waters: (a) spectra of mass-specific light absorption coefficients of humic acids (HA) and fulvic acids (FA) extracted from Gulf of Mexico waters (GMW; adapted from Carder et al. (1989)); (b) comparison of specific light absorption coefficients of marine humic $a_h^*(\lambda = 400\text{ nm})$ and fulvic $a_f^*(\lambda = 400\text{ nm})$ acids from the Gulf of Mexico with characteristic absorptions by these compounds from other environments. The various curves are given from various sources, for example, Zepp and Schlotzhauer (1981) and Hayase and Tsubota (1985).

are almost exponential, they can be well approximated by means of two-parameter relationships:

for humic acids (HA)

$$a_h^*(\lambda) = a_h^*(\lambda_0)e^{-S_h(\lambda - \lambda_0)}, \tag{4.4}$$

for fulvic acids (FA)

$$a_f^*(\lambda) = a_f^*(\lambda_0)e^{-S_f(\lambda - \lambda_0)}, \tag{4.5}$$

where S_h and S_f are the respective slopes (expressed in [nm^{-1}]) of the spectra (on a semi-logarithmic scale) for HA and FA, and $a_h^*(\lambda_0)$ and $a_f^*(\lambda_0)$ are the respective mass-specific light absorption coefficients of HA and FA for any wavelength λ_0 in the UV–VIS region. In this book we mostly use the value of $\lambda_0 = 400$ nm.

However, the absolute absorption coefficients and slopes of the spectra for both these compounds differ considerably: see Table 4.6, which gives the magnitudes of these parameters for HA and FA from different environments. As we can see from Figure 4.13 and Table 4.6, HA are much stronger absorbers than FA in sea water. This is a consequence of their chemical structures (Figure 4.12): in FA, the nuclei around which other compounds condense are usually aliphatic hydrocarbons, which are weak absorbers of radiation. In HA, by comparison, these nuclei are aromatic rings, often several times conjugated, which enhances the absorption capabilities of these acids. Thus, in the case illustrated in

TABLE 4.6. Parameters of the mass-specific light absorption spectra of humic and fulvic acids from the Gulf of Mexico and other environments.

Environment	a_h* (400 nm) a_f*(400 nm) [m^2g^{-1}]	S_h, S_f [nm^{-1}]	References
		Humic Acids	
Soil	1.52[a]	0.0090	Stuermer 1975,
	0.58 ± 0.08		Zepp & Schlotzhauer 1981
Sea (Gulf of Mexico)	0.228 ± 0.00005	0.0110 ± 0.00012	Carder et al. 1989
Bottom sediments	0.050–0.130	0.008–0.011	Hayase & Tsubota 1985, Zepp & Schlotzhauer 1981
		Fulvic Acids	
Soil	0.29 ± 0.25	0.011	Zepp & Schlotzhauer 1981
Bottom sediments	0.026–0.080	0.0125–0.017	Hayase & Tsubota 1985
River water (Mississippi delta)	0.018 ± 0.001	0.0194 ± 0.00044	Carder et al. 1989
Sea (Gulf of Mexico)	0.013 ± 0.001	0.0184 ± 0.00166	Carder et al. 1989

[a] The hypothetical value measured for humus substances extracted from coastal Atlantic water on the assumption that the dominant fraction of these substances was terrigenous.

Figure 4.13, the specific absorptions of radiation of wavelength $\lambda = 400$ nm by HA are about 20 times greater than by FA.

The second characteristic distinguishing the absorption of radiation from various spectral regions by HA and FA is the magnitude of the slopes of their absorption spectra S_h and S_f. These slopes are not as steep for HA (c. 0.008–0.011 nm^{-1}) as for FA (c. 0.011–0.020 nm^{-1}). This is an obvious consequence of the fact that the ratio of chromophores with large numbers of conjugated electrons (i.e., which absorb longer waves) to simple chromophores absorbing in the UV, is larger for HA than for FA molecules.

As we can see from Figure 4.13b and the data in Table 4.6, the differences between the absorption properties of HA and FA from other natural environments qualitatively resemble those among the properties of these acids from the Gulf of Mexico. Only the absolute values of the absorption spectra parameters are different. On the other hand, the quantitative proportions between these parameters for the two acids from different environments are interesting. We find a series of more or less complex regularities that, given our current understanding of the nature of HS, cannot all be explained in a simple manner.

We now attempt to explain the most important of the obvious regularities.

1. The mass-specific absorptions of HA and FA are higher when these substances are of pedological rather than of marine origin. This may indicate a greater accumulation of optically active chromophores in these acids formed in the soil.
2. The slopes S_h of the absorption spectra of HA are similar for all environments, usually varying in the range from 0.008 to 0.011 nm^{-1}. Thus, we

have the same proportions among the various chromophores absorbing radiation from different spectral ranges in the UV–VIS region, regardless of the environment in which HA are formed.

3. The slopes S_f of the absorption spectra of FA are steepest when these compounds are of marine origin (c. 0.018 – 0.020 nm^{-1}) and decrease to 0.011–0.013 nm^{-1} when they have come from the soil or bottom sediments. This could be evidence for higher relative contents of longwave chromophores in terrigenous FA, such that their optical properties approach those of HA.

We do not yet have a definitive explanation for why the mass-specific absorption coefficients of HA in marine bottom sediments are lower than the corresponding values in sea water. Some authors (e.g., Hayase and Tsubota (1985) and Carder et al. (1989)) are of the opinion that this is linked with the existence of hitherto unexplained relationships between the molecular masses of the various HS forming in the bottom sediments and in sea water on the one hand, and their absorption properties (such as the absolute absorption coefficients and spectral slopes). Empirical plots of these relationships are illustrated in Figure 4.14. A significant conclusion to be drawn from this figure is that the variations in the mass-specific absorption coefficients of HA and FA with changes in their molecular masses tend to be opposed: with increasing molecular mass, the absorptions of FA rise and those of HA fall (Figure 4.14a). At the same time, an increase in molecular mass causes the slopes of the absorption spectra of fulvic acids to drop to the values typical of humic acids (see Figure 4.14b). As a result, fractions of the large HA and FA molecules ($>10^5$ Daltons) have identical absorption properties and could very well be the same or extremely similar compounds. However, this is speculation: there is no evidence to substantiate this last statement.

Fulvic and humic acids (FA and HA) are present in sea water in various proportions (Harvey et al. 1983). On a global scale, they are primarily autogenous substances (c. 95%), forming as a result of the metabolic processes of phytoplankton, the decay of marine organisms, and the subsequent countless transformations of organic matter in the water itself (Duursma and Dawson 1981). The remaining 5% or so are usually allogenous substances, mostly terrigenous (from the soil), which are carried to the sea by rivers; smaller quantities will have been washed out of bottom sediments. These allogenous substances are characteristic of the coastal zones of seas and oceans. Their concentrations fall away rapidly with increasing distance from the shore and depth in the sea (Højerslev 1980, 1981; Blough and Del Vecchio 2002). We should therefore expect some variation in the absorption properties of natural humus associations in the sea water of different basins. This diversification will depend, among other things, on the individual properties of the FA and HA forming in a given sea and on the proportion and properties of the allogenous humus substances.

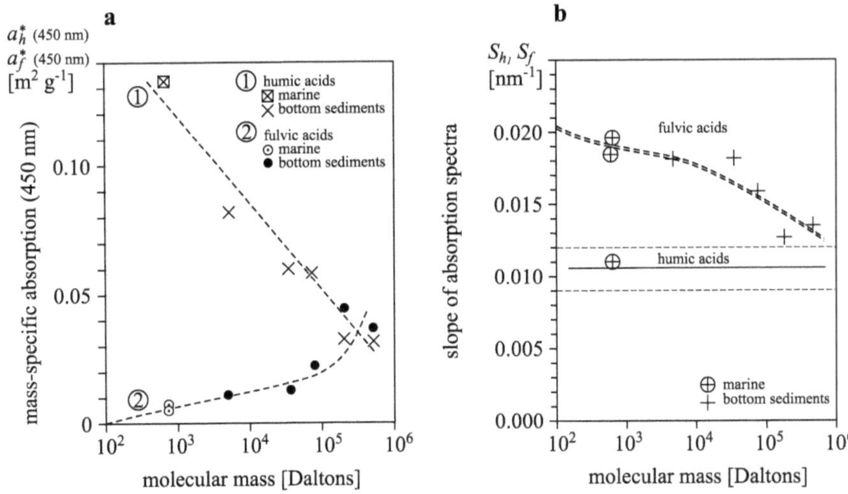

FIGURE 4.14. Approximate relationships between mass-specific absorption coefficients (a), slopes of absorption spectra (b), and the molecular masses of humic and fulvic acids. (Adapted from Carder et al. (1989).)

The influence of the proportions between the levels of FA and HA on the resultant absorption of radiation by HS in water from the Gulf of Mexico was analyzed by Carder et al. (1989). Assuming that in a given region of sea, the individual properties of the two humus components do not change, the resultant mass-specific absorption coefficient of humus acids a^*_H can be linked to the so-called degree of fulvization of humus substances f in accordance with the relationship:

$$a^*_H(\lambda) = f a^*_f(\lambda) + (1 - f)\, a^*_h(\lambda), \tag{4.6}$$

where by the degree of fulvization

$$f = C_f / C_H = C_f / (C_h + C_f), \tag{4.7}$$

is meant the relative proportion of the concentration of FA C_f in the total concentration of all humus acids $C_H = C_f + C_h$ (where C_h is the concentration of HA in the given water). The spectra of the specific absorption coefficients of all HS, calculated on this basis for various degrees of fulvization, are shown in Figure 4.15. It is obvious that the coefficients $a^*_H(\lambda)$ take values intermediate between those characteristic of FA and HA (i.e., $a^*_f < a^*_H < a^*_h$) and that they are strongly dependent on the degree of fulvization f.

The approximate ranges of variation in the values of the degree of fulvization f found in nature are given in Table 4.7. These data show that over the vast expanses of oligotrophic waters in the World Ocean, degrees of fulvization are high (c. 97%), so that the absolute specific light absorption coefficients of all

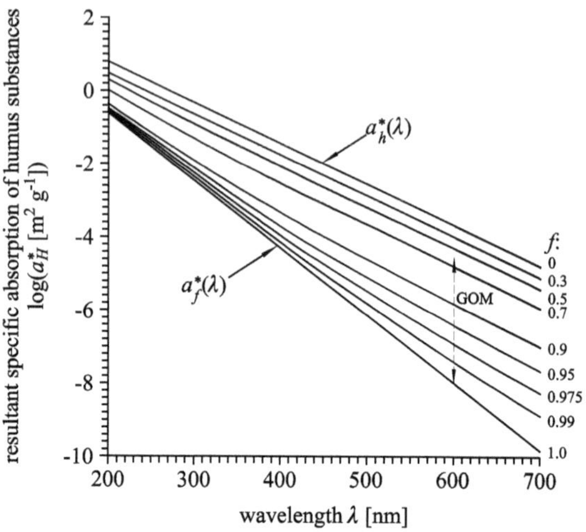

FIGURE 4.15. The effect of the degree of fulvization of humus substances f on the resultant specific absorption coefficients of humus substances $a_H{}^*$ (λ) typical of the Gulf of Mexico: spectra a^*_H (λ) calculated on the basis of Equations (4.6) and (4.4)–(4.5), assuming appropriate numerical values: a^*_f (400 nm) = 0.013 and a^*_h (400 nm) = 0.228 [m^2 g^{-1}], also S_f = 0.0184 and S_h = 0.011 [nm^{-1}]; GOM (Gulf of Mexico) illustrates the range of variability of a^*_H (λ) corresponding to the range of variability of the degree of fulvization found in the Gulf of Mexico; that is, 0.63 ≤ f ≤ 0.985 (see Table 4.7). (Adapted from Carder et al. (1989).)

TABLE 4.7. Ranges of variability of the degree of fulvization in the World Ocean.[a]

Basin	Concentration of chlorophyll *a* C_a [mg m^{-3}]	Degree of fulvization f	Source of data
-1-	*-2-*	*-3-*	*-4-*
Gulf of Mexico:			
– Mississippi delta	1.2–7.9	0.83–0.91	Harvey et al. 1983
– Cape San Blas	0.4– –1.06	0.63–0.93	
– Oligotrophic central areas	0.08–0.24	0.086–0.985	
Atlantic Ocean:			
– Sargasso Sea (surface waters)	Traces	0.97	Stuermer & Harvey 1974
– Sargasso Sea (abyssal waters 1500 m)	~0	0.78	
– Coastal waters (Woods Hole)	>2	0.85	

[a] For greater clarity, the chlorophyll concentration C_a, indicating the trophic type of sea, is also given.

humus acids are close to (although always higher than) the corresponding coefficients for FA.

On the other hand, in eutrophic waters, where the degree of fulvization is lower, $a^*_h(\lambda)$ can be much higher than the corresponding coefficients for FA (e.g., for the Gulf of Mexico the resultant mass-specific absorption coefficient for light of wavelength c. 400 nm of all HS in extreme cases is about eight times greater than for FA), making up over 40% of the mass-specific absorption coefficient of HA (see the plot marked GOM in Figure 4.15). Apart from affecting the absolute magnitude of the light absorption coefficients of all HS in the sea, the degree of fulvization also determines the slopes S_H of their resultant absorption spectra $a_H(\lambda)$ (or $a^*_H(\lambda)$). For remote-sensing purposes (see, e.g., Dera (2003)), the latter effect is of far greater significance than the former. Many remote-sensing algorithms, which enable the interpretation of remotely obtained optical data (for a review of such algorithms, see Kondratiev et al. (1990)), use the expression of light absorption by yellow substances in the form:

$$a_y(\lambda) \approx a_H(\lambda) = a_H(\lambda_0)e^{-S_h(\lambda - \lambda_0)}. \tag{4.8}$$

In view of the convention applied in this book, namely, that the terms "yellow substances" and "humus acids" are equivalent, we can assume that $a_y \equiv a_H$.

The accuracy with which the spectral slope S_H is established has a marked influence on that with which concentrations of optically active substances in sea water are estimated, for example, the concentration of chlorophyll determined on the basis of satellite measurements. It emerges from Figure 4.15 that the resultant spectra $a^*_H(\lambda)$ for various degrees of fulvization have different slopes S_H, which further differ with the wavelength of the light. The relationships of the mean spectral slopes S_H for two spectral ranges, determined for Gulf of Mexico waters are illustrated in Figure 4.16. This shows that even in quite a small area of sea, the spectral slopes S_H vary over a wide range: they almost double—from c. 0.0115 to c. 0.019 nm^{-1}—for degrees of fulvization from c. 0.6 to almost 1.0.

These absorption properties of marine HS, regarded as a bicomponental association of substances, refer to basins in the Gulf of Mexico and are thus unique. On their basis we can estimate with quite a good degree of precision the absolute absorption coefficients of marine humus acids in that we apply the concentrations of FA (C_f) and HA (C_h) determined for this part of the Atlantic by Harvey et al. (1983). However, we cannot apply these data to estimate the absorption properties of HS in other seas or oceans, because the nature and absorption capabilities of FA and HA forming in diverse environments are different (Zepp and Schlotzhauer 1981).

In the works of other authors (Zepp and Schlotzhauer 1981, Nyquist 1979, Stuermer 1975), in which the absorption capacities of HS were analyzed, the relevant measurements were performed on samples containing unseparated

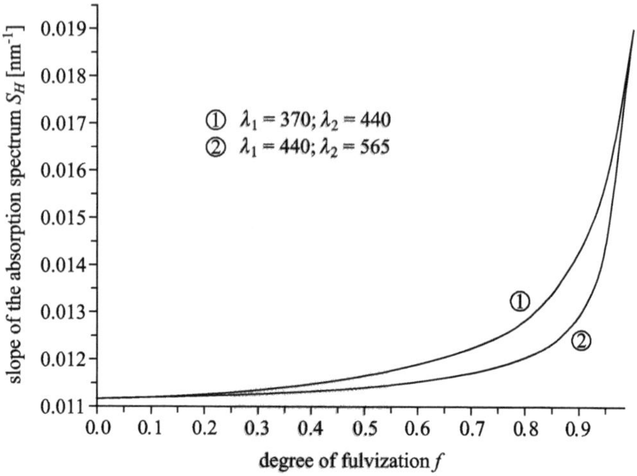

FIGURE 4.16. Relationship between the slopes of visible light absorption spectra $S_H(\lambda_1,\lambda_2)$ and the degree of fulvization of humus substances in the Gulf of Mexico. (Based on data from Carder et al. (1989).)

natural mixtures of FA and HA. The empirical spectra of the absorption coefficients of all HS $a^*_H(\lambda)$ determined by these authors are illustrated in Figure 4.17 and elsewhere. The mean parameters $a^*_H(400\text{ nm})$ and according to Equation (4.8) approximating these spectra for λ_0 = 400 nm are given in Table 4.8. For comparison, this table also sets out the range of variability of these parameters for different parts of the Gulf of Mexico. This range was estimated on the basis of the absorption capabilities of FA and HA discussed earlier, and of the degree of fulvization of these waters.

As can be gleaned from Table 4.8, the spectral mass-specific light absorption coefficients of natural associations of HS dissolved in various seas, recorded by different authors, lie within a range of values that stretches over two orders of magnitude. In our opinion, the scale of this variation has been overestimated, probably as a result of measurement errors and the application of different (nonstandard) chemical and optical measurement techniques. In particular, the very high absorption coefficients recorded by Stuermer (1975) in the coastal regions of the Atlantic[10] seem highly improbable, although not impossible.

Regardless of possible quantitative errors, the material presented in Table 4.8 and in Figure 4.17 point to a number of characteristic features of the absorption capabilities of HS that are specific to particular water types. So, the smallest spectral mass-specific absorption coefficients and at the

[10] Such values would be possible where the entire humus material was composed of polyaromatic compounds, such as those shown in Figure 4.12a.

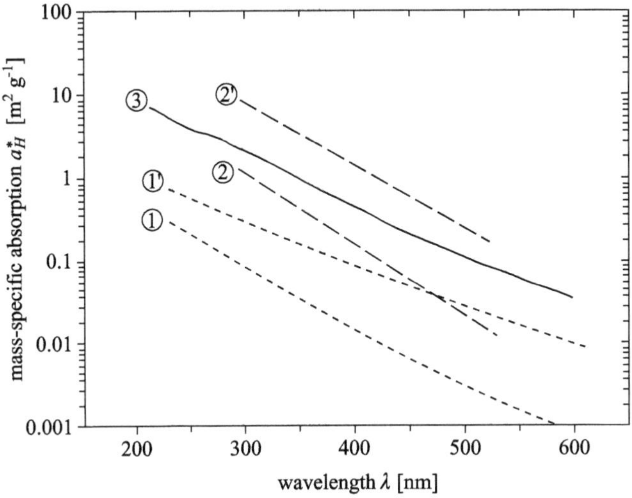

FIGURE 4.17. Approximate spectra of mass-specific light absorption coefficients of natural associations of humus substances dissolved in different seas: 1,1′. are the range of variability of coefficients typical of the Gulf of Mexico; 1. is the least polluted central part of the Gulf, 1′ is the vicinity of Cape San Blas (magnitudes estimated on the basis of Tables 4.6 and 4.7 after Carder et al. (1989) and Harvey et al. (1983); 2,2′. are the approximate range of variability recorded in various parts of the Atlantic; 2. is the NW Sargasso Sea, 2′. is the coastal waters off Woods Hole, Massachusetts (after Stuermer (1975)); 3. is the coastal waters of the Baltic Sea (after measurements by Nyquist (1979); see also Table 4.3).

TABLE 4.8. Approximate parameters of the spectra of the mass-specific light absorption coefficients of natural associations of humus substances dissolved in different waters.

Region	a_H^* (400 nm) [m^{2}g^{-1}]	Mean Slope $\bar{S}_H$ [nm^{-1}]	References
Atlantic (coastal region)	1.52	0.0175	Stuermer 1975
Baltic (littoral zone)	0.460	0.0140	Nyquist 1975
Inland waters (mean)	0.271	0.0111	Zepp & Schlotzhauer 1981
Sea waters (mean)	0.125	0.0114	Zepp & Schlotzhauer 1981
Sargasso Sea	0.165	0.020	Stuermer 1975
Gulf of Mexico[a]:			Carder et al. 1985,
– Cape San Blas	~0.030–0.100	0.0115–0.0148	Harvey et al. 1983
– Mississippi delta	~0.026–0.055	0.0128–0.0141	
– Oligotrophic central regions	~0.017–0.029	0.0133–0.0172	

[a] Variability intervals of the parameters calculated on the basis of specific coefficients of absorption by fulvic and humic acids (Table 4.6) and the degree of fulvization in the Gulf of Mexico (Table 4.7).

same time the steepest spectral slopes S_H are typical of unpolluted oligotrophic waters. This is because FA plays the dominant role in light absorption by humus acids in these waters. In contrast, in enclosed eutrophic basins or in the littoral zones of oceans, the mass-specific absorption coefficients of all HS increase, whereas the slopes of their spectra become gentler. This is without doubt the result of the substantial accumulation in these latter waters of allogenous HS, especially HA derived from the soil. The research of Pempkowiak (1989), for instance, has shown that the proportion of allogenous humus substances containing lignin fragments[11] in open Baltic waters is around 50% of the overall concentration of dissolved HS. In the Gulf of Gdańsk this percentage is higher still, c. 57%, and in the waters of the Vistula River (Wisła), it rises to 100%. We return to the question of the slopes of the spectra of light absorption by humus substances in Section 4.3.

At the present time, our knowledge of the nature and optical properties of marine HS is very fragmentary and does not suffice to make quantitative generalizations that would characterize the absorption capabilities of these substances in the various basins of the World Ocean. It is simply not possible to state, as some authors have done (e.g., Pempkowiak (1989) and Stuermer (1975)), that in vast areas of oceanic waters the quantities of humus acids and their absorption properties are known. The results of empirical investigations done up to the present do not provide sufficient evidence to make such a claim. To do so would require the standardization of the relevant chemical and optical research procedures, and on the other hand, that meticulous studies of these substances be carried out, at the very least on the scale of the investigations performed in the Gulf of Mexico.

4.3. The Total Absorption of UV–VIS Radiation by All Organic Substances Dissolved in Sea Water

The overall UV–VIS absorption spectra of all DOM in sea water can to a good approximation be described as the sum of the four group of absorbers discussed in the previous section:

$$a_{DOM}(\lambda) = \sum_{i=1}^{4} C_i a_i^*(\lambda), \tag{4.9}$$

where a_i* and C_i are the respective mass specific absorption coefficients and the concentrations of:

---

[11] As we said earlier, marine plants (phytoplankton), unlike terrestrial higher plants, do not biosynthesize lignin. That is why this component does not occur in autogenous humus substances.

Amino acids, their derivatives, and fragments of proteins together with parts of nonproteins (proteids) $a_1^* = a_{prot}$ and $C_1 = C_{prot}$
Various purine and pyrimidine compounds $a_2^* = a_{pp}^*$ and $C_2 = C_{pp}$
Lignins (only in shelf regions) $a_3^* = a_L^*$ and $C_3 = C_L$
Humus substances $a_4^* = a_H^*$ and $C_4 = C_H$

Inasmuch as the first three groups of substances absorb chiefly UV light, it is the practice in marine optics to denote the coefficient of the total VIS absorption of visible light by DOM variously by a_{DOM} or a_y (i.e., absorption by yellow substances), so that for the VIS spectral region we can write:[12]

$$a_{DOM}(\lambda) \equiv a_y(\lambda) \approx a_H(\lambda). \tag{4.10}$$

From our earlier discussion of these absorbers we know that DOM includes a vast number of different types of chromophoric groups capable of absorbing UV or VIS radiation. Nevertheless, this list of chromophores is to some extent restricted. Especially in oceanic waters, these substances are represented mainly by simple π-electron chromophores, consisting of small groups of atoms attached to single aromatic rings and their auxochromes, or at most two-ring purine groups. On the other hand, the high-molecular-mass, conjugated π-electron chromophores characteristic of, say, plant pigments or terrigenous, polyaromatic humus substances make up only a small percentage of the total DOM. On the basis of these assumptions, Brown (1974) attempted to predict statistically the spectral nature of absorption by DOM by analyzing the so-called "pseudo-spectrum" of absorption by organic substances.

We have produced a modified version of this pseudo-spectrum (Figure 4.18): it is a histogram of the frequency of occurrence (expressed on a probability density scale $\Delta p(\lambda)$) of some of the 539 organic chromophores listed by Scott (1964). They include all the simple chromophores, that is, functional groups attached to a single benzene ring, but no polyaromatic compounds. This classification was carried out with respect to the wavelengths of the principal absorption peaks of these chromophores, in spectral intervals of width $\Delta\lambda = 10$ nm. Because the simple π-electron chromophores all have quite similar specific absorption coefficients (c. 10–100 $m^2\ g^{-1}$; see Table 3.1 in Section 3.1) regardless of the spectral position of their absorption maxima, the pseudo-spectrum $\Delta p(\lambda)$ is presumed to have very much the same shape as the corresponding absorption spectrum of the chromophore mixture. This supposition is in fact correct.

As Figure 4.18 shows, one feature of this pseudo-spectrum is the presence of absorption peaks at around $\lambda \approx 280$ nm and, for even shorter wavelengths, in the

[12] The identity expressed by Equation (4.10) is a very rough approximation, because in addition to humus substances, the protein fragments contained in DOM make quite a considerable contribution to visible light absorption. This is illustrated in Figures 4.21 and 4.22 and in Table 4.10.

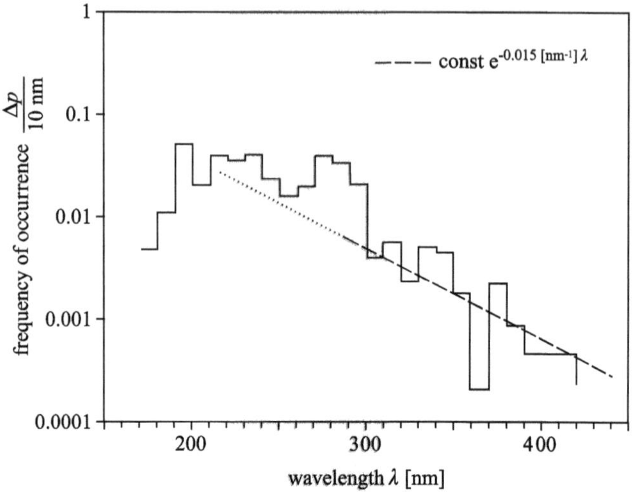

FIGURE 4.18. The relative frequency of occurrence of simple organic chromophores with given absorption properties as a function of the light wavelength. (the so-called "pseudo-spectrum" of light absorption by DOM; for explanations, see the text).

$\lambda < 250$ nm range. Notice that this overlaps the absorption band positions of the principal dissolved UV absorbers, that is, amino acids, purine and pyrimidine derivatives, and lignins (Figure 4.19). On the other hand, in the $\lambda > 300$ nm range, which extends into the visible region, $\Delta p(\lambda)$ decreases with wavelength λ almost monotonically with a mean exponential slope of c. 0.015 nm^{-1}. The slope of this absorption spectrum is thus similar to those for humus substances.

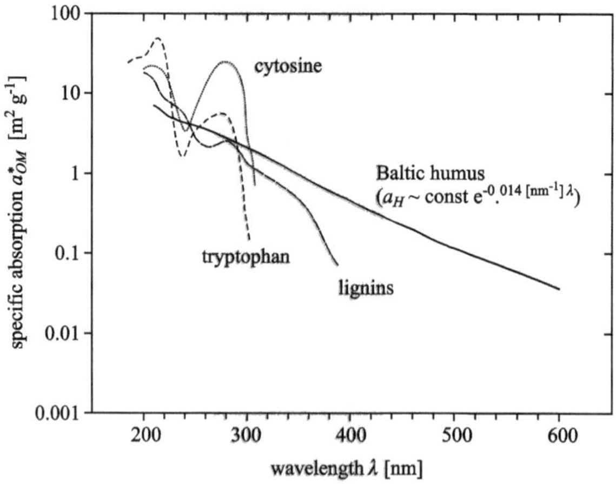

FIGURE 4.19. Spectra of the specific light absorption coefficients for examples of the four main groups of organic absorbers dissolved in sea water (after Table 4.3).

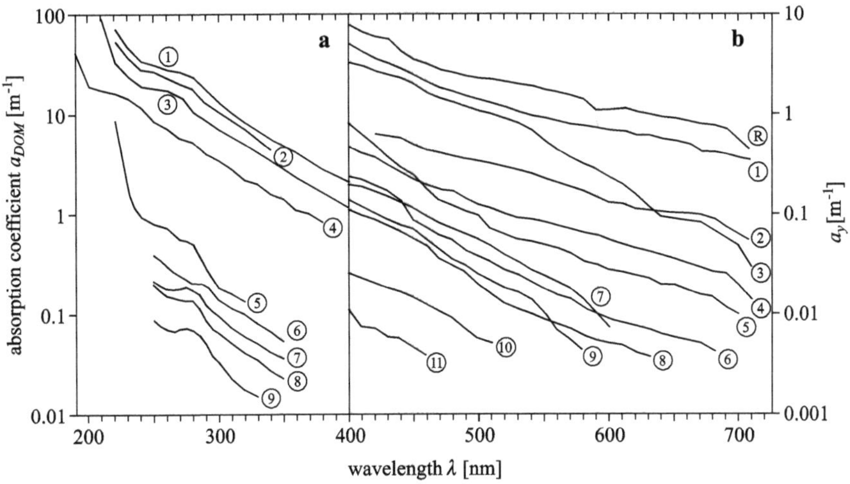

FIGURE 4.20. Approximate spectra of absorption coefficients ($a_{DOM}(\lambda)$ or $a_y(\lambda)$ in different waters: (a) in the UV region: 1–4. Baltic Sea, various parts of the Gulf of Riga (determined on the basis of data from Pelevin and Rutkovskaya (1980)); 5. surface waters of the Atlantic (estimated on the basis of data from Armstrong and Boalch (1961)); 6–9. eastern equatorial Atlantic, depths from 50 to 500 m (estimated on the basis of data from Ivanoff (1978)); (b) in the VIS region, on the basis of our own results and those of our colleagues: R, Lake Raduńskie (Poland); 1–3. Gulf of Gdańsk and Puck Bay; 4–6. open Baltic; 7–11. mid-Atlantic, surface waters.

The concordance of the shapes of the pseudo-spectrum $\Delta p(\lambda)$ with the overall light absorption spectra $a_{DOM}(\lambda)$ by all DOM is amazing. Figure 4.20 gives examples of such overall spectra in different types of natural waters.

The figure shows that, regardless of these similarities, the total absorption spectra of radiation by organic substances in different seas do display differences, which to varying degrees relate to:

1. The fine spectral structure, established mainly by the proportions among the contents of different groups of absorbers
2. The values of the absolute absorption coefficients and their associations with the total DOM content and the trophic type of waters C_a
3. The slopes of the absorption spectra S_H

We now discuss each of these questions in turn.

4.3.1 Fine Spectral Structure

Against a background of almost monotonic variations in light absorption by humus acids with respect to the wavelength, and regardless of the absolute values of the absorption coefficients, we find more or less distinct absorption bands (or inflexion points) in the UV in practically all spectra $a_{DOM}(\lambda)$.

To varying extents, they are due to the three groups of UV absorbers already mentioned (the absorption coefficients: a_{prot}, a_{pp}, a_L). The numerical proportions of these three sets of absorbers in the overall absorption of light by DOM vary in different seas and have not been subjected to close scrutiny. So, in order to obtain a rough idea of these proportions, we estimated the total absorption by DOM and its components with the aid of Equation (4.9) for two extreme, hypothetical types of sea water: type 2 "coastal-Baltic" waters (Case 2 waters; with lignins), and type 1 "oceanic" waters (Case 1 waters). Table 4.9 gives the approximate input data for these calculations, that is, the characteristic contents and specific absorption coefficients of the principal absorbers in these two hypothetical sea water types. In selecting these data, we took into account, as far as possible, the general trends of the concentrations (Table 4.2) and absorption properties (Tables 4.3 and 4.5) of the various groups of absorbers in the different sea water types. The estimated overall spectra of absorption by DOM and its components are shown in Figure 4.21.

TABLE 4.9. The concentrations and specific absorption coefficients of the principal organic absorbers dissolved in hypothetical sea water types.

Concentrations of Organic Forms:

	Total Concentration	Percentage Content			
	DOM [mg·dm^{-3}]	C_{prot}/DOM [%]	C_{pp}/DOM [%]	C_L/DOM [%]	C_H/DOM [%]
"Coastal-Baltic" type	10	8	2	18	55
"Oceanic" type	3	1.2	0.3	0	10

Specific Coefficients of Humus Substances and Lignins

"Coastal-Baltic" type	$a_H^*(\lambda)$ $a_L^*(\lambda)$	According to measurements by Nyquist in the Baltic (see Table 4.3.)
"Oceanic" type	$a_H^*(\lambda) = 0.125 \exp - [0.0114 (\lambda - 400)]$, based on measurements by Zepp and Schlotzhauser (Table 4.8)	

Specific Absorption Coefficients for Proteins and for Purine and Pyrimidine Compounds (in both sea water types)

$a^*_{prot}(\lambda) = 0.075 a^*_{OM,arom} + 0.93 \exp[-0.00768 (\lambda - 220)]$[a]

λ [nm]	210	220	230	240	250	260	270	280	290	300	310
(1) $a^*_{OM,arom}$ [m^2 · g^{-1}]	18.1	21.2	2.7	0.7	1.3	1.8	2.5	2.4	0.68	0.07	0
(2) a^*_{pp} [m^2 · g^{-1}]	12.9	12.0	6.3	9.6	12.6	15.5	17.2	13.7	9.1	2.1	0

[a] See Equation (4.3).

(1) $a^*_{OM,arom}$: Calculated as the mean specific absorptions for tyrosine, phenylalanine, and tryptophan (based on the data in Table 4.3).

(2) a^*_{pp}: Calculated as the mean specific absorptions for adenine, cytosine, and guanine (based on the data in Table 4.3).

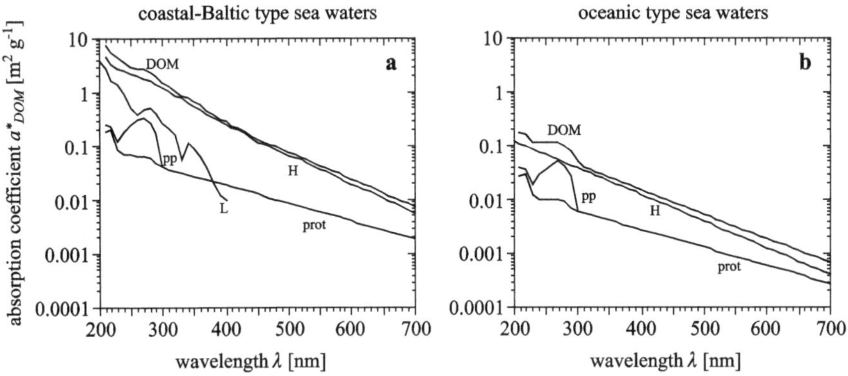

FIGURE 4.21. Dissolved organic matter absorption spectra, total (DOM) and components (prot: amino acids and fragments of proteins, pp: purine and pyrimidine compounds, L: lignins, H: humus acids): (a) in hypothetical "coastal-Baltic" sea waters; (b) in hypothetical "oceanic" sea waters.

In both types of sea water, UV absorbers (amino acids and their derivatives, compounds of purine and pyrimidine, and lignins in shelf waters) are responsible for enhanced absorption and the appearance of two more or less clearly defined absorption peaks in the $\lambda \approx 220$ nm and $\lambda \approx 280$ nm ranges. The intensity of these peaks makes up c. 30–50% of the total absorption in coastal waters and c. 30% in oceanic waters. In the former case, these features are due primarily to the lignin sulfates. In the latter, however, it is the amino acids that are principally responsible for the absorption band at $\lambda \approx 220$ nm, whereas the band at $\lambda \approx 280$ nm is due predominantly to absorption by purine and pyrimidine compounds. Table 4.10 gives the percentage shares of these absorbers in the total absorption.

The spectra of these percentage shares are illustrated in Figure 4.22. As Figure 4.21 clearly shows, spectra of the total absorption of visible light by DOM exhibit hardly any fine-structural features at all (except for the sporadic situations illustrated in Figure 4.23 and discussed later); they are characterized by a monotonic, exponential fall in the value of the absorption coefficient a^*_{DOM} with increasing wavelength λ. It also emerges from Figures 4.21 and 4.22 that, although visible light absorption by all DOM is indeed due mainly to humus substances, the proportion of this absorption due to protein fragments (along with the nonprotein parts, i.e., proteids) is probably equally significant. Thus, as we hinted in the introduction to this section, the identity expressed by Equation (4.10) is reasonable only if treated as an approximation. To separate the proportions of humus substances and proteins in the total absorption of light by yellow substances $a_y(\lambda)$ is, however, a complex problem, one that is impossible to unravel using the optical methods at present applied in oceanology or analytical chemical techniques.

TABLE 4.10. Percentage shares of the various groups[a] of UV and VIS absorbers in the total absorption of the light by DOM.

	Δ_{prot} [%]		Δ_{pp} [%]		Δ_L [%]	Δ_H [%]	
Wavelength λ[nm]	O[c]	B[b]	O[c]	B[b]	B[b]	O[c]	B[b]
-1-	-2-	-3-	-4-	-5-	-6-	-7-	-8-
210	16.1	2.5	22.0	3.4	36.1	61.9	58.0
230	10.8	1.9	16.0	2.8	32.0	73.3	63.2
250	8.6	2.3	32.3	8.5	17.0	59.1	72.3
270	8.5	2.4	44.3	12.8	17.4	47.2	67.4
290	9.1	2.3	34.9	8.8	20.0	56.0	69.0
310	13.8	2.7	0.0	1.0	16.2	86.0	80.4
330	14.7	3.6			6.1	85.3	90.3
350	15.7	3.9			13.6	84.3	82.5
400	18.3	6.2			3.0	81.7	87.9
450	21.3	9.9			0.0	78.7	90.1
500	24.5	11.6				75.5	88.4
550	28.1	14.0				71.9	86.0
600	32.0	17.3				68.0	82.7
650	36.2	21.1				63.8	78.9

[a] Δ_{prot}: amino acids and fragments of proteins, Δ_{pp}: purine and pyrimidine compounds, Δ_L: lignin sulfates, Δ_H: humus acids.
[b] In hypothetical "coastal-Baltic" sea waters.
[c] In hypothetical 'oceanic' sea waters.

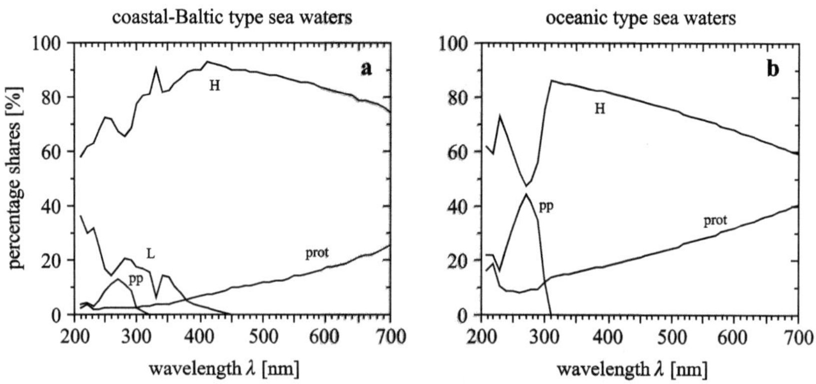

FIGURE 4.22. Percentage shares of the various groups of UV and VIS absorbers (prot: amino acids and fragments of proteins, pp: purine and pyrimidine compounds, L: lignins, H: humus acids): in the total absorption of the light by DOM (a) in hypothetical "coastal-Baltic" sea waters; (b) in hypothetical "oceanic" sea waters.

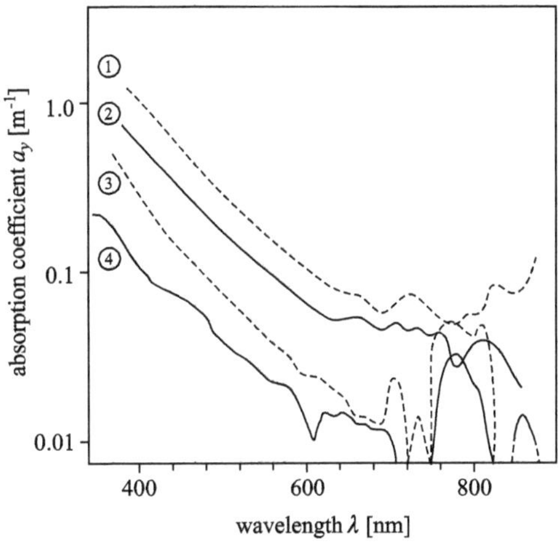

FIGURE 4.23. Spectra of light absorption by yellow substances in the Adriatic: 1, Po estuary; 2, Venice Lagoon; 3,4, coastal regions of the open Adriatic. (Adapted from Fischer et al. (1986).)

These fine-structural features of the UV (200–300 nm) absorption spectra $a_{DOM}(\lambda)$ are typical of practically all natural sea waters. In most oceanic waters, no other fine-structural characteristics are observed. Especially in the near-UV and the shortwave part of the VIS region, these spectra are exceptionally smooth, although their slopes can vary somewhat. Exceptions to this rule are the absorption spectra obtained from some shelf waters. For example, Sournia (1965) and Lyendekkers (1967) reported the existence of three fairly weak absorption bands in the 240–255 nm, 290–310 nm, and 330–390 nm ranges (maximum for λ = 370 nm), which were recorded in Indian Ocean waters off Madagascar. However, they were unable to ascribe these spectral features unequivocally to any particular functional groups or organic compounds.

In the longwave region, from 600 nm to the near-IR, the structure of the absorption spectrum $a_{DOM}(\lambda)$ is probably even more complex. An empirical study of this would be difficult owing to the low values of their absorption coefficients in comparison with the coefficients of the water itself. The spectra illustrated in Figure 4.23 give some idea of this complexity. These features were discovered with the aid of highly sensitive, high-resolution spectral apparatus in various coastal sea waters and estuarine waters in the Baltic, North Sea, and Adriatic, and are presented, for example, in Fischer et al. (1986), Reuter et al. (1986), and Diebel-Langohr et al. (1986). According to these authors, there are at least three sharply delineated absorption bands in the 600–800 nm range. They have not been examined in any detail; they could be due to humus–inorganic complexes with very large molecular structures, for example, with terrigenous metals. This is by no means certain, although the physical principles underlying the absorption of light by such complexes (Section 3.3) suggest that this may be so.

4.3.2 *Absolute Magnitudes of Absorption Coefficients*

As Figure 4.20 shows, the absolute light absorption coefficients of DOM in sea water for the same wavelengths vary over a range of two or three orders of magnitude. This is an enormous range of variation if we take into account the fact that in different basins, DOM concentrations vary by no more than 15–20 times (Figure 4.24).

The variability of these properties of natural waters (i.e., the DOM concentration and their absorption properties) in the context of various trophic types of waters[13] was subjected to a preliminary analysis by Woźniak (1990b). Some modified results of these analyses are compared in Figures 4.24 and Figure 4.25. These show that DOM concentrations in supereutrophic waters, such as those

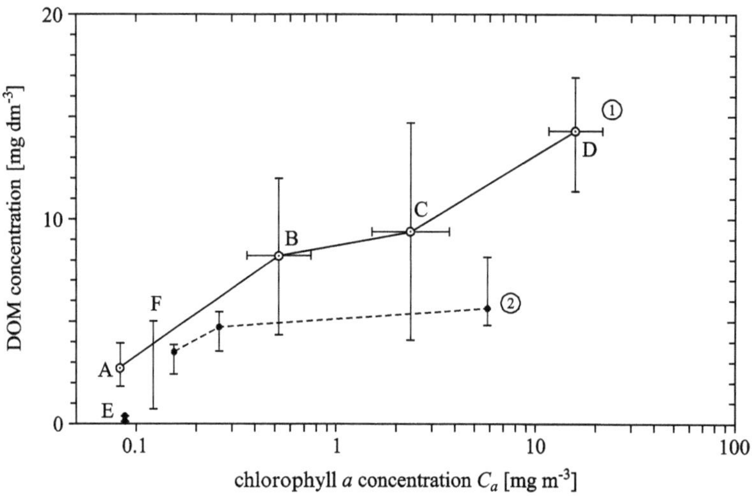

FIGURE 4.24. Relationships between the concentrations of DOM and chlorophyll *a* C_a in the euphotic zones of various basins. A: Atlantic (the present authors' own investigations; 12 measurements); B: southern Baltic (the present authors' own investigations; 61 measurements); C: Gulf of Gdańsk (the present authors' own investigations; 60 measurements); D: mouth of the Vistula River (the present authors' own investigations; 8 measurements); E: clean oceanic waters (only humus substances) (Stuermer 1975); F: range of variability in oceans (Popov et al. 1979); 1: averaged plot of interrelationships (the present authors' own investigations); 2: magnitudes of DOM typical of oligotrophic, mesotrophic and eutrophic regions of the oceans. (After Skopintsev et al. (1979); adapted from Woźniak (1990b).)

[13] The total concentration of chlorophyll a + phaeophytin (denoted C_a in our papers) is assumed to indicate the trophic type of water or basin.

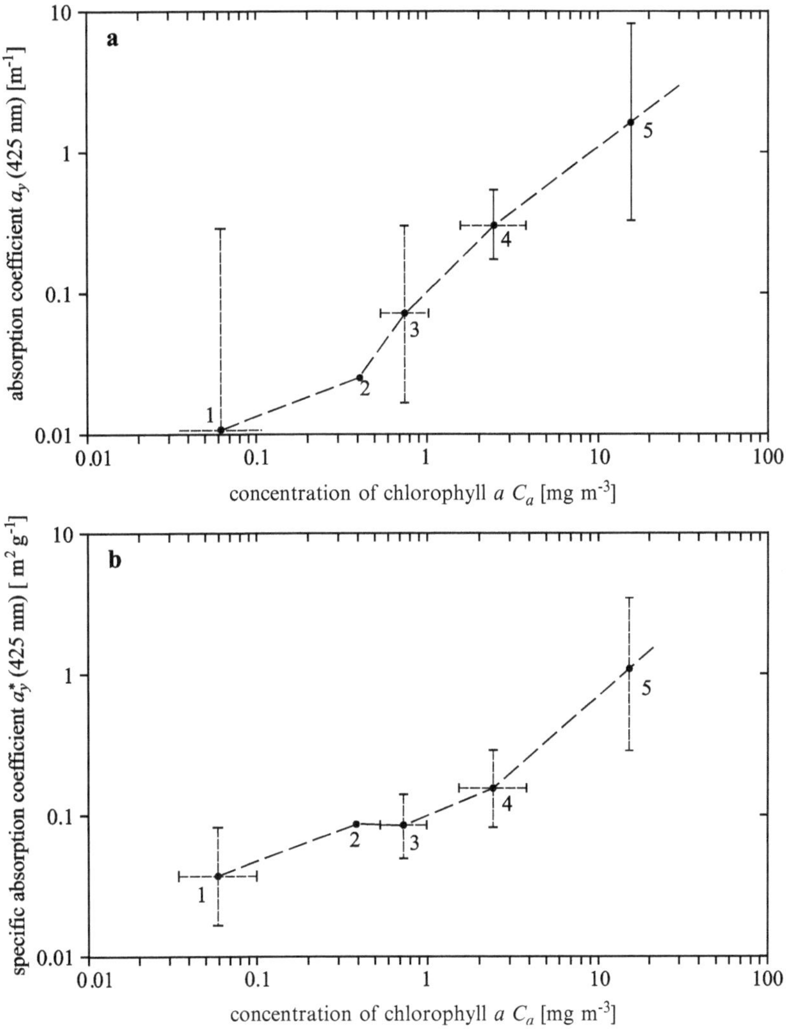

FIGURE 4.25. Observed relationships with the chlorophyll a concentration C_a: (a) light absorption coefficients at λ = 425 nm of yellow substances in sea water; (b) specific light absorption coefficients at λ = 425 nm. The respective means and standard deviations relate to measurements from: 1. the Atlantic; 2. the Baltic; 3. the Black Sea; 4. the Gulf of Gdańsk; 5. the mouth of the Vistula River. The respective vertical and horizontal sections depict the ranges of standard deviations of the recorded absorption coefficients a_y and a_y^* and the chlorophyll concentration C_a. (Adapted from Woźniak 1990b).)

off the mouth of the Vistula River, are four to five times as high as in oligotrophic basins of the Atlantic Ocean (Figure 4.24), whereas the corresponding absorbances a_y (425 nm) differ by a factor of c. 200 (Figure 4.25a). In view of this, the specific light absorption coefficients of DOM referred to the total DOM concentration

$$a_y^* \ (= a_{DOM}^*) = a_y / C_{DOM} \tag{4.11}$$

in eutrophic basins are from 10- to more than 100-fold times larger than in oligotrophic seas, and depend on the wavelength λ. The relationship among these specific coefficients and the chlorophyll a concentration (i.e., the trophic type of waters) is illustrated in Figure 4.25b. This state of affairs can be explained by the following two circumstances.

- As we can see in Table 4.2, the percentage share of humus substances in the total DOM concentration, that is, the degree of humification of the organic matter, varies in different waters. It is lowest, c. 10–15%, in open oceanic waters and rises to c. 23–25% in coastal areas and oceanic bights. Finally, in brackish sea waters and almost enclosed bays the degree of humification reaches values of 25–55% (Pempkowiak 1989). Inasmuch as humus substances are the main absorbers of radiation, and in the VIS region are practically the only absorbers of light, the specific absorption, described by Equation (4.11) also rises with increasing humification.
- A factor further enhancing the diversity of a_y^* values recorded in natural basins is that humus substances in oligotrophic waters are much weaker absorbers of light than their counterparts in coastal waters, which additionally contain considerable amounts of very strongly absorbing terrigenous humus substances (see Figures 4.13, 4.14, 4.17, and Tables 4.6, 4.8).

In this situation, there simply cannot exist an all-embracing relationship between light absorption coefficients a_y and concentrations C_{DOM} or C_{org} in different basins that would enable the content of organic compounds in the sea to be determined accurately by optical methods. In addition, the correlation coefficients of the dependences between a_y and C_{DOM} in the same basins, even at the same measuring stations at sea, are low. A certain correlation can be expected only with regard to water samples taken from different depths at the same sampling station (see, e.g., Kopelevitch (1983)). This is due, among other things, to the degree of fulvization of humus substances, which varies considerably in time and space, and also with depth in the sea (see Table 4.7, values of the degree of fulvization in the Sargasso Sea). Thus, at the present stage of research in marine optics, the measure of yellow substance concentrations in sea water is not the actual concentration C_{DOM} or C_{org} but the coefficient of light absorption $a_y(\lambda_0)$ at conventionally accepted wavelengths, for example, $\lambda_0 = 380$ nm (Baker and Smith 1982), $\lambda_0 = 390$ nm (Kopelevitch 1983, Pelevin and Rutkovskaya 1980), $\lambda_0 = 450$ nm (Højerslev 1980b) and so on. Moreover, the content of yellow substances is often defined in terms of their fluorescence, artificially induced and measured by standard immersion fluorimeters (e.g., Karabashev (1987)), or induced by laser and calibrated with respect to the Raman scattering of water (e.g., Brown (1980), Ivanoff (1978), Bristow and Nielsen (1981), and Blough and Del Vecchio (2002)).

4.3.3 The Slopes of Absorption Spectra

Determining the slopes of the light absorption spectra $a_y(\lambda)$ of DOM in various sea waters, especially in the VIS region, has become a very popular research topic (see the list of references in Table 4.11). This is because a knowledge of these slopes is an essential aspect of the algorithms for interpreting remote-sensing data, which are needed to define the optical and biological characteristics of marine and inland-water ecosystems; see, for example, the research programs of the European Space Agency (ESA) and the National Aeronautics and Space Administration (NASA) (e.g., Grassl et al. (1986a,b), Carder et al. (1989), Roesler et al. (1989), Moore et al. (1999), Arst (2003), Woźniak et al. (2004)).

The first comprehensive investigations into the spectral slopes of light extinction by organic substances extracted from diverse fresh waters, sea waters and ground waters were done by Kalle (1949, 1961, 1962, 1966). The slopes of the absorption spectra S_H, which Kalle calculated from extinction ratios $\varepsilon(\lambda_1)/\varepsilon(\lambda_2)$ (where λ_1 = 420 nm, λ_2 = 665 nm) measured for different natural associations of organic substances, lie between 0.007 and 0.014 nm^{-1}. Without devaluing Kalle's achievements—he pioneered research into yellow substances (Gelbstoff) in natural waters—Brown (1974) nevertheless questioned the credibility of the former's results. He demonstrated that as a result of the too-high density of the solutions of organic compounds with which Kalle experimented and the consequent concentration effects, the slopes of the extinction spectra determined by Kalle were correlated more closely with the concentrations of those solutions than with the real nature of the organic substances he was studying.

The first credible slope of a light absorption spectrum of yellow substances in the sea, $S_H = 0.015\ nm^{-1}$, emerges from data published by Jerlov (1968). This was later confirmed by results published by numerous authors from various natural basins (seas, lakes, and rivers): these results are reviewed in Table 4.11. The table shows that the spectral slopes S_H differ over a wide range of values, from c. 0.006 to c. 0.033 nm^{-1}. However, the most frequent values of this spectral slope for yellow substances exhibit a somewhat smaller scatter and lie in the middle part of the above-mentioned range. The relevant average for sea waters that we calculated from some 900 S_H values for spectra of $a_y(\lambda)$ in the 400–700 nm range is $\overline{S}_H = 0.0153\,nm^{-1}$ with a standard deviation of $\sigma_s = \pm 0.0046\ nm^{-1}$. These statistics include not only our own data and those referring to the seas stated in Table 4.11, but also similar data by different authors, gleaned from Blough and Del Vecchio (2002), and those from the papers by Babin et al. (2003a) and Kowalczuk (1999). Recorded in different marine basins, these slopes S_H clearly display a wide scatter. The reasons for this scatter are those we

TABLE 4.11. A review of the parameters S_H characterizing the exponential slopes of the UV and/or visible light absorption spectra of yellow substances in different basins.[a]

Basin or region	S_H [nm^{-1}]	Spectral range	References
-1-	*-2-*	*-3-*	*-4-*
	Lakes		
Lake Raduńskie (Poland)	0.0086	VIS	Our own studies
Ontario (Canada)	0.0091	VIS	Bukata et al. 1991
Ladoga (Russia)	0.0060–0.069	VIS	Kondratev et al. 1990
Coastal lakes	0.015	VIS	Kirk, 1976
Kizaki (Japan)	0.016	VIS	Kishino et al. 1984
Freshwater lakes	0.0187 (0.0151–0.0205)	280–460 nm	Davis-Colley and Vant 1987
Subarctic lakes	0.0151	295–400 nm	Laurion et al. 1997
Arctic lakes	0.0174	295–400 nm	Markager and Vincent 2000
Alps and Pyrences lakes	0.0128–0.0199	295–400 nm	Laurion et al. 2000
	Rivers		
Vistula (Poland)	0.0070–0.0109	VIS	Our own studies
Po (Italy)	0.0132	400–500 nm	Reuter et al. 1986
	0.0107	450–550 nm	Reuter et al. 1986
Brenta (Italy)	0.0119	400–500 nm	Reuter et al. 1986
	0.0089	450–550 nm	Reuter et al. 1986
Adige (Italy)	0.0098	400–500 nm	Reuter et al. 1986
	0.0085	450–550 nm	Reuter et al. 1986
Elbe (Germany)	0.0111	400–500 nm	Reuter et al. 1986
	0.0083	450–550 nm	Reuter et al. 1986
Orinoco (Venezuela)	0.0141	290–700 nm	Blough et al. 1993
	0.0112	400–500 nm	Battin 1998
Upper Orinoco (Ven.) Amazon	0.014–0.033	290–700 nm	Green and Blough 1994
	Baltic (Littoral Zones)		
Puck Bay	0.0089	VIS	Our own studies
Gulf of Gdańsk	0.0096 ± 0.0053	VIS	Our own studies (Samuła-Koszałka and Woźniak 1979)
Kattegat	0.0132	400–500 nm	Reuter et al. 1986
	0.0090	450–550 nm	Reuter et al. 1986
Skagerrak	0.0147	400–500 nm	Reuter et al. 1986
	0.0070	450–550 nm	Reuter et al. 1986
Bornholm Basin	0.0170	400–500 nm	Reuter et al. 1986
	0.0117	450–550 nm	Reuter et al. 1986
Fehmarn Belt	0.0133	400–500 nm	Reuter et al. 1986
	0.0090	450–550 nm	Reuter et al. 1986
Bays (South Baltic):			Kowalczuk 1999
Spring	0.019	350–600 nm	
Summer	0.019	350–600 nm	
Autumn/Winter	0.018	350–600 nm	

TABLE 4.11. A review of the parameters S_H characterizing the exponential slopes of the UV and/or visible light absorption spectra of yellow substances in different basins.—Cont'd.

Basin or region	S_H [nm^{-1}]	Spectral range	References
-1-	*-2-*	*-3-*	*-4-*
Coastal (South Baltic):			Kowalczuk 1999
Spring	0.020	350–600 nm	
Summer	0.020	350–600 nm	
Autumn/Winter	0.019	350–600 nm	
Eutrophic and Mesotrophic Sea Waters			
Open Baltic	0.014	VIS	Lundgren 1976
	0.018	VIS	Bricaud et al. 1981
Venice Lagoon	0.0151	400–500 nm	Reuter et al. 1986
	0.0110	450–550 nm	Reuter et al. 1986
Adriatic	0.0145	400–500 nm	Reuter et al. 1986
	0.0112	450–550 nm	Reuter et al. 1986
Pacific and Indian Oceans	0.012–0.015	<500	Kopelevitch 1983
Atlantic, San Juan I.	0.017±0.003	VIS	Roesler et al. 1989
North Sea (coastal)	0.0188	350–600 nm	Stedman et al. 2000
Open South Baltic:			Kowalczuk 1999
Spring	0.020	350–600 nm	
Summer	0.019	350–600 nm	
Autumn/Winter	0.021	350–600 nm	
Oligotrophic Waters			
Atlantic	0.016–0.020	VIS	Our own studies
Adriatic	0.0155	400–500 nm	Reuter et al. 1986
	0.0158	450–550 nm	Reuter et al. 1986
Pacific and Indian Oceans	0.015	<500	Kopelevitch 1983
Mauritania	0.015	VIS	Bricaud et al. 1981
Gulf of Guinea	0.014	VIS	Bricaud et al. 1981
Mediterranean	0.014	VIS	Bricaud et al. 1981
Eastern Pacific	0.017	VIS	Okami et al. 1982
	0.014	VIS	Kishino et al. 1984
Gulf of Mexico	0.014	VIS	Carder and Steward 1985
Crystal Sound (Antarctic open ocean)	0.023	260–330 nm	Sarpal et al. 1995
Abyssal Waters			
Pacific >100 m	0.017	<500	Kopelevitch 1983
Indian Ocean >100 m	0.019	<500	Kopelevitch 1983

[a] $a_y(\lambda) = a_y(\lambda_0)\exp[-S_H(\lambda - \lambda_0)]$.

have already mentioned: the different degrees of fulvization, the different proportions of auto- and allogenous humus substances, and the natural diversification of the absorption properties of humus acids produced in different waters. The scatter in S_H values could also be due to measurement errors. Moreover, S_H varies with the light wavelength (Reuter et al. 1986, Kopelevitch et al. 1988).

The utilization of such averaged data for practical purposes (e.g., in remote-sensing algorithms) has its limits and can only be regarded as a first approximation. An exact description of absorption properties in particular basins requires both regional and seasonal investigations, because it is characteristic of yellow substances, especially in coastal seas of the temperate zone, that the slopes of their absorption spectra vary widely over the annual cycle (Kowalczuk 1999, Babin et al. 2003a).

5 Light Absorption by Suspended Particulate Matter (SPM) in Sea Water

After the dissolved substances, the second group of optical admixtures in seas and other natural water bodies consists of various kinds of organic and inorganic suspended particles or, speaking generally, suspended matter[1] (see Jerlov (1975), Kirk (1994), and Dera (1992, 2003)). This and the next chapter describe the absorption by these substances of visible and near-UV radiation. We do not cover the absorption of IR radiation because the absorption of IR by suspended matter in the sea is incomparably weaker than its absorption by molecules of water and sea salt.

The UV absorption coefficients of suspended particles in the sea $a_p(\lambda)$ are usually much lower than their counterparts for dissolved organic matter (DOM), $a_{DOM}(\lambda)$ or yellow substances, $a_y(\lambda)$. This is because the mass concentrations of particulate organic matter (POM) are far smaller than those of DOM (Table 4.1; Figure 5.7d). On the other hand, inorganic suspensions are generally weak UV absorbers, and their concentrations are very much lower than those of dissolved inorganic substances.

Now, where visible light (VIS) is concerned, suspended particles in sea water usually exert a very strong influence on its overall absorption, one that equals or even exceeds the effect of yellow substances on that absorption. This is the case if the proportions of the visible light absorption coefficients of these two groups of substances, that is, $a_p(\lambda)$ and $a_y(\lambda)$, are compared in the context to the overall absorption $a(\lambda)$ in different marine basins. This is illustrated in Figure 5.1, which gives examples of the absolute values of these two component coefficients observed in various sea waters. Although these values resemble or differ from one another several times over, they do so within a range of no more than one order of magnitude. Moreover, they depend to a large extent on the type of basin and the depth of the water layer in question.

[1] In the oceanological literature *suspended matter* is also referred to as *suspensions*, *suspended particles*, *particulate matter*, or *hydrosols*.

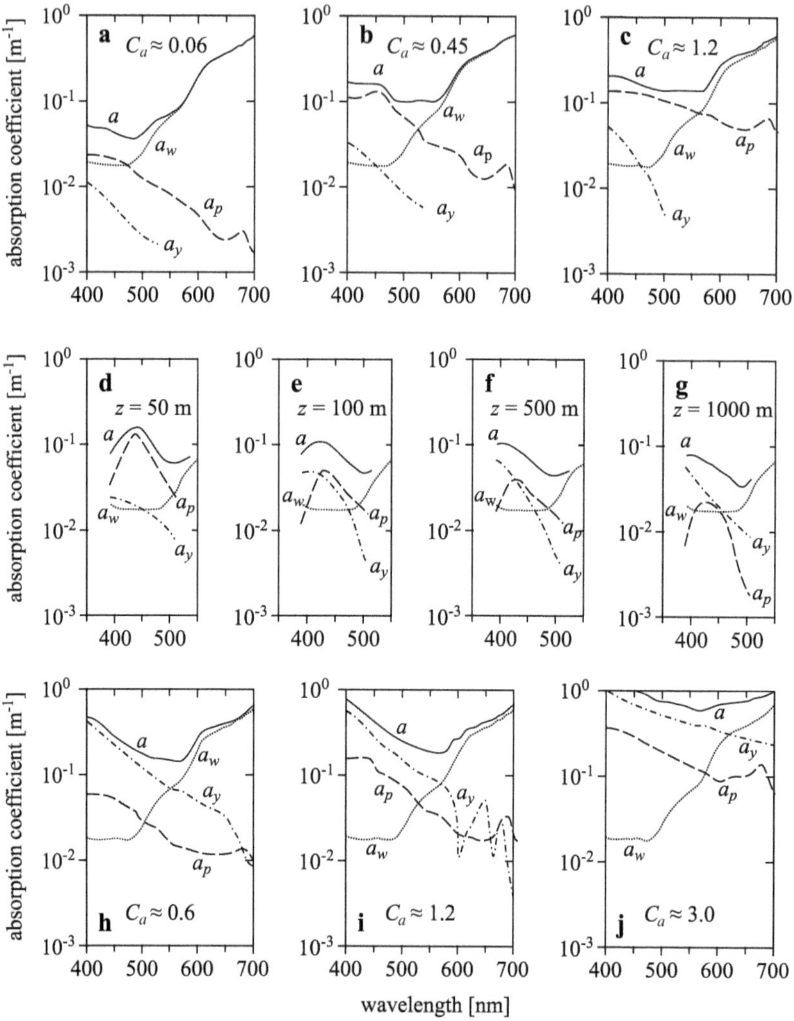

FIGURE 5.1. Spectra of light absorption coefficients: the overall absorption (a) and its component absorptions due to water (a_w), yellow substances (a_y), and suspended particles (a_p). a–c: in surface layers of Case 1 waters containing various concentrations of chlorophyll C_a [mg tot. chl a m^{-3}]; d–g: at various depths in the ocean; h–j: in Case 2 waters. Spectra plotted on the basis of empirical data: a,b: Atlantic Ocean (authors' own research); c: Pacific Ocean (after Kishino et al. (1984)); d–g: Atlantic Ocean (after Kopelevitch and Burenkov (1977)); h–j: Baltic (authors' own research).

To address these issues with respect to the multifarious conditions existing in the different basins of the World Ocean is a highly complicated matter and exceeds the scope of the present volume. We refer the interested reader to the article by Babin et al. (2003a). Among other things, this includes a quantitative analysis of the proportions between the light absorption coefficients of DOM and of different groups of suspended matter at more than 300 measurement stations in the seas around Europe, in both Case 1 waters (Atlantic) and Case 2 waters (Adriatic, Baltic, English Channel, and North Sea).

Suspended particles in the sea not only absorb light, they also scatter it. This scattering alters the direction of the light, thereby extending its optical pathway. In consequence, there is a greater probability that the photons will be absorbed by other suspended particles, or by molecules of water and the substances dissolved in it. Both processes—the absorption and scattering of light by suspended particles—depend quantitatively and spectrally on the chemical composition of the particles, their concentration in the water, and their physical properties.

One of the most important physical properties of particles in the context of their interaction with light is the real and imaginary part of refractive indices of the matter they consist of, that is, optical constants, as well as their shapes and dimensions. However, in order to understand these phenomena, the traditional quantum-mechanical principles for describing the interaction of photons with molecules are insufficient. We therefore begin this chapter by describing the theoretical principles underlying the interaction of light with optical inhomogeneities larger than molecules, such as marine suspended particles, that is, interactions with a dispersing medium (Section 5.1). Later (Section 5.2), we characterize the types and resources of suspended particles in the sea as well as their most prominent physical and chemical properties, determining the extent of their interaction with light. To end with (Section 5.3), we concentrate on the absorption properties of nonalgal particles (NAP) in sea water, on the basis of the empirical results obtained up to the present. The focus here is on NAP, because the absorption properties of phytoplankton, a separate, highly significant group of SPM in the sea, are described in detail in Chapter 6.

5.1 The Optical Properties of Dispersing Media: Theoretical Principles

It is perfectly usual for solid particles to be suspended in natural waters, which frequently occur in large numerical concentrations (10^3, 10^4 particles larger than 1 μm in 1 cm^3 of water; see, e.g., Parsons (1963, 1975) and Dera (2003)). Consequently, natural waters are not optically homogeneous or even quasi-homogeneous molecular solutions: they are discontinuous polydispersing media. This means they contain particles of various sizes, whose absolute refractive indices are different, usually greater than the refractive index of the surrounding water and the substances dissolved in it (see Section 5.2, which describes the physical and chemical properties of marine suspensions). In such optically inhomogeneous media we come across the so-called optical packaging effect. We now define this concept and explain the theory behind it.

5.1.1 *The Packaging Effect: What Is It and How Does It Manifest Itself?*

Already at the end of the nineteenth century (e.g., Love (1899)) it was known that the optical properties (absorption and scattering) of dispersing media differed fundamentally from those of homogeneous media with the same molecular composition.[2] This is due to the so-called packaging effect. Under natural conditions, for example, in the atmosphere or in the sea during the condensation of discrete molecules in aerosols and hydrosols, the packaging effect usually, but not always, causes: (1) light-scattering coefficients to increase in relation to molecular scattering (forward-scattering directions are particularly privileged (Love 1899, Mie 1908, Shifrin 1951, 1952), and (2) light absorption coefficients in the medium to drop, and to fall further as the inhomogeneities increase in size (Duysens 1956, Shifrin 1957). Moreover, the absorption and scattering properties of dispersing media are intimately linked: one could say they interact with each other. If the absorption coefficients of the suspended particle matter change, but their shapes, dimensions, and real refractive indices (which in the case of nonabsorbing suspensions determine the amount of light scattered) do not, then the scattering coefficients in the medium will change accordingly. These changes usually take place in opposite directions; that is, if the absorption coefficient of the particle matter increases, then the amount of light scattered by those particles will decrease, and vice versa.

Let us now examine how the packaging effect influences light absorption in a dispersing medium. This influence violates the Lambert–Beer law,[3] a basic law of optics applying to homogeneous media, for example, hypothetical solutions of absorbers in nonabsorbing solvents. According to this law, the light absorption coefficient of a particular substance for a given wavelength $a_{sol}(\lambda)$ depends only on its concentration C in the medium in question, in line with the formula

$$a_{sol}(\lambda) = a^*_{sol}(\lambda)\, C\,, \tag{5.1}$$

---

[2] For the sake of simplicity, we here use the term "homogeneous medium" in its broadest sense. By this we mean not only strictly homogeneous media, consisting of isotropically distributed atoms or molecules of the same chemical element or compound, but also well-mixed solutions or mixtures of gases in which the substance to be analyzed is distributed isotropically.

[3] The Lambert–Beer law states that the absorbancy of a homogeneous system, defined as $A = \log(I_0/I)$ (where I_0 and I are the respective intensities of radiation before and after traveling a distance z in the system), is proportional to the distance z, and the concentration of the absorber C, with the decimal specific absorption coefficient, $a^*_{10,sol}$, that is, $A = a^*_{10,sol}\, C\, z$. Starting out from this law and utilizing the terminology of marine optics, it is easily demonstrated that the absorption coefficient of this system $a_{sol} = (\ln 10)\, a_{10,sol}$ is equal to $a_{sol} = a^*_{sol}\, C$, where $a^*_{sol} = (\ln 10)\, a^*_{10,sol}$.

where a^*_{sol} is the specific light absorption coefficient of that substance (the subscript sol is derived from *solvent*).

The value of a^*_{sol} is determined by the laws of quantum mechanics, and for certain chemical structures remains practically constant, regardless of the concentration of the absorber in homogeneous media, as long as intermolecular interactions and the effects of the thermodynamic state parameters are neglected. In dispersing media, on the other hand, the Lambert– Beer law does not hold, because the absorption coefficients of these media, a_p, derived from molecules "packaged" in the particles, are usually smaller than the absorptions of the same molecules distributed isotropically; that is, $a_p < a_{sol}$ (the subscript p stands for *particles*). The upshot of this is that the absorption coefficients a_p are not linearly related to the mean concentration C of the absorber in the dispersing medium, because they depend additionally on the dimensions of its particles and also on the intraparticular concentration C_i.

It turns out, then, that although the laws of quantum mechanics are good enough to describe the absorption of radiation by dissolved substances, they are inadequate to explain absorption phenomena in dispersing media. The most we can do with them is define the coefficients a_{sol} (or a^*_{sol}). Although the values of these coefficients for substances in a homogeneous medium do affect light absorption by a suspension, they are not the only factor determining this quantity. A complete description of the absorption properties of a suspension that includes the effect of molecular packaging requires additional theories taking account of the "dispersivity" of the media and the interaction between absorption and scattering to be invoked. Such possibilities are offered by the classical electrodynamic theories resulting from Maxwell's equations, in particular, Mie's theory (Mie 1908, Hulst 1957, 1981, Deirmendjian 1969, Born and Wolf 1968).

In the following sections we outline the theoretical foundations of such a combined (quantum-mechanical and electrodynamic) description of the optical properties of polydispersing media, which we need for analyzing the absorption properties of suspended matter in the sea. Without going into the details of the mechanisms of these phenomena, we present formal mathematical relationships useful for grasping the quantitative differences between the light absorption coefficients of dissolved substances and of the substances contained in suspended matter particles, dispersed to varying degrees. We refer the interested reader to the literature describing in detail the influence of the packaging effect on the absorption properties of dispersing media and related issues. The first work in this field to apply to sea water was the one by Duysens (1956). But a real renaissance in theoretical and empirical studies in this field has taken place in the last 25 years (see, e.g., Morel and Bricaud (1981), Bohren and Huffman (1983), Bricaud et al. (1983, 1995), Bricaud and Morel (1986), Smirnov and Shifrin (1987), Zieliński et al. 1987, Mitchell and Kieffer (1988), Iturriga and Siegel (1988,1989), Bricaud and

Stramski (1990), Król (1985, 1991, 1998), Kirk (1994), Morel and Ahn (1990), Stramski and Kieffer (1990), Stramski and Reynolds (1993), Babin et al. (1993), Woźniak et al. (2000), Bernard et al. (2001), Quirantes and Bernard (2004, 2006), Babin and Stramski (2004), Ficek et al. (2004), Woźniak S.B. and Stramski (2004), and Olmo et al. (2006)).

5.1.2 Light Absorption in Polydispersing Media: A Quantum-Mechanical–Electrodynamic Description[4]

The optical properties (including light absorption and scattering) of complex media emerge from the elementary physical and chemical properties of the matter they contain: for example, the chemical composition, the concentration of components, the refractive indices, and the shapes and sizes of the suspended particles or clusters of molecules. The consequences of all this are shown in Figure 5.2, a conventional and much simplified diagram that points out the theories explaining and describing the relevant relationships. At the same time, it is a schematic diagram of the theoretical algorithm for determining the optical properties of different media on the basis of their physical and chemical properties with the aid of the mathematical formulas of quantum mechanics and classical electrodynamics.

Figure 5.2 shows that determining the optical properties of dispersing media takes place in two distinct stages, corresponding to the two separate parts of the schematic diagram Figure 5.2 Sector a and 5.2 Sector b. First, we determine the absorption properties of the homogeneous components of the medium (the quantum-mechanical stage), and only then do we define the resultant optical properties, taking account of the dispersivity of the media (the electrodynamic stage).

We presented the first, quantum-mechanical stage (Figure 5.2 Sector a) when we were describing the absorption properties of molecules of water and dissolved substances (Chapters 2 and 3). In this stage, using the input data (Sector Ia)—the chemical structure of the absorber molecule (block 1a) and its concentration C in the medium (block 2a)—and the mathematical

[4] The expression "quantum-mechanical–electrodynamic description" is imprecise, even tautological, because quantum mechanics itself is also an electrodynamic theory. It is, however, an unconventional theory, because the quantum laws in many places violate the fundamental laws of Maxwell's theory, such as the existence of discrete stationary states (orbits) in the Bohr model of the atom, which do not radiate energy continuously and are not incident on the nucleus, as predicted by classical electrodynamics. In view of this it would actually be more precise to term this description "unconventionally conventional" or "nonclassically classical", but these expressions are oxymorons. We trust, therefore, that the reader will bear with the above tautology.

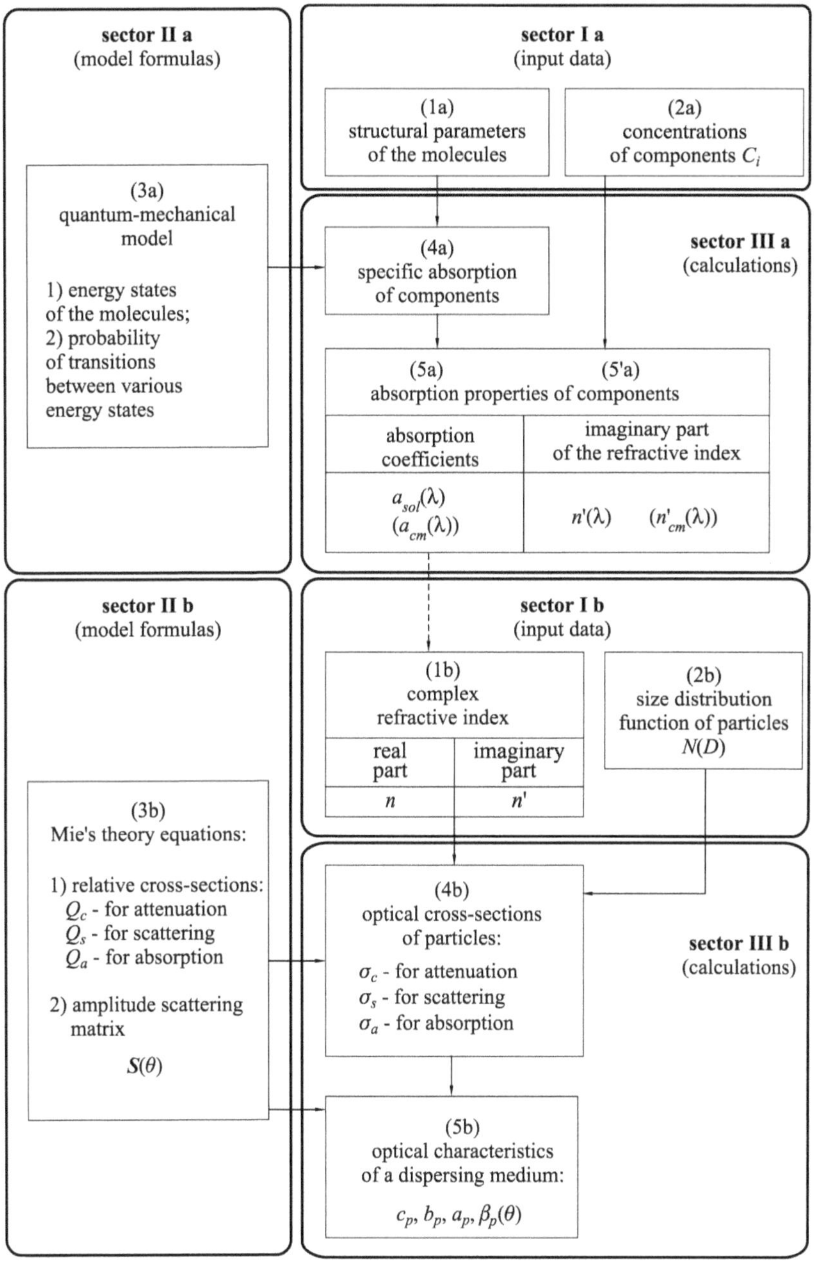

FIGURE 5.2. Schematic diagram of the quantum-mechanical–electrodynamic algorithm for determining the optical properties of dispersing media, including: the absorption properties of homogeneous components of the medium (sectors and blocks a); and the resultant absorption and scattering properties of a dispersing medium as a whole (sectors and blocks b).

apparatus of quantum mechanics (Sector IIa, block 3a), we can usually calculate (Sector IIIa) the absorption properties of homogeneous media as listed below, although this is often very difficult to do:

1. The specific absorption coefficients of the substance under scrutiny $a^*_{sol}(\lambda)$ (block 4a).
2. The absorption coefficients of this substance evenly distributed (dissolved) in the medium $a_{sol}(\lambda)$ (block 5a); to this end we also use the high concentration of this substance C_i and the Lambert–Beer law.
3. The imaginary parts of absolute refractive indices in the medium $n'(\lambda)$ (block 5′a), which, in the conventions applied in electrodynamics, describe the absorption of electromagnetic waves.[5] They are determined from known values of $a_{sol}(\lambda)$ or $a^*_{sol}(\lambda)$ and C_i, in accordance with the relationships:

$$n'(\lambda) = \frac{a_{sol}(\lambda) \cdot \lambda}{4\pi}. \tag{5.2a}$$

or

$$n'(\lambda) = \frac{a^*_{sol}(\lambda)\, C_i \cdot \lambda}{4\pi}. \tag{5.2b}$$

If, for the sake of simplicity, we assume that a polydispersing medium (i.e., in general, optically inhomogeneous) consists of particles that internally are optically homogeneous, we can use this first, quantum-mechanical stage to describe the internal absorption properties of the particle matter: intracellular in the case of phytoplankton, or intraparticular in the case of organic detritus or suspended mineral particles. These properties are determined from the relationships:
absorption by cellular or particulate matter:

$$a_{cm}(\lambda) \quad \text{or} \quad a_{pm}(\lambda) = a^*_{sol}(\lambda)\, C_i \tag{5.3}$$

the imaginary refractive index of the cellular or particulate matter:

$$n'(\lambda) = \frac{a_{pm}(\lambda)\lambda}{4\pi} \tag{5.4a}$$

or

---

[5] The electric vector $\boldsymbol{E}$ of an electromagnetic wave propagating in a homogeneous medium with a absolute complex refractive index $m = n - in'$ (where: n is the real part of complex refractive index of light, n′ is the imaginary part of this index and $i = \sqrt{-1}$ varies as $E = E_0 \exp[-kn'z]\exp[-ikc(t - nz/c)]$, where $k = 2\pi/\lambda$, wave number, z is the distance of wave propagation, t is the time, and c is the velocity of light in a vacuum.

The factor $\exp[-kn'z]$ describes the reduction in the amplitude of the electric field intensity with distance z as a result of absorption. Because the intensity of radiation as classically understood (see, e.g., Dera (2003, Section 4.3)) is proportional to the square of the electric vector, the relationship between the imaginary refractive index n' and the absorption coefficient a_{sol} (according to the Lambert–Beer law) is $a_{sol} = 2kn' = 4\pi n'/\lambda$.

$$n'(\lambda) = \frac{a^*_{pm}(\lambda)\, C_i\, \lambda}{4\pi}, \tag{5.4b}$$

where the specific absorptions $a^*_{sol}(\lambda)$ emerge from quantum mechanics, and C_i is the intraparticular or intracellular concentration of the absorber. The last of the coefficients mentioned above, the imaginary part of refractive index of the cellular matter n', is at the same time one element of the input data of the algorithm enabling the optical properties of polydispersing media to be found (see the arrow between the successive stages of the algorithm in Figures 5.2 Sectors a and b).

In contrast to the quantum-mechanical stage, which takes account only of the reactions of light quanta with the energy states of molecules, the second, electrodynamic, stage of the algorithm (Figure 5.2 Sectors b) describes the interaction of electromagnetic waves in a medium containing optical inhomogeneities whose dimensions exceed those of molecules. These inhomogeneities are suspended particles; by this we mean very generally defined areas of space with various complex refractive indices relative to the surrounding medium. Under the influence of the electric field of the electromagnetic waves passing through the particles, these become induced electric dipoles, emitting secondary electromagnetic waves in all directions in space (i.e., scattered light). These emitted secondary waves interfere with each other to a greater or lesser extent, thereby altering their spatial structure and the absolute quantities of energy emitted (scattered) at various angles to the direction of the primary, oscillation-inducing waves.

The changes in the wave field as a result of this interaction in the suspended particles (i.e., the distributions of intensities and other characteristics of the secondary waves with respect to the primary waves) are described by the appropriate coefficients and functions of the absorption, scattering, and attenuation of light (in general, electromagnetic waves) in dispersing media (Sector IIIb, Figure 5.2). These characteristics depend on the physical properties of the medium (Sector Ib), such as the complex refractive indices (block 1b), and the dimensions and shapes of the suspended particles (block 2b).

5.1.3 *Elements of Mie Theory*

A quantitative description of the attenuation, scattering, and absorption of light by a suspended particle as a function of the above-mentioned physical properties is obtained by solving the Maxwell equations for an electromagnetic field passing through an area of optical inhomogeneity. However, the geometrical shape of the particle itself, which sets the boundary conditions for such calculations, already limits the number of practicable solutions to the few cases of particles with regular shapes (spheres, cylinders, etc.). The possible solutions for irregular particles of any shape or size are extremely complex (see, e.g., Shifrin (1983a, b), Bohren and Huffman (1983), Voshchinnikov and Farafonov (1993), Wriedt (1998), and Quirantes and Bernard (2004, 2006)).

For calculating light scattering characteristics in the sea and atmosphere we usually use the set of solutions of the Maxwell equations for isotropic spherical particles of any size,[6] emerging from the so-called Mie theory and its later modifications (Figure 5.2, Sector IIb). Mie (1908) was the first to obtain such solutions for scattering, but nonabsorbing, spheres, that is, with only the real part of the complex refractive index $m = n$, and $n' = 0$. Expanding them to cover light-absorbing particles as well was the work of Hulst (1957, 1981) and Deirmendijan (1969), among others. In accordance with the convention applied by these and other authors, the most important optical, absorption, and scattering properties of a single particle are described by the characteristics:

The amplitude scattering matrix is $\boldsymbol{S}(\theta, \varphi)$.

The relative cross-sections of the particle, also known as optical efficiency factors:

Q_c is for attenuation,
Q_s is for scattering,
Q_a is for absorption.

We now discuss the significance of these characteristics.

The amplitude scattering matrix $\boldsymbol{S}(\theta, \varphi)$ links the combined amplitudes of the electric field intensities of the incident and scattered wave (where θ and φ are angles describing the scattering direction in the system of spherical coordinates r, θ, φ). Let a flat ray of light propagating along the z-axis be incident on a particle located at the origin of the system of coordinates (Figure 5.3).

The conversion of the electric vector of the incident light wave into the electric vector of a light wave scattered in direction $[\theta, \varphi]$ at a point $P(r, \theta, \varphi)$, sufficiently distant from the particle in comparison with the wavelength ($r >> \lambda$), is described by the matrix according to the relationship:

$$\begin{pmatrix} E_l \\ E_r \end{pmatrix} = [\boldsymbol{S}(\theta, \varphi)] \frac{e^{-ikr + kz}}{ikr} \begin{pmatrix} E_{l,0} \\ E_{r,0} \end{pmatrix}, \tag{5.5}$$

[6] Suspended matter in the sea consists of a set of particles of various sizes and irregular shapes. However, the observed optical properties of such a set of particles with statistically distributed spatial orientations resemble the optical properties of a set of spherical particles whose volumes (dimensions) are equal to the volumes (substitute dimensions) of particles with real shapes (see, e.g., Kullenberg (1974), Ivanoff (1978), and Shifrin (1983a)).

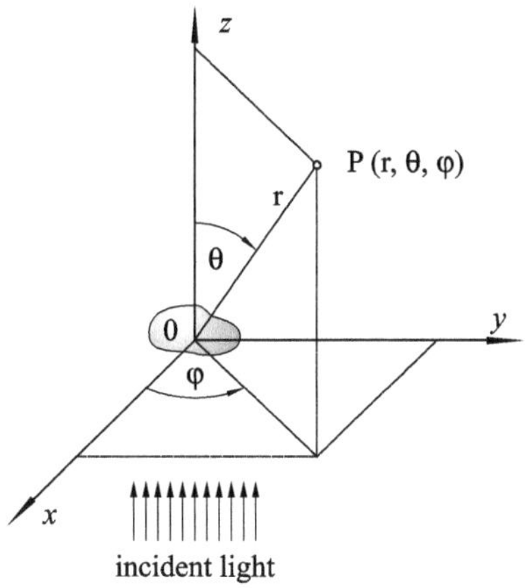

FIGURE 5.3. Geometrical sketch as an adjunct to the Mie theory of light scattering by a particle (see explanation in the text).

where $E_{l,0}$, $E_{r,0}$ are the respective parallel and perpendicular[7] (to the scattering plane $[z,r]$) components of the electric vector of the incident wave, E_l, E_r are the identical components of the electric vector of the scattered wave, and $k = 2\,\pi/\lambda$ is the wave number.

In the general case, that is, for an anisotropic particle of any shape, the scattering matrix $\boldsymbol{S}(\theta,\varphi)$ takes the form:

$$\boldsymbol{S}(\theta,\varphi) = \begin{pmatrix} S_2(\theta,\varphi) & S_3(\theta,\varphi) \\ S_4(\theta,\varphi) & S_1(\theta,\varphi) \end{pmatrix}. \tag{5.6}$$

But in the case of spherical and isotropic particles, this matrix is much simpler. Two of its terms, $S_3(\theta,\varphi)$ and $S_4(\theta,\varphi)$, take zero values, and the whole matrix is independent of the azimuthal angle φ; that is,

$$S_1(\theta,\varphi) \equiv S_1(\theta);\; S_2(\theta,\varphi) \equiv S_2(\theta).$$

The overall attenuation of light by absorption and scattering, and the separate processes of light scattering and absorption by a single scattering particle, are usually both treated as a straightforward screening of the light by the relevant cross-sections of the particle:

σ_c: cross-section for attenuation,
σ_s: cross-section for scattering,
σ_a: cross-section for absorption.

[7] The indices l and r introduced by Chandrasekhar (1960) are derived from the last letters of the English words *parallel* and *perpendicular*.

The respective relative cross-sections (optical efficiency factors) for the attenuation, scattering, and absorption of light are defined by the ratios:

$$Q_c = \sigma_c/S_g, \tag{5.7a}$$

$$Q_s = \sigma_s/S_g, \tag{5.7b}$$

$$Q_a = \sigma_a/S_g, \tag{5.7c}$$

where S_g is the geometrical cross-section of the particle (for spheres $S_g = \pi D^2/4 = \pi r_0^2$, where D is the particle diameter and r_0 is the particle radius).

The complete, formal mathematical apparatus of the algorithm for determining the optical, absorption, and scattering properties of dispersing media (both monodispersing and polydispersing), treated as a set of equivalent, isotropic spherical particles, is given in Table 5.1. It is based on Mie's theory, which describes the above optical characteristics ($\boldsymbol{S}(\theta)$, Q_c, Q_s, Q_a) as a function of two independent variables:

(a) x is the particle size parameter:

$$x = \frac{\pi D}{\lambda} \text{ or } x = \frac{\pi D}{\lambda_2} = \frac{\pi D n_2}{\lambda}; \tag{5.8}$$

that is, the ratio of the particle's circumference (*) to the wavelength of light in a vacuum λ if we are considering the optical properties of particles suspended in a vacuum or in a medium with absolute refractive index $n_2 = 1$, or (**) to the wavelength λ_2 in the medium surrounding the particle ($\lambda_2 = \lambda/n_2$) if we are considering a particle suspended in a medium with refractive index $n_2 > 1$;

(b) m is the complex refractive index of the particle matter with respect to the medium in which it is suspended:

$$m = n - in' \text{ or } m = \frac{1}{n_2}(n - in') \tag{5.9}$$

where n is the real part of the refractive index, equal to the ratio of the velocity of light in a vacuum to that in the particle matter; $n' = a_{pm}\lambda/4\pi$ is the imaginary part of the refractive index, where a_{pm} is the coefficient of light absorption by the particle matter. This means that in calculations involving the use of Mie's theory, we take as input data either the absolute complex refractive indices of the particle matter (if it is suspended in a vacuum) or the relative refractive indices (i.e., differing by the factor n_2^{-1}), if the particle is suspended in a medium with refractive index $n_2 > 1$ (e.g., in water, $n_2 \cong 1.34$ – for visible light).

The first three of these optical characteristics of a particle ($\boldsymbol{S}(\theta)$, Q_c, and Q_s) are described directly by Mie theory equations (Table 5.1, Equations (T-1)–(T-4). On the other hand, the relative cross-section for absorption Q_a is defined indirectly as the difference between the relative cross-sections for attenuation Q_c and scattering Q_s (Equation (T-5)). This relationship emerges from the assumption that scattering and absorption in a medium are additive. Now, if the dependences of these four elementary characteristics of a single particle on the parameters x and m, and on the physical properties of real dispersing media

TABLE 5.1. The algorithm for estimating the optical properties of dispersing media.

Sector Ib. Input parameters (blocks 1b and 2b in Figure 5.5.1):
$m = n - in'$ or $m = (n + in')/n_2$ (where n_2-refractive index of the medium in which particles are suspended): dimensionless complex refractive index of the particle matter: absolute or relative to the medium
N_v [number of particles m^{-3}]: numerical density of particles in the medium
D [μm]: substitute diameter of particles in a monodispersing medium
$N(D)$ [number of particles m^{-3} μm^{-1}]: size (substitute diameter) distribution function of particles in a polydispersing medium

Sector IIb. Model formulas from Mie's theory (after Hulst (1958) and Deirmendjian (1969); block 3b in Figure 5.5.1):

1. The amplitude scattering matrix $S(\theta,\varphi)$, as a function of the particle size parameter

$x = \frac{\pi D}{\lambda}$ or $x = \frac{\pi D n_2}{\lambda}$ and the complex refractive index $m = n - in'$ or $m = (n - in')/n_2$:

$$\boldsymbol{S}(\theta) = \begin{pmatrix} S_2(\theta) & 0 \\ 0 & S_1(\theta) \end{pmatrix} \tag{T-1}$$

$$S_1(\theta) = \sum_{n=1}^{\infty} \frac{2n+1}{n(n+1)} \left\{ A_n \frac{P_n^1(\cos\theta)}{\sin\theta} + B_n \frac{d}{d\theta} P_n^1(\cos\theta) \right\} \tag{T-2a}$$

$$S_2(\theta) = \sum_{n=1}^{\infty} \frac{2n+1}{n(n+1)} \left\{ B_n \frac{P_n^1(\cos\theta)}{\sin\theta} + A_n \frac{d}{d\theta} P_n^1(\cos\theta) \right\}, \tag{T-2b}$$

where $P_n^1(\cos\theta)$ is associated Legendre polynomials.

2. The relative cross-section as a function of x and m for attenuation

$$Q_c = \frac{2}{x^2} \sum_{n=1}^{\infty} (2n+1)\,\mathrm{Re}(A_n + B_n), \tag{T-3}$$

for scattering

$$Q_s = \frac{2}{x^2} \sum_{n=1}^{\infty} (2n+1)(|A_n|^2 + |B_n|^2), \tag{T-4}$$

for absorption

$$Q_a = Q_c - Q_s, \tag{T-5}$$

where $A_n = f(x,m)$ and $B_n = f(x,m)$ – Mie parameters, defined as functions of two variables: x, and $y = m\,x$:

$$A_n = \frac{\psi_n(x)\,\psi_n'(y)/\psi_n(y) - m\psi_n'(x)}{\xi_n(x)\,\psi_n'(y)/\psi_n(y) - m\xi_n'(x)} \tag{T-6a}$$

$$B_n = \frac{m\psi_n(x)\,\psi_n'(y)/\psi_n(y) - \psi_n'(x)}{m\xi_n(x)\,\psi_n'(y)/\psi_n(y) - \xi_n'(x)} \tag{T-6b}$$

and the auxiliary functions $\psi_n(z)$ and $\xi_n(z)$ are defined by Bessel functions of the first kind $J_{n+1/2}(z)$, of the order $(n + 1/2)$:

$$\psi(z) = \sqrt{\frac{\pi z}{2}}\, J_{n+1/2}(z) \tag{T-7a}$$

$$\xi(z) = \sqrt{\frac{\pi z}{2}}\, J_{n+1/2}(z) + (-1)^n i\, J_{-n-1/2}(z), \tag{T-7b}$$

where z is the variable x or y.

(*Continued*)

TABLE 5.1. The algorithm for estimating the optical properties of dispersing media. —Cont'd.

Sector IIIb. Formulas for calculating the optical properties of single particles and a dispersing medium (blocks 4b and 5b in Figure 5.2):

1. For a single particle: the optical cross-section as a function of the wavelength λ, the complex refractive index m, and the substitute diameter of a particle D (the dependence of the calculated optical functions on the wavelength λ and particle diameter D emerges from the dependence of these functions on the particle size parameter x):

$$\sigma_j = Q_j \pi D^2/4, \quad \text{(T-8)}$$

where $j = c$ for attenuation, $j = s$ for scattering, $j = a$ for absorption.

2. For a monodispersing medium (i.e., one containing N_v identical particles in unit volume): the volumetric light scattering function $\beta_p(\theta)$:

$$\beta_p(\theta) = \frac{N_v}{2k^3}[i_1(\theta) + i_2(\theta)], \quad \text{(T-9)}$$

the volumetric coefficient of light attenuation c_p, scattering b_p and absorption a_p:

$$c_p = N_v \sigma_c, \quad \text{(T-10a)}$$

$$b_p = N_v \sigma_s, \quad \text{(T-10b)}$$

$$a_p = N_v \sigma_a. \quad \text{(T-10c)}$$

For a polydispersing medium (i.e., one containing particles of various sizes described by the size distribution function $N(D)$):
the volumetric light scattering function $\beta_p(\theta)$:

$$\beta_p(\theta) = \frac{1}{2k^3} \int_{D_{min}}^{D_{max}} N(D)[i_1(\theta) + i_2(\theta)]\, dD, \quad \text{(T-11)}$$

the volumetric coefficient of light attenuation c_p, scattering b_p, and absorption a_p:

$$c_p = \int_{D_{min}}^{D_{max}} N(D)\, \sigma_c\, dD, \quad \text{(T-12a)}$$

$$b_p = \int_{D_{min}}^{D_{max}} N(D)\, \sigma_s\, dD, \quad \text{(T-12b)}$$

$$a_p = \int_{D_{min}}^{D_{max}} N(D)\, \sigma_a\, dD, \quad \text{(T-12c)}$$

where the wave number $k = 2\pi/\lambda$, and the components:

$$i_1(\theta) = |S_1(\theta)|^2, \quad \text{(T-13a)}$$

$$i_2(\theta) = |S_2(\theta)|^2, \quad \text{(T-13b)}$$

are so-called Mie intensity functions.

(Sector 1b of input data, Table 5.1), are known, their resultant optical properties (the coefficients of attenuation c_p, scattering b_p, and absorption a_p, as well as the volume scattering function of light by suspended particles $\beta_p(\theta)$)[8]

[8] Definitions of these functions can be found in the standard works on marine optics, e.g. Jerlov (1976), Kirk (1994), and Dera (2003).

and their spectral distributions can be determined. To do this, the formal relations between the optical properties of media and the optical characteristics of single inhomogeneities (particles), familiar from marine and atmospheric optics, have to be additionally employed (see the equations given in Sector III b, Table 5.1; for more details, see, e.g., Dera (2003)).

5.1.4 Some Theoretical Optical Characteristics of Suspended Particles

Mie's theory, as outlined above, allows one to estimate the influence of the basic physical properties of suspended particles (i.e., the complex refractive index and the particle dimensions) on their optical properties, especially on their absorption capabilities. It turns out that within the range of variability of these physical properties characteristic of various natural suspensions in seas and other natural water bodies, this influence is enormous. This is backed up by the plots in Figures 5.4–5.6. They illustrate the theoretical dependences,

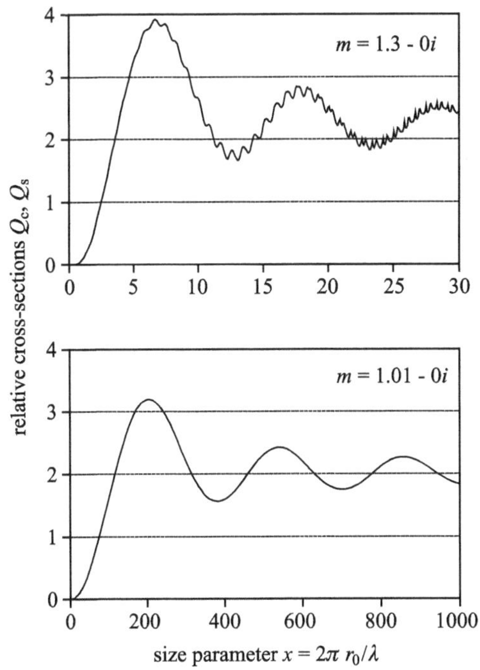

FIGURE 5.4. Relationships among the relative cross-sections for the attenuation Q_c and scattering of light Q_s, and the size distribution parameter x for particles with negligibly small imaginary part ($n' = 0$) and various real refractive indices n (shown in the figure). For these cases the relative cross-section for absorption $Q_a = 0$ and $Q_c = Q_s$. The upper curve applies to mineral suspensions with conventionally high (with respect to water) indices n, the lower one to some biological, practically nonabsorbing suspensions with low values of n.

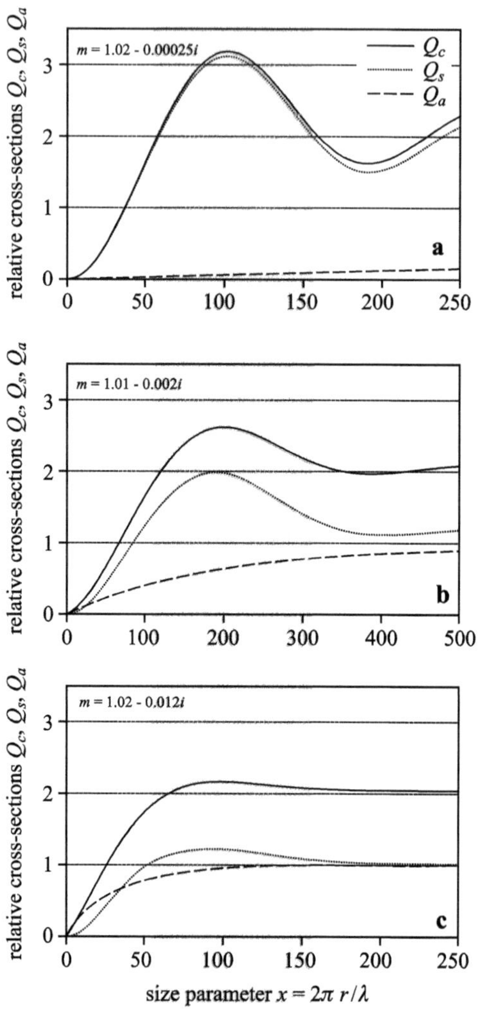

FIGURE 5.5. Relationships among the relative cross-sections for the attenuation Q_c, scattering Q_s, and absorption of light Q_a, and the particle size parameter x for optically soft suspensions with various complex refractive indices with respect to water. Typical relationships for: (a) weakly absorbing, pigment-free organic particles; (b) large, strongly absorbing organic particles, or weakly absorbing phytoplankton; (c) strongly absorbing phytoplankton with a high pigment content.

determined on the basis of Mie's theory, of the relative cross-sections for light attenuation Q_c and scattering Q_s, and also of their differences, that is, the cross-sections for absorption $Q_a = Q_c - Q_s$, on the particle size parameter $x = \pi D/\lambda$ for a few complex refractive indices $m = n - in'$. We selected the magnitudes of these indices, or to be more precise, of the various configurations of their component parts (real n and imaginary in'), on the basis of the extent to which they are representative of different types of natural suspensions.

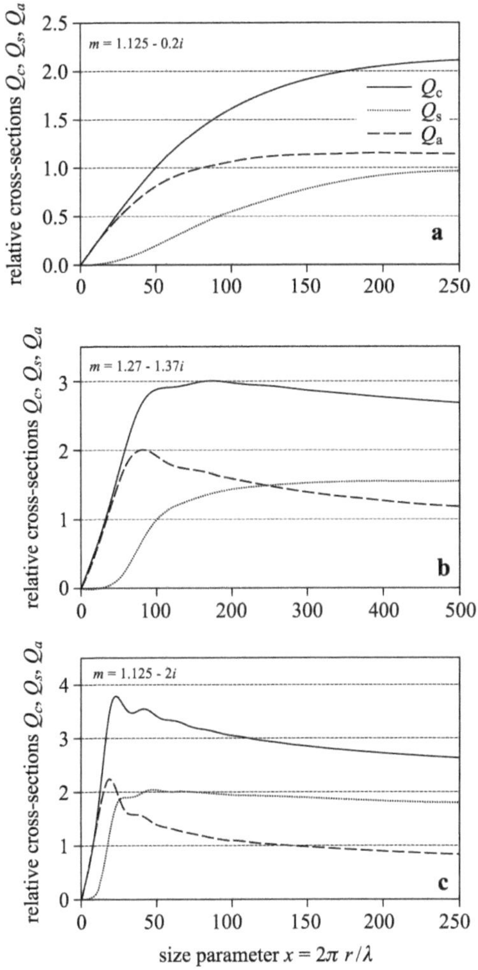

FIGURE 5.6. The same relationships as in Figure 5.5 for particles: (a) and (c) hypothetical internal substructures of phytoplankton; (b) containing iron (e.g., hematite). (Other explanations as for Figure 5.5.)

Taking into consideration the components n and n' of this index, it is possible to make a rough distinction between the main types of particles found in natural suspensions (covered in greater detail in Section 5.2):

- Particles with real refractive indices that are relatively high compared to that of water, up to $n \approx 1.3$, and with negligibly small imaginary part of refractive indices (i.e., $n' \approx 0$); this group includes lot of mineral particles, especially those not containing iron compounds, for example, hematite (Figure 5.4 upper curve).
- Optically "soft" particles, that is, with real relative refractive indices $n \approx 1$ and imaginary ones $n' \approx 0$; they include many organic particles not containing pigments or humus substances (Figure 5.4, lower curve; Figures 5.5a and 5.5b).

- Optically "soft" particles with small but nonzero imaginary part of refractive indices $1 >> n' > 0$; this group embraces the phytoplankton and many organic particles containing humus substances (Figure 5.5c).
- Particles whose real and imaginary parts of refractive indices are both relatively high (Figure 5.6).

The diversity in the values of these efficiency factors for attenuation (Q_c), scattering (Q_s), and absorption (Q_a), evident from an inspection of Figures 5.4 through 5.6, is characteristic of the various types of SPM with different physical properties including the different optical constants of the particles (i.e., their complex refractive indices) and their different sizes. This diversity shows how wide-ranging can be the influence of the packaging function on the optical properties of these particles, especially on their absorption properties in different basins and under different environmental conditions. Section 6.5 describes the effect of packaging on the light absorption coefficients of particles in greater and more vivid detail, the phytoplankton cells occurring in sea waters of different trophicities serving as an example.

5.2 Suspended Particulate Matter in the Sea: Nature, Origins, Chemical, and Physical Properties

The previous section described the theory underpinning the optics of dispersing media. On the basis of Mie's theory, it focused on the mathematics of the interactions of light with the optical inhomogeneities of a medium that are larger than chemical molecules, such as suspended particulate matter (SPM). It showed that these interactions vary for different types of particles and that they depend on the chemical composition of the particles and their fundamental physical properties. The most important of these properties are the optical constants of the substances making up these particles, (i.e., their real and imaginary spectral refractive indices of light), and the shapes and sizes of the particles.

We begin Section 5.2 by characterizing the origin, main types, and resources of organic and inorganic SPM in sea waters. Then we describe the most important chemical and physical properties of the main types of SPM that determine its optical properties (including absorption capacities); at the same time, they constitute sets of input data essential for calculating these optical properties of SPM using Mie's or other such theories.

5.2.1 Suspended Particulate Matter in the Sea: Main Types, Origins, and Resources

The fundamental chemical and physical properties of SPM governing its interaction with light vary greatly, depending on the type of SPM and its origin. So before discussing these properties, we first describe the main types of

SPM, their origin, and how they get into the sea. Following on from this, we characterize qualitatively, and where current knowledge allows, also quantitatively, the combined resources of SPM in the World Ocean and the regularities controlling the occurrence of various types of particles at different depths in different seas.

The Main Types and Sources of Suspended Particulate Matter in the Sea

Particles of matter suspended in sea water are exceedingly complex with respect not only to their shapes and sizes, but also to their chemical composition and physical (optical) properties. These particles can be divided into two main types: inorganic particles of mineral origin, and organic particles of biological origin. The latter type can be split into two groups: living microorganisms: bacteria, plankton, fungi, and others, although the various phyto- and zooplankton species are predominant; and dead organic matter (detritus), consisting mainly of the remains and the decomposition/metabolic products of marine plants and animals.

Henceforth we use the following abbreviations to denote these types and groups.

- SPM: Suspended particulate matter, that is, all suspended matter
- PIM: Particulate inorganic matter, that is, suspended inorganic (mineral) matter
- POM: Particulate organic matter, that is, all suspended organic matter
- LO: Living organisms, or the practically synonymous (LP) Living plankton, that is, living, particulate organic matter
- Od: Organic detritus, that is, dead particulate organic matter

These abbreviations, written in italics, also represent the concentrations in [g m^{-3}] of these groups of particulate matter.

The origins of particles suspended in sea water can be very diverse. Inorganic particles, for example, are derived primarily from minerals and occur in or reach the seas as a result of mechanical processes. The weathering of rocks produces countless such particles, many of which end up in the soil. These in turn are eroded by water; the mineral particles of the soil reach the rivers, which transport them to the sea. Mineral particles can also enter the sea via the atmosphere: winds gather them up from the land, mostly from desert areas, and deposit them in the sea as dusts and other types of atmospheric fallout. Volcanic ash[9] and cosmic dust also get into the sea from the atmosphere. The flux of mineral particles entering the seas and oceans is enormous, and their mass in the water—despite their ongoing deposition—is very high (see the data in Table 5.2). It has been estimated that rivers alone carry some 400 billion metric tons of particulate inorganic matter into the ocean every 24 hours (Romanovsky 1966, Dera 2003).

[9] Part of this mineral matter is also generated by underwater volcanoes.

TABLE 5.2. Estimated resources of suspended particulate matter in the World Ocean.

Form of occurrence (Symbol)		Mass		Ref. (Remarks)[b]
		10^9 t	% of SPM	
-1-		-2-	-3-	-4-
Suspended Particulate Matter (SPM)				(1)
	Total	1370	100	
Particulate Inorganic Matter (PIM)				(2)
	Total	1307	95.5	
Particulate Organic Matter[a] (POM)				(3)
	Total	62.4	4.5	
Living organisms (LO)				
	Total	~2.4	0.175	
	Phytoplankton	0.80	0.058	
	Zooplankton	1.48	0.108	
	Nekton	0.10	0.007	
	Bacteria	0.007	0.0005	
Organic detritus (Od)				
	Total	60.0	4.38	

[a] The total mass of POM includes the mass of phytoplankton, zooplankton, nekton, bacteria, and organic detritus, but not the biomasses of the phytobenthos and zoobenthos; [b] (1) data from Gershanovich and Muromtsev (1982); (2) the content of particulate inorganic matter (PIM) determined as the difference between the contents of suspended particulate matter (SPM) and particulate organic matter (POM), that is, PIM = SPM – POM; (3) the sources of the data were quoted in the caption to Table 4.1.

Winds and inflowing waters also carry certain amounts of organic particles to the sea (Romankevich 1977), mainly fragments of organisms present in the atmospheric dust, for example, bacteria, fungal spores, pollen grains, seeds, stem tissue, and ash (Pye 1987, Sokolik and Toon 1999). Even so, the vast majority of particulate organic matter in the World Ocean is associated with marine life and comes into being in the water itself. As we showed in Table 4.1 (see Chapter 4), the annual production of POM as a result of the photosynthesis (primary production) of phytoplankton and bacteria in all the seas and oceans amounts to some 60 billion metric tons dry mass. Apart from this primary production, there are many other processes occurring in the water that alter the composition and concentrations of SPM, for example, zooplankton development or detritus formation. We discuss these transformations of the various forms of suspended organic particles further in Section 5.2.4.

Notice that, in addition to organic particles, some mineral particles, such as fragments of the skeletons of organisms, are indirectly of biological origin, because they are derived from the decomposition of living organisms. Moreover, some suspended particles can have a hybrid nature, forming organic–inorganic aggregates as a result of processes combining matter of mineral and biological provenance. Such aggregates can form within the water itself, or can be carried into the sea from without.

The Total Content of Suspended Particulate Matter (SPM) in the World Ocean

The total resources of all types of SPM in the World Ocean—like those of other admixtures in sea water—can be calculated only very approximately. According to one such rough approximation (see Gershanovich and Muromtsev (1982)), the combined mass of SPM at all depths in all the seas and oceans is c. 1.37 10^{12} metric tons[10] (Table 5.2). If the approximate mass of all particulate organic matter (POM) in the oceans is equal to 62.4 10^9 metric tons (phytoplankton + zooplankton + nekton + bacteria + detritus; see table 5.2), then the mass of particulate inorganic matter will be c. 1.31 10^{12} metric tons. Making up c. 95.5% of the combined mass of SPM, PIM is thus predominant, because the mass of POM is barely 4.5%.

This predominance of PIM over POM is reflected in the data cited in Table 5.3. This gives examples of mass concentrations (means or ranges of variability) of suspended particles in the surface waters (mainly the euphotic zone) of various marine regions, as determined empirically by different authors. The data in this table show that the proportion of POM in the overall SPM concentration in the surface layers of seas is usually greater or very much greater than 4.5%; indeed, we show in due course (Table 5.5, Figure 5.7c) that in the eutrophic and supereutrophic regions of enclosed seas such as the Baltic or the Black Seas, the proportion of POM can reach 90% and more of the total SPM mass. These discrepancies between the percentage contents of POM and PIM, both in surface waters and on a World Ocean scale, emerge from the fact that at great depths in the oceans, mineral particles make up by far the greatest part of SPM. Durable and usually far heavier than water, they sink much more rapidly to the bottom (see Table 5.7, which gives, among other things, the densities of various mineral particles; Dera (2003)). In contrast, the densities of POM are usually about the same as or slightly greater than that of the surrounding water; they also break down quite quickly. Their rate of deposition is therefore small and they will usually have undergone mineralization even before they reach abyssal waters, for example, by spontaneous decomposition or bacterial degradation (see Romankevich (1977)).

It is clear from the data in Tables 5.2 and 5.3 that the mass concentrations of suspended matter found under natural conditions in the ocean, both the total SPM content and the component POM and PIM contents, vary in different regions and at different depths, and can do so by more than three orders of magnitude. This differentiation is due to a whole range of factors, both biotic and abiotic, and depends, among other things, on the geographical position of the marine region in question, its biological productivity, as

[10] This figure was calculated on the assumption that the mean concentration of suspended particulate matter in the World Ocean is probably close to 1 mg dm^{-3}, and that the combined volume of water in all the seas and oceans is 1.37 10^{21} dm^3.

TABLE 5.3. Mass concentrations of particulate organic and inorganic matter in various regions of the World Ocean.[a]

Region investigated	Total dry mass of particulate matter in mg dm^{-3} of water	Mass of POM as a percentage of the total dry mass [%]	Authors of investigations
-1-	-2-	-3-	-4-
Pacific, coastal zone	10.5	62	Fox et al. (1953)
Pacific Ocean	3.8	29	Fox et al. (1953)
NW Pacific	0.45–1	—	Hobson (1967)
North Sea	6.0	27	Postma (1954)
Wadden Sea (The Netherlands)	18.0	14	Postma (1954)
Bering Sea	2–4	—	Lisitzin (1959)
Mean value in the oceans	0.8–2.5	20–60	Lisitzin (1959)
Baltic Sea	0.2–12	—	Jonasz (1980)
Nepheloid layer at 3–5 km, North Atlantic (Nova Scotia Continental Rise)	~ 0.02 – ~ 7	—	McCave (1983)
Coastal waters around Europe:			
Adriatic	~ 0.21 – ~12	—	Babin et al. (2003a)
Atlantic	~ 0.019 – ~1.3	—	Babin et al. (2003a)
Baltic	~ 1.2 – ~6.5	—	Babin et al. (2003a)
English Channel	~ 0.35 – ~7	—	Babin et al. (2003a)
Mediterranean	~ 0.075 – ~17	—	Babin et al. (2003a)
North Sea	~ 0.51 – ~75	—	Babin et al. (2003a)
Baltic (central regions)	0.26 – 4.80	16.1 – 51.2	The present authors' own studies
Baltic (coastal regions and bays)	0.35 – 31.3	17.0 – 95.2	
Black Sea (central regions)	0.18 – 3.85	11.5 – 45.0	The present authors' own studies
Black Sea (coastal regions and bays)	0.26 – 18.0	13.0 – 61.1	The present authors' own studies

[a] Adapted from Dera (2003).

well as the dynamics of its water masses and of the air in the atmosphere above it. The problem is thus highly complex, one that has not yet been studied and described in detail in the context of SPM in general. A certain insight into the quantitative statistical regularities governing the occurrence of SPM in different types and regions of the seas in and around Europe is provided by the research results of Babin et al. (2003a,b) and of our own research team at Sopot. Both sets of results have demonstrated that connections indubitably exist between the total SPM content and the concentrations of its various constituents, and, the trophicity of the waters investigated. We now discuss these results.

How the Contents of SPM and Its Various Constituents Are Linked to the Trophicity of Sea Waters

Babin et al. (2003a,b) measured the optical properties of phytoplankton, nonalgal particles, and colored dissolved organic matter, as well as the total

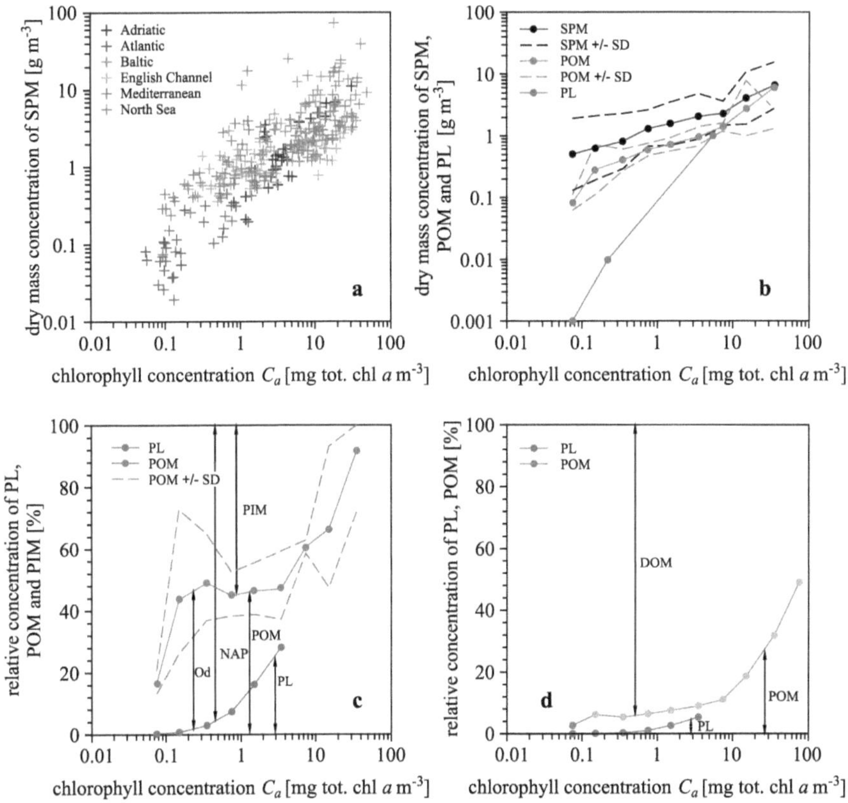

FIGURE 5.7. Concentrations of various marine suspended particles versus chlorophyll *a* concentration in sea waters: (a) measured in coastal waters around Europe (adapted from Babin et al. (2003a)); (b)–(d) mean values from all the data measured in the euphotic zones of the Baltic and Black Seas, and the Atlantic Ocean (Canary Islands region; based on the data bank of IO PAS Sopot). Abbreviations: SPM, suspended particulate matter; POM, particulate organic matter; DOM, dissolved organic matter; PIM, particulate inorganic matter; NAP, nonalgal particles; PL, phytoplankton; Od, organic detritus; SD, standard deviation. Lengths of the vertical arrows in (c) and (d) indicate the percentage contents of the various components. (See Colour Plate 4)

concentration of suspended particle matter *SPM* and the total chlorophyll *a* concentration in sea water C_a at about 350 stations in various coastal waters around Europe, including the English Channel, Atlantic, Adriatic Sea, Baltic Sea, Mediterranean Sea, and North Sea. These measurements show, among other things, that a distinct relationship exists between *SPM* and C_a concentrations: in the case of the coastal zones of the seas around Europe, these two quantities vary widely over a range of more than three orders of magnitude (see Figure 5.7a).

Despite the wide scatter of empirical points on this plot, the total SPM concentrations measured in these seas do tend to rise with increasing chlorophyll *a* concentration C_a in the sea.[11] Our own statistical researches (see Table 5.4) have confirmed the existence of such a relationship. At the same time, the chlorophyll *a* concentration C_a can be treated as the trophic index of waters, of which more in Section 6.1.1.

From statistical analyses of empirical data gathered by our team, we have been able to obtain more detailed relationships of this type, not just between the total *SPM* and C_a, but which also take into account the concentrations of the various groups or types of suspended particulate matter. This empirical material includes measurement data from the euphotic zones of different parts of the Baltic and Black Seas, as well as the Atlantic Ocean in the vicinity of the Canary Islands: 586 empirical points for the *SPM* versus C_a relation, 185 such points for the *POM* versus C_a relation, and a further 147 such points for the relationship between colored dissolved organic matter (*CDOM*) and C_a. Tables 5.4–5.6 give the more significant results of these statistical generalizations, and Figures 5.7b,c,d provide illustrations. We now proceed to discuss these results.

The statistical relationship that we have worked out between the concentration of total SPM and that of chlorophyll *a* C_a in the euphotic zones of the seas concerned is given in Table 5.4A and illustrated in Figure 5.7b, the SPM plot. This relationship is characterized by a broad scatter of experimental points, because a large proportion of the particulate matter in these semi-enclosed and shelf seas is made up of inorganic particles, the concentrations of which are governed by physical factors not directly connected with the chlorophyll *a* concentration C_a. In spite of this, the rise in mean SPM concentrations with increasing concentrations C_a is clearly discernible. The lowest SPM contents are typical of oligotrophic basins of types O3 and O4: mean values range from c. 0.50 to 0.63 g m^{-3} particulate dry mass, extreme values from 0.2 to 2.30 g m^{-3}. But in chlorophyll-rich eutrophic waters, for example, where C_a ranges from 20 to 50 mg tot. chl *a* m^{-3}, the SPM content is on average 6.5 g m^{-3}, and in extreme cases can exceed 30 g m^{-3}. The proportions of the various groups of suspended matter in the total quantities of SPM also vary in different trophic types of sea water (see below).

The mean concentration of the total SPM resources in the World Ocean is c. $\overline{SPM} \approx 1\,\mathrm{g\,m^{-3}}$. This is equivalent to the particulate matter concentration in mesotrophic waters, that is, where the chlorophyll concentration C_a lies within the 0.2–0.5 mg tot. chl. *a* m^{-3} range. In oligotrophic waters there is correspondingly less particulate matter; in eutrophic waters there is more. Unlike the case of dissolved organic matter (see Chapter 4), the overall mean concentration of SPM in the oceans is governed principally by its concentrations in moderately and highly productive seas.

---

[11] The division of marine basins into trophic types is given in Section 6.1 (Table 6.1).

TABLE 5.4. SPM concentrations measured in basins of different trophicity in the Baltic Sea, Black Sea, and Atlantic Ocean (off the Canary Islands).[a]

A. Dry-Mass Concentration of Total Suspended Particulate Matter (SPM)

Trophic type of basin	Range of chlorophyll *a* concentration C_a [mg tot. chl *a* m^{-3}]	SPM concentrations [g m^{-3}] Logarithmic mean[c]	Standard range of variability[c]	Range of variability in the field	Number of experimental points
-1-	*-2-*	*-3-*	*-4-*	*-5-*	*-6-*
Oligotrophic O3	0.05–0.1	0.50	0.13–1.92	0.20–2.15	20
O4	0.1–0.2	0.63	0.19–2.10	0.18–2.30	42
Mesotrophic M	0.2–0.5	0.82	0.29–2.30	0.35–2.60	76
Mesoeutrophic I	0.5–1.0	1.31	0.67–2.60	0.46–3.41	64
Eutrophic E1	1–2	1.57	0.70–3.52	0.52–3.85	160
E2	2–5	2.05	0.88–4.78	0.80–4.82	102
E3	5–10	2.30	1.47–3.58	1.28–4.25	46
E4	10–20	4.05	1.53–10.7	1.41–10.20	40
E5	20–50	6.50	2.75–15.4	2.05–31.3	36

B. Dry-Mass Concentration of Particulate Organic Matter (POM)[b]

Trophic type of basin	Range of chlorophyll *a* concentration C_a [mg tot. chl *a* m^{-3}]	POM Concentrations [g m^{-3}] Logarithmic mean[c]	Standard range of variability[c]	Range of variability in the field	Number of experimental points
-1-	*-2-*	*-3-*	*-4-*	*-5-*	*-6-*
Oligotrophic O3	0.05–0.1	0.082	0.062–0.11	0.050–0.12	8
O4	0.1–0.2	0.28	0.11–0.72	0.060–1.1	15
Mesotrophic M	0.2–0.5	0.40	0.27–0.60	0.18–1.3	27
Mesoeutrophic I	0.5– .0	0.59	0.47–0.74	0.48–0.80	20
Eutrophic E1	1–2	0.73	0.5–0.93	0.52–1.5	50
E2	2–5	0.96	0.68–1.4	0.56–1.8	24
E3	5–10	1.4	1.2–1.6	1.1–1.7	18
E4	10–20	2.7	1.0 –7.7	0.7–8.2	16
E5	20–50	6.0	1.3–27	1.2–21	7

[a] Values from the data bank of IO PAS, Sopot.
[b] POM concentrations estimated on the basis of measured concentrations of particulate organic carbon (POC), assuming that the ratio *POM*/*POC* is approximately equal to 2; that is, $POM \approx 2\,POC$.
[c] The logarithmic mean $\left(\langle x \rangle_{\log}\right)$ and standard range of variability (lower limit x_l and upper limit x_u) of the data set $[x_i]$ is defined here as $\langle x \rangle_{\log} = 10^{\bar{y}}$ and $x_l = 10^{\bar{y} - \sigma_y}$ and $x_u = 10^{\bar{y} + \sigma_y}$, where $\bar{y}$ and σ_y denote the arithmetic mean and standard deviation of sets of $[y_i]$, where $y_i = \log x_i$.

TABLE 5.5. Approximate proportions (% of dry mass) of the various groups of suspended particulate matter in the total dry mass of particulate matter in euphotic zones of the Atlantic Ocean (off the Canary Islands), the Baltic Sea, and the Black Sea.

		Proportion [% dry mass of total SPM concentration]					
			Organic particles				
Trophic type of basin	Range of chlorophyll *a* concentration C_a [mg tot. chl *a* m^{-3}]	Inorganic particles PIM	Total POM	Chloro-phyll *a* *Chl*	Phyto-plankton PL	Organic detritus Od (and zoo -plankton)	Nonalgal particles NAP
-1-	*-2-*	*-3-*	*-4-*	*-5-*	*-6-*	*-7-*	*-8-*
Oligotrophic							
O3	0.05–0.01	83.5	16.5	0.014	0.20	16.30	99.8
O4	0.1–0.2	56.3	43.7	0.022	0.71	42.99	99.3
Mesotrophic M	0.2–0.5	51.0	49.0	0.040	2.86	46.14	97.14
Mesoeutrophic I	0.5–1.0	55.1	44.9	0.054	7.3	37.6	92.7
Eutrophic E1	1–2	53.6	46.4	0.090	16.2	30.2	83.8
E2	2–5	53.0	47.0	0.16	28.1	18.9	71.9
E3	5–10	39.3	60.7	0.31	55.2	5.5	54.8
E4	10–20	33.5	66.5	0.35	—	—	—
E5	20–50	8.2	91.8	0.49	—	—	—

TABLE 5.6. Approximate proportions of different organic constituents in the total concentration of organic matter in euphotic zones of the Atlantic Ocean (off the Canary Islands), the Baltic Sea, and the Black Sea.

		Proportions [% of the total organic matter concentration]				
			Particulate organic matter POM			
Trophic type of basin	Range of chlorophyll *a* concentration C_a [mg tot. chl *a* m^{-3}]	Dissolved organic matter DOM	Total POM	Chlorophyll *a* *Chl*	Phyto-plankton PL	Organic detritus Od (and zoo -plankton)
-1-	*-2-*	*-3-*	*-4-*	*-5-*	*-6-*	*-7-*
Oligotrophic						
O3	0.05–0.01	97.3	2.7	0.0023	0.033	2.667
O4	0.1–0.2	93.8	6.2	0.0031	0.01	6.1
Mesotrophic M	0.2–0.5	94.6	5.4	0.0044	0.32	5.08
Mesoeutrophic I	0.5–1.0	93.6	6.4	0.0076	1.04	5.36
Eutrophic E1	1–2	92.5	7.5	0.015	2.62	4.88
E2	2–5	91.1	8.9	0.030	5.32	3.58
E3	5–10	88.9	11.1	0.056	10.1	1.0
E4	10–20	81.4	18.6	0.098	—	—
E5	20–50	68.2	31.8	0.17	—	—
E6	50–100	(51.0)	(49.0)	(0.27)	—	—

Particulate organic matter is much more strongly correlated with the chlorophyll concentration C_a than is the total SPM. The statistical relationship between the concentrations of *POM* and chlorophyll C_a in the euphotic zone of the seas investigated is presented in Table 5.4B and illustrated in Figure 5.7b, the POM plot. The standard deviation intervals for the relation $POM = f(C_a)$ are relatively narrower than for $SPM = f(C_a)$. More exact than for other optical constituents, the relationship between *POM* and chlorophyll concentration also exhibits a faster increase in organic particle concentration with increasing chlorophyll content in the water. The differences in POM concentrations recorded under various conditions in the investigated seas can reach two orders of magnitude, and in extreme cases are even greater. The corresponding mean POM concentrations in oligotrophic waters O3 and O4 range from 0.08 to 0.28 g m^{-3}. In biologically rich waters POM concentrations are around 100 times higher: for example, for chlorophyll concentrations C_a ~20–50 mg tot. chl *a* m^{-3} the mean *POM* amounts to c. 6 g m^{-3}, and in extreme cases $POM > 20$ g m^{-3}.

The mean POM concentration in the World Ocean (calculated from its overall total given in Table 5.2) is barely 0.045 g m^{-3}, which is far less than in the euphotic zone, and lower even than in the euphotic zones of oligotrophic waters. This is indicative of a considerable accumulation of POM in euphotic zones, that is to say, in the surface layers of seas and oceans. We look at the variability in concentration of the different groups of suspended particulate matter with depth in the sea later on in this section.

Although not dominant on a global scale, phytoplankton is an important constituent of particulate organic matter. It owes this significance to the fact that it is practically the sole producer of all the autogenic organic matter resources in the sea. We have thus attempted to estimate the contributions of this constituent to the total organic content, particulate and dissolved, in different types of waters.

Inasmuch as we did not have any experimental data at our disposal that would allow us to establish the same kinds of statistical relationships for the dependence of the phytoplankton concentration *PL* on the chlorophyll concentration C_a as in the case of *SPM* and *POM*, we were able to define the dependence $PL = f(C_a)$ (where *PL* is the concentration of the phytoplankton biomass) only approximately and indirectly. Hence, we assumed that $C_a/C_{org,pl}$, the ratio of the chlorophyll *a* content C_a to the organic carbon content in phytoplankton $C_{org,pl}$, is known and depends on the trophic index of the waters in question (i.e., the chlorophyll *a* concentration C_a in the sea.)

According to Koblentz-Mishke and Vedernikov (1977) this ratio is:[12]

[12] This is only an approximation, because in reality the ratio $C_a/C_{org,pl}$ depends on many biotic and abiotic factors. The values of this ratio, measured by various authors for different conditions in the sea and for cultivations of different species of algae, have been reviewed by Kirk (1994); see also the data on carbon and chlorophyll concentrations in the cells of different species of algae given by Stramski et al. (2001), for example.

- $C_a/C_{org,pl} = 0.15$ in oligotrophic waters ($C_a = 0.075$ mg tot. chl *a* m^{-3}).
- $C_a/C_{org,pl} = 0.045$ in mesotrophic waters ($C_a = 0.22$ mg tot. chl *a* m^{-3}).
- $C_a/C_{org,pl} = 0.011$ in eutrophic waters ($C_a = 5.5$ mg tot. chl *a* m^{-3}).

Having interpolated these data, and assuming that $PL = 2C_{org,pl}$, we defined the approximate relationship of $PL = f(C_a)$. The plot PL in Figure 5.7b illustrates this dependence. Clearly, then, the phytoplankton concentration increases strongly with increasing chlorophyll *a* content in the sea. In optically clear ocean waters, where $C_a \approx 0.08$ mg tot. chl *a* m^{-3}, the phytoplankton concentration is scarcely c. $PL \approx 0.001$ g m^{-3} dry mass. Moving into eutrophic seas, where the chlorophyll concentration is c. 5.5 mg tot. chl *a* m^{-3}, the phytoplankton concentration rises by around three orders of magnitude to reach values of $PL \approx 1.1$ g m^{-3} dry mass, whereas the chlorophyll concentration increases by only two orders. This considerable increase of *PL* with increasing C_a means that in euphotic waters phytoplankton is the predominant type of suspended particulate organic matter.

Knowledge of the statistical relationships between the total concentration of all SPM, and only POM or only the phytoplankton biomass PL, on the trophicity of a basin (see above and Figure 5.7b) allows the concentrations of other constituent types of particulate matter characteristic of these waters to be indirectly estimated as well:

- *PIM*, that is, the concentrations of inorganic particles, is obtained from the difference $PIM = SPM-POM$.
- *NAP* (nonalgal particles), that is, the combined concentrations of inorganic particles *PIM* and the organic particles minus the phytoplankton, is equal to the difference $NAP = SPM-PL$.
- NAP_{org}, that is, the nonalgal particles minus the phytoplankton, is equal to the difference $POM-PL$.

Notice that this last group of particles NAP_{org} is practically equivalent to the whole group of particles classified as Od (organic detritus) because the proportion of living nonalgal particles (mainly zooplankton) in this group of particles is usually very much smaller than the proportions of nonliving particles; hence, $NAP_{org} \approx Od$.

The percentage shares of these several forms of optical sea water constituents (concentrations estimated as above) in the total SPM concentrations are given in Table 5.5 and illustrated in Figure 5.7c. For comparison, the percentage shares of POM and DOM in the total organic resources in the sea are given in Table 5.6 and illustrated in Figure 5.7d. Here, we have additionally made use of the results of parallel analyses of the statistical relationships $DOM = f(C_a)$. The data given in Tables 5.5 and 5.6 refer to the respective mean proportions in the euphotic zones of different types of sea water. The regularities emerging from these data are convergent with those discussed earlier and do not require separate discussion.

The diversity of concentrations of the various optical admixtures in sea water, discussed in this section, is the prime cause of the diversity of optical properties of various types of waters. Nevertheless, it should be borne in mind, first, that the results we have presented here are very rough estimates, which may be fraught with serious errors, and second, that our analysis covers only certain basins in and around Europe. These results—merely examples and mainly qualitative—should therefore be approached with a considerable dose of caution: they are not universal regularities linking particulate matter concentrations with trophicity that are applicable to all parts of the World Ocean.

Vertical Distributions of Particulate Matter Concentrations in the Sea

The reader will recall that the concentrations of particulate matter, both the total amounts and the contents of the separate groups of particles, differ at different depths in the sea. Figure 5.8 depicts typical POM concentration changes with depth in the sea. The shapes of $POM(z)$ profiles in eutrophic waters resemble the vertical distributions of phytoplankton or chlorophyll *a* (see Figure 6.15 in Chapter 6). Their maxima are thus often observed at a depth approaching that of the photosynthesis peak. In oligotrophic seas, however, such analogies do not occur. The organic particle content tends to fall with increasing depth. Moreover, $POM(z)$ profiles resemble $DOM(z)$ profiles, except that the former have steeper gradients.

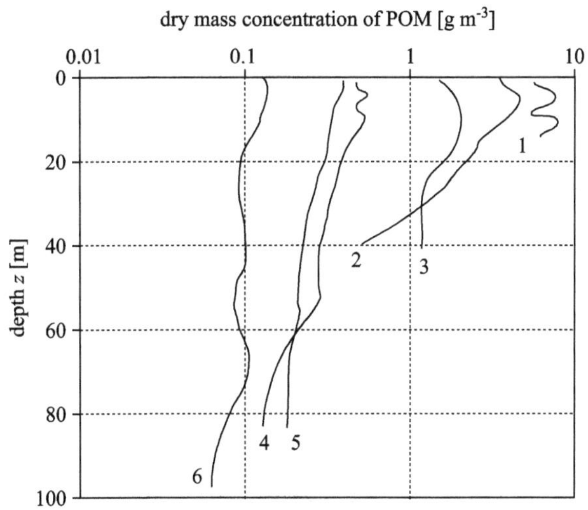

FIGURE 5.8. Examples of empirical vertical distributions of particulate organic matter (POM) concentrations in different sea basins: 1. Puck Bay (Baltic); 2. Gulf of Burgas (Black Sea); 3. Gulf of Gdańsk (Baltic); 4. Gdańsk Deep (Baltic); 5. Black Sea; 6. Atlantic (Canary Islands region). (Based on the data bank of IO PAS, Sopot.)

The vertical distributions of the SPM content are not as clearly defined as the *POM* profiles, and considerable diversity is the order of the day (see Figure 5.9a). It is therefore hard to make any quantitative generalizations. Usually, when SPM concentrations are small, they fall with increasing depth; when they rise, the profiles become more complex. *SPM*(z) profiles depend,

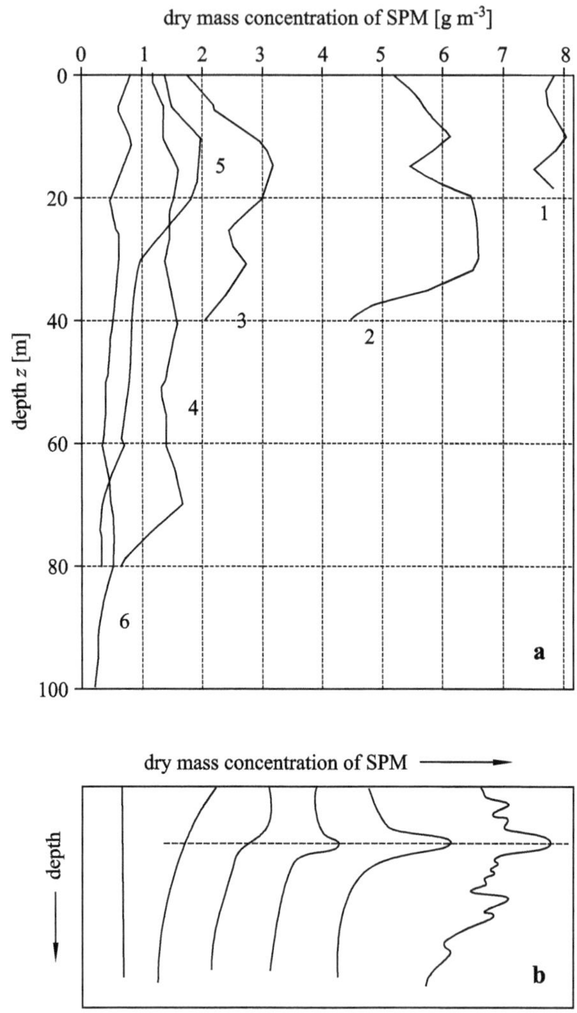

FIGURE 5.9. Vertical distributions of suspended particulate matter (SPM) concentrations in different sea basins. (a) Experimental examples of the concentrations in: 1. Puck Bay (Baltic); 2. Gulf of Burgas (Black Sea); 3. Gulf of Gdańsk (Baltic); 4. Gdańsk Deep (Baltic); 5. Black Sea; 6. Atlantic (Canary Islands region). (Based on data bank of IO PAS, Sopot.) (b) Idealized types of the most frequent vertical distributions in different seas. The horizontal broken curve depicts the positions of the pycnocline (the layer of rapidly decreasing water density). (Adapted from Ivanov (1975).)

for example, on the type and dimensions of suspended particles, and also on the process of deposition and on the dynamics of the waters in a given basin. They are, moreover, significantly affected by the presence of a pycnocline in the sea, in the vicinity of which particles can accumulate, thereby raising their concentration (Figure 5.9b). These problems are discussed in articles and monographs dealing with sedimentation processes, for example, Wassmann and Heiskanen (1988) and Wassmann et al. (1991); see also Dera (2003).

In summary, it has to be said that this outline of the characteristic vertical distributions of SPM concentrations in the sea does not suffice to make any quantitative generalizations, because the set of appropriate empirical data presently available is too small. However, it gives general information about the problem important for optical characterization of sea waters.

5.2.2 The Chemical Composition and Optical Constants of Mineral Particles

We begin our discussion of the fundamental physical properties of particulate matter, knowledge of which is indispensable for modeling the absorption and scattering properties of suspended particulate matter (SPM) in the sea using, for example, Mie's theory, by presenting and specifying the optical constants (i.e., the real and imaginary parts of the absolute complex refractive index of light), and describing their chemical compositions, which govern the optical constants. Section 5.2.2 deals with the optical constants of mineral particles, and Sections 5.2.3 and 5.2.4 with the optical constants of various organic particles. Section 5.2.5 focuses on the shapes and sizes of particles, further characteristics that are essential for modeling their interactions with light.

The Chemical Composition and Types of Minerals Present in Particulate Inorganic Matter in the Sea

The main source of particulate inorganic matter in the sea is, as we stated in the previous section, the mineral fallout from the atmosphere and soil material carried to the sea by rivers in the form of silt of various grain sizes. It therefore consists of mineral particles, derived from the erosion and breakdown of rocks and carried off by the wind into the atmosphere or by flowing water into the sea. Physically and chemically, the composition of this material is very complex. Consisting chiefly of amorphous or crystalline particles of matter, it can be divided into a number of size classes (after their origin in soil), for example, clays, silty clays, and sand. It is a mixture of hundreds of minerals, variously classified,[13] and too numerous to mention here. The particles can be homogeneous or aggregates of many minerals.

[13] There are several systems of soil classification based on different criteria (type, composition, texture, etc.). The details of these classifications can be found in, for example, Soil Survey Staff (1996); see also Webb et al. (1981), Sokolik and Toon (1999).

The most common minerals (or groups of them) present both in atmospheric dust (see, e.g., Pye (1987) and Sokolik and Toon (1999)) and in sea water (see, e.g., Shifrin (1983b, 1988) and the papers cited therein) are the following: quartz, the feldspar group, calcite, dolomite, gypsum, the mica group, kaolinite, illite, montmorillonite, palygorskite, the chlorite group, and hematite. Table 5.7 gives their principal chemical composition, together with their densities and real parts of the absolute refractive index in the VIS region. It shows that silicon, aluminum, and iron are the most common elements in these minerals, usually in the form of their oxides: SiO_2, Al_2O_3, and Fe_2O_3. Also present are calcium, magnesium, sodium, potassium, sulfur, carbon, and hydrogen, not to mention a whole range of other elements present as common impurities. These compounds not infrequently occur in hydrated form.

The proportions of these minerals in the form of particulate matter in the sea vary in different waters and cannot be defined unequivocally with respect to the entire World Ocean. Presumably, however, these proportions are similar to the proportions typical of atmospheric dust in different parts of the world; Table 5.8 gives approximate values of the latter.

As can be seen from these data, quartz (mean 37.2%)—the principal constituent of soil—is predominant for obvious reasons. Substantial quantities of illite, kaolinite, montmorillonite, and calcite are also present, the first three especially in the clay-size fraction, (see column 10 of Table 5.8A and column 9 of Table 5.8B). Less common, or present in smaller amounts are chlorite, hematite, and gypsum. The proportions of the other numerous minerals are much smaller and do not usually exceed 10%.

Apart from the eight major minerals mentioned above, that enter the sea via atmospheric fallout, and whose presence in the sea has to be reckoned with, there is often one further significant group of PIM constituents in the sea, particularly in shelf regions, namely, minerals from the mica group, which are present together with quartz (Egan and Hilgeman 1979). They enter the sea with the sediment loads of rivers.

The Imaginary Component n' of the Absolute Complex Refractive Index

The imaginary part of refractive index n' of the natural mixtures of minerals that one can expect to come across in the sea varies strongly, depending on their composition. Figure 5.10 illustrates this: it presents spectra of the effective (bulk) values of the imaginary part n' of the absolute complex refractive index in the UV–VIS–IR region.

They refer to the atmospheric dust in various geographical regions, which is composed of various minerals in different proportions. This variety of composition means that the absolute values of n' for natural dust of diverse provenance can vary from three- to c. tenfold, depending on the spectral range under scrutiny. On the other hand, the shapes of the spectra of $n'(\lambda)$ are similar in all cases. So all these spectra in the IR range above 2.5 μm exhibit a whole series of absorption bands, the intensity of which increases

TABLE 5.7. The principal chemical constituents and some physical properties[a] of mineral dust derived from surface soils that can occur in the sea.[b]

		Chemical Properties			Physical Properties	
No.	Constituent	Formula	Elements	Common impurities	Density ρ [$\times 10^3$ kg m^{-3}]	VIS ref. index n [Dimensionless]
-1-	-2-	-3-	-4-	-5-	-6-	-7-
1	Calcite	$Ca(CO_3)$	Ca, C, O	Mn, Fe, Zn, Co, Ba, Sr, Pb, Mg, Cu, Al, Ni, V, Cr, Mo	2.71	1.486–1.660 (1.573)
2	Chlorite group	$(Mg, Al, Fe)_{12}$, $(Si, Al_8)\ O_{20}\ (OH)_{16}$	Mg, Al, Fe, Si, O, H	Ni, Mn, Zn, Li, Ti	2.6–3.3 (2.95)	1.57–1.67 (1.62)
3	Dolomite	$CaMg(CO_3)$	Ca, C, Mg, O	Fe, Mn, Co, Pb, Zn	2.85–2.94 (2.89)	1.500–1.681 (1.590)
4	Feldspar group	X, $Al_{(1\text{-}2)}$, $Si_{(2\text{-}3)}$, O_8, where X may be Na, K, Ca	Al, Si, O, Na, K, Ca	Various	2.55–2.76	1.54 ± 0.03
5	Gypsum	$Ca(SO_4)\ 2(H_2O)$	Ca, S, O, H	—	2.32	1.519–1.530 (1.525)
6	Hematite	Fe_2O_3	Fe, O	Ti, Al, Mn, H_2O	4.9–5.3 (5.26)	Selective (Figure 5.12 and Table 5.9A)
7	Illite	$(K, H_3O)\ (Al, Mg, Fe)_2$ ... $(Si,Al)_4O_{10}[(OH)_2\ (H_2O)]$	Al, H, K, O, Si	—	2.6–2.9 (2.75)	1.535–1.605 (1.570)
8	Kaolinite	$Al_2Si_2O_5(OH_4)$	Al, H, O, Si	Fe, Mg, Na, K, Ti, Ca, H_2O	2.63	1.553–1.570 (1.561)
9	Montmorillonite	$(Na, Ca)_{0.33}\ (Al, Mg)_2$ ... $Si_4O_{10}(OH)_2 \cdot nH_2O$	Al, Ca, H, Mg, Na, O, Si	Fe, K	2.00–2.70 (2.35)	1.485–1.550 (1.517)
10	Muscovite (mica group)	$KAl[AlSi_3O_{10})\ (OH)_2]$	Al, K, H, O, Si	Cr, Li, Fe, V, Mn, Na, Cs, Rb, Ca, Mg, H_2O	2.77–2.88 (2.83)	1.552–1.616 (1.584)
11	Palygorskite	$(Mg, Al)_5(Si, Al)_8$ $O_{20}\ (OH)_2 \cdot 8H_2O$	Al, H, Mg, O, Si	Fe, K	2.1–2.2 (2.15)	1.5–1.57 (1.535)
12	Quartz	$Si\ O_2$	Si, O	Various	2.60–2.66 (2.63)	1.543–1.554 (1.548)

[a] This list gives the most probable values and/or the range limits of the values, quoted from various sources, and the corresponding mean densities (ρ) (column 6) and real parts of the complex absolute visible-light refractive index (VIS ref. index n,, column 7) for different minerals. In the case of hematite (see item 6) the refractive index is not given, because it is strongly selective with respect to the light wavelength (see Figure 5.12 and Table 5.9A). The mean values are given in parentheses.

[b] The information on the chemical and physical properties of these minerals has been taken from many different sources, for example, Lide (1997), Kerr (1977), Berry and Mason (1959), and Egan and Hilgeman (1979), and from data available on the Internet, for example, http://www.mindat.org, http://www.minweb.co.uk, http://webmineral.com/data, http://en.wikipedia.org/wiki, http://mineral.galleries.com/minerals).

TABLE 5.8. Mineralogical composition of bulk dust samples (Table A) and clay-size mode particles (Table B) from different geographical regions.[a]

A. Bulk Dust Samples

No.	Constituent	Sal Island	Barbados	Miami	Southern New Mexico	Sudan	Nigeria (1)	Nigeria (2)	Mean
-1-	-2-	-3-	-4-	-5-	-6-	-7-	-8-	-9-	-10-
1	Quartz	19.6	13.8	14.2	6.0	56.3	83.3	67.4	37.2
2	Illite	53.8	64.3	62.9	19.0	3.8	1.17	5.6	30.1
3	Kaolinite	6.6	8.3	6.3	20.0	N/R	12.0	22.6	10.8
4	Calcite	8.2	3.9	6.9	16.0	7.3	N/R	N/R	6.0
5	Montmorillonite	N/R	N/R	N/R	35.0	N/R	N/R	N/R	5.0
6	Chlorite	4.3	4.1	4.2	N/R	N/R	N/R	N/R	1.8
7	Hematite	N/R	N/R	N/R	N/R	6.23	N/R	N/R	0.9
8	Gypsum	N/R	N/R	N/R	N/R	5.0	N/R	N/R	0.7
9	Others	7.5	5.4	5.5	4.0	21.4	3.5	4.4	7.4

N/R, data not reported.

B. Clay-Size Mode Profiles

No.	Constituent	Chad	Egypt	Saudi arabia	Mali	Barbados (Saharan dust)	Australia	Mean
-1-	-2-	-3-	-4-	-5-	-6-	-7-	-8-	-9-
1	Kaolinite	11	30	55[b]	67.8	32	61.8	42.9
2	Illite	87	15	5	24.6	41	38.2	35.4
3	Montmorillonite	2	55	40	2.7	16	0	19.3
4	Others	0	0	0	3.1	11.0	0	2.4

[a] Values are in units of % by weight (based on data from Kiefert et al. (1996), Delany et al. (1967), Ganor and Foner (1996), Hoidale and Blanco (1969), Sharif (1995), and Adedokun et al. (1989); adapted from Sokolik and Toon (1999).

[b] Values given for kaolinite together with palygorskite.

with the wavelength of the radiation. They are due to vibrational transitions and restricted rotations (librations) in the molecules of the main constituent mineral oxides, that is, SiO_2, Al_2O_3, and Fe_2O_3, among others, in their solid state, and also in part to analogous energy transitions in the water molecules incorporated in these minerals.

Strong UV absorption due to the electronic transitions in all these mineral components is also observed. The VIS absorption coefficients of these minerals, however, decrease with increasing light wavelength; the upshot is that in the longwave part of the VIS and the near-IR as far as c. 2.5 μm the spectra of the imaginary part of the refractive index $n'(\lambda)$ display a broad absolute minimum. This characteristic spectral variability of $n'(\lambda)$ for natural mixtures of minerals is, of course, due to the combined effect of the absorption properties of the various constituent minerals in these mixtures (see Figure 5.11).

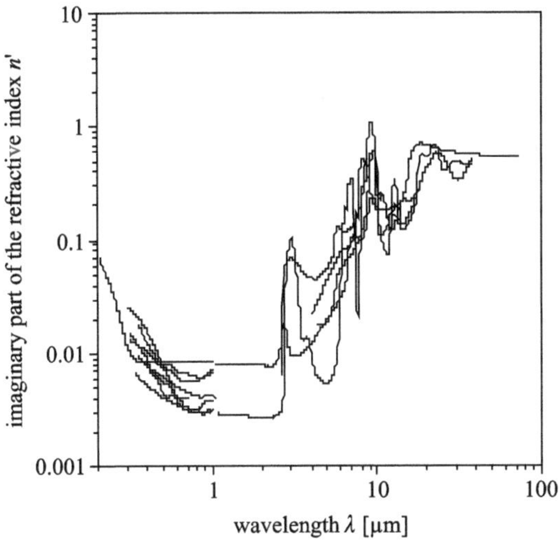

FIGURE 5.10. Spectra of the effective (bulk) values of the imaginary part n' of the absolute complex refractive index m in the wavelength range 200 nm–80 μm for atmospheric dust in different geographical regions. (Determined by various authors: Levin et al. (1980), Lindberg et al. (1976), Ivlev and Andreev (1986), Lindberg and Laude (1974), Patterson et al. (1977), Zuev and Krekov (1986), Volz (1973), and Carlson and Benjamin (1980); adapted from Sokolik et al. (1993).)

Figure 5.11a shows examples of individual spectra of $n'(\lambda)$ in the VIS and adjacent regions (200–1000 nm) for various minerals. We have limited ourselves here just to the most important of them, namely, the ones with the greatest influence on the absorption capacity of the total SPM. Having analyzed the available literature data regarding the optical constants of different minerals characteristic of visible light and adjacent regions of the electromagnetic spectrum (see, e.g., Querry et al. (1978), Egan and Hilgeman (1979), Peterson and Weinman (1969), and the data available at http://minerals.gps.caltech.edu/FILES/Visible/Index.htm, we found that of the very many minerals occurring naturally in the sea, the most important ones in this respect are the numerous varieties of quartz, along with the five minerals hematite, illite, montmorillonite, kaolinite, and mica.

The many natural varieties of quartz are very weak absorbers of visible light (see, e.g., Peterson and Weinman (1969)), with low values of $n'(\lambda)$; see plots I to IV in Figure 5.11a. That is why, for example, model calculations of the optical properties of discrete particles of pure quartz take account only of the real parts of the complex refractive index (i.e., $m(\lambda) \approx n(\lambda)$, but $n'(\lambda) \approx 0$). In other words, it is assumed that these particles do not absorb light at all; they only scatter it. Even so, quartz cannot be ignored when analyzing the effective (bulk) values of $n'(\lambda)$ for whole groups of PIM or for particles that

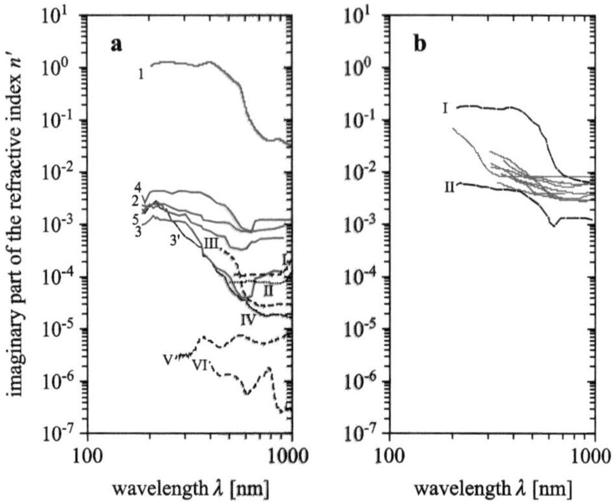

FIGURE 5.11. Example spectra of the imaginary part n' of the absolute complex refractive index m in the VIS range and adjacent regions: (a) pure forms of the most important minerals that may occur in suspended particles in the sea: 1. hematite, 2. illite, 3,3′. examples of two varieties of montmorillonite, 4. kaolinite, 5. mica, I–IV. different varieties of quartz; (b) particle aggregates: unbroken curves, atmospheric dusts in different geographical regions; broken curves (I, II), hypothetical dusts or particle aggregates containing hematite and illite in quantities corresponding to 15 and 85% of the particle volume (plot I), 0.3 and 99.7% of the particle volume (plot II). (Data taken from the following sources: curve 1 (a), Querry (1987), Querry et al. (1978); curves 2–5 (a), Egan and Hilgeman (1979); curves I–IV (a), calculations base of data from Internet (http://minerals.gps.caltech.edu/FILES/Visible/Index.htm); unbroken curves (b), various authors (see the caption to Figure 5.10); broken curves (b) authors' own calculations.)

are aggregates of quartz and other minerals: its presence in such associations or aggregates diminishes the effective (bulk) values of $n'(\lambda)$.

In contrast, the indices n' of the group of five minerals mentioned above are relatively high compared with those of other minerals (see Table 5.9B). Of these five, iron-rich hematite is the strongest absorber of light in the spectral range under scrutiny here. Its characteristic values of $n'(\lambda)$ are usually higher by two and more orders of magnitude than the corresponding values for the other minerals (see plot 1 in Figure 5.11a). So despite its relative rarity (with respect to other minerals; see Table 5.8), it has a marked effect both on the bulk values of n' for SPM in total and on the values of this index for the discrete particles that are aggregates of hematite and other minerals, even when their hematite content is quite low (see plots I and II in Figure 5.11b).

As regards the other four minerals in this group (illite, montmorillonite, kaolinite, and mica), their average values of $n'(\lambda)$ are higher than those of

TABLE 5.9. Examples of absolute complex refractive indices m (= $n - in'$) in the 185 nm $\leq \lambda \leq$ 1000 nm range for the most important mineral absorbers possibly occurring in the sea in the form of suspended particles.[a]

A. Real Part, *n* [Dimensionless]

Wavelength λ [nm]	Hematite	Illite	Montmorillonite	Kaolinite	Mica
-1-	*-2-*	*-3-*	*-4-*	*-5-*	*-6-*
185	—	1.448	1.544	1.491	1.478
190	—	1.444	1.543	1.494	1.478
200	—	1.441	1.542	1.496	1.478
210	1.222	1.438	1.541	1.498	1.478
215	1.328	1.434	1.540	1.501	1.478
220	1.429	1.431	1.539	1.503	1.477
225	1.508	1.427	1.539	1.506	1.477
233	1.629	1.424	1.538	1.508	1.477
240	1.724	1.420	1.537	1.511	1.477
260	1.947	1.417	1.536	1.513	1.459
280	2.114	1.411	1.534	1.506	1.456
300	2.214	1.401	1.523	1.514	1.447
325	2.327	1.404	1.523	1.512	1.437
360	2.334	1.406	1.524	1.509	1.438
370	2.381	1.415	1.524	1.500	1.438
400	2.657	1.423	1.525	1.490	1.437
433	2.917	1.420	1.525	1.491	1.436
466	3.051	1.418	1.526	1.492	1.435
500	3.083	1.415	1.526	1.493	1.433
533	3.094	1.414	1.524	1.493	1.431
566	3.162	1.412	1.522	1.493	1.429
600	3.065	1.411	1.520	1.493	1.429
633	2.949	1.407	1.522	1.494	1.429
666	2.870	1.403	1.523	1.496	1.430
700	2.815	1.399	1.525	1.497	1.430
817	2.701	1.395	1.527	1.499	1.430
907	2.668	1.391	1.528	1.501	1.430
1000	2.643	1.387	1.530	1.502	1.430

B. Imaginary Part, *n'* [Dimensionless]

Wavelength λ [nm]	Hematite	Illite	Montmorillonite	Kaolinite	Mica
-1-	*-2-*	*-3-*	*-4-*	*-5-*	*-6-*
185	—	2.24×10^{-3}	1.91×10^{-3}	9.55×10^{-4}	1.66×10^{-3}
190	—	2.19×10^{-3}	1.66×10^{-3}	1.05×10^{-3}	1.91×10^{-3}
200	—	2.29×10^{-3}	2.09×10^{-3}	1.20×10^{-3}	2.40×10^{-3}
210	1.13	2.34×10^{-3}	2.19×10^{-3}	1.41×10^{-3}	2.04×10^{-3}
215	1.16	2.40×10^{-3}	2.24×10^{-3}	1.26×10^{-3}	2.19×10^{-3}
220	1.19	2.69×10^{-3}	2.57×10^{-3}	1.38×10^{-3}	2.19×10^{-3}
225	1.20	2.40×10^{-3}	2.04×10^{-3}	1.23×10^{-3}	2.34×10^{-3}
233	1.22	2.34×10^{-3}	1.95×10^{-3}	1.20×10^{-3}	2.29×10^{-3}
240	1.23	2.45×10^{-3}	1.91×10^{-3}	1.23×10^{-3}	1.78×10^{-3}
260	1.22	2.04×10^{-3}	1.17×10^{-3}	1.17×10^{-3}	1.58×10^{-3}

(*Continued*)

TABLE 5.9. Examples of absolute complex refractive indices m (= $n - in'$) in the 185 nm ≤ λ ≤ 1000 nm range for the most important mineral absorbers possibly occurring in the sea in the form of suspended particles.[a] — Cont'd.

B. Imaginary Part, n' [Dimensionless]

Wavelength λ [nm]	Hematite	Illite	Montmorillonite	Kaolinite	Mica
-1-	-2-	-3-	-4-	-5-	-6-
280	1.19	1.86×10^{-3}	8.32×10^{-4}	1.17×10^{-3}	1.48×10^{-3}
300	1.16	1.82×10^{-3}	5.89×10^{-4}	1.07×10^{-3}	1.51×10^{-3}
325	1.08	1.66×10^{-3}	4.68×10^{-4}	8.13×10^{-4}	1.07×10^{-3}
360	1.11	1.45×10^{-3}	3.55×10^{-4}	3.98×10^{-4}	9.33×10^{-4}
370	1.16	1.23×10^{-3}	2.51×10^{-4}	2.75×10^{-4}	8.32×10^{-4}
400	1.22	1.17×10^{-3}	2.04×10^{-4}	2.04×10^{-4}	7.24×10^{-4}
433	1.03	1.07×10^{-3}	1.23×10^{-4}	1.45×10^{-4}	6.17×10^{-4}
466	8.10×10^{-01}	1.07×10^{-3}	1.07×10^{-4}	1.23×10^{-4}	5.62×10^{-4}
500	6.19×10^{-01}	1.02×10^{-3}	5.25×10^{-5}	9.55×10^{-5}	3.72×10^{-4}
533	5.05×10^{-01}	8.32×10^{-4}	4.27×10^{-5}	5.25×10^{-5}	3.47×10^{-4}
566	3.58×10^{-01}	7.08×10^{-4}	3.39×10^{-5}	3.89×10^{-5}	3.31×10^{-4}
600	1.48×10^{-01}	7.08×10^{-4}	3.63×10^{-5}	3.80×10^{-5}	3.39×10^{-4}
633	8.97×10^{-02}	6.92×10^{-4}	4.47×10^{-5}	4.17×10^{-5}	4.68×10^{-4}
666	6.48×10^{-02}	1.15×10^{-3}	9.33×10^{-5}	9.33×10^{-5}	4.57×10^{-4}
700	5.11×10^{-02}	1.20×10^{-3}	9.77×10^{-5}	1.00×10^{-4}	5.01×10^{-4}
817	4.05×10^{-02}	1.23×10^{-3}	1.29×10^{-4}	1.29×10^{-4}	5.50×10^{-4}
907	3.83×10^{-02}	1.20×10^{-3}	1.26×10^{-4}	1.23×10^{-4}	5.50×10^{-4}
1000	2.91×10^{-02}	1.23×10^{-3}	1.58×10^{-4}	1.58×10^{-4}	6.03×10^{-4}

[a] Data selected from Querry et al. (1978; hematite) and Egan and Hilgeman (1979; illite, montmorillonite, kaolinite and mica).

most other minerals, even though they absorb light far less strongly than hematite does (see plots 2–5 in Figure 5.11a). They are also quite common in natural associations of suspended particulate matter in the sea. This is the reason why we feel that their direct involvement in the absorption of light from the visible and neighboring regions of the spectrum by the totality of SPM should not be neglected (the same applies to quartz, which in large measure affects these absorptions only indirectly).

It is worth drawing attention to the fact that among the spectra of $n'(\lambda)$ for pure minerals shown here (see Figure 5.11a) there is not one that lies within the natural diversity of the effective (bulk) value of $n'(\lambda)$ characteristic of the mineral associations naturally present in atmospheric dusts (see Figure 5.11b, unbroken plots). This is because values of n' for hematite are usually from a few to over a dozen times higher than n' for atmospheric dust. On the other hand, the values of n' in the visible and adjacent regions of the spectrum for the other pure minerals are usually lower (in some cases by several orders of magnitude) than n' for natural dusts. This means that all natural atmospheric dusts (and hence the natural associations of particulate matter in the sea) are heterogeneous in that they consist of mixtures of different minerals.

Nevertheless, they always contain one mineral that elevates their absorption capabilities—hematite—or, less commonly, other iron-rich constituents, for instance, goethite, which contains FeOOH. Because iron is a strong absorber, this admixture need not be large, as the positions of the $n'(\lambda)$ spectra in Figure 5.11b testify; these were calculated for hypothetical dusts containing Fe_2O_3 (i.e., hematite) in the following proportions: 15% by volume (broken curve I) and 0.3% by volume (broken curve II). The distance between these two curves I and II covers and even exceeds by far the natural range of diversity of n' for natural dust. The connections linking the absorption capabilities of natural associations of mineral particles with their hematite content, or directly with their iron content, have been analyzed in much greater detail by, for example, Sokolik and Toon (1999), Babin and Stramski (2004), and Marra et al. (2005).

The Real Part n of the Absolute Complex Refractive Index

The great majority of natural minerals (with the exception of hematite and a number of rare but very iron-rich minerals; see below) display very similar values of the real part n of the absolute refractive index in the range from the near-UV to the near-IR from $\lambda < 2.5$ μm (see, e.g., Kerr (1977), Egan and Hilgeman (1979), and S.B. Woźniak and Stramski (2004)). Moreover, in this spectral range, the values of n are practically independent of the wavelength λ. Figure 5.12 (plots 2–5) presents example spectra of $n(\lambda)$ for some of the principal minerals occurring in atmospheric dusts and in particulate

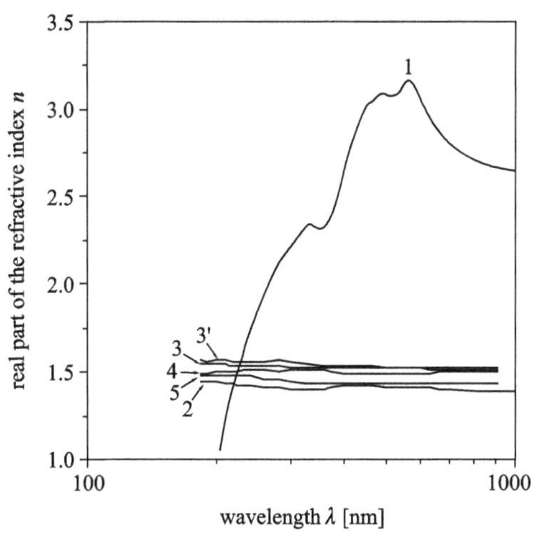

FIGURE 5.12. Example spectra of the real part n of the absolute complex refractive index m in the VIS range and adjacent regions for pure forms of the most important minerals that may occur in suspended particles in the sea: 1. hematite; 2. illite; 3 and 3′. examples of two varieties of montmorillonite; 4. kaolinite; 5. mica. (Data taken from: plot 1. Querry (1987), Querry et al. (1978); plots 2–5. Egan and Hilgeman (1979).)

mineral matter in the sea, and Table 5.9A (columns 3–6) provides selected values of n for these minerals in the spectral range from 185 nm to 1 μm.

A review of the literature data shows that the mean VIS refractive index n for all the minerals (except hematite) represented in Table 5.7 is $n \approx 1.58$, the amplitude of values ranging from $n \approx 1.48$ to $n \approx 1.68$. This range of values of n does not appear to reflect the actual limits of this diversity; it is more likely the result of imperfections in the various techniques for measuring the refractive index. In reality, therefore, this range of natural diversity is probably narrower. Some authors have linked the value of n of different minerals with their density ρ. For instance, S.B. Woźniak and Stramski (2004) have drawn attention to the general tendency for ρ to increase with n_{rel}, which is described by the following regression:

$$n_r = 0.1475\ 10^{-6}\ \rho + 0.7717,$$

where $n_r = n/1.34$ is the real part of the relative refractive index of a mineral with respect to water (1.34 is a typical value of the real part of the absolute refractive index for water); the density ρ is expressed in [g m^{-3}].

As far as the values of n and their variability with light wavelength are concerned, among the natural minerals possessing similar features, hematite and other, rarer, but iron-rich minerals stand out. One of the heaviest of the natural constituents of soil, hematite ($\rho \approx 5.26\ 10^3$ kg m^{-3}) exhibits high values of $n(\lambda)$, selective with respect to wavelength, with a maximum of $n_{max} \approx 3.16$ in the visible light region at $\lambda \approx 566$ nm. Plot 1 in Figure 5.12 shows the spectrum of $n(\lambda)$ for hematite, and Table 5.9A (column 2) sets out its spectral values in the range from 210 nm to 1 μm.

5.2.3 The Chemical Composition and Optical Constants of the Planktonic Components of Organic Particles

Now that we have discussed the chemical compositions and optical constants of mineral particles in the sea, we can move on to describe the present state of knowledge regarding the properties of organic particles occurring in sea water. This problem is a highly complex one, and the material we present in this context is copious. We therefore deal with it in two parts: Section 5.2.3 concentrates on the chemical and biochemical compositions and the optical constants of living particles (mainly phyto- and zooplankton), whereas Section 5.2.4 focuses on the optical constants of organic detritus.

Major Metabolites of Marine Organisms and Their Optical Constants

The chemical composition of living organisms is highly complex: the thousands of different organic and inorganic substances they contain (apart from water) possess very diverse physical and chemical properties, and fulfill a multitude of biological functions in them. To go into the fine detail of this issue, however, is exceedingly difficult and would go beyond the scope of this volume. Matters become somewhat simpler if these metabolites, as they are known, are divided into a few groups of structurally related compounds. Hand in hand with the chemical similarities of these compounds go the similarities in their physical properties, among others, their densities and optical constants.

The most important metabolites of marine organisms are proteins, carbohydrates, fats (lipids), and, where plants are concerned, pigments: the chlorophylls, carotenoids, and phycobilins (biliprotenoids) (Parsons et al. 1977, Romankevich 1977). Apart from these organic constituents, certain inorganic compounds also play an essential part in organisms by lending rigidity to their structures: the skeletons of animals, and the multifarious shells and scales of some of the phytoplankton, especially the classes *Bacillariophyceae* and *Prymnesiophyceae* or *Haptophyceae* (coccolithophorids, in particular). These compounds are usually two types of minerals: calcium carbonate $CaCO_3$ in the form of amorphous calcite or anhydrous crystalline, and silicon dioxide SiO_2, mostly in the form of hydrated amorphous silica (i.e., $SiO_2 \cdot nH_2O$), otherwise known as opal.

The most important physical properties of the principal organic metabolites (proteins, carbohydrates, fats, and pigments) and inorganic compounds (calcite, silica, and water) relevant to this description are set out in Tables 5.10 and 5.11 (see also Figure 5.13). They are the density of these compounds in the dry state ρ (column 4 in Table 5.10), and the real n (column 5 in Table 5.10) and

TABLE 5.10. Marine organisms: Principal chemical constituents and selected physical properties.

			Physical properties		
No.	Constituent	Chemical formula	Density ρ^a $[10^3 kg\ m^{-3}]$	VIS real ref. index $n^{(b)}$ [Dimensionless]	Imag ref. index n^b [Dimensionless] at the spectral band peak λ_{max} [nm]
-1-	*-2-*	*-3-*	*-4-*	*-5-*	*-6-*
1	**Algal Pigments: In Total**		**1.12 ± 0.06**	**1.50 ± 0.04**	**1.77 (440 nm)**
2	*-Chlorophylls a*	$(C_{55}H_{72}MgN_4O_5)$	1.11	1.52	2.72 (436 nm)
3	*-Chlorophylls b*	$(C_{55}H_{70}MgN_4O_6)$	1.13	1.52	3.51 (453 nm)
4	*-Chlorophylls c*	$(C_{35}H_{28\text{ or }30}MgN_4O_5)$	1.31	1.54	3.47 (460 nm)
5	*-α-Carotene*	$(C_{40}H_{56})$	~1.00	1.451	—
6	*-β-Carotene*	$(C_{40}H_{56})$	~1.00	1.453	—
7	*-Xanthophylls*	$(C_{40}H_{56}O_2)$	~1.06	1.448	—
8	*-Photosynthetic carotenoids* (in total)	$(C_{40}H_{56}$ and $C_{40}H_{56}O_2)$	(~1.03)	(1.451)	1.43 (494 nm)
9	*-Photoprotecting carotenoids* (in total)	$(C_{40}H_{56}$ and $C_{40}H_{56}O_2)$	(~1.03)	(1.451)	3.27 (464 nm)

(Continued)

TABLE 5.10. Marine organisms: Principal chemical constituents and selected physical properties.—Cont'd.

			Physical Properties		
No.	Constituent	Chemical formula	Density ρ^a [10^3kg m^{-3}]	VIS real ref. index $n^{(a)}$ [Dimensionless]	Imag ref. index n'^b [Dimensionless] at the spectral band peak λ_{max} [nm]
-1-	*-2-*	*-3-*	*-4-*	*-5-*	*-6-*
10	**Proteins: in Total**		**1.22 ± 0.02**	**1.57 ± 0.01**	**6.18×10^{-2} (220 nm)**
11	**Carbohydrates: In Total**		**1.53 ± 0.04**	**1.55 ± 0.02**	~0 (Weak absorbing)
12	*-Cellulose, crystalline*	$[(C_6H_{10}O_5)_n]$	1.27–1.61	1.55	”
13	*-Cellulose, amorphous*	$[(C_6H_{10}O_5)_n]$	1.482–1.489	—	”
14	*-Cellulose, fibers*	$[(C_6H_{10}O_5)_n]$	1.48–1.55	1.563–1.573	”
15	*-Starch*	$[(C_6H_{10}O_5)_n]$	1.50–1.53	1.51–1.56	”
16	*-Sucrose*	$(C_{12}H_{22}O_{11})$	1.581–1.588	1.538–1.560	”
17	*-Glucose*	$(C_6H_{12}O_6)$	1.544–1.562	1.55	”
18	**Fats (Lipids): In Total**		**0.93 ± 0.02**	**1.47 ± 0.01**	~0 (Weak absorbing)
19	**Silica: In Total**		**2.07 ± 0.1**	**1.48 ± 0.07**	~0 (Weak absorbing)
20	*-Opal*	$(SiO_2.nH_2O)$	1.73–2.20	1.406–1.46	”
21	*-Opal in diatoms*	$(SiO_2.nH_2O)$	2.00–2.70	1.42–1.486	”
22	*-Vitreous quartz*	(SiO_2)		1.485	”
23	*-Crystal quartz*	(SiO_2)	2.64–2.66	1.547	”
24	**Calcite: In Total**		**2.71 ± 0.1**	**1.60 ± 0.01**	~0 (Weak absorbing)
25	*-Calcite*	$(CaCO_3)$	2.71–2.94	1.601–1.677	”
26	*-Aragonite*	$(CaCO_3)$	2.93–2.95	1.632–1.633	”
27	**Oceanic water**	(H_2O) + 3.5% salt	**1.025**	**1.339**	See Table 5.11

[a] Mean or typical values (or ranges of variability): the densities of dry matter of the various constituents ρ (column 4) and the real parts n of the complex absolute visible-light refractive index (column 5) were gleaned mainly from Aas (1996) and are based on the research findings of numerous authors, e.g., Ketelaar and Hanson (1937), Weast (1981), Euler and Jansson (1931), Barer (1966), Ross (1967), Armstrong et al. (1947), Oncley et al. (1947), McCrone et al. (1967), Stecher (1968), Treiber (1955), Davies et al. (1954), Gibbs (1942), Chamot and Mason (1944), Forsythe (1954), Levin (1962 a,b), Jenkins and White (1957), and Hodgson and Newkirk (1975).
[b] The values of the imaginary part of the absolute refractive index n' are typical of the pigments and groups of pigments in oceanic phytoplankton and of the proteins in marine plankton; they were derived by the authors (see Woźniak et al. (1999, 2005a,b)). The most typical values are indicated by the bold print.

imaginary n' (Table 5.11) parts of the absolute complex refractive index of the various forms of these major constituents of organisms. Determined by various authors using a variety of techniques (see Aas (1996) and the papers cited therein, and Woźniak et al. (1999,2005a,b)), they must be taken into consideration in any model of the optical properties of the planktonic

TABLE 5.11. Imaginary part n' [dimensionless] of refractive index of light in range 200–700 nm for important absorbers present in marine organisms.[a,b]

Wavelength λ [nm]	Chl*a*	Chl*b*	Chl*c*	PSC	PPC	Pig-ocean	Pr	Wt
-1-	*-2-*	*-3-*	*-4-*	*-5-*	*-6-*	*-7-*	*-8-*	*-9-*
200	—	—	—	—	—	—	—	4.89×10^{-08}
210	—	—	—	—	—	—	5.53×10^{-02}	3.33×10^{-08}
220	—	—	—	—	—	—	6.18×10^{-02}	2.29×10^{-08}
230	—	—	—	—	—	—	2.73×10^{-02}	1.70×10^{-08}
240	—	—	—	—	—	—	2.27×10^{-02}	1.38×10^{-08}
250	—	—	—	—	—	—	2.33×10^{-02}	1.11×10^{-08}
260	—	—	—	—	—	—	2.37×10^{-02}	9.46×10^{-09}
270	—	—	—	—	—	—	2.47×10^{-02}	8.02×10^{-09}
280	—	—	—	—	—	—	2.39×10^{-02}	6.42×10^{-09}
290	—	—	—	—	—	—	1.92×10^{-02}	4.96×10^{-09}
300	—	—	—	—	—	—	1.70×10^{-02}	3.37×10^{-09}
350	6.71×10^{-01}	1.83×10^{-01}	3.11×10^{-03}	1.12×10^{-14}	1.25×10^{-02}	—	1.34×10^{-02}	5.68×10^{-10}
380	1.12	2.02×10^{-01}	4.90×10^{-01}	1.15×10^{-09}	1.68×10^{-01}	—	1.15×10^{-02}	3.44×10^{-10}
400	1.06	2.11×10^{-01}	2.07	1.03×10^{-06}	5.82×10^{-01}	7.13×10^{-01}	1.04×10^{-02}	2.11×10^{-10}
410	1.31	2.15×10^{-01}	2.40	1.90×10^{-05}	9.35×10^{-01}	9.09×10^{-01}	9.87×10^{-03}	1.54×10^{-10}
420	1.71	2.25×10^{-01}	2.19	2.51×10^{-04}	1.36	1.18	9.37×10^{-03}	1.52×10^{-10}
430	2.18	3.76×10^{-01}	2.04	2.37×10^{-03}	1.80	1.51	8.88×10^{-03}	1.69×10^{-10}
440	2.38	1.06	1.12	1.59×10^{-02}	2.17	1.77	8.42×10^{-03}	2.22×10^{-10}
450	4.93×10^{-01}	3.29	4.35×10^{-01}	7.60×10^{-02}	2.58	1.43	7.97×10^{-03}	3.30×10^{-10}
460	1.10×10^{-01}	2.55	3.63×10^{-01}	2.58×10^{-01}	3.19	1.43	7.55×10^{-03}	3.59×10^{-10}
470	6.39×10^{-02}	2.41	2.78×10^{-01}	6.27×10^{-01}	3.09	1.42	7.14×10^{-03}	3.97×10^{-10}
480	4.49×10^{-02}	1.63	1.38×10^{-01}	1.09	2.84	1.30	6.75×10^{-03}	4.85×10^{-10}
490	3.95×10^{-02}	5.89×10^{-01}	5.01×10^{-02}	1.40	3.02	1.25	6.38×10^{-03}	5.85×10^{-10}
500	4.19×10^{-02}	2.58×10^{-01}	3.21×10^{-02}	1.38	2.41	1.03	6.03×10^{-03}	8.12×10^{-10}
510	4.87×10^{-02}	2.18×10^{-01}	5.49×10^{-02}	1.17	1.19	6.54×10^{-01}	5.70×10^{-03}	1.32×10^{-09}

(*Continued*)

TABLE 5.11. Imaginary part n' [dimensionless] of refractive index of light in range 200–700 nm for important absorbers present in marine organisms.[a,b]—Cont'd.

Wavelength λ [nm]	Chl*a*	Chl*b*	Chl*c*	PSC	PPC	Pig-ocean	Pr	Wt
-1-	*-2-*	*-3-*	*-4-*	*-5-*	*-6-*	*-7-*	*-8-*	*-9-*
520	5.81×10^{-02}	2.13×10^{-01}	1.06×10^{-01}	1.01	4.21×10^{-01}	4.11×10^{-01}	5.38×10^{-03}	1.69×10^{-09}
530	6.95×10^{-02}	2.09×10^{-01}	1.90×10^{-01}	9.28×10^{-01}	1.48×10^{-01}	3.22×10^{-01}	5.08×10^{-03}	1.83×10^{-09}
540	8.25×10^{-02}	2.04×10^{-01}	3.07×10^{-01}	8.27×10^{-01}	5.94×10^{-02}	2.85×10^{-01}	4.79×10^{-03}	2.04×10^{-09}
550	9.70×10^{-02}	1.99×10^{-01}	4.51×10^{-01}	6.44×10^{-01}	2.38×10^{-02}	2.45×10^{-01}	4.52×10^{-03}	2.47×10^{-09}
560	1.13×10^{-01}	1.97×10^{-01}	6.02×10^{-01}	4.20×10^{-01}	8.77×10^{-03}	2.03×10^{-01}	4.26×10^{-03}	2.76×10^{-09}
570	1.30×10^{-01}	2.09×10^{-01}	7.28×10^{-01}	2.27×10^{-01}	2.93×10^{-03}	1.70×10^{-01}	4.02×10^{-03}	3.15×10^{-09}
580	1.48×10^{-01}	2.65×10^{-01}	8.01×10^{-01}	1.01×10^{-01}	8.89×10^{-04}	1.56×10^{-01}	3.79×10^{-03}	4.14×10^{-09}
590	1.68×10^{-01}	3.95×10^{-01}	7.99×10^{-01}	3.72×10^{-02}	2.44×10^{-04}	1.65×10^{-01}	3.57×10^{-03}	6.35×10^{-09}
600	1.87×10^{-01}	5.67×10^{-01}	7.24×10^{-01}	1.13×10^{-02}	6.09×10^{-05}	1.83×10^{-01}	3.36×10^{-03}	1.06×10^{-08}
610	2.07×10^{-01}	6.89×10^{-01}	5.95×10^{-01}	2.82×10^{-03}	1.38×10^{-05}	1.95×10^{-01}	3.16×10^{-03}	1.28×10^{-08}
620	2.27×10^{-01}	7.60×10^{-01}	4.44×10^{-01}	5.84×10^{-04}	2.82×10^{-06}	2.07×10^{-01}	2.98×10^{-03}	1.36×10^{-08}
630	2.47×10^{-01}	9.31×10^{-01}	3.01×10^{-01}	9.95×10^{-05}	5.24×10^{-07}	2.29×10^{-01}	2.80×10^{-03}	1.46×10^{-08}
640	2.65×10^{-01}	1.28	1.85×10^{-01}	1.40×10^{-05}	8.82×10^{-08}	2.67×10^{-01}	2.63×10^{-03}	1.58×10^{-08}
650	2.98×10^{-01}	1.60	1.03×10^{-01}	1.62×10^{-06}	1.35×10^{-08}	3.15×10^{-01}	2.48×10^{-03}	1.76×10^{-08}
660	5.52×10^{-01}	1.60	5.24×10^{-02}	1.56×10^{-07}	1.87×10^{-09}	4.17×10^{-01}	2.33×10^{-03}	2.15×10^{-08}
670	1.33	1.23	2.41×10^{-02}	1.23×10^{-08}	2.34×10^{-10}	6.84×10^{-01}	2.19×10^{-03}	2.34×10^{-08}
675	1.53	9.85×10^{-01}	1.58×10^{-02}	3.22×10^{-09}	8.01×10^{-11}		2.12×10^{-03}	2.41×10^{-08}
680	1.35	7.42×10^{-01}	1.01×10^{-02}	8.02×10^{-10}	2.67×10^{-11}	6.45×10^{-01}	2.06×10^{-03}	2.52×10^{-08}
690	5.94×10^{-01}	3.71×10^{-01}	3.83×10^{-03}	4.32×10^{-11}	2.76×10^{-12}	2.98×10^{-01}	1.93×10^{-03}	2.83×10^{-08}
700	3.50×10^{-01}	1.80×10^{-01}	1.32×10^{-03}	1.92×10^{-12}	2.58×10^{-13}	1.66×10^{-01}	1.82×10^{-03}	3.48×10^{-08}

[a] Chl*a*: chlorophylls *a*; Chl*b*: chlorophylls *b*; Chl*c*: chlorophylls *c*; PSC: photosynthetic carotenoids, PPC: photoprotecting carotenoids; Pig-ocean: the set of all pigments typical for the oceanic phytoplankton; Pr: a set of proteins typical for the plankton; Wt: water.
[b] After Woźniak et al. (1999, 2005a,b) data.

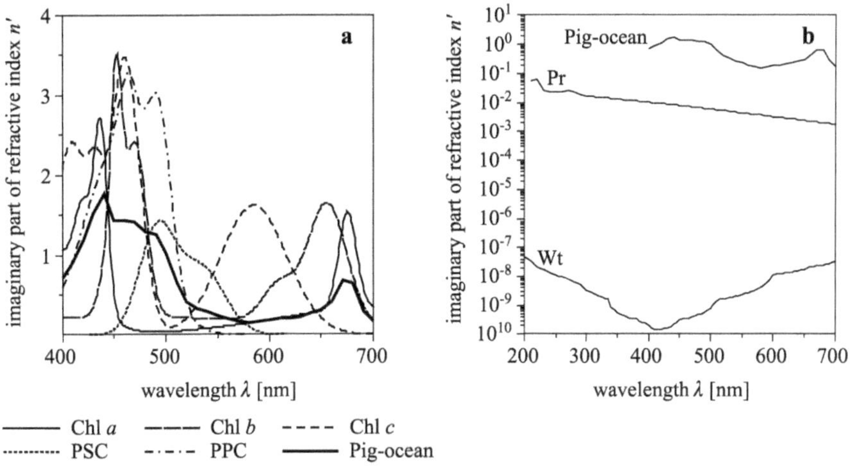

FIGURE 5.13. Spectra of the imaginary part of the absolute complex refractive index for selected major constituents of marine organisms: (a) chlorophylls *a* (Chl *a*), chlorophylls *b* (Chl *b*), chlorophylls *c* (Chl *c*), photosynthetic carotenoids (PSC), photoprotecting carotenoids (PPC), the set of all pigments occurring in the proportions typical of oceanic phytoplankton (Pig-ocean); (b) the set of all pigments (as above – Pig-ocean), proteins (Pr) and water (Wt).

components of particulate matter. Because the real indices n of these compounds practically do not change with light wavelength (Aas 1996), we give just one value of this parameter for each compound, typical of the VIS range and adjacent regions of the light spectrum.

In contrast, the imaginary indices n' of these compounds are mostly strongly selective with respect to light wavelength. In this case, therefore, apart from the values of n' given in Table 5.10, we have also plotted their changes over a broad spectral range (see Table 5.11 and Figure 5.13), determined on the basis of the known densities of these compounds ρ (Table 5.10, column 4) and their spectral mass-specific absorption coefficients of light a^* (see earlier Equation (4.3) and Tables 4.3 and 4.9 for proteins, and also Equation (6.3) and Table 6.8 for algal pigments) using the well-known relationship $n' = a^*\rho\lambda/4\pi$. We give only the values of $n'(\lambda)$ for algal pigments and proteins, inasmuch as the level of light absorption by the other principal constituents is very low and can be safely ignored when modeling the optical properties of suspended particulate matter in the sea.

The reader should bear in mind that all the numerical data in Tables 5.10 and 5.11 are approximate, particularly those referring to plant pigments; very little information about their physical properties, for example, ρ, n, and n', is given in the literature. We ourselves calculated the values of n' for the pigments in these tables from their mass-specific absorption coefficients (also determined by us). The values of these absorption coefficients (see Figure 6.11 in Chapter 6), and hence the values of n' given by other authors differ somewhat from those stated here; our values of n' should therefore be treated as examples.

As the data in Table 5.10 (column 5) show, the real parts of the refractive indices of all the major organic and inorganic constituents of marine organisms are fairly similar, from c. 1.47 (fats) to c. 1.60 (calcite). Nevertheless, these values are 10–20% greater than the corresponding value n_w for water (e.g., for seawater of salinity 35 psu, at a temperature of 20°C, and for a light wavelength of 589 nm, $n_w = 1.339$). Their refractive indices relative to water ($n_r = n/n_w$) thus range from 1.10 to 1.19.

The situation with respect to n', the imaginary part of absolute complex refractive index of these principal constituents, is quite different; its values vary widely (see Tables 5.10 (column 6) and 5.11). Plant pigments exhibit the very highest values of n' for VIS light: around 1 and more, they exceed the values of n' of the other constituents by several, sometimes as many as ten, orders of magnitude. Pigments are very strong VIS absorbers because of the multiple π, electron conjugations forming in their molecules (see Chapter 3, e.g., Sections 3.2.1 and 3.2.3) These conjugated π–electron bonds are responsible for the complex absorption bands, differently shaped for different pigments; Figure 5.13a illustrates this variability with regard to the spectra of $n'(\lambda)$. The structure of the absorption spectra of these pigments is described in Chapter 6.

Almost all the other important constituents of organisms are very weak absorbers of light and can be omitted from models of the optical properties of particulate suspended matter; their values of n' are assumed to be zero. Proteins are the exception here. It is true that their values of n' are quite low in the VIS region, but they rise with diminishing light wavelength, reaching relatively high values (10^{-2}–10^{-1}) in the far UV (see Figure 5.13b). This is due to the presence in proteins of amino acids, especially the aromatic ones (tryptophan, tyrosine, and phenylalanine), which are very strong UV absorbers. We discussed this in detail in Chapter 4 (Section 4.2.1).

The Biochemical Composition of Marine Organisms

The bulk optical constants of marine organisms are governed not only by the individual optical constants of their constituents, but also by the proportions between the latter. These proportions differ widely with respect even to the same plant or animal species living under different conditions in the sea; they also depend on the age of the organism. Now, little research effort has been devoted to these complexities, so we are unable to quote any mean values or ranges of variability of these proportions, either for particular species, or for the higher taxonomic units to which they are assigned. We can speak only about typical proportions. Table 5.12 gives such typical proportions of the constituents (the mean biochemical composition) of different marine

TABLE 5.12. The biochemical composition of marine organisms.

	Relative content as % of the mass												
	Wet matter			Dry matter					Organic matter				
Organism	Water[a]	Inorganic matter	Organic matter	Inorganic matter	Proteins	Carbo-hydrates	Fats (Lipids)	Pigments	Proteins	Carbohydrates	Fats (lipids)	Pigments	References
-1-	*-2-*	*-3-*	*-4-*	*-5-*	*-6-*	*-7-*	*-8-*	*-9-*	*-10-*	*-11-*	*12-*	*-13-*	*-14-*
Phytoplankton	**80**	**9**	**11**	45	29	20	5	1	53	36	9	2	(1′)
-Bacillariophyceae				36	41.2	14.3	6.4	2.2	64.4	22.3	9.9	3.4	(2)
				1–67 (Silica only)	21–64	1–36	2–13	1–4	39–88	1.5–43	5.7–21	1.8–6	
-Dinophyceae									65.0	23.5	9.5	2	(3)
									35–84	8–45	3–23	1–3	
-Prymnesiophyceae (coccolithophorids)				23 (Calcite only)	54	17	5	1	54	17	5	1	(4)
-Cyanobacteria									44	38	16	2	(5)
-Chlorophyceae									39.1	37.4	20.8	2.7	(6)
									15–58	21–49	4–34	2–3	
-Rhodophyceae									34.5	53	10.5	2	(7)
									26–40	45–65	8–14	1–3	
Phytobenthos	**80**	**5**	**15**	**25**	**13.5**	**58.5**	**0.5**	**2.5**	**18**	**78**	**1**	**3**	(1′)
Zooplankton	**80**	**2**	**18**	**10**	**60**	**15**	**15**	**~0**	**67**	**16**	**17**	**~0**	(1)
Zoobenthos	**63**	**23**	**14**	**62**	**27**	**8**	**3**	**~0**	**71**	**21**	**8**	**~0**	(1)
Nekton	**73**	**3**	**24**	**11**	**70**	**4**	**15**	**~0**	**78**	**5**	**17**	**~0**	(1)

References: (1) in accordance with the assessment of Romankevich (1977); (1′) in accordance with the assessment of Romankevich (1977) and modified by the authors to take account of the presence of plant pigments; (2) means and ranges of variability estimated by the authors on the basis of 18 datasets for various species; gleaned from Brandt and Raben (1920), Parsons et al. (1961), McAllister et al. (1961), and Antia et al. (1963); (3) means and ranges of variability estimated by the authors on the basis of 9 datasets for various species; gleaned from Brandt and Raben (1920) and Parsons et al. (1961); (4) example data for *Syracosphaera carierae*, after Parsons et al. (1961); (5) example data for *Agmenellum quadruplicatum*, after Parsons et al. (1961); (6) means and ranges of variability estimated by the authors on the basis of 8 datasets for various species; gleaned from Parsons et al. (1961), Ketchum and Redfield (1949), and Collyer and Fogg (1955); (7) means and ranges of variability estimated by the authors on the basis of 4 datasets for various species; gleaned from Collyer and Fogg (1955).

[a] typical quantities of water contained in organisms (in accordance with the assessment of Romankevich (1977)). The most typical values are indicated by the bold print.

organisms (phytoplankton, phytobenthos, zooplankton, zoobenthos, and nekton), in accordance with the assessment of Romankevich (1977). In view of the particular significance of marine algae, the strongest light absorbers of all organisms, Table 5.12 also provides the corresponding data for some of the most important classes of algae, namely, the *Bacillariophyceae*, *Dinophyceae*, *Prymnesiophyceae*, *Cyanobacteria*, *Chlorophyceae*, and *Rhodophyceae*. These data are based on experimental material obtained by numerous authors (see the footnotes to this table), collected and analyzed by Aas (1996).

With the aid of the data in Table 5.12 and the optical constants of the individual constituents of marine organisms given in Tables 5.10 and 5.11, it is possible to estimate the values of their real n and imaginary n' refractive indices. Aas (1996), among others, calculated n for phytoplankton, and we ourselves (Woźniak et al. 2005a,b, 2006) calculated values of n' for different organisms and for organic detritus. We discuss some of these values in this and the next section.

The Real Part of the Bulk Absolute Complex Refractive Index

The bulk refractive index n_p of particles of matter made up of several constituents (e.g., live organisms) can be estimated from known relative volumes v_i of the constituents of the particles and their individual refractive indices n_i, in accordance with the relationship:

$$n_p = \sum_i n_i v_i. \tag{5.10}$$

It is assumed that these constituents are reasonably well mixed, in other words, that the particle is isotropic. Equation (5.10) is easily reduced to a form describing the relationship between this bulk index n_p for a particle and parameters, such as those given in Tables 5.10 and 5.12, that is, the density of the constituents ρ_i, their individual refractive indices n'_i and their relative concentrations μ_i in the particle (in this case, in the organism). This relationship takes the form:

$$n_p = \rho_p \sum_i \frac{n_i \mu_i}{\rho_i}, \tag{5.11a}$$

where ρ_p denotes the bulk density of the particle, which is given by the expression:

$$\rho_p = \left(\sum_i \frac{\mu_i}{\rho_i} \right)^{-1}. \tag{5.11b}$$

The bulk densities ρ_p and bulk indices n_p of various marine organisms, calculated using Equations (5.10) and (5.11) together with the data in Tables 5.10 and 5.12, are given in Table 5.13. Part A of this table presents the values of these parameters as determined by Aas (1996) for selected major classes of phytoplankton; referring to both dry (columns 2 and 5) and wet matter, the values are typical of such cells in the natural live state (columns 3, 4, 6, and 7).

TABLE 5.13. Bulk density ρ_p and bulk real absolute refractive index n_p calculated for the dry and wet matter of marine organisms.

A. For Phytoplankton Classes[a]

	Bulk ρ_p			Bulk n_p		
		Wet matter[b]			Wet matter[b]	
Organisms	Dry matter	$v_w = 0.6$	$v_w = 0.8$	Dry matter	$v_w = 0.6$	$v_w = 0.8$
-1-	*-2-*	*-3-*	*-4-*	*-5-*	*-6-*	*-7-*
Bacillariophyceae	1.536	1.229	1.076	1.538	1.415	1.377
Dinophyceae	1.240	1.111	1.068	1.552	1.420	1.379
Prymnesiophyceae (coccolithophorids)	1.426	1.185	1.105	1.562	1.424	1.381
Cyanobacteria	1.252	1.116	1.070	1.541	1.416	1.377
Chlorophyceae	1.229	1.107	1.066	1.535	1.414	1.376
Rhodophyceae	1.317	1.142	1.083	1.545	1.418	1.378
Total mean	**1.342**	**1.152**	**1.088**	**1.542**	**1.417**	**1.377**
Standard deviation	0.046	0.015	0.011	0.023	0.009	0.004

[a] Adapted from Aas (1996).
[b] The calculations were performed on the assumption that from c. 60% ($v_w = 0.6$) to c. 80% of the cell volume ($v_w = 0.8$) of these organisms consists of sea water.

B. For Other Organisms[a]

	Bulk ρ_p		Bulk n_p	
Organisms	Dry matter	Wet matter[b]	Dry matter	Wet matter[b]
-1-	*-2-*	*-3-*	*-4-*	*-5-*
Phytobenthos	1.634	1.108	1.506	1.370
Zooplankton	1.269	1.067	1.548	1.379
Zoobenthos	1.869	1.230	1.572	1.395
Nekton	1.247	1.077	1.551	1.389

[a] The authors' own calculations.
[b] The calculations were performed on the assumption that the organisms contain the quantities of sea water given in Table 5.12.

From c. 60% ($v_w = 0.6$) to c. 80% of the cell volume ($v_w = 0.8$) of these organisms in vivo was assumed to consist of sea water. Part B of this table gives our own estimates of ρ_p and n_p for dry (columns 2 and 4) and wet (columns 3 and 5) matter of other marine organisms: phytobenthos, zooplankton, zoobenthos, and nekton. In this case only one value of the water content was taken to be representative of each of these organisms in the natural live state (see Table 5.12 column 2): $\mu_w = 0.8$ (phytobenthos), $\mu_w = 0.8$ (zooplankton), $\mu_w = 0.63$ (zoobenthos), and $\mu_w = 0.73$ (nekton).

The data in both parts of Table 5.13 make it clear that the real parts of the absolute refractive index n_p of all planktonic components, and of other organisms that are potential sources of suspended particular matter in the sea, are predictably very similar to or only slightly higher than the value for water. Practically all the values of n_p in the VIS range for marine organisms lie within a narrow band, from c. 1.37 to c. 1.43. This corresponds to the range of differentiation of the relative values (relative to sea water) of coefficients $n_{p,r}$ from c. 1.02 to c. 1.07. This is corroborated by the data in Figure 5.14, which presents empirical spectra of $n_p(\lambda)$ in the VIS range for selected phytoplankton cultures, and also for all particles naturally suspended in the sea (cases when planktonic components predominated over other forms of organic and inorganic suspended matter). Most of these empirical spectra of $n_p(\lambda)$ lie within the calculated limits, as do in fact nearly all the empirical values of $n_p(\lambda)$ for phytoplankton (in the VIS range and adjacent regions) that are available in the literature (e.g., Aas (1996)).

Table 5.14 (column 3) lists some of these values of $n_{p,r}$(550 nm), the real part of the relative (to water) complex refractive index for light of λ = 550 nm, determined empirically for a number of planktonic components; they range from $n_{p,r}$ = 1.024 to $n_{p,r}$ = 1.063. On converting them to absolute values of the real part of the absolute complex refractive index (i.e., n_p, relative to a vacuum), we obtain empirical values from n_p = 1.371 to n_p = 1.423, which lie within the range emerging from theoretical calculations.

That this range of variability of n_p for phytoplankton is narrow seems obvious in view of the fact that all the major constituents of these organisms, both organic and inorganic, have similar values of these coefficients. Perusal of Table 5.10 will confirm this. As numerous authors have demonstrated (e.g., Aas (1996), Stramski (1999), and DuRand et al. (2002) and the

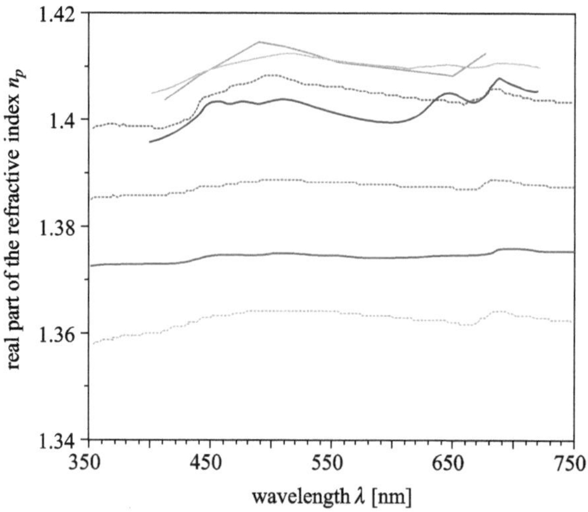

FIGURE 5.14. Some empirical spectra of the real part of the absolute complex refractive index for: selected phytoplankton species: *Synechocystis* (black and gray unbroken lines; after Stramski et al. (1988)), *Cyanobacteria* (blue unbroken line; after Morel and Bricaud (1986)), *Micromonas pusilla* (green unbroken line; after DuRand et al. (2002)); particulate matter in southern Benguela Atlantic Ocean (red, orange, and violet broken lines; after Bernard et al. (2001)). (See Colour Plate 5)

TABLE 5.14. Real $n_{p,r}(\lambda)$ and imaginary $n'_{p,r}(\lambda)$ parts of the complex refractive index (relative to water) for selected planktonic components; determined experimentally by various authors using different techniques.[a,b]

Label	Planktonic component	$n_{p,r}$ (550 nm)	$n'_{p,r}$ (440 nm)	$n'_{p,r}$ (675 nm)	References
-1-	-2-	-3-	-4-	-5-	-6-
viru	Viruses	1.050	0	0	1
hbac	Heterotrophic bacteria	1.055	5.09×10^{-04}	5.70×10^{-05}	2
proc	*Prochlorococcus* strain MED	1.055	2.33×10^{-02}	1.38×10^{-02}	3
	Average of *Prochlorococcus* strains NATL and SARG	1.046	1.38×10^{-02}	6.69×10^{-03}	3
syne	*Synechococcus* strain MAX41 (*Cyanophyceae*)	1.047	5.42×10^{-03}	2.91×10^{-03}	3
	Synechococcus strain MAX01 (*Cyanophyceae*)	1.049	4.51×10^{-03}	2.55×10^{-03}	3
	Synechococcus strain ROS04 (*Cyanophyceae*)	1.049	4.52×10^{-03}	2.15×10^{-03}	3
	Synechococcus strain DC2 (*Cyanophyceae*)	1.050	4.25×10^{-03}	2.38×10^{-03}	3
	Synechococcus strain WH8103 (*Cyanophyceae*)	1.062	9.25×10^{-03}	4.67×10^{-03}	4
syma	Synechocystis (*Cyanophyceae*)	1.050	4.53×10^{-03}	1.91×10^{-03}	5
	Anacystis marina (*Cyanophyceae*)	1.060	8.46×10^{-03}	3.60×10^{-03}	5
ping	*Pavlova pinguis* (*Haptophyceae*)	1.046	4.18×10^{-03}	2.71×10^{-03}	6
pseu	*Thalassiosira pseudonana* (*Bacillariophyceae*)	1.045	9.23×10^{-03}	7.40×10^{-03}	7
lluth	*Pavlova lutheri* (*Haptophyceae*)	1.045	5.77×10^{-03}	2.40×10^{-03}	6
galb	*Isochrysis galbana* (*Haptophyceae*)	1.056	7.67×10^{-03}	5.10×10^{-03}	5
huxl	*Emilianina huxleyi* (*Haptophyceae*)	1.050	5.01×10^{-03}	2.95×10^{-03}	5
crue	*Porphyridium cruentum* (*Rhodophyceae*)	1.051	3.35×10^{-03}	2.44×10^{-03}	6
frag	*Chroomonas fragarioides* (*Cryptophyceae*)	1.039	4.28×10^{-03}	2.90×10^{-03}	5
parv	*Prymnesium parvum* (*Haptophyceae*)	1.045	2.16×10^{-03}	1.33×10^{-03}	6
bioc	*Dunaliella bioculata* (*Chlorophyceae*)	1.038	1.05×10^{-02}	7.84×10^{-03}	5
tert	*Dunaliella tertiolecta* (*Chlorophyceae*)	1.063	6.26×10^{-03}	5.08×10^{-03}	8
curv	*Chaetoceros curvisetum* (*Bacillariophyceae*)	1.024	2.88×10^{-03}	1.48×10^{-03}	6
elon	*Hymenomonas elongata* (*Haptophyceae*)	1.046	1.39×10^{-02}	7.59×10^{-03}	5
mica	*Prorocentrum micans* (*Dinophyceae*)	1.045	2.47×10^{-03}	1.71×10^{-03}	5

[a] Adapted from Stramski et al. (2001).
[b] References (those supplying data are also indicated): 1. Stramski and Kiefer (1991); 2. Stramski and Kiefer (1990); 3. Morel et al. (1993); 4. Stramski et al. (1995); 5. Ahn et al. (1992); 6. Bricaud et al. (1988); 7. Stramski and Reynolds (1993); 8. Stramski et al. (1993).

references cited therein), there is a strong correlation between the indices n_p for phytoplankton and the relative volumes v_w of water they contain or the intracellular concentration C_c of organic carbon; they depend only slightly, if at all, on the proportions between the individual major cell constituents. This dependence of indices n_p on the relative volume of water v_w hinges on the inverse proportionality between n_p and v_w and is expressed approximately by the equation (after Aas (1996)):

$$n_p = 1.533 - 0.194\, v_w. \qquad (5.12)$$

In contrast, n_p is directly proportional to the carbon concentration C_c; that is, the higher the value of n_p, the greater is the value of C_c. This latter relation can be expressed by the following regression equation, which we derived on the basis of the empirical data of various authors:

$$n_p = 0.000209 C_c + 1.352, \qquad (5.13)$$

where C_c is expressed in units of [kg m^{-3}] and the correlation coefficient $R^2 = 0.609$. The plot of this relationship is depicted in Figure 5.15.

The results presented in Figures 5.14 and 5.15 refer to the bulk real part of the refractive index of phytoplankton cells. Unfortunately, the literature gives no empirical bulk refractive indices for other marine organisms, that is, the phytobenthos, zooplankton, zoobenthos, and nekton. Nonetheless, according

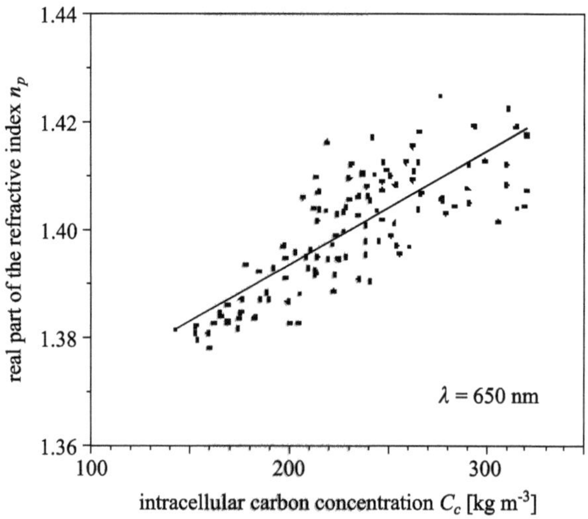

FIGURE 5.15. Relationship between the real part of the absolute refractive index n_p (λ = 650 nm), and the intracellular carbon concentration C_c for phytoplankton. The 115 points correspond to the measurements of these properties in various phytoplankton species by numerous authors (e.g., Stramski et al. (1995, 2002), DuRand and Olson (1998), Stramski and Reynolds (1993), Stramski (1999), and DuRand et al. (2002); see also the other authors mentioned in the references to Table 5.14). The continuous line is the plot of the linear regression expressed by Equation (5.13).

to the estimates given in Part B of Table 5.13, these indices are similar in value to those of phytoplankton. Moreover, the dependences of these n_p for the other organisms on the volumes of water v_w they contain or on their carbon concentrations C_c are surely similar, resembling Equations (5.12) and (5.13) for phytoplankton.

The Imaginary Part of the Bulk Absolute Complex Refractive Index for Phytoplankton and Its Direct Degradation Products

Unlike the real parts of the refractive indices of marine organisms, the imaginary parts of these indices n'_p vary widely in value (see Figures 5.16–5.19). This applies in particular to these indices in the VIS region for photosynthesizing organisms, that is, principally the phytoplankton and phytobenthos, and also heterotrophic bacteria. Among these organisms, the strongest absorbers of visible light thanks to their pigment content, there is a c. 20-fold differentiation in the value of n'_p This emerges, for example, from Table 5.14, which, among other data, gives values of the imaginary part of the relative (to water) refractive index $n'_{p,r}$ of selected planktonic components for light of wavelengths 440 nm and 675 nm, which are roughly equivalent to the positions of the two most common absorption peaks among plants. The highest value of $n'_{p,r}$ (440 nm) = 1.39 10^{-2}, measured for *Hymenomonas elongata,* is over 20 times greater than $n'_{p,r}$ (440 nm) = 5.09 10^{-4} for heterotrophic bacteria.

The values of this imaginary refractive index for plant organisms in the sea exhibit not only quantitative differentiation but also considerable spectral diversity, as the plots in Figure 5.16 show. This figure illustrates empirical spectra of the imaginary part of the absolute refractive index $n'_p(\lambda)$ for selected phytoplankton species. The characteristic features of these spectra of $n'_p(\lambda)$ and their spectral diversity are, of course, due to the differences in the pigment composition of these organisms, a matter that is dealt with in great detail in Chapter 6.

Like the real refractive indices n_p of suspended particulate matter, the imaginary refractive indices n'_p can also be calculated from their chemical compositions. The corresponding dependences of the bulk indices n'_p on the relative volumes v_i or the relative mass contents μ_i of their various constituents and their individual indices n'_i are analogous to those for n_p described above by Equations (5.10) and (5.11):

$$n'_p = \sum_i n'_i v_i \tag{5.14}$$

and

$$n'_p = \rho_p \sum_i \frac{n'_i \mu_i}{\rho_i}, \tag{5.15}$$

where ρ_i is the densities of the separate constituents, and ρ_p is the bulk density of the particle, given by Equation (5.11b).

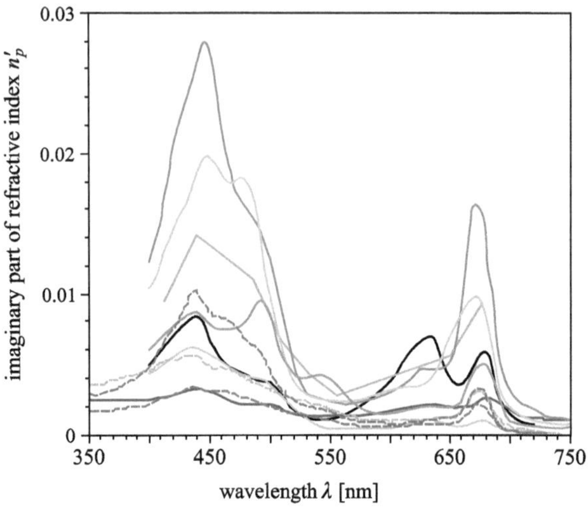

FIGURE 5.16. Empirical spectra of the imaginary part of the absolute complex refractive index, defined by various methods: for selected phytoplankton species: *Synechocystis* (unbroken black and gray lines; after Stramski et al. (1988)), *Cyanobacteria* (unbroken blue line; after Morel and Bricaud (1986)), *Micromonas pusilla* (unbroken pale green line; after DuRand et al. (2002)), *Prochlorococcus* (unbroken dark green and pale blue lines; after Morel et al. (1993)), *Synechococcus* (unbroken brown line; after Morel et al. (1993)); for particulate matter in southern Benguela Atlantic Ocean (broken red, orange and violet lines; after Bernard et al. (2001)). (See Colour Plate 6)

We used these relations to estimate, among other things, the complete range of variability of the possible spectral values of $n'_p(\lambda)$ for phytoplankton under different conditions in the oceans (Woźniak et al. 2005a). In this respect we took account not only of possible variations in intracellular organic matter and pigment concentrations, but also the changes in pigment compositions brought about by the photo- and chromatic adaptation of phytoplankton to the ambient light conditions in the water.[14] We also first estimated the ranges of variability of the spectral indices $n'(\lambda)$ for a number of detrital forms resulting directly from the degradation of phytoplankton cells, for example, particles containing dry organic matter consisting solely of phytoplankton pigments, and particles containing dry organic matter consisting of all the organic substances in phytoplankton. These results are listed in Table 5.15 and illustrated in Figure 5.17b. The areas of dependence marked on this figure depict the ranges of possible differentiation in the spectral indices $n'_p(\lambda)$ of

[14] Photo- and chromatic adaptation, the processes whereby phytoplankton adapts to the light conditions in the sea, are discussed in detail in Chapter 6 (e.g., Section 6.4).

TABLE 5.15. Values of the imaginary part of the absolute refractive index n'_p of visible light of selected wavelengths for different types of particles 1–4; explanations are given beneath the table.[a]

Particle Type[b]	Imaginary Part of the Refractive Index n'_p for Wavelength λ [nm] 400	440	480	500	520	540	560
-1-	-2-	-3-	-4-	-5-	-6-	-7-	-8-
1	8.95×10^{-01}	2.41	1.95	1.64	4.84×10^{-01}	2.69×10^{-01}	1.83×10^{-01}
2	3.05×10^{-02}	7.18×10^{-02}	5.80×10^{-02}	4.90×10^{-02}	1.64×10^{-02}	1.00×10^{-02}	7.35×10^{-03}
3	2.18×10^{-02}	5.13×10^{-02}	4.14×10^{-02}	3.50×10^{-02}	1.17×10^{-02}	7.17×10^{-03}	5.25×10^{-03}
4	9.01×10^{-04}	1.63×10^{-03}	5.98×10^{-04}	4.24×10^{-04}	2.29×10^{-04}	2.06×10^{-04}	2.11×10^{-04}
Particle Type[b]	**Imaginary Part of the Refractive Index n'_p for Wavelength λ [nm] 580**	**600**	**620**	**640**	**660**	**675**	**700**
-1-	-9-	-10-	-11-	-12-	-13-	-14-	-15-
1	1.40×10^{-01}	1.51×10^{-01}	1.63×10^{-01}	1.95×10^{-01}	3.35×10^{-01}	7.43×10^{-01}	1.67×10^{-01}
2	5.91×10^{-03}	5.98×10^{-03}	6.13×10^{-03}	6.84×10^{-03}	1.06×10^{-02}	2.19×10^{-02}	5.63×10^{-03}
3	4.22×10^{-03}	4.27×10^{-03}	4.38×10^{-03}	4.88×10^{-03}	7.56×10^{-03}	1.56×10^{-02}	4.02×10^{-03}
4	2.20×10^{-04}	2.23×10^{-04}	2.13×10^{-04}	2.11×10^{-04}	3.29×10^{-04}	7.51×10^{-04}	1.95×10^{-04}

[a] After Woźniak et al. (2005a).

[b] The separate values of this index n'_p refer to various types of particle. Type 1: particles containing dry organic matter consisting solely of phytoplankton pigments with a composition typical of the highest irradiances and oligotrophic waters O1; type 2: particles containing dry organic matter consisting of all the organic substances in phytoplankton with a pigment composition typical of the highest irradiances and oligotrophic waters O1; type 3: phytoplankton cells with the highest possible intracellular concentrations of organic carbon (C_c = 500 kg m^{-3}) and chlorophylls *a* (C_{chl} = 10 kg m^{-3}), with a composition typical of the highest irradiances and oligotrophic waters O1; type 4: phytoplankton cells with the lowest possible intracellular concentrations of organic carbon (C_c = 25 kg m^{-3}) and chlorophylls *a* (C_{chl} = 0.5 kg m^{-3}) with a pigment composition typical of eutrophic waters E5.

living phytoplankton cells, and also of phytoplanktonlike particles (i.e., with a similar organic matter content but different water content) and particles of phytoplanktonic origin; these last, if they consist purely of pigments, achieve the highest possible values of $n'_p(\lambda)$. The lower boundary of these latter two areas is delineated by the indices n'_w for water.

Recapitulating, we can state that these calculations demonstrate the approximate range of natural differentiation of the absolute values of the imaginary refractive index n'_p (λ) of the matter contained in phytoplankton cells. This more than 20-fold differentiation is in excess of one order of magnitude, a result confirmed by empirical data; see Figure 5.17a. Furthermore, the dataset of $n'_{p,r}$ (after conversion to n'_p) for the various systematic units of phytoplankton given in Table 5.14 coincides perfectly with the range of differentiation of this index illustrated in Figure 5.17b. This range, however, is not large in comparison with that of the indices $n'_p(\lambda)$ for phytoplanktonlike particles, that is, particles containing similar organic matter. As a result of the hydration and aeration, as well as the dehydration and de-aeration of dead

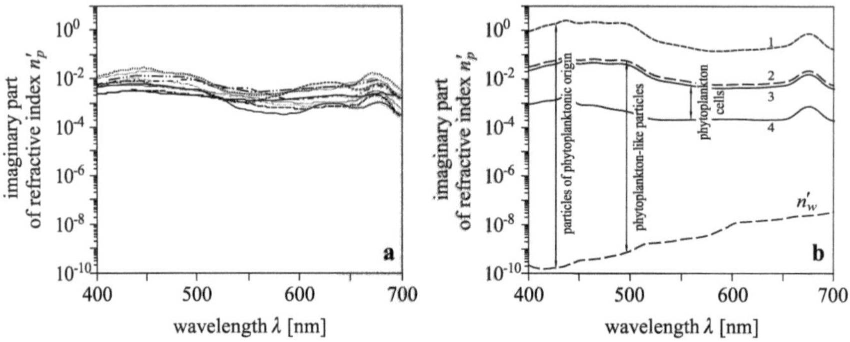

FIGURE 5.17. Spectra of the imaginary part of the absolute complex refractive index (a) The same empirical spectra for selected phytoplankton species and particulate matter as are depicted clearly in Figure 5.16 (b) Comparison of the probable ranges of natural differentiation of the most common spectra of the imaginary parts of the absolute refractive index for oceanic phytoplankton cells (area between plots 3 and 4), their associated phytoplankton-like particles (area between plots 2 and n'_w), and particles of phytoplanktonic origin (area between plots 1 and n'_w). The plots present model spectra of this index for particles of types 1–4 (explanations in the footnote to Table 5.15) and n'_w for water particles (i.e. pure water) (after Woźniak et al. 2005a).

algae, the suspended particles these processes produce have a much wider diversity of absorption properties than living phytoplankton. These properties, expressed by values of n'_p, vary over several orders of magnitude: around eight orders for blue light ($\lambda = 420$ nm) and around five orders for red light ($\lambda = 650$ nm; see the area between plot 2 and the spectrum of $n'_w(\lambda)$ in Figure 5.17b).

The break-up of phytoplankton cells can also be a source of suspended particles with an organic matter content different from that discussed earlier. Therefore, some of the particles derived from phytoplankton may contain organic matter that absorbs light more strongly than the natural mixture of the various organic compounds of which phytoplankton is constituted, for instance, chloroplast fragments composed of pigments. In effect, the range of differentiation of $n'_p(\lambda)$ is further widened by nearly two orders of magnitude (see the area between plot 1 and the spectrum of $n'_w(\lambda)$ in Figure 5.17b). The range of differentiation of these indices for any particle of phytoplanktonic origin is thus around ten orders of magnitude for blue light ($\lambda = 420$ nm) and around seven orders for red light ($\lambda = 650$ nm).

So far we have been talking about the spectral indices $n'_p(\lambda)$ of phytoplankton and nonliving particles derived directly from it with respect to VIS light. Phytoplankton also absorbs considerable amounts of UV radiation, which is the result of its containing proteins. The plots in Figure 5.18 (see also Table 5.16, columns 2–4) illustrate this by depicting spectra of $n'_p(\lambda)$ over the

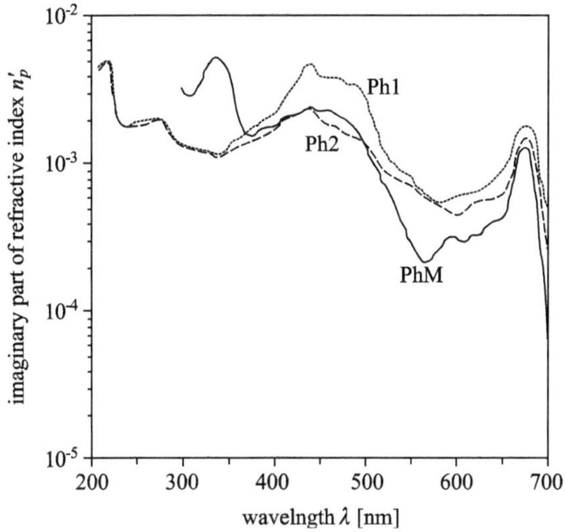

FIGURE 5.18. Calculated possible spectra of the imaginary part of the absolute refractive index in the UV-VIS region for three different types of phytoplankton: oceanic (Ph1), Baltic (Ph2) and polar (PhM) (based on data taken from Woźniak et al. 2005b)

entire UV–VIS range for three types of phytoplankton: oceanic (which, besides other cell constituents, contain sets of pigments, the properties and quantitative compositions of which are typical of algae living in open oceans; see Woźniak et al. (1999)); "Baltic" (containing, among other constituents, sets of pigments, the properties and quantitative compositions of which are typical of algae living in the Baltic Sea; see Ficek et al. (2004)); and polar (see Moisan and Mitchell 2001). These spectra were determined on the basis of the typical contents of the various chemical constituents of cells. In the case of polar phytoplankton, we assumed that apart from "ordinary" proteins, in which the strongest absorbers are the aromatic amino acids (these absorb inter alia UV light in the 180–300 nm range; see the absorption bands in Figure 4.8a), this phytoplankton may also contain certain amounts of mycosporine-like amino acids (MAAs). In Chapter 4 (Figure 4.8b) we showed that these MAAs are very strong UV absorbers in the 310–370 nm range and serve to protect the organisms from this lethal UV radiation (Garcia-Pichel 1994, Vernet and Whitehead 1996).

Figure 5.18 shows that the spectra of $n'_p(\lambda)$ for "ordinary" oceanic and "Baltic" phytoplankton (i.e., not containing MAAs) are usually characterized by two peaks in the UV region at wavelengths c. 220 nm and c. 280 nm, due to UV absorption by the aromatic amino acids in the proteins. The values of n'_p for UV light in these bands are comparable with the values of this index for visible light c. 440 nm and 675 nm in wavelength. In contrast, the

TABLE 5.16. Modeled spectra of the imaginary part of the absolute complex refractive index of UV–VIS light for oceanic phytoplankton (Ph1), Baltic phytoplankton (Ph2), polar phytoplankton (PhM), zooplankton (Zoo), zoobenthos (Zoo-b), and nekton (Nek).

Wavelength λ[nm]	Ph1	Ph2	PhM	Ph-b	Zoo	Zoo-b	Nek
-1-	-2-	-3-	-4-	-5-	-6-	-7-	-8-
210	4.34×10^{-03}	4.34×10^{-03}	—	1.35×10^{-03}	5.80×10^{-03}	5.57×10^{-03}	9.22×10^{-03}
220	4.85×10^{-03}	4.85×10^{-03}	—	1.51×10^{-03}	6.48×10^{-03}	6.23×10^{-03}	1.03×10^{-02}
230	2.14×10^{-03}	2.14×10^{-03}	—	6.68×10^{-04}	2.86×10^{-03}	2.75×10^{-03}	4.55×10^{-03}
240	1.79×10^{-03}	1.79×10^{-03}	—	5.57×10^{-04}	2.39×10^{-03}	2.29×10^{-03}	3.79×10^{-03}
250	1.83×10^{-03}	1.83×10^{-03}	—	5.71×10^{-04}	2.44×10^{-03}	2.35×10^{-03}	3.89×10^{-03}
260	1.86×10^{-03}	1.86×10^{-03}	—	5.82×10^{-04}	2.49×10^{-03}	2.39×10^{-03}	3.96×10^{-03}
270	1.94×10^{-03}	1.94×10^{-03}	—	6.06×10^{-04}	2.59×10^{-03}	2.49×10^{-03}	4.12×10^{-03}
280	1.88×10^{-03}	1.88×10^{-03}	—	5.86×10^{-04}	2.51×10^{-03}	2.41×10^{-03}	3.99×10^{-03}
290	1.51×10^{-03}	1.51×10^{-03}	—	4.71×10^{-04}	2.02×10^{-03}	1.94×10^{-03}	3.20×10^{-03}
300	1.33×10^{-03}	1.33×10^{-03}	3.28×10^{-03}	4.17×10^{-04}	1.78×10^{-03}	1.71×10^{-03}	2.83×10^{-03}
310	1.26×10^{-03}	1.26×10^{-03}	2.96×10^{-03}	—	1.69×10^{-03}	1.62×10^{-03}	2.68×10^{-03}
320	1.21×10^{-03}	1.21×10^{-03}	3.52×10^{-03}	—	1.61×10^{-03}	1.55×10^{-03}	2.57×10^{-03}
330	1.15×10^{-03}	1.15×10^{-03}	4.83×10^{-03}	—	1.54×10^{-03}	1.48×10^{-03}	2.45×10^{-03}
340	1.10×10^{-03}	1.10×10^{-03}	5.12×10^{-03}	—	1.47×10^{-03}	1.41×10^{-03}	2.34×10^{-03}
350	1.25×10^{-03}	1.19×10^{-03}	4.27×10^{-03}	—	1.40×10^{-03}	1.35×10^{-03}	2.23×10^{-03}
360	1.41×10^{-03}	1.28×10^{-03}	2.35×10^{-03}	—	1.34×10^{-03}	1.28×10^{-03}	2.12×10^{-03}
370	1.58×10^{-03}	1.37×10^{-03}	1.58×10^{-03}	—	1.27×10^{-03}	1.22×10^{-03}	2.02×10^{-03}
380	1.75×10^{-03}	1.47×10^{-03}	1.59×10^{-03}	—	1.21×10^{-03}	1.16×10^{-03}	1.92×10^{-03}
390	1.93×10^{-03}	1.57×10^{-03}	1.76×10^{-03}	—	1.15×10^{-03}	1.10×10^{-03}	1.83×10^{-03}
400	2.12×10^{-03}	1.68×10^{-03}	1.84×10^{-03}	3.15×10^{-03}	1.09×10^{-03}	1.05×10^{-03}	1.74×10^{-03}
410	2.58×10^{-03}	1.95×10^{-03}	2.03×10^{-03}	3.93×10^{-03}	1.04×10^{-03}	9.95×10^{-04}	1.65×10^{-03}
420	3.22×10^{-03}	2.12×10^{-03}	2.14×10^{-03}	5.02×10^{-03}	9.83×10^{-04}	9.44×10^{-04}	1.56×10^{-03}
430	4.01×10^{-03}	2.23×10^{-03}	2.26×10^{-03}	6.35×10^{-03}	9.32×10^{-04}	8.95×10^{-04}	1.48×10^{-03}
440	4.62×10^{-03}	2.33×10^{-03}	2.42×10^{-03}	7.39×10^{-03}	8.83×10^{-04}	8.48×10^{-04}	1.40×10^{-03}
450	3.77×10^{-03}	2.02×10^{-03}	2.31×10^{-03}	5.98×10^{-03}	8.36×10^{-04}	8.04×10^{-04}	1.33×10^{-03}
460	3.75×10^{-03}	1.88×10^{-03}	2.27×10^{-03}	5.96×10^{-03}	7.92×10^{-04}	7.61×10^{-04}	1.26×10^{-03}

470	3.71×10^{-03}	1.76×10^{-03}	2.16×10^{-03}	5.91×10^{-03}	7.49×10^{-04}	7.20×10^{-04}	1.19×10^{-03}
480	3.41×10^{-03}	1.57×10^{-03}	1.97×10^{-03}	5.43×10^{-03}	7.09×10^{-04}	6.81×10^{-04}	1.13×10^{-03}
490	3.28×10^{-03}	1.49×10^{-03}	1.71×10^{-03}	5.22×10^{-03}	6.70×10^{-04}	6.44×10^{-04}	1.07×10^{-03}
500	2.73×10^{-03}	1.35×10^{-03}	1.34×10^{-03}	4.32×10^{-03}	6.33×10^{-04}	6.08×10^{-04}	1.01×10^{-03}
510	1.80×10^{-03}	1.11×10^{-03}	9.90×10^{-04}	2.79×10^{-03}	5.98×10^{-04}	5.74×10^{-04}	9.51×10^{-04}
520	1.20×10^{-03}	9.20×10^{-04}	7.49×10^{-04}	1.80×10^{-03}	5.65×10^{-04}	5.42×10^{-04}	8.98×10^{-04}
530	9.71×10^{-04}	8.31×10^{-04}	5.69×10^{-04}	1.43×10^{-03}	5.33×10^{-04}	5.12×10^{-04}	8.47×10^{-04}
540	8.71×10^{-04}	7.68×10^{-04}	4.05×10^{-04}	1.27×10^{-03}	5.03×10^{-04}	4.83×10^{-04}	7.99×10^{-04}
550	7.64×10^{-04}	7.06×10^{-04}	2.93×10^{-04}	1.10×10^{-03}	4.74×10^{-04}	4.56×10^{-04}	7.54×10^{-04}
560	6.52×10^{-04}	6.28×10^{-04}	2.26×10^{-04}	9.29×10^{-04}	4.47×10^{-04}	4.30×10^{-04}	7.11×10^{-04}
570	5.62×10^{-04}	5.66×10^{-04}	2.19×10^{-04}	7.88×10^{-04}	4.22×10^{-04}	4.05×10^{-04}	6.70×10^{-04}
580	5.20×10^{-04}	5.08×10^{-04}	2.54×10^{-04}	7.27×10^{-04}	3.97×10^{-04}	3.82×10^{-04}	6.32×10^{-04}
590	5.34×10^{-04}	4.88×10^{-04}	3.13×10^{-04}	7.58×10^{-04}	3.74×10^{-04}	3.60×10^{-04}	5.95×10^{-04}
600	5.70×10^{-04}	4.51×10^{-04}	3.11×10^{-04}	8.25×10^{-04}	3.52×10^{-04}	3.39×10^{-04}	5.60×10^{-04}
610	5.92×10^{-04}	4.72×10^{-04}	2.90×10^{-04}	8.69×10^{-04}	3.32×10^{-04}	3.19×10^{-04}	5.28×10^{-04}
620	6.14×10^{-04}	5.35×10^{-04}	3.26×10^{-04}	9.13×10^{-04}	3.12×10^{-04}	3.00×10^{-04}	4.97×10^{-04}
630	6.60×10^{-04}	5.62×10^{-04}	3.52×10^{-04}	9.96×10^{-04}	2.94×10^{-04}	2.82×10^{-04}	4.67×10^{-04}
640	7.47×10^{-04}	5.67×10^{-04}	4.11×10^{-04}	1.15×10^{-03}	2.77×10^{-04}	2.66×10^{-04}	4.40×10^{-04}
650	8.59×10^{-04}	5.99×10^{-04}	4.34×10^{-04}	1.34×10^{-03}	2.60×10^{-04}	2.50×10^{-04}	4.13×10^{-04}
660	1.10×10^{-03}	8.50×10^{-04}	6.46×10^{-04}	1.75×10^{-03}	2.45×10^{-04}	2.35×10^{-04}	3.89×10^{-04}
670	1.75×10^{-03}	1.41×10^{-03}	1.22×10^{-03}	2.83×10^{-03}	2.30×10^{-04}	2.21×10^{-04}	3.65×10^{-04}
675	—	1.51×10^{-03}	1.29×10^{-03}	—	2.23×10^{-04}	2.14×10^{-04}	3.54×10^{-04}
680	1.65×10^{-03}	1.40×10^{-03}	1.07×10^{-03}	2.67×10^{-03}	2.16×10^{-04}	2.08×10^{-04}	3.44×10^{-04}
690	7.96×10^{-04}	6.92×10^{-04}	3.49×10^{-04}	1.25×10^{-03}	2.03×10^{-04}	1.95×10^{-04}	3.23×10^{-04}
700	4.71×10^{-04}	2.58×10^{-04}	6.65×10^{-05}	7.18×10^{-04}	1.91×10^{-04}	1.83×10^{-04}	3.03×10^{-04}

spectra of $n'_p(\lambda)$ for polar plankton may exhibit, in addition to these features, a further strong peak at c. 340 nm. This exceeds by far the values of n'_p for visible light and may constitute the absolute maximum of this spectrum. The presence of this strong absorption band in the 310–370 nm range was demonstrated experimentally in a culture of *Phaeocystis antarctica* and investigated in detail by Moisan and Mitchell (2001).

To conclude this discussion of the imaginary refractive indices of the cellular matter of phytoplankton and of particles derived from the direct degradation of phytoplankton, we emphasize that the principal factor governing the values of these indices for visible light are the intracellular concentrations of pigments, the strongest VIS absorbers of all the chemical constituents of phytoplankton. Many authors (e.g., Stramski (1999), DuRand et al. (2002) and others cited therein) have demonstrated empirically the existence of strong correlations between indices n'_p and intracellular concentrations of chlorophyll *a* in phytoplankton cells C_{chl} This relationship is particularly striking for red light at c. 675 nm, where the main absorber of radiation is this very chlorophyll. To a good approximation, it can be expressed by the following regression equation, which we derived on the basis of empirical data taken from a large number of papers of by different authors:

$$n' = 0.001026C_{chl} + 0.000589, \quad (5.16)$$

where C_{chl} is expressed in [kg·m^{-3}] and the correlation coefficient $R^2 = 0.893$. The plot of this relationship is shown in Figure 5.19.

The Imaginary Part of the Bulk Complex Refractive Index for Other Marine Organisms

In the world literature, we did not come across any reports whatsoever of empirical studies of the bulk imaginary part of the refractive index by marine organisms, apart from the earlier-mentioned investigations with respect to phytoplankton and bacteria. We can therefore present the values of the bulk indices n'_p for the phytobenthos, zooplankton, zoobenthos, and nekton only by giving typical examples of the $n'_p(\lambda)$ spectra of these organisms, calculated with the aid of Equations (5.14) and (5.15), the data on their chemical composition (Table 5.12) and the imaginary spectral coefficients n'_i of these constituents (given in Table 5.11). These spectra of $n'_p(\lambda)$ are given in Table 5.16 (columns 5 to 8) and are illustrated graphically in Figure 5.20.

As can be seen from the data in Figure 5.20, the strongest VIS light absorber of all these organisms is, for obvious reasons, the phytobenthos (Ph-b), which contains plant pigments. Its imaginary refractive indices n'_p in this range are comparable with the corresponding indices for phytoplankton. In contrast, the values of n'_p are several times smaller for the other, animal organisms; they, of course, do not contain any light-absorbing pigments.

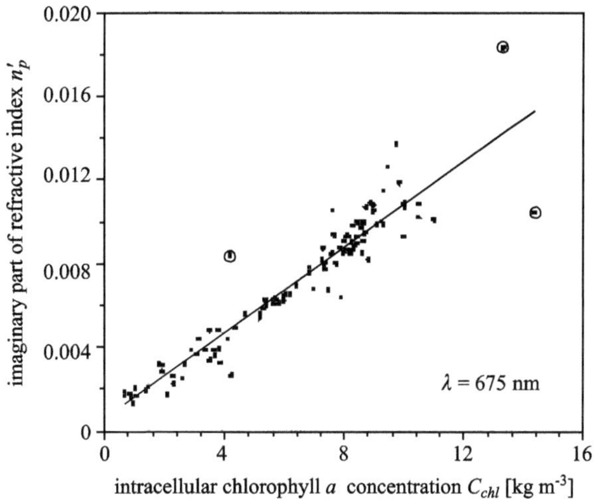

FIGURE 5.19. Relationship between the imaginary part of the absolute refractive index $n'_p(\lambda = 675 \text{ nm})$ and the intracellular chlorophyll a concentration C_{chl} for phytoplankton. The 108 points correspond to measurements of these properties in various phytoplankton species, performed by the various authors mentioned in the caption of Figure 5.15. The unbroken line is the regression curve described by Equation (5.16). The circled points were treated as serious errors and were left out of the calculation of the regression curve.

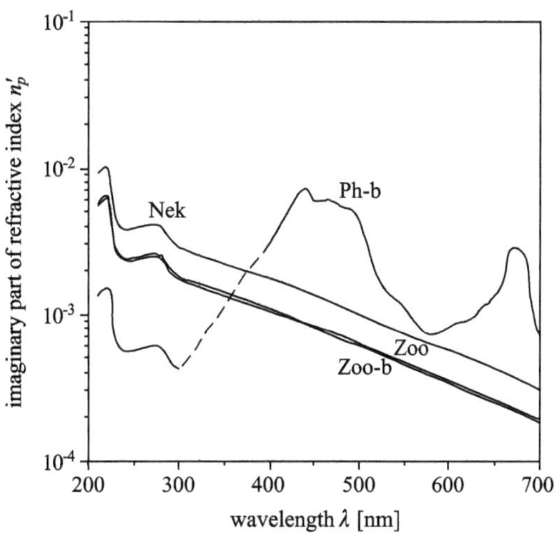

FIGURE 5.20. Possible, calculated spectra of the imaginary part of the absolute refractive index of UV–VIS light by the phytobenthos (Ph-b), zooplankton (Zoo), zoobenthos (Zoo-b), and nekton (Nek). The dashed part of the Ph-b curve indicates that data are lacking. (Based on data from Woźniak et al. (2005b).)

But matters are otherwise in the UV spectral region. There the values of n'_p for the animal organisms are quite considerable, thanks to the absorption bands of the aromatic amino acids in the proteins. On the other hand, n'_p for the phytobenthos may be several times lower than those of the animals, chiefly because of the much smaller protein content in the phytobenthos than in the other marine organisms (see columns 6 and 10 in Table 5.12).

In conclusion, we remind the reader that the spectra of $n'_p(\lambda)$ for the phytobenthos, zooplankton, zoobenthos, and nekton that we have presented in this section were obtained solely by approximate calculations. Although undoubtedly reflecting the general qualitative trends in the relations prevailing among them, they should be treated purely as examples. In reality, such spectra may well be highly diverse, lying within greater or smaller ranges of differentiation.

5.2.4 The Chemical Composition and Optical Constants of Organic Detritus

It is a far more complex problem to establish the chemical composition and basic physical properties (densities and optical constants) of organic detritus than of living organisms. For one thing, this is because detritus particles display a much greater diversity of chemical and morphological forms; for another, the methods for directly or indirectly determining the optical constants of detritus are more limited in scope than those available for studying plankton. As we showed in the previous section, current knowledge of the optical constants of planktonic components is largely the result of laboratory investigations done on cultures of certain phytoplankton species. Such methods are not applicable in optical studies of organic detritus.

As a result of these difficulties, the optical constants of organic detritus in the sea are not as well known as those of living organisms, even though the former usually make up the lion's share of POM in the sea. This paucity of knowledge applies in particular to spectra of the imaginary part of the absolute refractive index $n'_p(\lambda)$. A search of the literature revealed just one single spectrum of the relative (to water) index $n'_{p,r}(\lambda)$ (Stramski et al. 2001). Referring to natural organic detritus in the sea, it was obtained on the basis of microspectrophotometric measurements done in the Sargasso Sea[15] (Iturriga and Siegel 1989):

$$n'_{p,r}(\lambda) = 0.010658\exp(-0.007186\lambda), \tag{5.17}$$

[15] Another paper by Stramski and Woźniak (2005) has recently been published, in which a few more spectra of $n'_{p,r}(\lambda)$ for Sargasso Sea POM are presented, albeit only in graphic form.

where the index $n'_{p,r}(\lambda)$ is nondimensional and the wavelength λ is expressed in [nm]. Of course, this expression (5.17) is not generally applicable: it merely describes the averaged spectrum of $n'_{p,r}(\lambda)$ for a whole range of POM particles differing in origin, shape, size, and chemical structure, and occurring in the quantitative proportions that happen to prevail in the one basin investigated by Iturriga and Siegel (1989). Presumably, then, the indices $n'_p(\lambda)$ for the various types of suspended particles and their configurations in different seas, at different depths and under different conditions, will be highly dissimilar.

In view of the lack of appropriate, in situ empirical data on the structure, chemical composition, and physical properties (densities and optical constants) of organic detritus, we have attempted to define these chemical and physical properties approximately by way of theoretical speculations based on a general knowledge of marine physics, chemistry, and biology. We achieved this by means of multiphase modeling, a procedure that we have discussed in detail in a series of articles (Woźniak et al. 2005a, b, Woźniak et al. 2006). In the first phase we distinguished, for instance, hypothetical model groups of organic matter of which real suspended particulate matter in the sea could be composed. To simplify things, we called these groups "model chemical classes of particles." We drew this distinction on the basis of available knowledge of organic substances occurring in the sea, on the assumption that the same chemical substances (or their derivatives) are present in organic detritus as in living organisms or in the water as DOM. Then, starting out from the known spectral mass-specific absorption coefficients of light for these chemical constituents of the particles (the spectra of these coefficients for the various organic substances present in the sea are given in Chapters 3, 4, and 6), we estimated first the bulk values of the light absorption of particulate material $a_{pm}(\lambda)$, and from these, their imaginary refractive indices $n'_p(\lambda)$, in accordance with the relationship $n'_p(\lambda) = a_{pm}(\lambda)\lambda/4\pi$.

In the second phase of the modeling we availed ourselves of what information there is in the literature on the structure, formation, and possible evolution of organic detritus particles in the sea. On this basis we were able to distinguish more than 20 different morphological forms of these particles consisting of the various possible configurations of the model chemical classes of particles mentioned earlier. We treated these classes as constituent elements (fragments) of the morphological forms. On this basis, we then determined the bulk imaginary parts of the refractive index n'_p characterizing these morphological forms as mixtures of particles from different model chemical classes. Finally, using the relationship between the indices n_p (real part of the refractive index) and the intracellular concentration of carbon C_c (described in Section 5.2.3), we calculated the approximate values of indices n_p in the VIS region for these morphological forms of organic detritus.

In the present section, we describe the most important phases and results of this modeling of optical constants for organic detritus, which are discussed in detail in the above-mentioned cycle of articles (Woźniak et al. (2005a, b) and Woźniak et al. (2006)). This description may be useful, for example, in calculating the optical properties (e.g., absorption) of organic detritus particles in the sea on the basis of Mie's theory. The reader should, however, always remain aware that the numerical data given here characterizing the optical constants of detritus are by no means certain, because they are solely the result of theoretical speculations and have not yet been subjected to rigorous empirical verification.

At the end of this section we also cite some literature values of the real and imaginary parts of the absolute refractive index for a few crude oil derivatives. These substances may occur in the sea in the form of emulsions or suspended particles, and so are formally classified as a type of organic detritus.

The Chemical Composition and Imaginary Part of Refractive Index of Hypothetical Particulate Matter Contained in Organic Particles

Our model is subject to a number of limitations and simplifying assumptions. We have analyzed only those suspended organic particles that contain strong UV and VIS absorbers, in addition to other possible organic and inorganic admixtures. Nevertheless, it may be assumed that such are the vast majority of organic particles in the sea. The light-absorbing molecules in them are various unsaturated organic compounds, among them, two groups of very strongly UV-absorbing low- and medium-molecular weight nitrogen-containing compounds, and four groups of UV and/or VIS absorbing medium- or high-molecular weight compounds:

1. *Amino acids and their derivatives*, including peptides and protein fragments (strong UV absorbers)
2. *Purine and pyridine compounds*, including nucleic acid fragments (strong UV absorbers)
3. *Natural proteins* (UV and VIS absorbers)
4. *Plant pigments* (strong VIS absorbers)
5. *Humus acids*, that is, humic and fulvic acids and their mixtures (strong UV and VIS absorbers)
6. *Lignins* (UV and VIS absorbers; UV absorption is particularly strong)

The first five groups include organic compounds produced in biochemical and biogeochemical reactions, as well as in the various biological processes taking place in all natural terrestrial and marine ecosystems, and in bottom sediments (Hansell and Carlson 2002). It can therefore be taken as read that all such compounds, be they of marine or terrestrial origin, are potential constituents of suspended particulate matter (SPM) in the sea. In open oceanic waters these substances are in large measure formed in the sea itself; that is, they are autogenous products of marine ecosystems. But in shelf regions, especially near river mouths, they may also to some extent be allogenous (terrigenous) substances. Entirely allogenous are the substances in the sixth group of absorbers, the lignins, which are organic compounds produced only by the higher plants (trees), and are therefore terrigenous (Meyers-Schulte and Hedges 1986). Large amounts of lignins can therefore be expected, especially in suspended particle form, in the coastal waters of oceans and in enclosed seas, particularly shallow ones such as the Baltic (Nyquist 1979, Pempkowiak 1989).

With these premises in mind, and assuming, moreover, that every suspended, strongly UV–VIS absorbing, organic particle contains at least one or a combination of the six groups of substances mentioned above (in addition to other, nonabsorbing constituents), we attempted a chemical classification of these particles. To simplify the analysis, we made the additional assumption that POM consists of the same organic compounds and their associations in more or less the same proportions as in living plants and animals as in DOM. This assumption is to a large extent justified by the fact that organic matter in the sea consists of substances produced by living organisms and of their metabolism and degradation products. POM in the sea is thus composed of these organisms, together with the particle s produced from them, and the particles formed as a result of the coagulation of dissolved matter. Of course, it is quite likely that there exist in the sea organic particles with chemical compositions different from those we have assumed; we are unable to define them, however.

Having made these simplifying assumptions, we therefore identified 26 model chemical classes of POM of which real organic particles suspended in sea water may be composed. Table 5.17 lists these 26 classes; it also specifies their chemical compositions and/or the origins of the organic substances they contain. With respect to the number of constituents, these particles can be divided into the three types:

- Single-component organic particles[16] containing (see Table 5.17A): (1) aromatic amino acids or protein fragments with these amino acids, which are some of the strongest absorbers of UV, especially in the 200–300 nm range; these are class A1 particles; (2) mycosporinelike amino acids, UV absorbers in the 300–400 nm range. class A2; (3) natural proteins, class A3; (4) phytoplankton pigments with a composition typical of oceanic phytoplankton, class P1; (5) phytoplankton pigments with a composition typical of Baltic phytoplankton, class P2; (6) purine and pyridine compounds, and nucleic acid fragments, class N; (7) lignin fragments, class L.
- Particles containing organic humus matter, which because of the complex nature of humus (see, e.g., Aiken et al. (1985), Rashid (1985), and Spitzy and Ittekkot (1986)), can be classified as both quasi-single component matter and multicomponent matter. These particles contain: (1) humus substances from various marine basins, classes H1 to H9 particles, and (2) particles

[16] The term "single-component matter" is used conventionally and does not imply a single chemical compound. Here, it defines matter containing chemical compounds classified in the same group of compounds; for example, in a particle containing class A1 compounds various aromatic amino acids, such as a mixture of tryptophan, tyrosine, and phenylalanine, may occur simultaneously.

TABLE 5.17. Model chemical classes of particulate organic matter (POM) in the sea.

A. Particles Containing One Organic Component[a]

Symbol	Name of class (Kind of organic substance)	Composition and/or origin of organic substances in the suspended particle
-1-	*-2-*	*-3-*
A1	Aromatic amino acids (protein fragments)	Mixture of compounds: tryptophan, tyrosine, and phenylalanine in proportions by weight 1:1:1
A2	Mycosporinelike amino acids (MAAs)	Natural mixture of mycosporine-glycine, shinorine, and mycosporine-glycine valine, typical of Phaeocystis antarctica (Moisan and Mitchell 2001)
A3	Natural proteins	Mixture of various amino acids present in the sea, containing 7.5% by weight of aromatic amino acids
P1	Ocean phytoplankton pigments	Natural mixture of plant pigments typical of oceanic phytoplankton (Woźniak et al. 1999)
P2	Baltic phytoplankton pigments	Natural mixture of plant pigments typical of Baltic phytoplankton (Ficek et al. 2004)
N	Purine and pyridine compounds	Mixture of adenine, guanine and cytosine in proportions by weight 1:1:1
L	Lignins	Natural mixture of lignins isolated from Baltic waters (Nyquist 1979)

B. Particles Consisting of Organic Humic Matter

Symbol	Name of class (Kind of organic substance)	Composition and/or origin of organic substances in the suspended particle
-1-	*-2-*	*-3-*
H1	Baltic humus	Natural mixture of various humus substances isolated from Baltic waters (Nyquist 1979)
H2	Atlantic coastal humus	Natural mixture of various humus substances isolated from Atlantic coastal waters (Stuermer 1975)
H3	Humus from inland waters	Natural mixture of various humus substances, typical of inland waters (Zepp and Schlotzhauer 1981)
H4	Marine humus	Natural mixture of various humus substances, typical of sea waters (Zepp and Schlotzhauer 1981)
H5	Sargasso Sea humus	Natural mixture of various humus substances isolated from Sargasso Sea waters (Stuermer 1975)
H6, H7 and H8	Humus from the Gulf of Mexico	Mixtures of natural fulvic and humic acids isolated from waters of the central Gulf of Mexico in proportions by weight: 0:1 (H6); 1:0 (H7); 0.9:0.1 (H8) (Carder et al. 1989)
H9	Fulvic acids from Mississippi estuarine waters	Natural mixture of various fulvic acids isolated from Mississippi estuarine waters in the Gulf of Mexico (Carder et al. 1989)

TABLE 5.17. Model chemical classes of particulate organic matter (POM) in the sea.—Cont'd.

B. Particles Consisting of Organic Humic Matter

Symbol	Name of class (Kind of organic substance)	Composition and/or origin of organic substances in the suspended particle
-1-	*-2-*	*-3-*
H10	Soil humic acids	Natural humic acids isolated from the soil (Zepp and Schlotzhauer 1981)
H11	Soil fulvic acids	Natural fulvic acids isolated from the soil (Zepp and Schlotzhauer 1981)
H12	Humic acids from bottom sediments	Natural humic acids isolated from bottom sediments (Hayase and Tsubota 1985)
H13	Fulvic acids from bottom sediments	Natural fulvic acids isolated from bottom sediments (Hayase and Tsubota 1985)

C. Multicomponent Particles of Organic Matter

Symbol	Name of class (Kind of organic substance)	Composition and/or origin of organic substances in the suspended particle
-1-	*-2-*	*-3-*
Ph1	Oceanic phytoplankton and phytoplanktonlike particles	Proteins (46.9%), chlorophyll a (1%), plus the set of pigments typical of oceanic phytoplankton (2.1%); the remaining organic matter does not absorb light (50%); note that slightly different proportions between proteins and pigments are also possible, but their overall content remains the same
Ph2	Baltic phytoplankton and phytoplanktonlike particles	Proteins (44.7%), chlorophyll a (1%) plus the set of pigments typical of Baltic phytoplankton (4.3%); the remaining organic matter does not absorb light (50%); note that slightly different proportions between proteins and pigments are also possible, but their overall content remains the same
PhM	Polar phytoplankton and phytoplanktonlike particles	Composition typical of Phaeocystis antarctica (Moisan and Mitchell 2001)
Z	Zooplankton and zooplankton like and/or nektonlike particles	Proteins (70%); the remaining organic matter does not absorb light (30%)
D1	Oceanic DOM-like particles	Proteins (1.2%), purine and pyridine compounds (0.3%), Sargasso Sea humus (10%); the remaining organic matter does not absorb light (88.5%)
D2	Baltic DOM-like particles	Proteins (8%), purine and pyridine compounds (2%), Baltic humus (55%), lignins (18%); the remaining organic matter does not absorb light (17%)

[a] Adapted from Woźniak et al. (2005a,b).

formed in soils and bottom sediments, classes H10 to H13. Table 5.17B gives the exact environmental origins of these substances.

- *Multicomponent organic particles*, including those with an organic composition: (1) as in phytoplankton or resembling that of phytoplankton in the oceans (we call these "oceanic phytoplankton and phytoplanktonlike particles") and the Baltic ("Baltic phytoplankton and phytoplanktonlike particles"), classes Ph1 and Ph2 particles; (2) resembling that of phytoplankton species growing, for example, under elevated levels of UV (polar regions) and containing large amounts of mycosporinelike amino acids (MAAs) (polar phytoplankton and phytoplanktonlike particles), class PhM; (3) zooplankton and zooplankton-and/or nektonlike particles, class Z; (4) oceanic DOM-like particles, class D1, and Baltic DOM-like particles, class D2. In all these cases, the particulate matter also contains weakly absorbing organic substances. Table 5.17C states the approximate proportions among the various components of this POM, or the origin of natural associations of these components.

Woźniak et al. (2005b) quotes bulk values of the imaginary parts of absolute refractive index n'_p for these 26 model chemical classes of particles, with reference to their various possible physical types. These indices were determined from the concentrations and mass-specific light absorption coefficients of the constituents[17] (matter in the unpackaged state) of the particles. Here we give the spectral indices $n'_{p,dry}$, just for "pure heavy particles" (purely organic particles), that is, with densities exceeding that of the surrounding water and containing pure organic matter, without any interstitial water or air.[18] These particles have a decidedly negative buoyancy, so that in the absence of other forces acting on them in the water (a condition not normally fulfilled in the sea), they will sink freely to the bottom. For simplicity, we therefore assumed that the total concentration of organic matter C_{OM} in all such particles is identical and approximately equal to the density of the dry (or pure) organic matter ρ_{OM};[19] that is,. $C_{OM} = \rho_{OM} \approx 1400$ kg m^{-3}. The indices $n'_{p,dry}$ for a wide spectral range (200–700 nm) are given in Table 5.18 and illustrated in Figure 5.21.

---

[17] The absorption coefficients of these various constituents are given in the present work, for example, in Tables 4.3, 4.8, 4.9, 6.8, and Equation (6.3) or are available in the publications of the other authors cited in Table 5.17.

[18] POM (particulate organic matter) can be divided into a number of physical types and subtypes as a consequence of their various densities, which are due to the different proportions of gas (air) and water that they contain besides organic substances. Apart from the pure heavy type mentioned here (purely organic particles), other types are distinguishable, such as neutral particles (density approaching that of water), dry particles (containing organics and gas), and wet particles (additionally containing water), as well as light particles (density less than that of sea water), which can be either wet or dry. This issue is accorded detailed treatment in Woźniak et al. (2005a, b).

[19] In fact this density of dry organic matter differs slightly for different organic substances: this is evident from the data in column 4 of Table 5.10.

TABLE 5.18. The imaginary part of the absolute refractive index of UV–VIS light of selected wavelengths for (1) 26 model chemical classes of POM occurring in the form of pure organic matter $n'_{p,dry}$ (for $\rho_{OM} \approx 1400$ kg m^{-3}) and (2) water particles n'_w [a].

Particle	Imaginary part of the refractive index n′ for wavelength λ [nm]:													
type	200	230	280	300	340	400	420	440	500	550	600	650	675	700
-1-	-2-	-3-	-4-	-5-	-6-	-7-	-8-	-9-	-10-	-11-	-12-	-13-	-14-	-15-
A1	–	6.92×10^{-02}	7.49×10^{-02}	2.34×10^{-03}	(0)	(0)	(0)	(0)	(0)	(0)	(0)	(0)	(0)	(0)
A2	–	–	–	$9.26\times10^{+00}$	$1.36\times10^{+01}$	–	(0)	(0)	(0)	(0)	(0)	(0)	(0)	(0)
A3	–	2.73×10^{-02}	2.39×10^{-02}	1.70×10^{-02}	1.40×10^{-02}	1.04×10^{-02}	9.37×10^{-03}	8.42×10^{-03}	6.03×10^{-03}	4.52×10^{-03}	3.36×10^{-03}	2.48×10^{-03}	2.12×10^{-03}	1.82×10^{-03}
P1	(0)	(0)	(0)	(0)	–	7.13×10^{-01}	$1.18\times10^{+00}$	$1.77\times10^{+00}$	$1.03\times10^{+00}$	2.45×10^{-01}	1.83×10^{-01}	3.15×10^{-01}	7.01×10^{-01}	1.66×10^{-01}
P2	(0)	(0)	(0)	(0)	–	3.16×10^{-01}	4.30×10^{-01}	4.88×10^{-01}	2.74×10^{-01}	1.31×10^{-01}	8.00×10^{-02}	1.23×10^{-01}	3.45×10^{-01}	4.66×10^{-02}
N	–	1.61×10^{-01}	4.27×10^{-01}	7.02×10^{-02}	(0)	(0)	(0)	(0)	(0)	(0)	(0)	(0)	(0)	(0)
L	4.72×10^{-01}	2.02×10^{-01}	8.73×10^{-02}	5.01×10^{-02}	2.42×10^{-02}	2.45×10^{-03}	(0)	(0)	(0)	(0)	(0)	(0)	(0)	(0)
H1	–	1.31×10^{-01}	9.05×10^{-02}	7.35×10^{-02}	4.55×10^{-02}	2.05×10^{-02}	1.64×10^{-02}	1.31×10^{-02}	6.68×10^{-03}	4.04×10^{-03}	2.34×10^{-03}	–	–	–
H2	–	–	–	2.92×10^{-01}	1.65×10^{-01}	6.77×10^{-02}	5.01×10^{-02}	3.70×10^{-02}	1.47×10^{-02}	–	–	–	–	–
H3	5.56×10^{-02}	4.58×10^{-02}	3.20×10^{-02}	2.75×10^{-02}	2.00×10^{-02}	1.21×10^{-02}	1.02×10^{-02}	8.52×10^{-03}	4.97×10^{-03}	3.14×10^{-03}	1.97×10^{-03}	1.22×10^{-03}	9.63×10^{-04}	7.56×10^{-04}
H4	2.72×10^{-02}	2.22×10^{-02}	1.53×10^{-02}	1.31×10^{-02}	9.38×10^{-03}	5.57×10^{-03}	4.66×10^{-03}	3.88×10^{-03}	2.23×10^{-03}	1.39×10^{-03}	8.55×10^{-04}	5.24×10^{-04}	4.09×10^{-04}	3.19×10^{-04}
H5	–	–	–	4.07×10^{-02}	2.08×10^{-02}	7.35×10^{-03}	5.18×10^{-03}	3.63×10^{-03}	1.24×10^{-03}	–	–	–	–	–
H6	–	–	2.66×10^{-02}	2.29×10^{-02}	1.67×10^{-02}	1.02×10^{-02}	8.56×10^{-03}	7.20×10^{-03}	4.23×10^{-03}	2.68×10^{-03}	1.69×10^{-03}	1.06×10^{-03}	8.33×10^{-04}	6.56×10^{-04}
H7	–	–	3.69×10^{-03}	2.74×10^{-03}	1.49×10^{-03}	5.79×10^{-04}	4.21×10^{-04}	3.05×10^{-04}	1.15×10^{-04}	5.04×10^{-05}	2.19×10^{-05}	9.46×10^{-06}	6.20×10^{-06}	4.06×10^{-06}
H8	–	–	5.98×10^{-03}	4.75×10^{-03}	3.01×10^{-03}	1.54×10^{-03}	1.24×10^{-03}	9.95×10^{-04}	5.26×10^{-04}	3.14×10^{-04}	1.89×10^{-04}	1.14×10^{-04}	8.88×10^{-05}	6.92×10^{-05}
H9	–	–	5.76×10^{-03}	4.19×10^{-03}	2.18×10^{-03}	8.02×10^{-04}	5.71×10^{-04}	4.06×10^{-04}	1.44×10^{-04}	6.01×10^{-05}	2.48×10^{-05}	1.02×10^{-05}	6.52×10^{-06}	4.17×10^{-06}
H10	–	–	–	–	–	2.58×10^{-02}	2.27×10^{-02}	1.98×10^{-02}	1.31×10^{-02}	9.21×10^{-03}	6.41×10^{-03}	4.43×10^{-03}	3.67×10^{-03}	3.04×10^{-03}
H11	–	–	–	–	–	1.29×10^{-02}	1.09×10^{-02}	9.16×10^{-03}	5.38×10^{-03}	3.41×10^{-03}	2.15×10^{-03}	1.34×10^{-03}	1.06×10^{-03}	8.34×10^{-04}
H12	–	–	–	–	–	4.01×10^{-03}	3.48×10^{-03}	3.02×10^{-03}	1.94×10^{-03}	1.33×10^{-03}	9.00×10^{-04}	6.06×10^{-04}	4.96×10^{-04}	4.06×10^{-04}
H13	–	–	–	–	–	2.36×10^{-03}	1.85×10^{-03}	1.44×10^{-03}	6.75×10^{-04}	3.55×10^{-04}	1.85×10^{-04}	9.61×10^{-05}	4.77×10^{-04}	4.95×10^{-05}
Ph1	–	*1.87×10^{-02}*	*1.70×10^{-02}*	*8.50×10^{-03}*	*7.00×10^{-03}*	2.70×10^{-02}	4.10×10^{-02}	5.88×10^{-02}	3.47×10^{-02}	9.72×10^{-03}	7.26×10^{-03}	1.09×10^{-02}	2.50×10^{-02}	6.00×10^{-03}
Ph2	–	*1.87×10^{-02}*	*1.70×10^{-02}*	*8.50×10^{-03}*	*7.00×10^{-03}*	2.14×10^{-02}	2.70×10^{-02}	2.96×10^{-02}	1.72×10^{-02}	8.99×10^{-03}	5.74×10^{-03}	7.63×10^{-03}	1.93×10^{-02}	3.28×10^{-03}
Z	–	1.91×10^{-02}	1.67×10^{-02}	1.19×10^{-02}	9.81×10^{-03}	7.28×10^{-03}	6.56×10^{-03}	5.89×10^{-03}	4.22×10^{-03}	3.16×10^{-03}	2.35×10^{-03}	1.73×10^{-03}	1.49×10^{-03}	1.27×10^{-03}
D1	–	3.04×10^{-03}	3.10×10^{-03}	1.72×10^{-03}	1.11×10^{-03}	6.82×10^{-04}	5.77×10^{-04}	4.88×10^{-04}	2.95×10^{-04}	1.93×10^{-04}	1.26×10^{-04}	–	–	–
D2	–	1.14×10^{-01}	7.59×10^{-02}	5.22×10^{-02}	3.05×10^{-02}	1.25×10^{-02}	1.00×10^{-02}	8.05×10^{-03}	4.16×10^{-03}	2.59×10^{-03}	1.56×10^{-03}	9.37×10^{-04}	7.29×10^{-04}	5.68×10^{-04}
PhM	–	*1.87×10^{-02}*	*1.70×10^{-02}*	4.17×10^{-02}	6.52×10^{-02}	2.35×10^{-02}	2.73×10^{-02}	3.08×10^{-02}	1.70×10^{-02}	3.73×10^{-03}	3.96×10^{-03}	5.53×10^{-03}	1.65×10^{-02}	8.46×10^{-04}
n_w'	4.89×10^{-08}	1.70×10^{-08}	6.42×10^{-09}	3.37×10^{-09}	8.79×10^{-10}	2.11×10^{-10}	1.52×10^{-10}	2.22×10^{-10}	8.12×10^{-10}	2.47×10^{-09}	1.06×10^{-08}	1.76×10^{-08}	2.41×10^{-08}	3.48×10^{-08}

[a] After Woźniak et al. (2005b).

[b] The values of the index n'_p were not defined; (0), the values of the index $n'_{p,dry}$ are much smaller than in other spectral intervals; extrapolated values are given in italics.

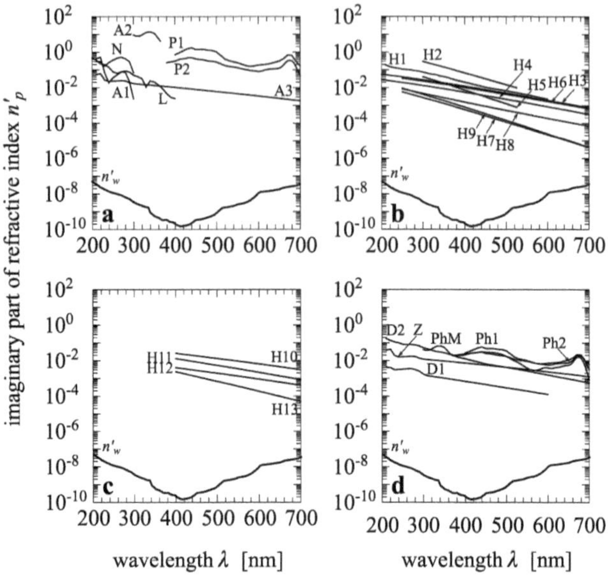

FIGURE 5.21. Hypothetical spectra of the imaginary part of the absolute refractive index for various chemical classes of pure organic matter of density $\rho_{OM} \approx 1400$ kg m^{-3}: (a) single-component organic particles; (b) particles containing organic humus matter originating from different marine basins; (c) particles containing organic humus matter produced in the soil and bottom sediments; (d) multicomponent organic particles. The solid line in the graphs illustrates the spectrum of the absolute imaginary refractive index of water $n'_w(\lambda)$. The symbols on the plots correspond to the model chemical classes of POM defined in Table 5.17.

The data in Table 5.18 and Figure 5.21 show that the indices $n'_{p,dry}$ of these pure model groups of organic matter exceed by many orders of magnitude the imaginary refractive indices of water, and they are also greater than these indices for minerals (except hematite) present in marine SPM (see Figure 5.11). The spectral structures of some of these groups are complex, especially those of particles containing one organic component (Figure 5.21a), whereas the $n'_{p,dry}(\lambda)$ spectra of others (humus substances; see Figures 5.21 b,c) run a more gentle course, without marked extremes. These spectral properties of POM are inextricably bound up with the molecular structure of the organic compounds; inasmuch as we discussed this aspect in Chapters 3 and 4, we do not analyze it further here. The practical utility of the results is, however, worth examining.

The spectral values of the imaginary refractive indices $n'_{p,dry}$ of all 26 chemical classes of particles containing pure organic matter (i.e., the physical type of purely organic matter) (see Table 5.18) can be used to define these indices n'_p of practically all physical types of particles that occur numerously in the sea. All that is needed to do this is knowledge of the intraparticulate concentration of organic matter C_{OM}. The imaginary part of the refractive indices of various physical types of particles can then be determined from the approximate relationship:

$$n'_p \approx n'_{p,dry} \frac{C_{OM}}{\rho_{OM}} , \qquad (5.18)$$

where $n'_{p,dry}$ and n'_p are the respective imaginary refractive indices of particles containing dry organic matter (data in Table 5.18) and of particles with any intraparticulate concentration of organic matter; C_{OM}, ρ_{OM} is the density of dry organic matter assumed in this model (ρ_{OM} = 1400 kg m^{-3}).

In Equation (5.18), the effect of the absorption properties of water on the value of n'_p of particles containing water, being negligible, was omitted. Nevertheless, these indices for the physical types of particles occurring most frequently in water, that is, neutral wet particles and the like, and also heavy particles containing water, take values that are directly proportional to their organic matter content C_{OM} and do not exhibit any conspicuous dependence on the absorption properties of water, even though these formally exist. The explanation for this is that the light absorption by organic matter (even though there is not much of it) exceeds by far the light absorption by water. This emerges from the obvious fact that the mass-specific absorption coefficients of UV–VIS light for water are many orders of magnitude smaller than the mass-specific absorption coefficients of the organic substances under scrutiny here. Light absorption by water of concentration C_w contained in particles may therefore dominate the bulk absorption by these particles only when the organic matter content is extremely small, for example, for $C_{OM}/C_w \approx 10^{-7}$ and less.

The spectra, presented above, of the imaginary refractive index $n'_{p,i}(\lambda)$ for the model chemical classes of POM can also be used to determine such spectra characteristic of different morphological groups of suspended particles $n'_{p,morph}(\lambda)$, which are mixtures of organic substances from several chemical classes. These bulk $n'_{p,morph}(\lambda)$ for a given morphological form can be determined on the basis of known relative volumes v_i or the relative mass contents μ_i of their various constituents with the aid of relationships analogous to Equations (5.14) and (5.15):

$$n'_{p,morph}(\lambda) = \sum_i n'_{p,i}(\lambda) v_i \qquad (5.19)$$

and

$$n'_{p,morph}(\lambda) = \rho_{p,morph} \sum_i \frac{n'_{p,i}(\lambda)\mu_i}{\rho_{p,i}} \qquad (5.20)$$

where $\rho_{p,i}$ is the density of the separate constituents i, and $\rho_{p,morph}$ is the bulk density of a particle of this particular morphological form.

Below we also discuss the practical application of these modeled indices n'_p for the model chemical classes of POM for determining these indices of particles of organic matter belonging to different morphological groups.

The Formation, Structure, Composition, and Optical Constants of Different Morphological Groups of Marine Particles

Particles of organic detritus can be classified into a number of morphological types and subtypes. They differ not only in size and external structure, but also in chemical composition, and hence they exhibit diverse optical properties. These differences result from the processes of their formation and subsequent evolution in the water. Figure 5.22 illustrates a simplified division of marine POM into morphological groups, with special emphasis on organic detritus. It also gives the approximate size ranges of these groups of particles and their typical relative proportions in the oceans. In contrast, Figure 5.23 is a simplified diagram showing how these various morphological groups of particles formed and evolved. Although conceived by us, both figures (i.e., 5.22 and 5.23) are based on the relevant knowledge available in the literature (e.g., monographs and collective works: Romankevich (1977), Wassmann and Heiskanen (1988), Wassmann et al. (1991); see also Woźniak et al. (2006) and

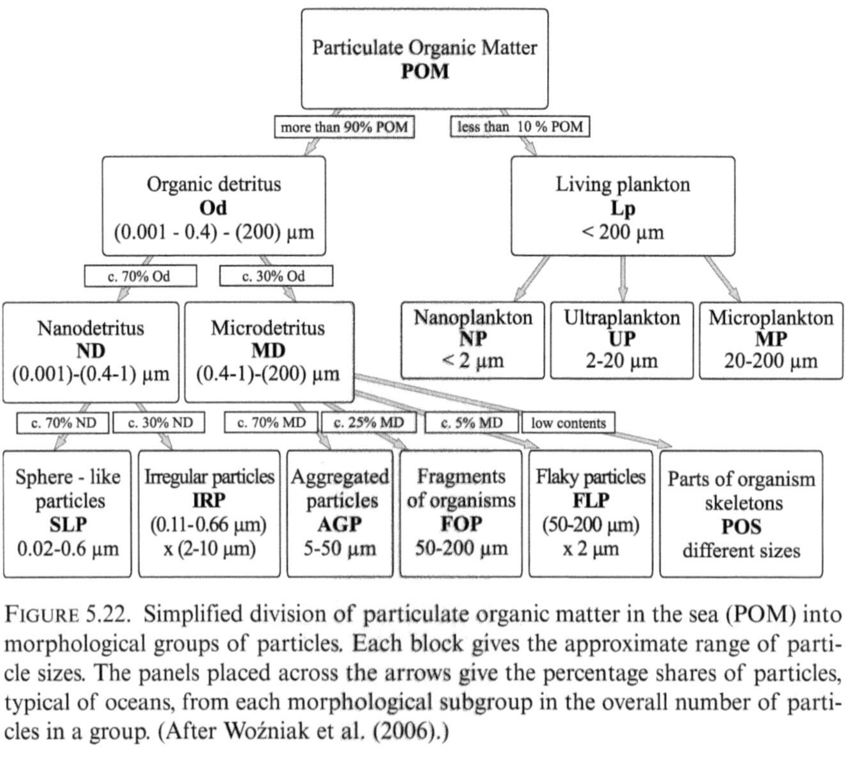

FIGURE 5.22. Simplified division of particulate organic matter in the sea (POM) into morphological groups of particles. Each block gives the approximate range of particle sizes. The panels placed across the arrows give the percentage shares of particles, typical of oceans, from each morphological subgroup in the overall number of particles in a group. (After Woźniak et al. (2006).)

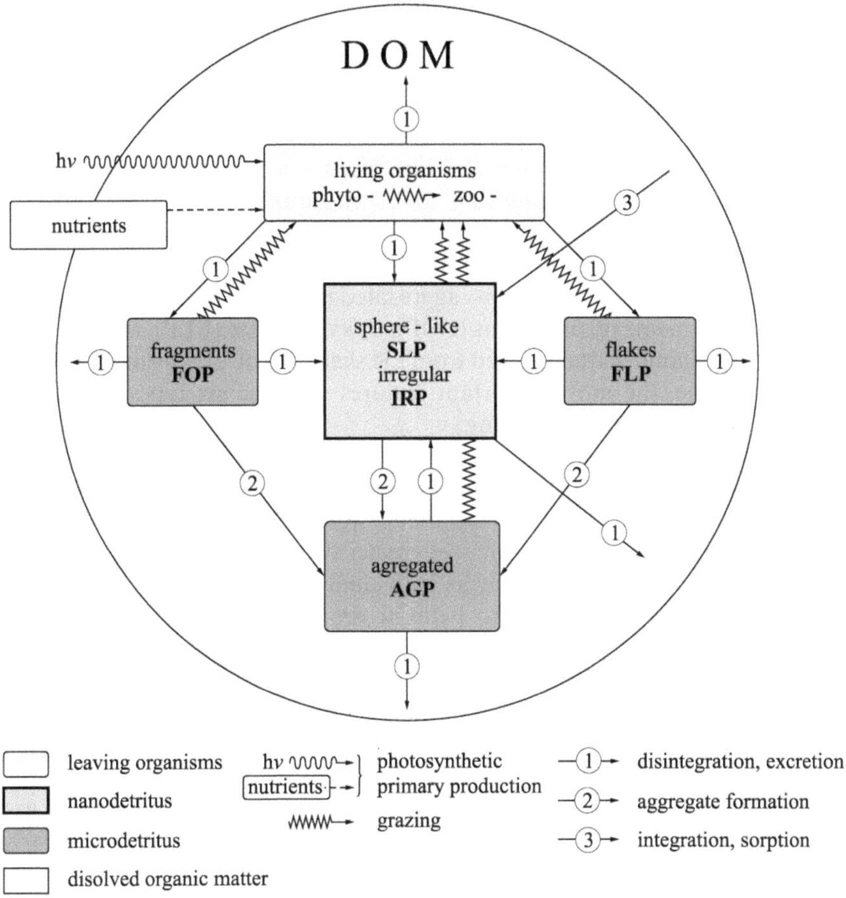

FIGURE 5.23. Simplified diagram of the formation and evolution of the various morphological forms of POM (after Woźniak et al. 2006).

the works of various authors cited therein). The terminology of the various morphological groups was taken from Romankevich (1977).

Figure 5.22 shows that POM in the sea contains both living and nonliving components. The living part of POM consists primarily of phyto- and zooplanktonic organisms Lp, which usually make up less than 10% of the overall mass of POM in the sea. The second preponderant part of POM (usually 90% and more of the mass of POM) consists of organic detritus. With respect to particle size, this component of POM usually breaks down into two morphological groups. One comprises the fraction of fine-grained organic detritus less than 1 μm in size, which, by analogy to the nanoplankton, can be referred to as nanodetritus (ND). The other morphological group of organic detritus consists of particles from 1 to 200 μm in size; this can be called microdetritus (MD). These two principal groups of organic detritus can be further subdivided into numerous smaller groups of particles differing not only in their size ranges and shapes, but also, as it turns out, in their chemical and physical properties, origin, and numbers. Because these particles are complex in nature and have been researched to only a limited extent, we decided to distinguish, to a first approximation, six types of organic detritus particles: two types of fine-grained particles differing in shape—spherelike particles (SLP) and irregular particles (IRP)—and four types of coarse-grained particles—aggregated particles (AGP), particles consisting of fragments of organisms (FOP), flaky particles (FLP), and particles containing organic matter derived from the skeletons of organisms (POS). To our knowledge, the most important features of these six types of organic detritus particles are the following:

- *Spherelike particles* (SLP) make up some 70% of the mass of organic nanodetritus (ND). Usually resembling spheres, from 0.02 to 0.6 μm in diameter and with molecular masses greater than 10,000 Daltons, these suspended particles occur not only as individual spheres, but also as chains of a few spheres, aggregates up to 1 μm in size consisting of many spherical forms, and all sorts of clusters of fibrous aggregates 4–10 μm in size made up of 0.01–0.16 μm diameter spheres. A number of spheres may combine to form clouds, which take on all manner of bizarre shapes. These spherical forms are built up of various organic compounds; the discrete spheres, however, are usually homogeneous: they consist of the same substance.
- *Irregular particles* (IRP) make up c. 30% of the mass of ND. 0.11–0.66 μm in diameter and 2–10 Å in thickness; they comprise a diversity of irregularly shaped suspended particles, among which transparent, thin-layered forms are common. Spherical structures are absent. As in the case of SLP, IRP are mostly homogeneous chemically, containing many of the different organic compounds present in sea water.
- *Aggregates* (AGP) make up c. 70% of the mass of coarse-grained organic detritus (MD). Highly diverse in internal structure, these are heterogeneous forms with sizes varying widely, from 1 μm and less to 200 μm and more, but usually lying within the 5–50 μm range. Some numerous forms smaller than 5 μm or greater than 50 μm can also be classified among the AGP. They are conglomerations of smaller, wet and dry fragments of organic matter consisting of both plant[20] and animal parts, and of fragments of such matter not derived directly from organisms, for example, certain amounts of humus, and interstices, which may be filled with bacteria and mineral admixtures. The proportions between the components—the smaller

[20] These may also be live cells of, for example, diatoms: the fact that AGP particles fluoresce in the red region is evidence for the presence in them of "live" chlorophyll.

substructures of aggregates—vary in different seas and at different depths. In most cases, however, their organic content is not identified as being the immediate remains of organisms. It is also likely that the major part of this organic matter consists of carbohydrates and their derivatives. Some authors (e.g., Bogdanov and Lisitzin (1968), Silver et al. (1991), and Lin et al. (1989)) have distinguished among the AGP yet other subtypes differing with respect to origin, morphology, dimensions, and content: for example, larger yellow-gray particles, less spherical, with a fairly irregular structure, and yellow-brown particles, quite spherical with a waxlike structure, usually less than 2.5 μm in size.

- *Fragments of organisms* (FOP) comprise c. 25% of MD. These are variously shaped wet particles, from 50 to 200 μm in size, derived from zoo- or phytoplanktonic organisms, the nekton, or their feces.
- *Flaky particles* (FLP) generally make up no more than 5% of the mass of MD. Like the two previous subgroups of coarse-grained detritus, this one is also directly biogenic. FLP are dry (anhydrous) fragments of animal or plant organisms, which may additionally contain some inorganic compounds. Occurring in the form of flakes, they are generally homogeneous, 50–200 μm in size and c. 2 μm thick. FLP of smaller dimensions (15–35 μm) have on occasion also been recorded.
- *Organic matter in skeletons* (POS). This last group of coarse-grained organic detritus consists of organic matter enclosed, as it were, in the remains of the skeletons of organisms or in various kinds of cocoons and shells. Because it occurs in very much smaller amounts (sporadically) than the other subgroups, we have not analyzed it.

The optical properties, in particular the optical constants, of the above-mentioned morphological groups of organic detritus depend on the compositions and concentrations of the various organic compounds they contain. The chemical composition of organic detritus in the sea conceived in this way has not yet been experimentally researched to any great extent, therefore we have attempted to reproduce an approximate version of it on the basis of the fragmentary information available on the formation and evolution of this detritus in the conditions obtaining in the sea. Although to produce a reasonably exact description of the mechanisms of the formation, evolution, and disappearance of the different forms of POM, taking into account the various factors governing these processes and all the interrelationships among them, is an exceedingly complex undertaking, it is nevertheless a feasible one. This is because these elements constitute one of the links in the biogeochemical cycle of organic carbon; the subject of analysis for many years now; this has been modeled quantitatively on different scales with varying degrees of success (see, e.g., Libes (1992)).

Nonetheless, the diagrams of this cycle published in the accessible literature (see, e.g., Lancelot and Billen (1985), Leppanen (1989), Pempkowiak (1997), and Lee (2005)) usually show organic detritus lumped together as a single entity, with no distinction being made for its various morphological forms, and no account being taken of the flow of organic matter among them. Hence, for the purposes of the present volume, we have drawn a simplified diagram, based on empirical knowledge gleaned from Romankevich (1977), Wassmann and Heiskanen (1988), and Wassmann et al. (1991), showing the formation and transformation of all the main morphological types of POM, with particular emphasis on organic detritus (see Figure 5.23). It applies to only one link in the multistage organic carbon cycle in the marine ecosystem, namely, the formation and flow of organic carbon through POM. Restricted, moreover, to the main streams of carbon, the most abundant ones, the diagram does not pretend to preserve the zero balance of matter flowing through the ecosystem. It does, however, state premises, important with regard to the aims of this book, enabling the approximate compositions and intraparticulate concentrations of organic matter in particles belonging to the various morphological forms of Od to be established. To this end, then, the following assumptions were made:

- Individual particles of the various morphological types of organic detritus (Od) are formed as a result of:
  - The degradation of and/or secretions by living forms of POM and nekton;
  - The degradation or aggregation of other morphological types of Od;
  - The sorption, deposition, and precipitation of DOM.
- The formation and transformation of particles take place in accordance with the diagram in Figure 5.23 and in the order shown in this diagram.[21]
- In many cases the composition of the organic matter and the concentration of given portions of it (not necessarily whole particles), transferred from one morphological type to another as a result of particle transformation, changes insignificantly, if at all.
- Nevertheless, these two properties (the composition and intraparticulate concentration of organic matter) of "whole" particles, belonging to two adjacent morphological types in this chain of cause and effect, may also differ. This situation may occur, for instance, when:
  - New particles come into existence as a result of the selective decomposition (or secretion) of the source particle into fragments with organic

[21] This order is the following. (1) Initially, mainly as a result of only the degradation of living organisms, particles that are fragments of organisms (FOP) and flaky particles (FLP), belonging to the microdetritus, are formed. (2) The degradation of organisms also gives rise to spherelike particles (SLP) and irregular particles (IRP), which belong to the nanodetritus. The process by which SLP and IRP are generated is reinforced by the decomposition of other detritus particles and also by the sorption of DOM. (3) Finally, the microdetritus of the aggregated particle type (AGP) is not directly biogenic, coming into existence chiefly by the aggregation of all the other morphological types of detritus.

matter compositions or concentrations differing from these properties averaged for the whole source particle;

— New particles come into existence as a result of the aggregation of fragments with differing physical or chemical properties, and/or the sorption of DOM;

— Particles coming into existence as a result of aggregation are able to "imprison" certain amounts of water or gas, or other inorganic substances.

- We have not taken into consideration possible modifications to particles during their existence in a given morphological form due to chemical, biochemical, and physicochemical processes, such as chemical decomposition, bacterial degradation, humification, hydration or dehydration, and aeration or deaeration. These "internal" processes, however, appear to exert a far smaller influence on the chemical and physical diversity of suspended particles than the processes during which they come into existence.

On the basis of the diagram showing the formation and evolution of organic detritus particles (Figure 5.23) and the above-mentioned simplifying assumptions, we have distinguished 25 variants of this detritus, belonging to different morphological groups. Moreover, we have estimated the probable organic matter compositions and the probable intraparticulate concentrations C_{OM}[22] of these variants. The details of this procedure are explained in Woźniak et al. (2006). Table 5.19 presents the results, giving the organic matter composition of each of the 25 distinct morphological variants of detritus as the proportion of the organic matter content in the model chemical classes (see Table 5.17). In defining the values of C_{OM} given in the table for these variants of detritus particles, the possible aeration and dehydration of these particles during their evolution has been accounted for, as have the proportions among these components, data on the densities of various organic substances, individually or in groups, in the dry state, as well as other data given, inter alia, in Tables 5.10, 5.12, and 5.13. Table 5.19 also gives the approximate values for these particles of n_p, the real part of the absolute refractive index in the VIS region. They were determined approximately or by the relevant summation, weighted for relative volume or the relative mass content (see Equations (5.10) and (5.11) in Section 5.11), of the individual particle components with known indices n_i, or else using the empirical formula (5.13) (Section 5.2.3) on the basis of known average concentrations of carbon in the particles C_c. These concentrations were assumed to be approximately half the value of C_{OM} ($C_c \approx 0.5\ C_{OM}$). The last step in the model was to determine the bulk imaginary refractive indices for these 25 morphological variants of detritus particles. This was done using Equations (5.18)–(5.20) in combination with data on the composition and intraparticulate concentration of all

[22] There may, of course, be more such variants of detritus with different compositions of the earlier-mentioned model chemical classes.

TABLE 5.19. Selected chemical and physical characteristics of model morphological groups of POM.[a,b]

A. Nanodetritus, Spherelike Particles (SLP) and Irregular Particles (IRP)

Symbol	Composition of organic matter	Physical type[c]	C_{OM} [kg m^{-3}]	$n'_{p,VIS}$ [Dimensionless]
-1-	-2-	-3-	-4-	-5-
SLP(A1), IRP(A1)	As chemical class A1	PH	1220	1.57
SLP(A2), IRP(A2)	As chemical class A2	PH	1220	1.57
SLP(A3), IRP(A3)	As chemical class A3	PH	1220	1.57
SLP(P1), IRP(P1)	As chemical class P1	PH	1120	1.50
SLP(P2), IRP(P2)	As chemical class P2	PH	1120	1.50
SLP(N), IRP(N)	As chemical class N	PH	1530	1.55
SLP(L), IRP(L)	As chemical class L	PH	1220	1.57
SLP(H1), IRP(H1)	As chemical class H1	PH	1530	1.55
SLP(H4), IRP(H4)	As chemical class H4	PH	1530	1.55
SLP(D1)	As chemical class D1	PH	(1400)	(1.54)
IRP(D2)	As chemical class D2	PH	(1400)	(1.54)

B. Microdetritus, Aggregated Particles (AGP)

Symbol	Composition of organic matter	Physical type[c]	C_{OM} [kg m^{-3}]	$n'_{p,VIS}$ [Dimensionless]
-1-	-2-	-3-	-4-	-5-
AGP(D1)	As chemical class D1	PH	(1400)	(1.54)
AGP(D2)	As chemical class D2	PH	(1400)	(1.54)
AGP(D1/Ph1/Z)	As mixture of morphological groups IRP(D1)-FOP(Z)-FLP(Z)-FOP(Ph1)-FLP(Ph1) Be found in proportions, respectively: 0.8861; 0.0854; 0.0171; 0.0095; 0.0019	NW	(862)	(1.44)
AGP(D2/Ph2/Z)	As mixture of optico - morphological groups IRP(D)-FOP(Z)-FLP(Z)-FOP(Ph2)-FLP(Ph2) Be found in proportions, respectively: 0.8861; 0.0475; 0.0095; 0.0475; 0.0095	NW	(760)	(1.43)
AGP(Ph1/Z)	As mixture of optico - morphological groups FOP(Z)-FLP(Z)-FOP(Ph1)-FLP(Ph1) Be found in proportions, respectively: 0.7498; 0.1501; 0.0834; 0.0167	NW	(210)	(1.40)

TABLE 5.19. Selected chemical and physical characteristics of model morphological groups of POM.[a,b]—Cont'd.

B. Microdetritus, Aggregated Particles (AGP)

Symbol	Composition of organic matter	Physical type[c]	C_{OM} [kg m^{-3}]	$n'_{p,VIS}$ [Dimensionless]
-1-	*-2-*	*-3-*	*-4-*	*-5-*
AGP(Ph2/Z)	As mixture of optico - morphological groups FOP(Z)-FLP(Z)-FOP(Ph2)-FLP(Ph2) Be found in proportions, respectively: 0.4167; 0.0833; 0.4167; 0.0833	NW	(150)	(1.40)

C. Microdetritus, Fragments of Organisms Particles (FOP)

Symbol	Composition of organic matter	Physical type[c]	C_{OM} [kg m^{-3}]	$n'_{p,VIS}$ [Dimensionless]
-1-	*-2-*	*-3-*	*-4-*	*-5-*
FOP(Ph1)	As chemical class Ph1	NW	(110)	(1.38)
FOP(Ph2)	As chemical class Ph2	NW	(110)	(1.38)
FOP(PhM)	As chemical class PhM	NW	(110)	(1.38)
FOP(Z)	As chemical class Z	NW	(200)	(1.38)

D. Microdetritus, Flake Particles (FLP)

Symbol	Composition of organic matter	Physical type[c]	C_{OM} [kg m^{-3}]	$n'_{p,VIS}$ [Dimensionless]
-1-	*-2-*	*-3-*	*-4-*	*-5-*
FLP(Ph1)	As chemical class Ph1	PH	1300	1.54
FLP(Ph2)	As chemical class Ph2	PH	1300	1.54
FLP(PhM)	As chemical class PhM	PH	1300	1.54
FLP(Z)	As chemical class Z	PH	1250	1.55

[a] Adapted from Woźniak et al. (2006);
[b] C_{OM}, intraparticulate concentration of organic matter; $n'_{p,VIS}$, imaginary part of the absolute refractive index in the VIS range
[c] Physical types: PH, pure heavy particles; NW, neutral (free) wet particles or similar particles.

organic matter in these particles, given in Table 5.19. Results of this calculation are given in Table 5.20 and depicted graphically in Figure 5.24. For comparison, this figure also shows the earlier-mentioned empirical spectra of the imaginary refractive index $n'_{p,SS}(\lambda)$ for organic detritus in the Sargasso Sea (after Stramski et al. 2001), along with the spectra of $n'_w(\lambda)$ for water.

It is clear from the data in Table 5.20 and Figure 5.24 that the theoretically estimated spectral values of indices n'_p for the morphological variants of organic detritus distinguished here are quite widely differentiated, particularly in the case of SLP and IRP from the nanodetritus (ND) group. The highest values, of the order of $n'_p \approx 10$ and more in the UV region, and of

TABLE 5.20. The imaginary part of the absolute refractive index at selected wavelengths of UV–VIS light for model morphological groups of POM, n'_p.[a, b]

Particle groups	Imaginary part of the refractive index n'_p for wavelength λ [nm]:												
	200	230	280	300	340	400	420	440	500	550	600	650	675	700
-1-	*-2-*	*-3-*	*-4-*	*-5-*	*-6-*	*-7-*	*-8-*	*-9-*	*-10-*	*-11-*	*-12-*	*-13-*	*-14-*	*-15-*
SLP(A1), IRP(A1)	–	6.03×10^{-02}	6.53×10^{-02}	2.04×10^{-03}	(0)	(0)	(0)	(0)	(0)	(0)	(0)	(0)	(0)	(0)
SLP(A2), IRP(A2)	–	–	–	$8.07\times10^{+00}$	$1.19\times10^{+01}$	–	(0)	(0)	(0)	(0)	(0)	(0)	(0)	(0)
SLP(A3), IRP(A3)	–	2.38×10^{-02}	2.08×10^{-02}	1.48×10^{-02}	1.22×10^{-02}	9.06×10^{-03}	8.17×10^{-03}	7.34×10^{-03}	5.25×10^{-03}	3.94×10^{-03}	2.93×10^{-03}	2.16×10^{-03}	1.85×10^{-03}	1.59×10^{-03}
SLP(P1), IRP(P1)	(0)	(0)	(0)	(0)	–	5.70×10^{-01}	9.44×10^{-01}	$1.42\times10^{+00}$	8.24×10^{-01}	1.96×10^{-01}	1.46×10^{-01}	2.52×10^{-01}	5.61×10^{-01}	1.33×10^{-01}
SLP(P2), IRP(P2)	(0)	(0)	(0)	(0)	–	2.53×10^{-01}	3.44×10^{-01}	3.90×10^{-01}	2.19×10^{-01}	1.05×10^{-01}	6.40×10^{-02}	9.84×10^{-02}	2.76×10^{-01}	3.73×10^{-02}
SLP(N), IRP(N)	–	1.76×10^{-01}	4.67×10^{-01}	7.67×10^{-02}	(0)	(0)	(0)	(0)	(0)	(0)	(0)	(0)	(0)	(0)
SLP(L), IRP(L)	4.11×10^{-01}	1.76×10^{-01}	7.61×10^{-02}	4.37×10^{-02}	2.11×10^{-02}	2.14×10^{-03}	(0)	(0)	(0)	(0)	(0)	(0)	(0)	(0)
SLP(H1), IRP(H1)	–	1.43×10^{-01}	9.89×10^{-02}	8.03×10^{-02}	4.97×10^{-02}	2.24×10^{-02}	1.79×10^{-02}	1.43×10^{-02}	7.30×10^{-03}	4.42×10^{-03}	2.56×10^{-03}	–	–	–
SLP(H4), IRP(H4)	2.97×10^{-02}	2.43×10^{-02}	1.67×10^{-02}	1.43×10^{-02}	1.03×10^{-02}	6.09×10^{-03}	5.09×10^{-03}	4.24×10^{-03}	2.44×10^{-03}	1.52×10^{-03}	9.34×10^{-04}	5.73×10^{-04}	4.47×10^{-04}	3.49×10^{-04}
IRP(D1)	–	3.04×10^{-03}	3.10×10^{-03}	1.72×10^{-03}	1.11×10^{-03}	6.82×10^{-04}	5.77×10^{-04}	4.88×10^{-04}	2.95×10^{-04}	1.93×10^{-04}	1.26×10^{-04}	–	–	–
IRP(D2)	–	1.14×10^{-01}	7.59×10^{-02}	5.22×10^{-02}	3.05×10^{-02}	1.25×10^{-02}	1.00×10^{-02}	8.05×10^{-03}	4.16×10^{-03}	2.59×10^{-03}	1.56×10^{-03}	9.37×10^{-04}	7.29×10^{-04}	5.68×10^{-04}
AGP(D1)	–	3.04×10^{-03}	3.10×10^{-03}	1.72×10^{-03}	1.11×10^{-03}	6.82×10^{-04}	5.77×10^{-04}	4.88×10^{-04}	2.95×10^{-04}	1.93×10^{-04}	1.26×10^{-04}	–	–	–
AGP(D2)	–	1.14×10^{-01}	7.59×10^{-02}	5.22×10^{-02}	3.05×10^{-02}	1.25×10^{-02}	1.00×10^{-02}	8.05×10^{-03}	4.16×10^{-03}	2.59×10^{-03}	1.56×10^{-03}	9.37×10^{-04}	7.29×10^{-04}	5.68×10^{-04}
AGP(D1/ Ph1/Z)	–	–	–	–	–	1.02×10^{-03}	–	–	6.71×10^{-04}	3.73×10^{-04}	2.68×10^{-04}	–	–	–

AGP(D2/ Ph2/Z)	–	–	–	–	–	6.92×10^{-03}			2.66×10^{-03}	1.62×10^{-03}	9.99×10^{-04}	7.41×10^{-04}		4.14×10^{-04}
AGP(Ph1/Z)	–	–	–	–	–	1.39×10^{-03}	1.50×10^{-03}	1.68×10^{-03}	1.09×10^{-03}	5.73×10^{-04}	4.26×10^{-04}	3.98×10^{-04}		2.62×10^{-04}
AGP(Ph2/Z)	–	–	–	–	–	1.54×10^{-03}	1.80×10^{-03}	1.90×10^{-03}	1.15×10^{-03}	6.51×10^{-03}	4.33×10^{-03}	5.02×10^{-03}	1.11×10^{-03}	2.44×10^{-03}
FOP(Ph1)	–	*1.07×10^{-03}*	*9.40×10^{-04}*	*6.65×10^{-04}*	*5.50×10^{-04}*	2.12×10^{-03}	3.22×10^{-03}	4.62×10^{-03}	2.73×10^{-03}	7.64×10^{-04}	5.70×10^{-04}	8.59×10^{-04}	1.91×10^{-03}	4.71×10^{-04}
FOP(Ph2)	–	*1.07×10^{-03}*	*9.40×10^{-04}*	*6.65×10^{-04}*	*5.50×10^{-04}*	1.68×10^{-03}	2.12×10^{-03}	2.33×10^{-03}	1.35×10^{-03}	7.06×10^{-04}	4.51×10^{-04}	5.99×10^{-04}	1.51×10^{-03}	2.58×10^{-04}
FOP(PhM)	–	–	–	3.28×10^{-03}	5.12×10^{-03}	1.84×10^{-03}	2.14×10^{-03}	2.42×10^{-03}	1.34×10^{-03}	2.93×10^{-04}	3.11×10^{-04}	4.34×10^{-04}	1.29×10^{-03}	6.65×10^{-05}
FOP(Z)	–	2.73×10^{-03}	2.39×10^{-03}	1.70×10^{-03}	1.40×10^{-03}	1.04×10^{-03}	9.37×10^{-04}	8.42×10^{-04}	6.03×10^{-04}	4.52×10^{-04}	3.36×10^{-04}	2.48×10^{-04}	2.12×10^{-04}	1.82×10^{-04}
FLP(Ph1)	–	*1.27×10^{-02}*	*1.11×10^{-02}*	*5.29×10^{-03}*	*6.50×10^{-03}*	2.51×10^{-02}	3.81×10^{-02}	5.46×10^{-02}	3.22×10^{-02}	9.03×10^{-03}	6.74×10^{-03}	1.01×10^{-02}	2.24×10^{-02}	5.57×10^{-03}
FLP(Ph2)	–	*1.27×10^{-02}*	*1.11×10^{-02}*	*5.29×10^{-03}*	*6.50×10^{-03}*	1.99×10^{-02}	2.51×10^{-02}	2.75×10^{-02}	1.60×10^{-02}	8.35×10^{-03}	5.33×10^{-03}	7.09×10^{-03}	1.79×10^{-02}	3.05×10^{-03}
FLP(PhM)	–	–	–	3.87×10^{-02}	6.05×10^{-02}	2.18×10^{-02}	2.54×10^{-02}	2.86×10^{-02}	1.58×10^{-02}	3.46×10^{-03}	3.68×10^{-03}	5.14×10^{-03}	1.53×10^{-02}	7.86×10^{-04}
FLP(Z)	–	1.71×10^{-02}	1.49×10^{-02}	1.06×10^{-02}	8.76×10^{-03}	6.50×10^{-03}	5.86×10^{-03}	5.26×10^{-03}	3.77×10^{-03}	2.82×10^{-03}	2.10×10^{-03}	1.54×10^{-03}	1.33×10^{-03}	1.13×10^{-03}

[a] Adapted from Woźniak et al. (2006).

[b] (-), the values of the index n'_{p} were not defined; (0), the values of the index n'_p are much smaller than in other spectral intervals; extrapolated values are given in italics.

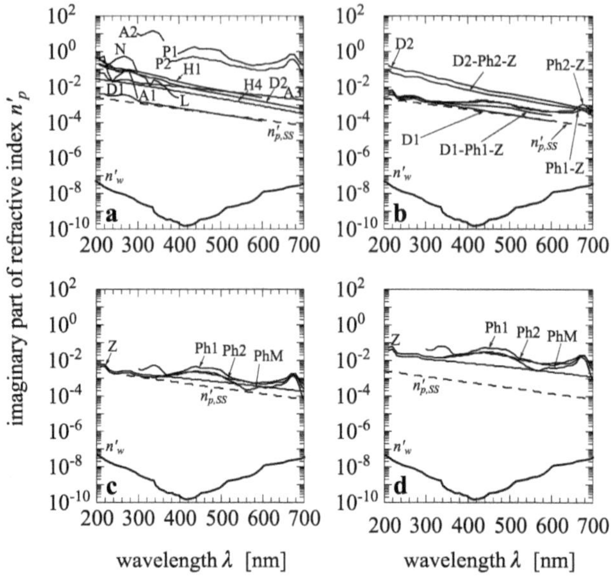

FIGURE 5.24. Possible spectra of the imaginary part of the absolute refractive index for selected morphological groups of POM: (a) spherelike particles (SLP) and irregular particles (IRP); (b) aggregated particles (AGP); (c) fragments of organisms (FOP); (d) flaky particles (FLP). The symbols on the plots stand for the model chemical classes of POM contained in the morphological forms of particles in the proportions given in Table 5.19. The solid line in the graphs denotes the spectrum of the absolute imaginary refractive index of water $n'_w(\lambda)$ (or of a particle composed exclusively of water). The dashed line represents the spectrum $n'_{p,SS}(\lambda)$ for Sargasso Sea detritus (after Stramski et al. (2001); see Equation (5.17)). (Adapted from Woźniak et al. (2006).)

the order of $n'_p \approx 1$ and more in the VIS region, are characteristic of particles made up respectively of pure MAAs (see A2 in Figure 5.24a) and of pure phytoplankton pigments (P1 in Figure 5.24a). In contrast, the lowest values of n'_p, some 3–4 orders of magnitude smaller than the values given above, are typical of particles comprising DOM solidified in sorption processes such as deposition and precipitation (see D1 in Figure 15.24a). Such low values of n'_p are also exhibited by aggregated particles (AGP) or fragments of organisms (FOP; Figures 15.24b,c), especially those produced by the nonselective degradation and immediate aggregation of the matter of which the organisms consist (see, e.g., particles AGP(D1), AGP(D1/Ph1/Z), etc. and, e.g., particles FOP(Z), FOP(Ph1), etc.) Their indices $n'_p(\lambda)$ take values approaching those of $n'_{p,SS}(\lambda)$ determined for detritus on the basis of empirical data from the Sargasso Sea.

The almost perfect match between the indices $n'_p(\lambda)$ determined theoretically for nanodetritus containing solidified DOM (see, e.g., D1 in Figure 5.24a) and the empirical values of $n'_{p,SS}(\lambda)$ is worthy of note. Particles of this type must therefore be present in far greater numbers than all the other morphological variants of organic detritus taken together. This conclusion is supported by the fact that DOM is the most abundant substrate in the sea from which SPM can form.

Moreover, the estimated values of $n'(\lambda)$ of all the morphological variants of POM distinguished here are higher than, or at the very least comparable with, the empirical values of $n'_{p,SS}(\lambda)$. This is because the calculations presented in this section took into account only groups of POM consisting of at least one of the substances classified as strong UV or VIS absorbers. The indices $n'_p(\lambda)$ given in Table 5.20 and in Figure 5.24 can therefore be used to determine directly, with the aid of Mie's theory, the optical properties of only those fractions of POM most strongly absorbing radiation. In reality, there are many other kinds of organic particles in the sea with values of n'_p lower than those given here: they are composed solely of weakly absorbing organic substances, or contain only very tiny quantities of strong absorbers. It seems, however, that their influence on the bulk absorption capacity of POM in general is small and can, to a first approximation, be neglected.

In conclusion, we stress that the above analyses and the results of our calculations are of a preliminary nature and hypothetical, because they have not yet been empirically verified to a sufficient extent. The problem requires further study.

Optical Constants of Crude Oil and its Derivatives

In sea waters one can also expect to find suspended particles consisting wholly or partially of crude oil, or its derivatives produced by evaporation, oxidation, and the other transformations to which crude oil is subject. These particles may have come from natural sources (e.g., oil seeps from undersea deposits) or arisen through human agencies (spillage from ships, pipelines, undersea oil wells, etc.). On entering the marine environment, crude oil usually first forms an oil film and a water-in-oil emulsion on the sea surface. As a result of various chemical and physical processes, some of the oil or its derivatives (fractions) subsequently sinks into the water in the form of droplets, sometimes in quantities that are significant with regard to the light field. Moreover, some of the crude oil or its derivatives may be deposited on various particles suspended in the bulk of water. In this way, crude oil and its derivatives become part of the group of multicomponental suspended particles, which may also contain inorganic components. Thus, we have drawn attention to a problem little researched so far, although we are unable to describe it in detail.

In Table 5.21 and in Figure 5.25 we give the values of both components of the optical constants, that is, the real and imaginary refractive indices (in the VIS region, at various temperatures) for two types of crude oil used, among other things, as marine engine fuel: *Romashkino* and *Petrobaltic*. Because

TABLE 5.21. Absolute complex refractive indices m (350 nm ≤ λ ≤ 750 nm) at various temperatures for *Romashkino* and *Petrobaltic* crude oil.[a]

A. Real Part n [Dimensionless]

Wavelength	Romashkino			Petrobaltic		
λ [nm]	0°C	10°C	20°C	0°C	10°C	20°C
-1-	-2-	-3-	-4-	-5-	-6-	-7-
350	1.5212	1.5178	1.5144	1.4959	1.4918	1.4877
360	1.5176	1.5142	1.5107	1.4948	1.4907	1.4865
370	1.5143	1.5108	1.5074	1.4936	1.4895	1.4854
380	1.5113	1.5078	1.5044	1.4925	1.4884	1.4843
390	1.5086	1.5051	1.5017	1.4914	1.4873	1.4832
400	1.5061	1.5027	1.4992	1.4904	1.4863	1.4822
410	1.5039	1.5005	1.4970	1.4894	1.4853	1.4812
420	1.5020	1.4985	1.4951	1.4885	1.4844	1.4802
430	1.5002	1.4968	1.4934	1.4875	1.4834	1.4793
440	1.4987	1.4953	1.4918	1.4867	1.4825	1.4784
450	1.4974	1.4939	1.4905	1.4858	1.4817	1.4776
460	1.4962	1.4928	1.4893	1.4850	1.4809	1.4768
470	1.4952	1.4918	1.4883	1.4842	1.4801	1.4760
480	1.4944	1.4909	1.4875	1.4835	1.4794	1.4753
490	1.4936	1.4902	1.4868	1.4828	1.4787	1.4746
500	1.4930	1.4896	1.4862	1.4821	1.4780	1.4739
510	1.4925	1.4891	1.4857	1.4815	1.4774	1.4733
530	1.4918	1.4884	1.4849	1.4803	1.4762	1.4721
550	1.4914	1.4879	1.4845	1.4793	1.4752	1.4711
570	1.4912	1.4877	1.4843	1.4785	1.4744	1.4703
590	1.4911	1.4877	1.4842	1.4778	1.4736	1.4695
610	1.4912	1.4877	1.4843	1.4772	1.4731	1.4690
630	1.4913	1.4878	1.4844	1.4768	1.4727	1.4686
650	1.4914	1.4879	1.4845	1.4765	1.4724	1.4683
670	1.4915	1.4880	1.4846	1.4764	1.4723	1.4682
690	1.4915	1.4881	1.4847	1.4764	1.4723	1.4682
710	1.4915	1.4881	1.4847	1.4766	1.4725	1.4683
730	1.4915	1.4881	1.4846	1.4769	1.4728	1.4687
750	1.4914	1.4880	1.4845	1.4773	1.4732	1.4691

B. Imaginary Part of n' [Dimensionless]

Wavelength	Romashkino			Petrobaltic		
λ [nm]	0°C	10°C	20°C	0°C	10°C	20°C
-1-	-2-	-3-	-4-	-5-	-6-	-7-
350	0.015689	0.015387	0.015085	0.006026	0.005999	0.005972
360	0.015220	0.014927	0.014634	0.002926	0.002913	0.002900
370	0.014670	0.014388	0.014106	0.001744	0.001736	0.001728
380	0.014057	0.013786	0.013516	0.001166	0.001161	0.001156
390	0.013395	0.013138	0.012880	0.000840	0.000836	0.000832
400	0.012700	0.012456	0.012211	0.000637	0.000634	0.000631
410	0.011983	0.011753	0.011522	0.000502	0.000500	0.000498
420	0.011256	0.011040	0.010823	0.000407	0.000405	0.000404
430	0.010530	0.010327	0.010125	0.000338	0.000336	0.000335

TABLE 5.21. Absolute complex refractive indices m (350 nm ≤ λ ≤ 750 nm) at various temperatures for *Romashkino* and *Petrobaltic* crude oil.[a]—Cont'd.

B. Imaginary Part of n' [Dimensionless]

Wavelength	Romashkino			Petrobaltic		
λ [nm]	0°C	10°C	20°C	0°C	10°C	20°C
-1-	*-2-*	*-3-*	*-4-*	*-5-*	*-6-*	*-7-*
440	0.009812	0.009623	0.009435	0.000286	0.000284	0.000283
450	0.009111	0.008936	0.008761	0.000245	0.000244	0.000243
460	0.008434	0.008272	0.008110	0.000213	0.000212	0.000211
470	0.007786	0.007637	0.007487	0.000187	0.000186	0.000185
480	0.007173	0.007035	0.006897	0.000166	0.000165	0.000164
490	0.006597	0.006470	0.006343	0.000148	0.000147	0.000147
500	0.006062	0.005945	0.005829	0.000133	0.000132	0.000132
510	0.005570	0.005463	0.005355	0.000120	0.000120	0.000119
530	0.004719	0.004628	0.004538	0.000100	0.000099	0.000099
550	0.004048	0.003970	0.003892	0.000084	0.000084	0.000084
570	0.003549	0.003480	0.003412	0.000072	0.000072	0.000071
590	0.003207	0.003146	0.003084	0.000062	0.000062	0.000062
610	0.003003	0.002945	0.002887	0.000054	0.000054	0.000054
630	0.002909	0.002853	0.002798	0.000048	0.000047	0.000047
650	0.002899	0.002844	0.002788	0.000042	0.000042	0.000042
670	0.002943	0.002887	0.002830	0.000037	0.000037	0.000037
690	0.003013	0.002955	0.002897	0.000033	0.000033	0.000033
710	0.003081	0.003022	0.002963	0.000029	0.000029	0.000029
730	0.003127	0.003067	0.003006	0.000026	0.000026	0.000026
750	0.003133	0.003073	0.003013	0.000023	0.000023	0.000023

[a] Data kindly made available by Dr. Zbigniew Otremba, private communication.

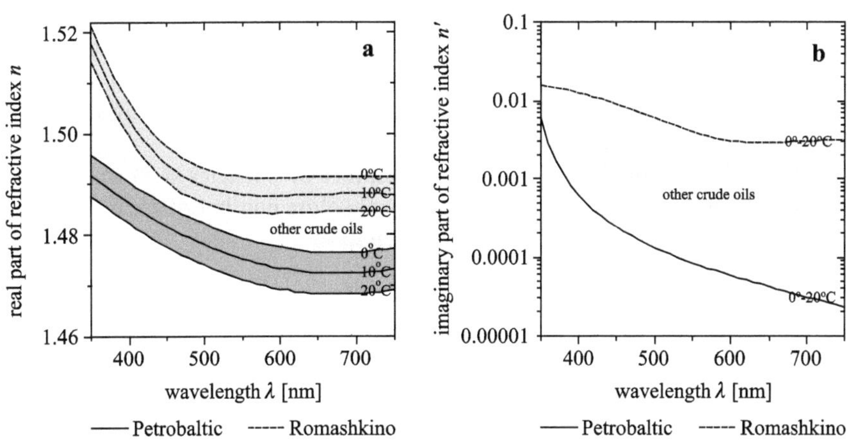

FIGURE 5.25. Spectra of the real (a) and imaginary (b) parts of the absolute refractive index in the VIS region for two types of unrefined crude oil with widely different optical properties (adapted from Otremba 2000).

their optical properties differ in the extreme, we can assume (see Figure 5.25) that their optical constants roughly delineate the envelope containing the possible real and imaginary refractive indices of most other types of natural crude oil and their derivatives.

The data in Table 5.21 and Figure 5.25 show that the real refractive indices $n(\lambda)$ of crude oil for visible light lie within the limits typical of organic substances, and that they approach the corresponding indices of the organic detritus particles mentioned earlier in this section and of the planktonic components of organic particles discussed in Section 5.2.3. On the other hand, the imaginary refractive indices $n'(\lambda)$ of these crude oils are of the same order of magnitude as, for example, those of living particles (with the exception of phytoplankton and phytobenthos of which the n'_p are greatest) or of the very numerous, strongly absorbing nanodetritus formed from solidified DOM. Only the indices $n'(\lambda)$ of a few types of crude oil (resembling *Petrobaltic* crude oil) are low enough for their absorption capacities, in comparison with those of POM, to be ignored. We refer readers interested in the problems of crude oil optical constants and in modeling various optical properties on their basis, such as absorption, using Mie's theory and the like, to the relevant literature (e.g., Otremba (2004) and Otremba and Piskozub (2001) and the papers cited therein).

5.2.5 Sizes and Shapes of Particles

In Sections 5.2.2 and 5.2.3 we discussed the optical constants (i.e., the complex refractive indices of the matter contained in suspended particles), knowledge of which is necessary in theoretical descriptions of the optical properties of such particles based on Mie's theory. Moreover, as we showed in Section 5.1, it is also essential to know the particle sizes (diameters D), their numerical concentration N in a monodispersing medium, or the total numerical concentration N_{tot} of particles with all possible values of diameter D, and their size distribution functions (SDF) in a polydispersing medium, such as the sea and the atmosphere.

The SDF is a function of the relative number of particles $n(D)$ with diameters in the interval $D \pm 1/2\, dD$. Hence, the absolute number of particles $N(D)$ in this diameter interval is given by the equation

$$N(D) = \mathrm{C}\, n(D). \tag{5.21}$$

The several magnitudes in this equation are expressed in the following units: D [μm], N [$\mathrm{m^{-3}\mu m^{-1}}$], and C [dimensionless], n [$\mathrm{m^{-3}\mu m^{-1}}$]. C is a constant dependent on N_{tot}, in accordance with the relationship:

$$C = N_{tot}\left(\int_{D_{\min}}^{D_{\max}} n(D)\,dD\right)^{-1}, \tag{5.22}$$

where N_{tot} [m^{-3}] is the total numerical concentration of particles in the medium:

$$N_{tot} = \int_{D_{min}}^{D_{max}} N(D)\,dD, \quad (5.23)$$

where D_{min} and D_{max}, respectively, denote the minimum and maximum diameters of the particles present in this medium.

The size distribution of particles given by Equation (5.21) can be regarded as a "differential distribution," because it describes the number of particles $N(D)$ (in unit volume of the medium) with dimensions from $D - 1/2\ dD$ to $D + 1/2\ dD$, that is, in the size interval corresponding to the differential dD. In practice, it is also convenient to use the so-called integral size distribution of particles. This describes the cumulative numerical concentration of particles $N_{int>D}(D)$, that is, the total number of particles (in unit volume of the medium) with diameters equal to and greater than D, in other words, in the interval from D to D_{max}. In practice, these will be the size intervals from D to ∞, because the upper size limit of particles, particularly in the sea, is usually undefined. The size distribution referred to such cumulative numbers of particles is linked to SDF, that is, $n(D)$, by the relationship

$$N_{int>D}(D) = C \int_{D}^{\infty} N(D)\,dD \quad (5.24)$$

Defining these characteristic sizes, numbers, and SDF of marine particles is an exceedingly complex task. First, this is because the shapes of particles suspended in the sea are generally irregular and one cannot talk about their diameters in the literal sense (see the geometrical characteristics of suspended mineral and detritus particles in Tables 5.22 and 5.24, and also the shapes of planktonic organisms illustrated in Figure 5.29). Secondly, the sizes of these different types of particles, both organic and mineral, cover an extremely broad range of values (Figure 5.26), whose upper and lower limits are difficult to establish objectively.

To resolve the first difficulty, arising from the intricacies of particle shapes, their sizes are established with the aid of so-called equivalent diameters, which are defined according to the method used to measure them (Zalewski 1977, Dera 2003). On examining particles under the microscope, one usually sees the projection of a randomly situated particle onto a plane. One way of assessing the particle's diameter is to take the distance between two tangents to the outline of the particle's projection that are perpendicular to the base of the microscope image; this is the so-called Feret diameter (Figure 5.27a).

Another equivalent particle diameter is the Martin diameter, which is the length of a section of straight line parallel to the base of the particle's image cutting the surface of its projection at half height (Figure 5.27a). In contrast, when investigating the sizes of suspended particles by means of the time-honored electronic conduction technique (the Coulter counter), and also by

TABLE 5.22. Sizes and shapes of mineral particles.

A. Size Ranges of Grains (According to the British Soil Classification System) and Their Usual Shapes

Type-Group	Grain width *d* [μm]	Shape
-1-	*-2-*	*-3-*
Clay	< 2	Flaky
Silt:		Flaky, elongated, angular, irregular, rounded
Fine	2–6	
Medium	6–20	
Coarse	20–60	
Sand:		Rounded, spherical
Fine	60–200	
Medium	200–600	
Coarse	600–2000	

B. Approximate Geometrical Characteristics[a] of Particulate Inorganic Matter (PIM) Commonly Occurring in Sea Waters

Mineral (Group)	Shape	Grain width *d* [μm]	Thickness	Equivalent spherical diameter *D*
-1-	*-2-*	*-3-*	*-4-*	*-5-*
Montmorillonite (clay)	Flaky	0.01–1.0	~0.01 *d*	~0.25 *d*
Illite (clay)	Flaky	0.2–2.0	~0.1 *d*	~0.53 *d*
Kaolinite (clay)	Flaky	0.3–2.0	~0.2 *d*	~0.67 *d*
Fine-grained quartz	Rounded	0.06–2.0	~*d*	~*d*
Coarse-grained quartz and other minerals from the and and silt group	Spherical	>2.0	*d*	*d*

[a] The approximate parameters characterizing particle shapes and sizes were estimated on the basis of data and information available on http://fbe.uwe.ac.uk/public/geocal/SoilMech/classification/soilclas.htm

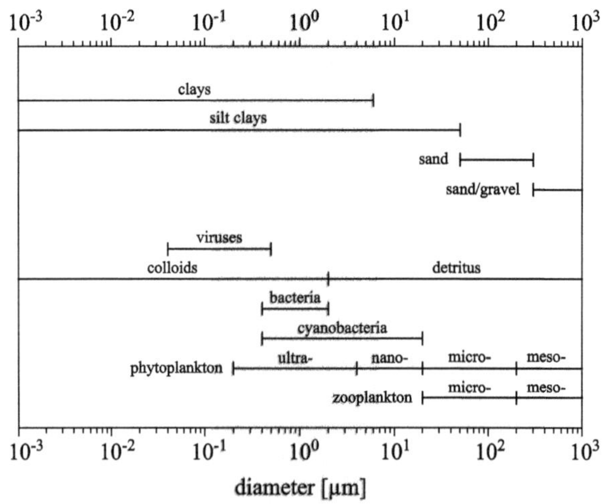

FIGURE 5.26. Approximate size ranges of various particles in sea water. The boundaries between these ranges are a matter of convention: their values have been calculated according to various criteria and may vary slightly in the works of different authors.

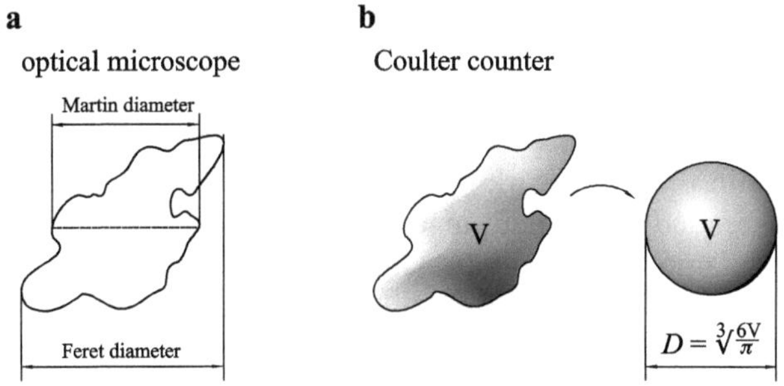

FIGURE 5.27. Diagrams showing how the diameters of irregularly shaped particles in marine suspended matter can be defined. (a) Outline of the microscopic image of a particle showing the Martin and Feret diameters. (b) The equivalent diameter used in the Coulter technique, that is, the diameter of a sphere with the same volume as the particle under scrutiny. (After Dera (2003); with the kind permission of PWN Publishers.)

a battery of more modern electronic techniques and indirect methods of assessing mean particle sizes on the basis of optical parameters,[23] the equivalent diameter is taken to be the diameter of a sphere having the same volume as that of the particle in question; see Figure 5.27b. It is this version of the equivalent diameter that we have in mind when in this description we talk about sizes or diameters of suspended particles.

The second difficulty with descriptions of size distributions of particles is, as we have already mentioned, the impossibility of establishing their lower (D_{min}) and upper (D_{max}) size limits.

Specifying the lower size limit of particles suspended in sea water is necessary, because the number of particles usually increases steeply with ever-decreasing size, so that their lower size limit can only be a matter of convention. This is because the degree of comminution of particles is theoretically unlimited—until they have actually broken down into individual molecules—and in water this process depends on the solubility of the particle matter. The smallest, submicroscopic organic particles in the sea include colloids, viruses, and bacteria; submicroscopic mineral particles are represented by various clays. In practice, then, the lower size limit is usually taken

[23] The Coulter counter measures particle diameters over a fairly limited range, usually from c. 1–2 μm to c. 40 μm (Coulter 1973, Sheldon et al. 1972, Zalewski 1977, Dera 2003). For studying size distributions and other properties of larger particles optical techniques are also applied, for example, underwater microphotography (Eisima et al. (1990), size range from c. 4 to 650 μm) and underwater holography (Carder and Meyers (1979), Malkiel et al. (1999), size range from c. 3 to 3000 μm). The sizes of small particles and colloids can be defined, for example, by image analyses of samples, obtained from transmission electron microscopy (Wells and Goldberg (1994), size range 0.01 to 0.2 μm), or by electronic, resistive pulse particle counters (Koike et al. (1990), Yamasaki et al. (1998), size range c. 0.4 to 1 μm).

to be the smallest size of particles D_{min} detectable with the available measurement techniques. Notice, however, that this value of D_{min} cannot be too high: the omission from analyses of an excessively large number of small particles with diameters smaller than this D_{min} may lead to serious errors in calculations within the framework of Mie's theory regarding the interaction of the total suspended particulate matter with, for example, light. Less problematical is the upper size limit D_{max} of suspended particles: it is defined, as it were, by their contribution to the process we are examining, inasmuch as the number of large particles falls rapidly with their increasing size. Suspended particles larger than 50 μm make up only a tiny fraction of one percent of their total number. Their contribution to optical processes in the sea is therefore correspondingly tiny and can be safely ignored (Dera 2003).

In the remainder of this section we discuss the sizes and shapes of mineral particles in the sea, the sizes and shapes of planktonic components (including viruses and bacteria), and the sizes and shapes of detritus and colloidal particles. Then we move on to the size distributions of particles recorded in different seas and their mathematical descriptions.

The Sizes and Shapes of Mineral Particles

As we have already mentioned, inorganic particles in the sea are usually terrigenous and take the form of grains of diverse sizes and shapes; see Table 5.22A. In size, these particles can vary over several orders of magnitude, from fractions of micrometers (clays) to several millimeters (coarse sand and gravel). In shape, they can be rounded, irregular (irregular shape with round edges), angular (flat faces and sharp edges), flaky (thickness small compared to length and breadth), elongated (length > breadth and thickness), and flaky-elongated (length > breadth > thickness). But among the water-worn or air-worn minerals present in atmospheric dusts and water bodies (see Table 5.8), two shapes predominate: round and flaky (Table 5.22B). Both fine-grained and coarse-grained quartz, and also other mineral particles from the sand and silt group are usually rounded, and the larger of these particles are almost ideally spherical. On the other hand, the smaller particles from the clay group, numerous in dust and water bodies (composed of montmorillonite, illite, and kaolinite) are usually of a flaky shape. Particles of mica schist are also flaky. Table 5.22B (columns 2–4) gives the characteristic parameters describing these shapes and the typical sizes of these particles. Column 5 of this table gives the approximate relations between the equivalent spherical diameters of these grains D, determined from their volumes, and the average grain width d for the plane of their maximum area cross-section. It is clear from these data that for analyzing the optical properties of sufficiently large mineral grains (sands and the larger clay fractions) Mie's theory for spherical particles is wholly adequate, because their real and equivalent spherical diameters are roughly the same (D similar to or not much smaller than d).

Certain errors may arise, however, when this theory (and others assuming the sphericity of particles) is applied to small particles, for example, montmorillonite clays. In such cases the solutions offered by the Maxwell equations are recommended: although resembling the solutions provided by the Mie theory, described in Section 5.1, they apply to nonspherical particles. Numerous solutions for nonspherical particles, for example, spheroids, ellipsoids, and finite cylinders, are available in the world literature (e.g., Asano and Yamamoto (1975), Asano (1979), Hulst (1981), Bohren and Hoffman (1983), Voshchinnikov and Farafonov (1993), Kurtz and Salib (1993); see also the review by Wriedt (1998)).

Further structural parameters of polydispersing media affecting their optical properties are the size distributions of the particles suspended in them. The separate distributions of particulate inorganic matter in the sea are difficult to establish empirically. Most experimental techniques used to investigate particle sizes register the total number of particles, both organic and mineral. On the other hand, many authors cite the size distributions of atmospheric dusts or atmospheric aerosols in general at various heights in the free troposphere (e.g., Junge (1963), Kent et al. (1983), Hoppel et al. (1990), Raes et al. (1997), Murugavel et al. (2001), Reid et al. (2003)). Containing relatively small quantities of water droplets, sea salt, and organic matter, such atmospheric aggregations of particles consist mainly of terrigenous mineral matter. They may also contain other inorganic components originating from industries or fires, such as flaky ash and black carbon (Hasler and Nussbaumer 1998, Clarke et al. 2004, Koukouzas et al. 2005).

Figure 5.28 shows examples of size distributions in such atmospheric aggregations. These atmospheric aerosols are the principal source of mineral particles in the ocean (apart from riverborne sediments). It is to be expected, therefore, that the SDFs of these particles in the sea are to a first approximation similar to the values characteristic of mineral particles in the atmosphere. In shelf seas, moreover, there may be slightly larger numbers of coarse-grained particles such as sand (e.g., Shirazi et al. (2001)), other soil fragments derived, for example, from rock erosion, and aggregations of clay particles (e.g., Vogel et al. (2005)), that have been carried to the sea by rivers or raised from the sea bed.

Quantitatively, the size distributions of atmospheric particles shown in Figure 5.28 vary a great deal. This is because they were recorded at different heights and in different geographical regions, at varying distances from the deserts that are their prime source. These quantities will therefore have been strongly modified by a variety of factors and dynamic processes, such as sinking or horizontal transport in the atmosphere (e.g., Maring et al. (2003) and Tunved et al. (2005)).

Apart from this quantitative differentiation, however, the size distribution functions (SDF) of airborne minerals are much the same. The relative numbers of particles drop very sharply with increasing particle size. For an increase in particle size D from c. 0.1 μm to c. 10 μm, that is, three orders of

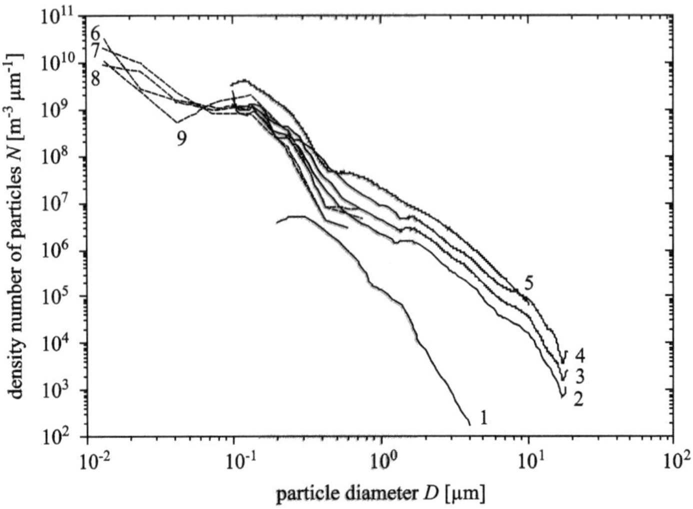

FIGURE 5.28. Examples of aerosol or dust-size distribution in different regions of the atmosphere: 1. aerosol typical of the continental free troposphere (after Patterson et al. (1980)); 2–4. examples of Saharan dust (after Reid et al. (2003)); 5. Asian–Pacific dust (Iwakuni, Japan) (after Clarke et al. (2004)); 6–9. examples of aerosols over the Indian Ocean (Mauritius; after Murugavel et al. (2001).)

magnitude, this drop in numbers amounts to almost five orders of magnitude. Typical of these distributions is also the tendency for the largest numbers of particles to occur in the range of small diameters D, c. 0.1 μm. Presumably, the SDFs of PIM exhibit similar features.

Sizes and Shapes of Planktonic Components (Including Viruses and Bacteria)

The composition of particulate organic matter in the sea is far more diverse than that of PIM, with respect both to its optical properties (i.e., the optical constants discussed in Section 5.2.3) and its sizes and shapes. POM consists of multifarious living (mainly plankton, bacteria, and viruses) and nonliving forms (detritus). Nevertheless, the supreme forms of organic particles are the living single-celled plant organisms, which are grouped into classes, for example, *Bacillariophyceae* (diatoms), *Dinophyceae* (dinoflagellates), and *Haptophyceae* (inter alia coccolithophorids).[24] The most numerous of these classes, the diatoms, which consists mainly of various forms of marine phytoplankton, alone contains around 10,000 species. The zooplankton, the

[24] The fundamentals of the taxonomy of marine phyto- and zooplankton can be found in, for example, Parke and Dixon (1968) and Nawell and Nawell (1966). See also Parsons et al. (1977) and other monographs on marine biology and biological oceanography.

second-largest group of living organisms in the ocean, is represented by almost as many species as the phytoplankton. The shapes of 16 species of marine phyto- and zooplankton are illustrated in Figure 5.29.

Even from this tiny sample, the rich diversity of shapes is evident; establishing their sizes in some unambiguous manner, for example, by means of the equivalent spherical diameter, is a complex matter. To this end, the shapes

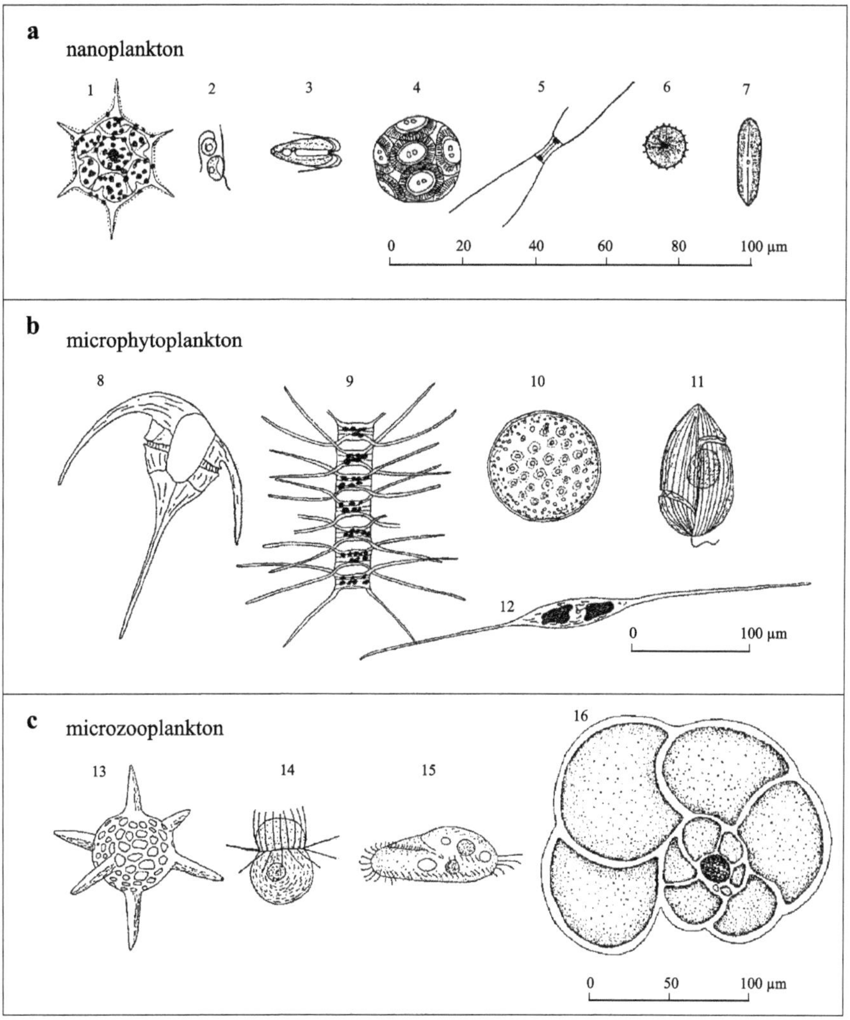

FIGURE 5.29. Examples of shapes of marine plankton cells: (a) nanoplankton: flagellates (1. Distephanus; 2. Thalassomonas; 3. Tetraselmis; 4. Coccolithus); diatoms (5. Chaetoceros; 6. centrate; 7. pennate); (b) microphytoplankton: dinoflagellates (8. Ceratium, 11. Gyrodinium); diatoms (9. Chaetoceros; 10. Coscinodiscus; 12. Nitzschia); (c) microzooplankton: radiolarians (13. Hexastylus); ciliates (14. Mesodinium; 15. Amphisia); foraminiferans (16. Pulvinulina). (Adapted from Parsons et al. (1977).)

of these particles can be approximated in calculations by the use of geometrical figures whose equivalent volume can be determined by a known mathematical method. This operation is performed on the basis of specialist catalogues and other compilations, which help to identify the species in question and give the shape of the equivalent figure of the identified cell together with a relationship defining its equivalent volume (e.g., Butcher (1959) and Starmach (1966)). To give an example: such a catalogue was compiled for the plankton most commonly and most numerously occurring in the Baltic in the form of a recommendation by Working Group No. 9, appointed for the purpose by the Baltic Marine Biologists (Edler 1979).

One should bear in mind, however, that employing such approximate sizes of particles to calculate their optical properties in terms of Mie's theory may, in the case of plankton cells with shapes very unlike spheres, lead to serious errors. In such cases we recommend, as we did with regard to nonspherical mineral particles, that the Maxwell equations—analogous to the Mie theory equations—be used for nonspherical and optically inhomogeneous organic particles. This is because the optical properties of the internal structures of organic particles (especially phytoplankton) are to a high degree inhomogeneous (see, e.g., Król (1998)), which is not the case with the more or less homogeneous mineral particles. Several methods of determining the optical properties of inhomogeneous nonspherical particles are described in the review by Wriedt (1998), mentioned above, and by Quirantes and Bernard (2004, 2006).

The sizes of this living fraction of POM vary over a wide range (Figure 5.26): from viruses ($D \approx$ a few to over a dozen hundredths of a μm), through bacteria, cyanobacteria, and ultraphytoplankton ($D \approx$ a few tenths of a μm to a few μm), and nanophytoplankton ($D \approx$ a few μm to c. 20 μm), to the larger phyto- and zooplanktonic forms, classified formally as microplankton (D from c. 20 μm to c. 200 μm) and macroplankton ($D > 200$ μm). The largest single organisms or their colonies, classified among the macrozoo- and macrophytoplankton, achieve sizes of even a few millimeters (e.g., *pouchetii* of the class *Haptophyceae* has a diameter of 1000–2000 μm), and other small animals are larger still.[25]

Table 5.23 lists examples of equivalent spherical diameters D, calculated experimentally under laboratory conditions by different authors, for the various representatives of the living POM mentioned above. These values of D should be treated as average values. In reality these sizes are very diverse, whereby their individual size distributions are random, approaching normal distributions or other functional expressions of a similar character (we return

[25] Obviously, marine organisms such as the macroplankton (size 2–20 cm), megaplankton (20–200 cm), and especially the higher animals (nekton), achieve even greater or very much greater sizes (the Blue Whale *Balaenoptera musculus* is c. 31 m long!). But because of the relatively small number of these large-size forms and their negligible share of the total biomass of the marine biosphere, they do not directly affect the optical properties of sea water and can be omitted from such analyses.

TABLE 5.23. Some equivalent spherical diameters D empirically determined for selected planktonic components.

Label	Planktonic component	D [μm]	Ref[a]	Label	Planktonic component	D [μm]	Ref[a]
-1-	-2-	-3-	-4-	-1-	-2-	-3-	-4-
viru	Viruses	0.07	1	pseu	*Thalassiosira pseudonana* (Bacillariophyceae)	3.99	7
hbac	Heterotrophic bacteria	0.55	2	luth	*Pavlova lutheri* (Haptophyceae)	4.26	6
proc	*Prochlorococcus* strain MED	0.59	3	galb	*Isochrysis galbana* (Haptophyceae)	4.45	5
	Average of *Prochlorococcus* strains NATL and SARG	0.70	3	huxl	*Emilianina huxleyi* (Haptophyceae)	4.93	5
syne	*Synechococcus* strain MAX41 (Cyanophyceae)	0.92	3	crue	*Porphyridium cruentum* (Rhodophyceae)	5.22	6
	Synechococcus strain MAX01 (Cyanophyceae)	0.94	3	frag	*Chroomonas fragarioides* (Cryptophyceae)	5.57	5
	Synechococcus strain ROS04 (Cyanophyceae)	1.08	3	parv	*Prymnesium parvum* (Haptophyceae)	6.41	6
	Synechococcus strain DC2 (Cyanophyceae)	1.14	3	bioc	*Dunaliella bioculata* (Chlorophyceae)	6.71	5
	Synechococcus strain WH8103 (Cyanophyceae)	1.14	4	tert	*Dunaliella tertiolecta* (Chlorophyceae)	7.59	8
syma	*Synechocystis* (Cyanophyceae)	1.39	5	curv	*Chaetoceros curvisetum* (Bacillariophyceae)	7.73	6
	Anacystis marina (Cyanophyceae)	1.43	5	elon	*Hymenomonas elongata* (Haptophyceae)	11.77	5
ping	*Pavlova lutheri* (Haptophyceae)	3.97	6	mica	*Prorocentrum micans* (Dinophyceae)	27.64	5

Adapted from Stramski et al. (2001).

[a] 1: Stramski and Kiefer (1991); 2: Stramski and Kiefer (1990); 3: Morel et al. (1993); 4: Stramski et al. (1995); 5: Ahn et al. (1992); 6: Bricaud et al. (1988); 7: Stramski and Reynolds (1993); 8: Stramski et al. (1993).

to this question later). The dispersions of these distributions depend on the individual properties of the given species and also on the diverse growing conditions in the sea and oceans. The half-width of the size distributions may achieve values that are 10–20%, even ~50%, of their most probable or mean values. Figure 5.30 gives examples of such distributions for viruses, heterotrophic bacteria, and selected plankton species.

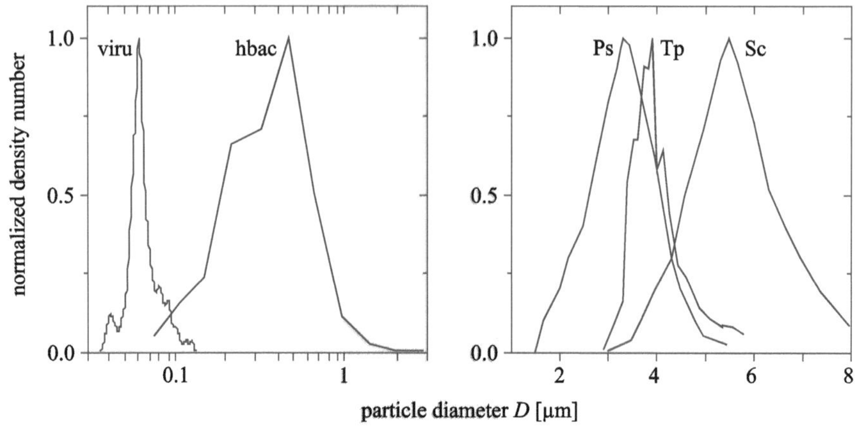

FIGURE 5.30. Examples of normalized size distributions of some marine planktonic components: viru (viruses, after Stramski and Kiefer (1991)), hbac (heterotrophic bacteria, after Jonasz et al. (1997)), Ps (*Platymonas suecica*, Prasinophyceae after Bricaud and Morel (1986)), Tp (*Thalassiosira pseudonana,* Bacillariophyceae, after Stramski and Reynolds (1993)), Sc (*Skeletomena costatum*, Bacillariophyceae, after Bricaud and Morel (1986)).

The size distributions of particles exemplified in Figure 5.30 refer to separate representatives of the living fraction of POM. Where, however, the natural associations of marine planktonic components—containing all the species growing at a particular time and place in the ocean—are concerned, their overall size distributions cannot be measured directly, as we stated earlier. Such distributions are usually measured with respect to all suspended particulate matter, that is, the combined total of living organisms, organic detritus, and PIM. Indirect analyses have implied, however, that size distributions of the living planktonic fraction of POM can be roughly divided into two kinds. One kind represents the situation when at any given instant and place in the sea many different planktonic components are growing, their overall size distributions are practically monotonic and smooth, the number of particles falls dramatically with increasing size, and there may be a single maximum in the small particle region (see later plot 1 in Figure 5.33c). The other kind comes into play when a suitable combination of abiotic factors (temperature, irradiance, availability of nutrients) gives rise to a bloom of one or just a few species of phytoplankton. Then in the total size distribution of POM particles there appear one or several bulges, or even fairly strong local maxima, in the region of diameters D characteristic of the blooming species (see, e.g., plot 2 in Figure 5.33f and also Figure 5.33e).

These specific and subtle features linking plankton size distributions are determined solely by phytoplankton blooms, because the similar size distributions for zooplankton do not usually exhibit those features to such an extent. We discuss in the last part of this section the influence of these phenomena on the total size distributions of all SPM.

Sizes and Shapes of Detritus and Colloids

The other component of POM in the sea comprises the nonliving organic particles, that is, detritus ($D > 1$ μm) and colloids ($D < 1$ μm), also known jointly as nanodetritus. We discussed the convention regarding the segregation of this group of particles into detritus and nanodetritus in Section 5.2.4; for the sake of clarity, we henceforth refer to all such particles as "detritus". As already outlined in Section 5.2.1 (e.g., Figure 5.7c, and Tables 5.5 and 5.6), the mass concentrations of this detritus, and therefore the numerical concentrations (i.e., the number of particles in unit volume of the medium) greatly exceed, and sometimes vastly so, the concentrations of planktonic components in most marine basins, but especially in oligo- and mesotrophic seas. Be that as it may, the sizes and shapes of the diverse morphological forms of such particles have been less intensively studied than the sizes and shapes of living organisms. What is known is that the enormous diversity and irregularity of the shapes of these particles are difficult to relate to their sizes and other morphological features.

Table 5.24 gives the approximate geometrical characteristics of different morphological types of organic detritus, and is based on our assessment of the data regarding these particles, which we discussed in Section 5.2.4. Evidently, the shapes of a great many of these particles are far from spherical. Among the small particles there are some particles of nanodetritus (colloids) with intermediate sizes ($d \approx 0.11$–0.66 μm), which are classified as irregular particles (IRP). For these detritus particles the ratio D/d (i.e., the equivalent spherical diameter D to grain width d) is small, amounting only to c. 0.15 or even less.[26] Among the larger-sized fractions, the shapes of flaky particles (FLP) are also quite "unspherical"; this includes mainly the fraction of large FLP (50–200 μm), for which $D/d \approx 0.3$, and to a lesser extent the fraction of smaller FLP (15–35 μm), for which $D/d \approx 0.5$. It seems, therefore, that invoking Mie's theory for analyzing the optical properties of these highly

TABLE 5.24. Approximate geometrical characteristics of different morphological types of organic detritus (Od).

Group and type of detritus	Shape	Grain width d [μm]	Thickness	Equivalent spherical diameter D
-1-	*-2-*	*-3-*	*-4-*	*-5-*
Nanodetritus (ND):				
Spherelike particles (SLP)	Spherical	0.01–0.16	~d	~d
Irregular particles (IRP)	Various	0.11–0.66	~0.002 d	<0.15 d
Microdetritus (MD):				
Aggregated particles (AGP)	Various	0.5–50	<d	<d
Fragments of organisms (FOP)	Various	50–200	<d	<d
Flake particles (FLP):				
Small	Flaky	15–35	~0.08 d	~0.5 d
Large	Flaky	50–200	~0.016 d	~0.3 d

[26] For ideally spherical particles, the ratio $D/d = 1$.

nonspherical particles may give rise to serious errors. Maxwell's equations for nonspherical particles (mentioned earlier) are therefore called for. Much more spherical, in contrast, are some other types of small nanodetritus particles, ranging in size from 0.01 to 0.16 μm, referred to as sphericlike particles (SLP), also aggregated particles (AGP), that is, microdetritus 0.5–50 μm in size, and fragments of organisms (FOP), with a size range from 50 to 200 μm. For SLP, AGP, and FOP, the ratio D/d ranges from c. 0.5 to slightly less than unity. In these latter cases, then, the application of Mie's theory for analyzing their optical properties seems perfectly reasonable.

It is also the case that the individual size distributions of organic detritus particles have not been examined by direct measurement. Indirect analyses have shown, however, that the numerical concentrations of such particles drop sharply with their increasing size and can be described with the aid of monotonic (more or less homogeneous) functions. We look at this question in greater detail in the last part of this section (see Formulas Approximating SDF).

Examples of Total Particle Size Distributions Recorded in Different Seas

The total size distributions of natural mixtures of all these types of particles recorded at different depths in different seas and in different seasons are widely differentiated. Figure 5.31 illustrates examples of such distributions over a broad range of sizes (from submicronic to several tens of millimeters). These plots show that the quantitative differentiation of suspended particles of the same sizes in different marine regions can extend over several orders of magnitude. Presenting typical cumulative numbers of particles with diameters in excess of 1 μm in 1 m³ sea water $N_{int>1\mu m}$, recorded in different regions of the World Ocean, Table 5.25 gives some idea of this diversity. It emerges from these and other literature data (e.g., Brun-Cottan (1971), Gordon and Brown (1972), Carder and Schlemmer (1973), Ackelson et al. (1988), and Carder (1970)) that the typical values of $N_{int>1\mu m}$ in various seas usually lie within the 10^8–10^{11} m^{-3} range. The lower values from this range ($N_{int>1\mu m} \approx 10^8$ m^{-3}) are characteristic mainly of very clear oceanic waters, principally superoligotrophic. The upper values ($N_{int>1\mu m} \approx 10^{11}$ m^{-3}) are usually recorded only in highly eutrophic waters, chiefly in shelf seas and enclosed basins. Rarely, in extreme cases of severely polluted waters, such as around river mouths and in shallow bays, these numerical concentrations can reach even higher values, of the order of 10^{13} particles per 1 m³ of water (Zalewski 1977, Jonasz 1980, Dera 2003).

Apart from the very considerable quantitative differentiation of particles in different seas, the nature of the relationship between the numerical concentration of these particles N and their size D is equally strongly differentiated. Figure 5.31 shows that in all cases the number of particles N drops very sharply with increase in size D. This drop in the number of particles N for an increase in size D from c. 10^{-2} μm to 10^2 μm (i.e., a range of 4 orders of magnitude of the size D) amounts to some 15 orders of magnitude. Nevertheless,

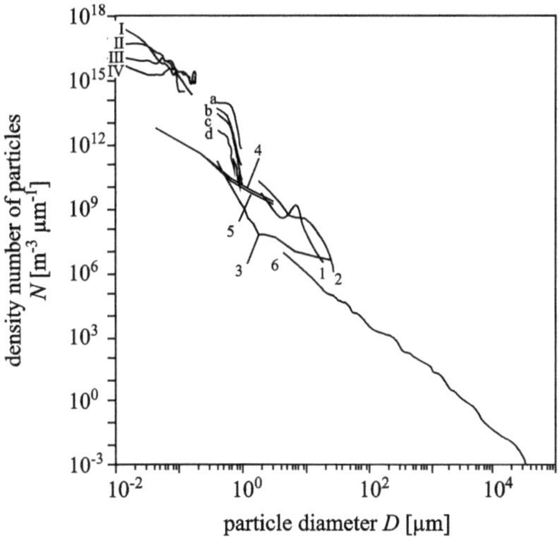

FIGURE 5.31. Examples of total particle size distributions measured in various size ranges at different depths in different seas: I, Scotian Shelf; II, Sargasso Sea; III,IV, Southern Ocean; a,b,c,d, Pacific (shelf waters off Japan); 1,2. Baltic; 3. Pacific (Rarotonga); 4,5. Gulf of Mexico (deep waters at 600 and 1800 m); 6. Pacific (Gulf of Monterey). Data gleaned from Yamasaki et al. (1998), plots a,b,c,d; Wells and Goldberg (1994), plots I, II, III,IV; Jonasz (1980), plots 1,2; Shifrin et al. (1974), plot 3; Harris (1977), plots 4,5; Jackson and Burd (1998), plot 6.

TABLE 5.25. Examples of cumulative numerical concentrations of suspended particles with diameters > 1 μm ($N_{int>1\mu m}$) recorded in the surface layers of various seas.

Region (Sea)	$N_{int>1\mu m}$ [10^9 m^{-3}]	Source of data[a]
-1-	*-2-*	*-3-*
North Atlantic:		1
Depth 1 m (29°22′N; 64°03′W)	~0.16	
Depth 1 m (41°40′N; 64°03′W)	~0.45	
Depth 5 m (36°25′N; 74°43,5′W)	~0.9	
Pacific (Rarotonga)	~1.76	2
Central Atlantic		
(Mean of 28 samples)	11.5	3
Central Baltic		
(Mean of 102 samples)	22.3	3
Baltic (Gulf of Gdańsk)		
(Mean of 67 samples)	44.4	3
South Atlantic (Ezcurra inlet)		
(Mean of 83 samples)	52.5	3

[a] Numerical concentrations $N_{int>1\mu m}$ estimated on the basis of data obtained from: 1. Sheldon et al. (1972); 2. Shifrin et al. (1974); 3. Jonasz (1980).

with respect to various random sets of particles suspended in diverse types of sea water, and with respect to the different size ranges of particles, this drop in the number of large particles can vary a great deal and be strongly non-monotonic. This applies in particular to particle size distributions in waters where one or several species of phytoplankton happen to be blooming. Then, the dominance resulting from such a bloom is reflected in bulges, inflexions, or even local peaks of the size distribution function $N = f(D)$ in the size intervals (ranges) characteristic of the species blooming in the given waters. These features are particularly conspicuous in differential size distributions, although much less strongly marked in integral distributions; see Figure 5.32.

In order to facilitate the analysis and modeling of the optical properties of marine suspensions and atmospheric aerosols in terms of Mie's theory and suchlike, researchers have long been attempting to describe in mathematical terms the size distribution functions (SDF) of these particles, not just complete sets of them, but also particular components of these sets. These attempts have been aimed mainly at particles with sizes from tenth parts of a micrometer to tens of micrometers, simply because the sizes of smaller particles, such as the smallest colloidal particle fractions, are empirically rather poorly understood. Nevertheless, this particular size range of particles is the most significant one as regards the influence of these types of particle on light absorption: with the largest overall volume and mass concentration; their share in the processes of light absorption by all suspended particulate matter in the sea is the dominant one.[27] Figure 5.33 presents size

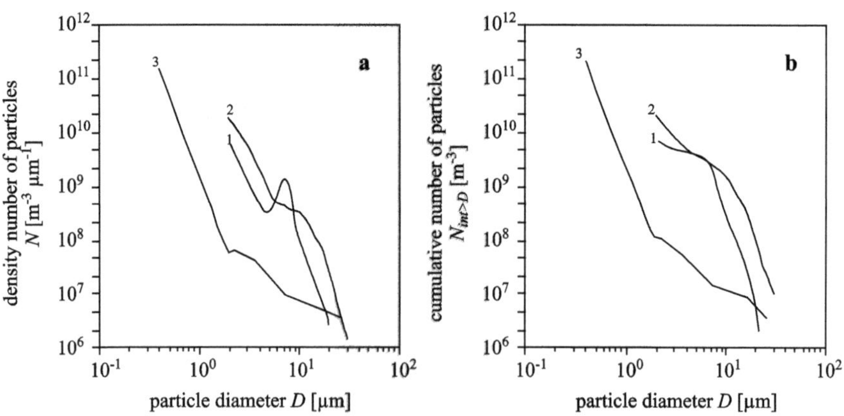

FIGURE 5.32. Differential (a) and integral (b) size distributions of marine particles in the Baltic Sea (plots 1 and 2, after Jonasz (1980)) and in the Pacific–Rarotonga (plot 3, after Shifrin et al. (1974)).

[27] But this does not apply to the same extent to light scattering. Models not taking account of light scattering by small colloidal particles may be encumbered with serious error (see Stramski and S.B. Woźniak (2005)).

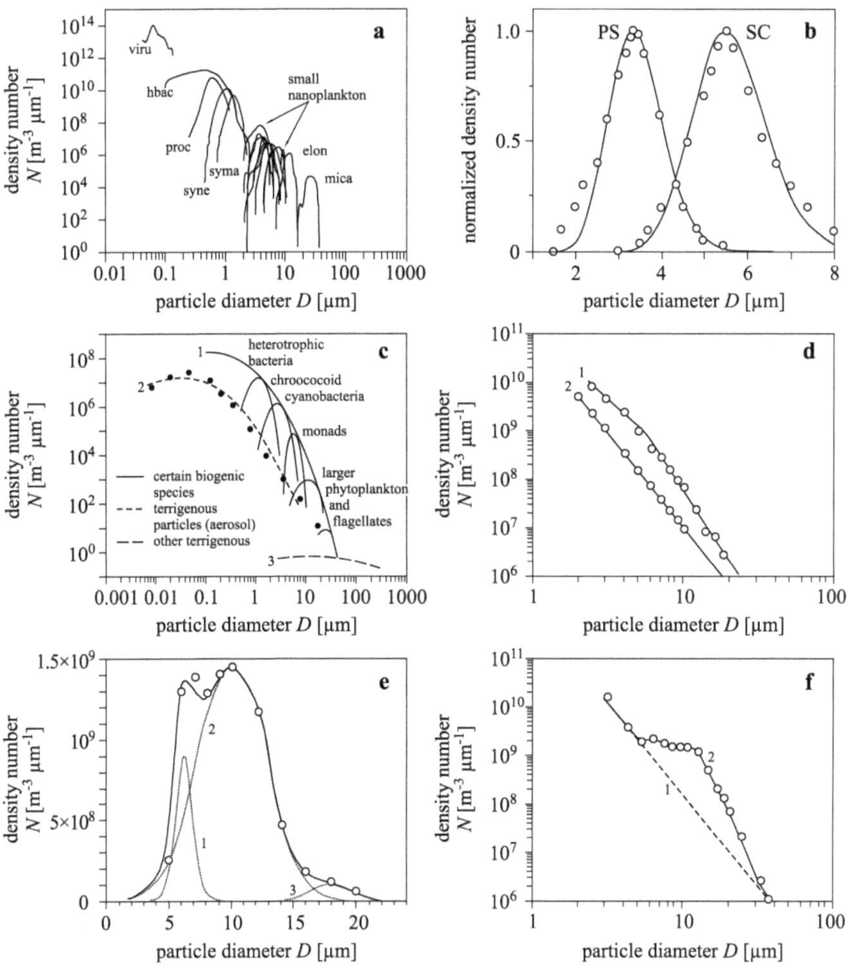

FIGURE 5.33. Modeled and/or empirical size distributions of various kinds of particles in the sea: (a) 18 planktonic components (for symbols, see Table 5.23) (adapted from Stramski et al. (2001)); (b) Ps, *Platymonas suecica* (*Prasinophyceae*) and Sc, *Skeletomena costatum* (*Bacillariophyceae*): points are measured data from Bricaud and Morel (1986), plots approximated with the gamma distribution (Equation (5.26)) (adapted from Risovic (1993)); (c) 1. particles of certain biogenic species; approximated with the generalized gamma distribution (Equation (5.28)) (adapted from Risovic (1993)); 2. mineral particles: points are measured data from Junge (1963), plot approximated with the generalized gamma distribution (Equation (5.28)) (adapted from Risovic (1993)); 3. schematic presentation of possible components of suspended particulate matter in coastal regions, mainly sand (the authors' own suggestion on the basis of information from Hwang et al. (2002) and Bigelow et al. (2004)); (d) 1. marine particles in the Baltic: points are measured data, plot approximated with the bisegmental hyperbolic distribution (Equation (5.29)) (adapted from Jonasz (1980)); 2. for marine particles in the Atlantic:

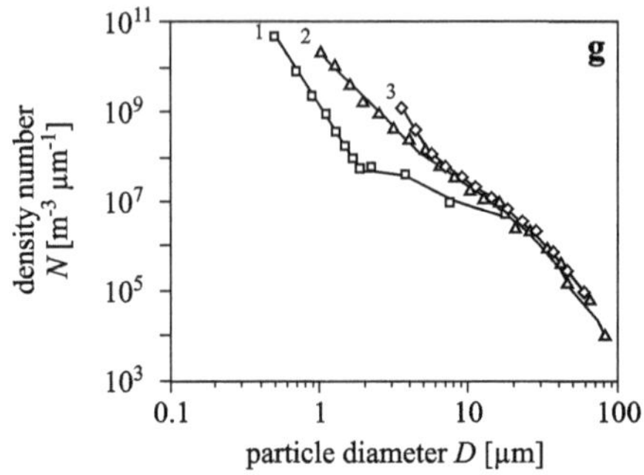

FIGURE 5.33. (Continued) points are measured data, plot approximated with the monosegmental hyperbolic distribution (Equation (5.25)) (adapted from Jonasz (1980)); (e) for three phytoplankton species in the Baltic during the bloom: points are measured data; dashed plots 1,2,3 approximated with the normal distribution (Equation (5.27)); continuous plot, sum of plots 1,2,3 (adapted from Jonasz (1980)); (f) for marine particles in the Baltic during the bloom of three phytoplankton species (see Figure e): continuous plot approximated with the hyperbolic-normal distribution (Equation (5.31)) (adapted from Jonasz (1980)); (g) for marine particles in the Pacific (Rarotonga) (plot 1; data from Shifrin et al. (1974)), North Pacific (plot 2; data from Sheldon et al. (1972)) and Sargasso Sea (plot 3; data from Del Grosso (1978)): points are measured data; plots approximated with the bisegmental generalized gamma distribution (Equation (5.30)) (adapted from Risovic (1993)). Note: on (c) the results are only examples and the proportions between the density numbers $N(D)$ of the various components have not been maintained.

distributions of various kinds of particles measured in the sea together with their modeled approximations. We now briefly discuss the most frequently applied mathematical formulas approximating size distribution functions along with their ranges of applicability.

Formulas Approximating SDF

The principal mathematical formulas used to describe SDFs of particles suspended in sea waters (see the symbol $n(D)$ [m^{-3} μm^{-1}] in Equation (5.21)) are as follows:

- *The single-parameter hyperbolic distribution* (Bader 1970), also known as Junge's distribution (Junge 1963) is used chiefly for describing SDFs for particles of minerals and organic detritus:

$$n(D) = D^{-k}, \qquad (5.25)$$

  where D [μm] is the equivalent diameter of the particle, k [dimensionless] is a constant taking values from 2.5 to 6, depending on the size range of the particles under scrutiny and the origin of the particle sample (where particles larger than 1 μm have been measured with a Coulter counter, values of k usually lie between 2.5 and 4.5).

- *The two-parameter gamma distribution*, used by Deirmendijan (1964) and Green et al. (1971) for describing the size distribution of aerosol and fog particles is suitable for describing the size distribution of plankton monocultures (see the empirical distributions in Figure 5.33a and the modeled ones in Figure 5.33b):

$$n(D) = \begin{cases} 0 & D < 0 \\ D^{\mu} e^{-bD} & D \geq 0; \mu > -1; b > 0 \end{cases}, \tag{5.26}$$

  where μ and b are parameters of the distribution.
- *The two-parameter normal distribution* is used, for example, by Jonasz (1980, 1983) for describing the components of particles derived from living organisms and also the hard skeletons from dead phytoplankton cells (see Figure 5.33e):

$$n(D) = e^{-b(D - D_{max})^2}, \tag{5.27}$$

  where b and D_{max} are parameters of the distribution.
- *The three-parameter generalized gamma distribution* (Risovic 1993), which, among other things, provides a description of the superposition of several plankton monocultures (see Figure 5.33c, plot 1) and of the size distributions of terrigenous particles (aerosol, dust, kaolin particles dissolved in water; Figure 5.33c, plot 2):

$$n(D) = \begin{cases} 0 & D \leq 0 \\ D^{\mu} e^{-bD^{\gamma}} & D > 0 \end{cases}, \tag{5.28}$$

  where μ, b, and γ are parameters of the distribution.

Using any one of the above functions, it is possible to describe with satisfactory accuracy the size distributions of marine suspended particles only when one characteristic biogenic or terrigenous component is dominant, and often only within a limited range of sizes (see the example approximating the real size distribution of particles in the sea by means of just one hyperbolic function, Figure 5.33d, plot 2). To describe the more complex, real size distributions of marine particles over a wide range of sizes, complex models have to be applied. The most common of these are as follows:

- *The bisegmental hyperbolic distribution*, applied by, for example, Harris (1977) and Jonasz (1980, 1983); it takes two different slopes in different particle size intervals into consideration (see, e.g., Figure 5.33d, plot 1):

$$n(D) = \begin{cases} C_A D^{-k_A} & D \le D_P \\ C_B D^{-k_B} & D > D_P \end{cases}, \tag{5.29}$$

where k_A, k_B, D_P are parameters of the distribution, and factors C_A and C_A are weights reflecting the proportions of the components of the distributions in the total number of particles N_{tot}.

- *The bisegmental generalized gamma distribution*, that is, the bicomponental distribution postulated by Risovic (1993), in which each component is described by a generalized gamma distribution with fixed values of parameters μ and b, with only parameter γ varying in both components (see the examples in Figure 5.33g):

$$n(D) = C_A D^{\mu_A} e^{-b_A(D/2)^{\gamma_A}} + C_B D^{\mu_B} e^{-b_B(D/2)^{\gamma_B}}, \tag{5.30}$$

where C_A and C_B are weights reflecting the proportions of the components of the distributions in the total number of particles, and the suggested values of the parameters are $\mu_A = 2$, $b_A = 52$, $\mu_B = 2$, $b_B = 17$, $\gamma_A = 0.145$–0.195, and $\gamma_B = 0.192$–0.322.

- *The multisegmental "hyperbolic-normal" distribution,* postulated by Jonasz ((1980, 1983); see also Jonasz and Fournier (1996)) consists of one or two hyperbolic segments describing the distributions of nonliving particles and a few "normal" segments describing plankton distributions (see the example in Figure 5.33f):

$$n(D) = \sum_i C_i e^{-b_i (D - D_{max,i})^2} + C_A D^{-k_A} + C_B D^{-k_B}, \tag{5.31}$$

where b_i, $D_{max,i}$, k_A, k_B denote the parameters of the distribution, and factors C_i, C_A, C_B are weights dependent on the proportion of particles in the several segments in the total number of particles N_{tot}.

These formulas approximating SDFs by no means exhaust the list of attempts undertaken by various authors at a mathematical description of the size distributions of the various kinds of particles in the sea and the atmosphere, and of the particles of various minerals and organisms. Many such formulas have been given by Shifrin (1983a, 1988), Risovic (1993), Jonasz and Fournier (1996), Hwang et al. (2002), and the papers cited in these publications. Unfortunately, there are no universal formulas describing real total SDFs for all seas among those mentioned above or in the literature. Modeling the size distributions of suspended particulate matter in the sea and their optical properties therefore comes up against numerous difficulties; any attempt must take into account the specifics of the basin, its geographical position, the depth at which the particles to be examined are suspended, and also the time of year to which the modeling refers.

5.3 Light Absorption Properties of Nonalgal Particles: Results of Empirical Studies

In Section 5.2 we discussed the nature and origin of the major types and subtypes (groups) of suspended particulate matter in sea waters. We also presented the current state of knowledge as regards the most important chemical and physical properties of the various different types of SPM and their associations occurring naturally in different sea waters. Our objective in Section 5.3 is to characterize the principal absorption properties of these natural associations of SPM.

The absorption properties (and also other optical properties) of marine SPM can be determined theoretically, for example, with the aid of Mie's theory, presented in Section 5.1. To this end, data are required on the concentrations of these particles, their size distributions, and their fundamental physical properties (optical constants and the like), which we described in Section 5.2. The reader can find examples of such model calculations of the optical properties of discrete particles with given dimensions, or of their variously sized associations, in numerous publications. Performed using both Mie's theory and other mathematical models for nonspherical particles, these calculations have been applied to mineral particles, such as those present in atmospheric dusts and in sea waters (see, e.g., Olmo et al. (2006), Plaza et al. (1997, 2002), Babin and Stramski (2004), Stramski et al. (2004), Woźniak S.B. and Stramski (2004), Sokolik and Toon (1999), and Marra et al. (2005)), and also to plankton cells and other organic suspended matter (see, e.g., Quirantes and Bernard (2004, 2006), Bernard et al. (2001), Stramski et al. (2001), and Król (1998)).

But the results of all these theoretical calculations do not usually refer to all SPM, that is, to entire associations of suspended particles in the sea: either particular groups of SPM were selected as the focus of study, or, where whole natural associations of SPM were indeed the focus, they were examined only in the context of a given set of local environmental conditions. These results do not therefore give the full picture of the natural diversity of the absolute light absorption coefficients of SPM in different sea waters, neither do they define the proportions between the contributions made by the various types of these particles to the total absorption of light by all SPM. Nonetheless, it is possible to obtain an approximation of this picture from the published results of relevant empirical research; it is this approximate picture that we present in this section. We describe in detail the absorption properties of nonalgal particles, that is to say, of all inorganic (mineral) particles and of all organic particles except phytoplankton. We deal with the absorption of light by phytoplankton pigments separately and in considerable detail in Chapter 6.

The optical characteristics of SPM discussed here refer to the visible light spectrum.

5.3.1 Light Absorption Spectra of All Suspended Particulate Matter (SPM) and Nonalgal Particles: General Characteristics

In this section we state more precisely the range of differentiation of the absolute visible light absorption coefficients of all SPM in different seas and describe their major spectral features. We then go on to discuss the state of knowledge regarding the proportions of the two principal absorbers among the SPM, that is, phytoplankton pigments and nonalgal particles, in the total absorption of light by all SPM in different marine basins. We wind up this section by describing the range of natural differentiation and spectral characteristics of the absolute visible light absorption coefficients of nonalgal particles. We describe in detail these aspects with regard to phytoplankton pigments in Chapter 6.

Absorption Spectra of SPM: Absolute Values and Structure

Figure 5.34 illustrates empirical spectra of the visible light absorption coefficients of SPM $a_p(\lambda)$ typical of the euphotic zones of different marine basins, from the optically clear waters of the oceans to the highly productive waters of Baltic Sea bays. The absolute values of these coefficients $a_p(\lambda)$ in particular, and to a lesser extent their spectral structure, display considerable diversity in the World Ocean. In fact, this differentiation is even greater than that depicted in Figure 5.34, varying as it does over more than four orders of magnitude, from exceedingly small values, of the order of 10^{-4} m^{-1}, characteristic of the clear, ultra-oligotrophic waters of ocean centers (see, e.g., $a_p(\lambda)$ for light from the middle range of the visible spectrum, plot 1 in Figure 5.34), to values of 1 m^{-1} (see plot 23 in Figure 5.34) and more in extreme cases, in the supereutrophic, coastal waters of oceans and in enclosed seas (Babin et al. 2003a).

In outline, the shapes of the spectra of $a_p(\lambda)$ for almost all types of SPM are similar, usually displaying two principal, wide, more or less intense absorption bands. One of these bands is wider and taller, and lies in the blue region with a maximum usually at c. 435–445 nm; its half-width is normally >100 nm. Resembling the so-called Soret band in the absorption spectrum of chlorophyll, this band is due to light absorption by chlorophyll *a* and most of the other phytoplankton pigments, the strongest visible light absorbers of all the organic compounds present in SPM. This was exemplified already in Section 5.2 by the relevant imaginary refractive indices. The other principal absorption band, a narrower and weaker one, with a half-width of c. 20–30 nm, is situated in the red with a maximum at c. 675 nm. It is due mainly to the principal phytoplankton pigment, chlorophyll *a*. These very distinctly two-banded spectra of $a_p(\lambda)$ are further differentiated by their fine structure, different in various marine basins. This consists of a number of "smaller" and randomly distributed irregularities: narrow, although not very intensive,

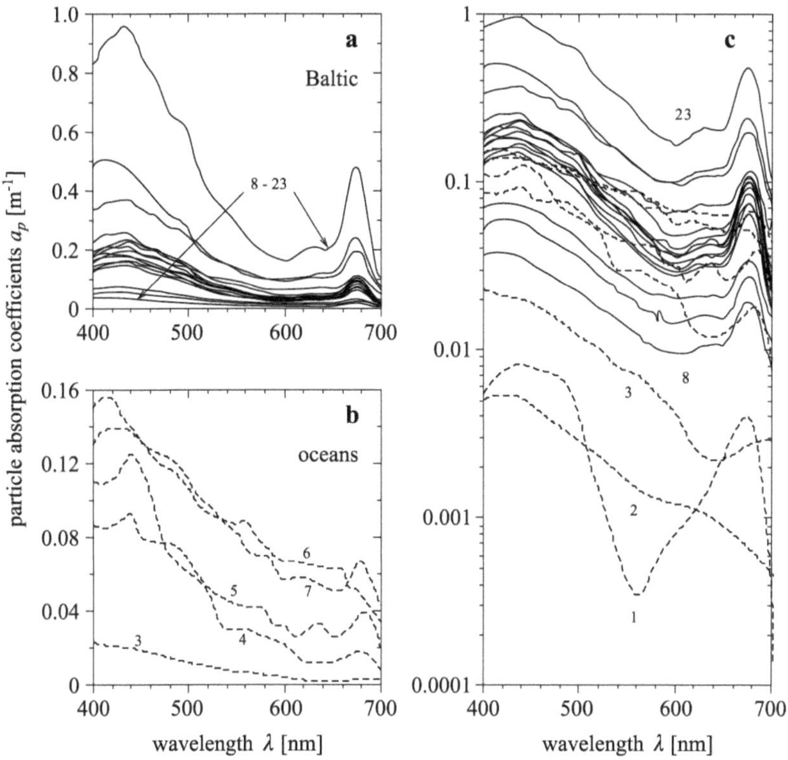

FIGURE 5.34. Empirical spectra of the total absorption of light by suspended particulate matter (SPM) in sea water, $a_p(\lambda)$: (a) and (b) – coefficients $a_p(\lambda)$ on a linear scale, (c) – coefficients $a_p(\lambda)$ on a logarithmic scale. The numbers of the plots denote spectra of $a_p(\lambda)$ for various marine regions: 1, 2 – Sargasso Sea; 3, 4 – South Atlantic; 5-7 – North-west Pacific; 8-23 – Southern Baltic (different areas in the open sea and in the Gulf of Gdańsk). (from the data bank of IO PAS Sopot).

local maxima and minima, manifested as a rule by slight bulges or inflexion points in the overall spectral structure.

The spectral values of the light absorption coefficients of all SPM $a_p(\lambda)$ generally correlate weakly not only with the total SPM concentration, but also with the concentrations of its various subgroups, for example, with POM or with C_a, a good index of the phytoplankton content in the sea. This is quite easily explained. The above-mentioned features of the total absorption spectra of $a_p(\lambda)$ for all SPM under different conditions in the seas and oceans and the quantitative differentiation in the absorption coefficients $a_p(\lambda)$ are due to the fact that the spectra of $a_p(\lambda)$ are the sums of light absorptions by diverse groups of suspended particles, present in sea waters in very different concentrations and in very different proportions.

Principal Components of Light Absorption by SPM in the Sea

The methodology underlying empirical studies[28] of the absorption properties of SPM assumes its division into two groups of particles: the phytoplankton (together with their light-absorbing pigments), with an absorption coefficient here denoted by $a_{pl}(\lambda)$, and nonalgal particles (NAP), the absorption coefficient of which is denoted by $a_{NAP}(\lambda)$. Such a division leads to the following equation:

$$a_p(\lambda) = a_{pl}(\lambda) + a_{NAP}(\lambda). \tag{5.32}$$

The term $a_{NAP}(\lambda)$ covers the combined effects of light absorption not only by mineral particles ($a_{PIM}(\lambda)$), but also by organic detritus along with all the "living" particles not included among the phytoplankton (i.e., zooplankton) and nonliving organic matter. Denoting the absorption coefficient of organic detritus so defined by $a_{Od}(\lambda)$, we can therefore write the expression for the total absorption by NAP as:

$$a_{NAP}(\lambda) = a_{PIM}(\lambda) + a_{Od}(\lambda), \tag{5.33}$$

and the expression for the total absorption of light by all SPM as:

$$a_p(\lambda) = a_{pl}(\lambda) + a_{Od}(\lambda) + a_{PIM}(\lambda). \tag{5.34}$$

Figure 5.35 shows a number of empirical examples of this division of the total light absorption coefficient by all SPM $a_p(\lambda)$ into its principal components (see Equation (5.32)), that is, light absorption by phytoplankton pigments $a_{pl}(\lambda)$ and by nonalgal particles $a_{NAP}(\lambda)$, in different seas and oceans. The spectral components $a_{pl}(\lambda)$ and $a_{NAP}(\lambda)$ clearly differ considerably in structure. We describe the characteristic spectral features of $a_{NAP}(\lambda)$ and $a_{pl}(\lambda)$ in due course; those of the former at the end of this section, and those of the latter in Chapter 6. Figure 5.35 also shows that the proportions of a_{pl} and a_{NAP} in the total coefficient a_p vary in different waters, a_{NAP} generally being smaller than a_{pl}. This emerges from the well-known fact that the absorption capacities of phytoplankton pigments far exceed those of other organic constituents. Hence, even though autogenic NAP (derived from phytoplankton) may be present in the water in much greater concentrations than the

[28] Such studies are usually carried out in the laboratory and generally involve two measurements of the light absorption spectrum of the filtered particles (in the absence of sea water). The first measurement is of the total absorption by the natural mixture of particles in the sample $a_p(\lambda)$. The second one is performed on the same particles after their decolorization and yields the absorption $a_{NAP}(\lambda)$ (i.e., that of nonalgal particles). To a good approximation, the difference between $a_p(\lambda)$ and $a_{NAP}(\lambda)$ is equal to the light absorption by the phytoplankton pigments: $a_{pl}(\lambda) = a_p(\lambda) - a_{NAP}(\lambda)$. The methodology of these measurements is complex; it is described in greater detail in Chapter 6 (Section 6.6.1).

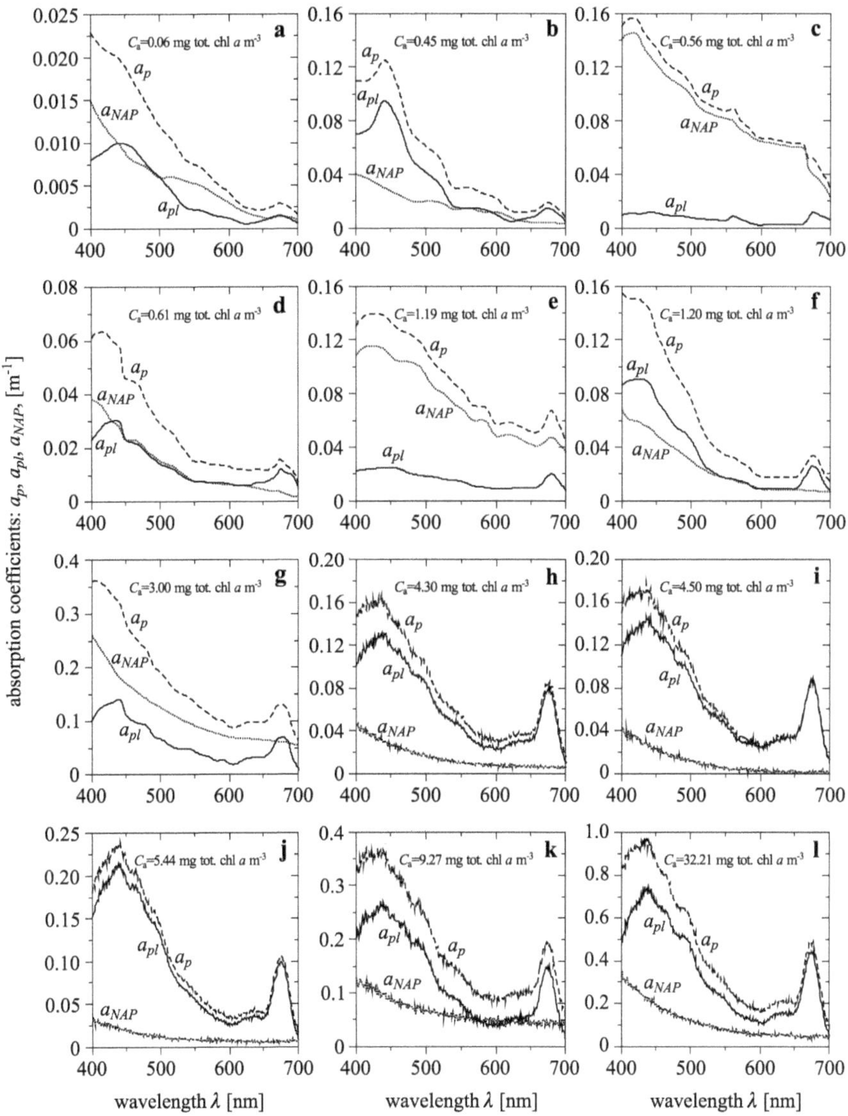

FIGURE 5.35. Empirical light absorption spectra of suspended particulate matter (SPM) and its components in the euphotic zones of various seas and oceans: (a_p) – total absorption by SPM; (a_{pl}) – absorption by phytoplankton pigments; (a_{NAP}) – absorption by non-algal particles; a,b – Atlantic; c,d – Pacific (coastal regions); f,g – central Baltic; h-l – Baltic Sea (Gdańsk Deep and Gulf of Gdańsk); all plots give the concentration of chlorophyll *a* C_a in the water samples examined for the absorption properties of the SPM they contained. (from the data bank of IO PAS Sopot).

phytoplankton itself, (see, e.g., Figure 5.7), their absorption coefficients a_{NAP} are lower than a_{pl}. Under certain conditions, however, this situation may be reversed: then light absorption by NAP exceeds, sometimes by a large margin, that by phytoplankton. This happens mostly in parts of seas and oceans where Case 2 waters prevail, when substantial amounts of allogenic particles, both mineral and organic, enter the sea (see Figure 5.35, plots c,d,g).

For many years now, attempts have been made to characterize the quantitative contribution of all SPM in the total absorption of light in sea water, and the contribution of the various types of SPM in the total absorption of light by all SPM in different marine basins (see, e.g., Woźniak (1973, 1977), Dera et al. (1974, 1978), Gohs et al. (1978), Kishino et al. (1984, 1986), Morrow et al. (1989), Roesler et al. (1989), Bricaud et al. (1995, 1998), and Babin et al. (2003a)). Most of this research was limited to investigating light absorption budgets, thus conceived, in certain marine basins and seasons.

The work of Bricaud et al. (1998), however, took a universal approach: their results are applicable to all the various types of Case 1 waters. On the basis of a statistical analysis of some 300 sets of empirical spectra of $a_p(\lambda)$, $a_{pl}(\lambda)$, and $a_{NAP}(\lambda)$, these authors developed a unique parameterization of these spectra for all the trophic types of Case 1 waters. This parameterization is founded on the statistical associations established between these spectral coefficients $a_p(\lambda)$, $a_{pl}(\lambda)$, and $a_{NAP}(\lambda)$ (in the VIS region) and the chlorophyll *a* concentration (C_a) in the basin; in all our publications, including the present book, we treat this as the trophic index (trophicity) of waters (see Section 6.1 Table 6.1). In Sections 5.3.3 and 6.7.3 we describe in greater detail some of the results from Bricaud et al. (1995, 1998).

Here we show how their parameterization can be applied to work out the mean statistical contributions of phytoplankton and nonalgal particles to the total absorption of light by all SPM in different trophic types of Case 1 waters. The absorption coefficients a_p, a_{pl}, and a_{NAP} in these waters calculated for light of wavelength $\lambda = 443$ nm are set out in Table 5.26A (items 1, 2, and 3). We also used this parameterization to determine, for the same waters, the mean proportions of phytoplankton ($\Delta_{pl} = a_{pl}/a_p$) and nonalgal particles ($\Delta_{NAP} = a_{NAP}/a_p$) in the total absorption of light by all SPM (Table 5.26A, items 4 and 5) and the corresponding mean ratios of the absorption of light ($\lambda = 443$ nm) by NAP and phytoplankton ($a_{NAP}/a_{pl} = \Delta_{NAP}/\Delta_{pl}$). Figure 5.36 illustrates some of these results.

According to the data presented in Table 5.26A and in Figure 5.36, the absorption coefficients for NAP a_{NAP}(443 nm) in different Case 1 waters are always markedly lower than the absorption coefficients for phytoplankton pigments a_{pl}(443 nm), and the ratio of these two coefficients is more or less stable. Hence, the proportion of nonalgal particles in the total light absorption by all SPM in oligotrophic waters O1 ($C_a = 0.035$ mg tot. chl *a* m^{-1}) is c. 23% and rises only slightly with increasing trophic index (i.e., C_a), achieving values of nearly 29% in eutrophic waters E4 ($C_a = 15$ mg tot. chl *a* m^{-3}). In contrast, the proportion of light ($\lambda = 443$ nm) absorbed by phytoplankton

TABLE 5.26. Averaged statistical parameters characterizing the absorption of light ($\lambda = 443$ nm) by suspended particles in the sea.[a]

A. In Different Trophic Types of Case 1 Waters[b]
Trophic index of waters C_a [mg tot.chl*a* m^{-3}]

No.	Parameter	O1 0.035	O2 0.075	O3 0.015	M 0.35	I 0.7	E1 1.5	E2 3.5	E3 7.0	E4 15
-1-	-2-	-3-	-4-	-5-	-6-	-7-	-8-	-9-	-10-	-11-
1	a_p [m^{-1}]	6.23×10^{-3}	1.03×10^{-2}	1.55×10^{-2}	6.23×10^{-2}	4.35×10^{-2}	6.54×10^{-2}	1.11×10^{-1}	1.84×10^{-1}	2.76×10^{-1}
2	a_{NAP} [m^{-1}]	1.46×10^{-3}	2.48×10^{-3}	3.84×10^{-3}	6.70×10^{-3}	1.13×10^{-2}	1.75×10^{-2}	3.04×10^{-2}	5.14×10^{-2}	7.90×10^{-2}
3	a_{pl} [m^{-1}]	4.77×10^{-3}	7.82×10^{-3}	1.17×10^{-2}	1.96×10^{-2}	3.22×10^{-2}	4.79×10^{-2}	8.07×10^{-2}	1.32×10^{-1}	1.97×10^{-1}
4	Δ_{NAP} [%]	23.5	24.1	24.8	25.5	26.1	26.7	27.4	28.0	28.6
5	Δ_{pl} [%]	76.5	75.9	75.2	74.5	73.9	73.3	72.6	72.0	71.4
6	a_{NAP}/a_{pl}	0.306	0.317	0.328	0.342	0.353	0.364	0.377	0.389	0.401

B. In Different Types of Case 2 Waters[c]
Study area

No.	Parameter	Adriatic	Eng. Channel	Baltic	Mediterranean	North Sea	Baltic[d] ($C_a = 9.61$)	Baltic[e] ($C_a = 1.4$)
-1-	-2-	-3-	-4-	-5-	-6-	-7-	-8-	-9-
1	Number of Data	39	75	54	45	85	18	28
2	a_p [m^{-1}]	—	—	—	—	—	1.84×10^{-1}	6.20×10^{-2}
3	a_{NAP} [m^{-1}]	—	—	—	—	—	3.4×10^{-2}	2.17×10^{-2}
4	a_{pl} [m^{-1}]	—	—	—	—	—	1.50×10^{-1}	4.03×10^{-2}
5	Δ_{NAP} [%]	27.8	31.1	38.5	42.9	50.0	18.5	35.0
6	Δ_{pl} [%]	72.2	68.9	61.5	57.1	50.0	81.5	65.0
7	a_{NAP}/a_{pl}	0.385	0.451	0.626	0.751	1.00	0.227	0.538

[a] a_p, a_{NAP} and a_{pl}: respective coefficients of light absorption by all SPM, nonalgal particles, and phytoplankton pigments; Δ_{NAP} and Δ_{pl}: respective proportions of nonalgal particles and phytoplankton pigments in the total absorption of this light by all SPM; a_{NAP}/a_{pl}:ratio of the absorption of this light by NAP to that by phytoplankton pigments.

[b] According to the parameterization by Bricaud et al. (1998).

[c] Columns 3–7 based on data gleaned from Babin et al. (2003a); Columns 8 and 9 from the data bank of IO PAS Sopot.

[d] Data gathered on cruises of r/v Oceania in the Gdańsk Deep and Gulf of Gdańsk during the phytoplankton bloom in spring 2004.

[e] Data gathered on a cruise of r/v Professor Shtockman in the Gdańsk Deep and Gulf of Gdańsk in January 1990; in both cases (d and e) the average chlorophyll *a* concentrations C_a in the study areas, expressed in [mg tot. chl *a* m^{-3}], are also given.

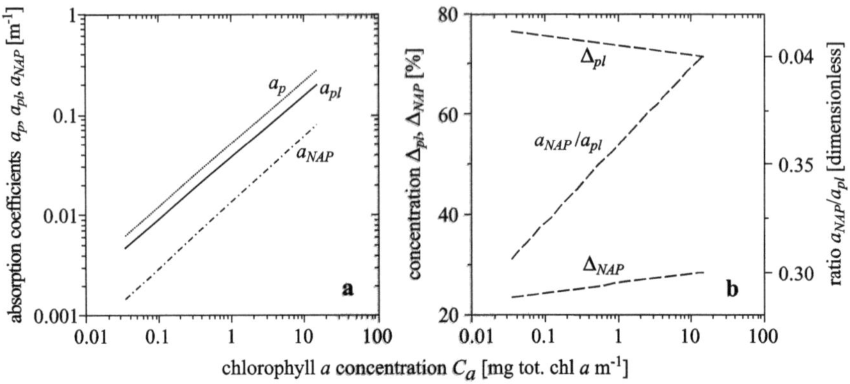

FIGURE 5.36. Relationship between the absorption of light (λ = 443 nm) by SPM, averaged for Case 1 waters, and their trophic index, calculated using the parameterization of Bricaud et al. (1998) (the trophic index of water is taken to be the chlorophyll a concentration C_a): (a) for the absorption coefficients of all SPM (a_p), phytoplankton pigments (a_{pl}) and non-algal particles (a_{NAP}); (b) for the proportions of phytoplankton pigments (Δ_{pl}) and non-algal particles (Δ_{NAP}) in the total absorption by all SPM and for the ratio of the absorption of this light by non-algal particles to that by phytoplankton pigments (a_{NAP}/a_{pl}).

pigments in the total absorption by all SPM diminishes somewhat with increasing trophic index: from c. 77% in O1 waters to c. 71% in E4 waters. The ratio of these two absorption coefficients, a_{NAP}/a_{pl}, in these oligotrophic waters is therefore c. 0.3, rising to c. 0.4 in eutrophic waters.

But such a stable situation, when the proportions between light absorption by NAP and phytoplankton are roughly the same and when a_{pl}(443 nm) is some three times greater than a_{NAP}(443 nm), occurs only in Case 1 waters, in which all suspended organic particles originate directly or indirectly from the phytoplankton. On the other hand, these proportions in Case 2 waters can vary a great deal. Table 5.26B presents some characteristics of light (λ = 443 nm) absorption by suspended matter in the Case 2 waters off the coasts of Europe (data gleaned from Babin et al. (2003a) and our own research). As a result of all kinds of allogenic organic and mineral suspended matter entering such Case 2 waters, the light absorption coefficients of NAP–a_{NAP} (and hence those of all SPM–a_p) are greater than the corresponding coefficients in Case 1 waters of the same trophicity. The ratio a_{NAP}/a_{pl} thus increases, sometimes by a considerable amount, to values above the c. 0.3–0.4 (for λ = 443 nm) typical of Case 1 waters (see Table 5.26B, item 7). Such an increase, however, is by no means always the rule. This is because in some Case 2 waters, the ratio a_{NAP}/a_{pl} can also reach values lower than those in Case 1 waters with the same C_a. This can happen in Case 2 waters, for instance, during a massive bloom of phytoplankton, when its content in the water rises sharply in a short time and this increase is way ahead of the increase in autogenic NAP. We recorded such a situation during the spring phytoplankton bloom in the Baltic (the Gdańsk Deep and Gulf of Gdańsk) in 2004 (see Table 5.26B, column 8).

On the basis of current knowledge of Case 1 waters, it is possible to state with satisfactory accuracy the magnitudes and proportions between the light absorption coefficients of phytoplankton pigments a_p, nonalgal particles a_{NAP}, and all SPM. But with respect to Case 2 waters, these characteristics are purely random values, which cannot be parameterized in such a simple way as for Case 1 waters. This is a complex issue that needs to be researched on a local scale.

Up to the present moment, we have not been able to gather sufficient empirical data to estimate the proportions of the mineral and organic constituents of NAP in the absorption of light by all NAP for either Case 1 or Case 2 waters.

Light Absorption Spectra of Nonalgal Particles: Absolute Values and Structure

Figure 5.37 illustrates some empirical VIS absorption spectra of nonalgal particles $a_{NAP}(\lambda)$ recorded in the euphotic zones of various oceanic regions and Baltic Sea basins. Clearly, the spectral coefficients a_{NAP} vary a good deal in value; in reality, however, this diversity is even greater than that suggested by the figure, covering as it does more than four orders of magnitude, analogously to the absorption coefficients $a_p(\lambda)$ of all SPM. The lowest values of a_{NAP} are characteristic of *NAP* from central oceanic regions: for red light (c. 700 nm) they may be less than 10^{-4} m^{-1}. In contrast, the highest values of c. 1 m^{-1} and more are typical of blue light and are from time to time recorded in river estuaries and the bays of shallow seas.

The structure of $a_{NAP}(\lambda)$ spectra is a relatively simple affair in comparison with that of $a_p(\lambda)$ spectra. It is evident from Figure 5.37 that, if we ignore the fine structure (the small narrow peaks and troughs are due mainly to apparatus "noise"), in the great majority of cases a_{NAP} falls monotonically in value with increasing light wavelength. The nature of this decrease is well described by exponential functions (the slopes of such functions approximating the absorption spectra of NAP are analyzed in Section 5.3.2). These spectra therefore resemble the absorption spectra of colored dissolved organic matter (CDOM) in sea water ($a_y(\lambda)$), which we discussed in Chapter 4. However, such a monotonic drop in absorption with increasing light wavelength typifies not only the absorption spectra of suspended organic particles, but also those of all NAP in a particular sea, regardless of whether more organic or inorganic particles are present. Only in a very few cases (more often in lakes than in seas) have some authors (e.g., Morel and Prieur (1977), Bukata et al. (1981a, b, 1983), Gallie and Murtha (1992), and Prieur and Sathyendranath (1981)) recorded enhanced absorption in the red light region,

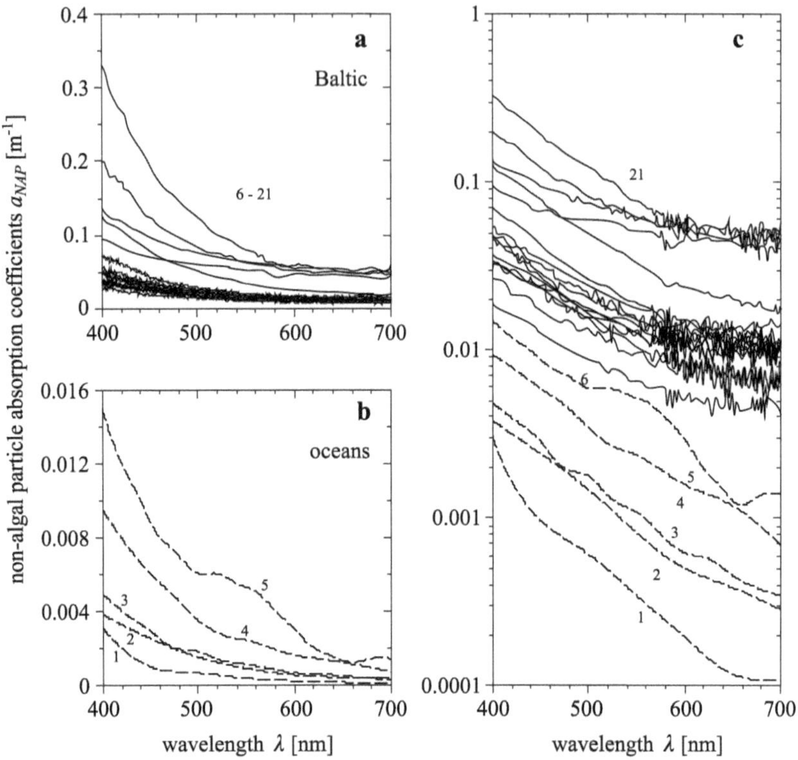

FIGURE 5.37. Empirical light absorption spectra of non-algal particles $a_{NAP}(\lambda)$: (a) and (b) – coefficients $a_{NAP}(\lambda)$ shown on a linear scale; (c) – coefficients $a_{NAP}(\lambda)$ shown on a logarithmic scale. The numbers of the plots denote the spectra of $a_{NAP}(\lambda)$ for various seas: 1, 2 – Sargasso Sea; 3, 4 – NW Pacific; 5 – South Atlantic; 6-21 – Southern Baltic (various areas of the open sea and the Gulf of Gdańsk). (from the data bank of IO PAS Sopot).

an increase due possibly to the presence in the suspended matter of rare minerals or organometallic complexes with sufficiently high maxima in the red and infrared.

5.3.2 *Spectra of the Mass-Specific Light Absorption Coefficients of Nonalgal Particles*

In the previous section we characterized in a general way the light absorption spectra of all suspended particulate matter $a_p(\lambda)$ and of nonalgal particles $a_{NAP}(\lambda)$ in the sea. The absolute values of a_p and a_{NAP} do not depend only on the properties of discrete particles; they depend above all on their concentrations in the water. In the present section we describe the mass-specific absorption coefficients of these suspended particles, $a_p^*(\lambda)$ and $a_{NAP}^*(\lambda)$, in other words, the ratios of the absorption coefficients $a_p(\lambda)$ and $a_{NAP}(\lambda)$ to the SPM concentration in the water:

$$a_p^*(\lambda) = a_p(\lambda)/SPM \tag{5.35}$$

$$a_{NAP}^*(\lambda) = a_{NAP}(\lambda)/SPM. \tag{5.36}$$

Now these mass-specific coefficients no longer depend on the SPM concentration in the water; they are solely properties of the particles themselves. That is to say, they depend on the chemical and physical properties of the particles, for example, their chemical compositions, sizes and shapes, and optical constants. Figure 5.38 illustrates some empirical spectra of these mass-specific coefficients.

The following topics are handled in the remainder of this section: the ranges of variability of a_p^* and a_{NAP}^*, with particular emphasis on the latter

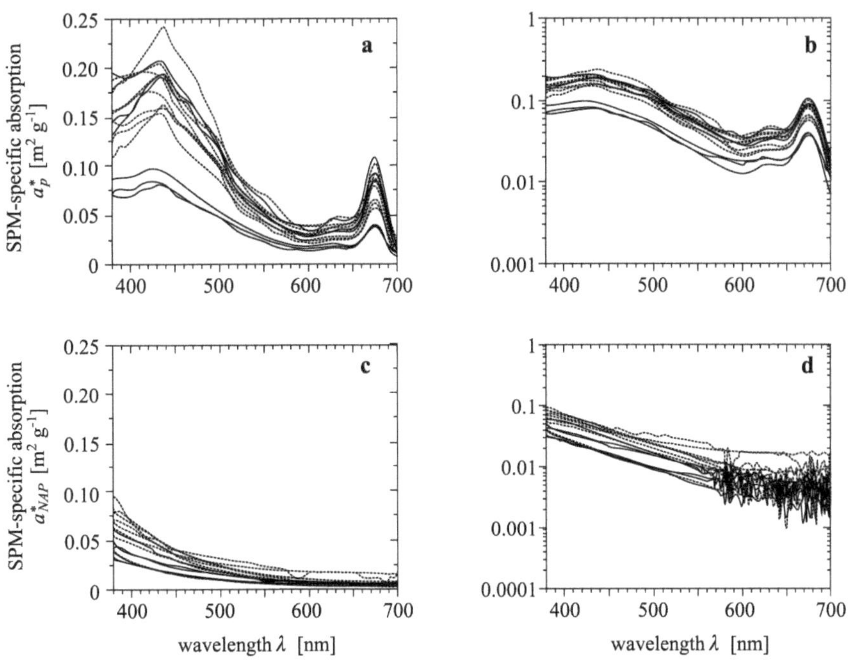

FIGURE 5.38. Spectra of the mass-specific absorption coefficients of all suspended particulate matter (SPM) $a_p^*(\lambda)$ – (a and b), and of non-algal particles $a_{NAP}^*(\lambda)$ – (c and d) recorded in the Baltic Sea (the Gdańsk Deep and the Gulf of Gdańsk) in spring 2004; Figures a and c are plotted on a linear scale, Figures b and d on a logarithmic scale. Dotted lines – spectra of the cases when $POM \geq 0.75\ SPM$; Continuous lines – other cases, i.e. when $POM < 0.75\ SPM$ (based on the data bank of IO PAS Sopot).

coefficient; the spectral slope of absorption $a_{NAP}(\lambda)$; and finally, an attempt to discover whether and how the spectra of the mass-specific absorption coefficients of natural associations of organic particles differ from those of mineral particle associations.

Absolute Magnitudes and Ranges of Variability of a_p^* and a_{NAP}^*

Research has shown that the mass-specific light absorption coefficients of all SPM (a_p^*) and of NAP (a_{NAP}^*) are not constant; they differ considerably in value at different depths in various natural basins, and at different times of the year (see, e.g., Babin et al. (2003a), Bowers and Binding (2006), S. B. Woźniak et al. (2006)). This emerges from the diverse make-up of mineral and organic particles suspended in the water, their various dimensions and chemical compositions. Even within the same basins a_p^* and a_{NAP}^* can vary by several factors. The spectra of $a_p^*(\lambda)$ and $a_{NAP}^*(\lambda)$ illustrated in Figure 5.38 exemplify this point. These spectra were measured on samples taken from the Gdańsk Basin (the Gdańsk Deep and the Gulf of Gdańsk) during one month in 2004 (April 17–May 22), during the phytoplankton bloom. With the proportions of the SPM constituents in this basin (phytoplankton, other organic particles, and mineral particles) changing, largely because of the phytoplankton bloom, the values of both a_p^* and a_{NAP}^* altered by some two to three factors in the same spectral intervals.

The real ranges of variability of a_p^* and a_{NAP}^* in different seas, rivers, and lakes are undoubtedly even wider that those exemplified in Figure 5.38 for the Gdańsk Deep and Gulf of Gdańsk regions of the Baltic Sea. As yet, however, there are too few empirical data in the subject literature to be able to define the scale of this variability more precisely. Moreover, the methods of measuring SPM concentrations,[29] knowledge of which is indispensable for determining a_p^* and a_{NAP}^*, are encumbered with considerable errors. Aside from these possible errors, the fragmentary data set out in Table 5.27 (see also selected data in Tables 5.28 and 5.29) suggest that the range of variability of these mass-specific coefficients may be even greater than one order of magnitude. For instance, the mass-specific light absorption coefficient at 400 nm for nonalgal particles, $a_{NAP}^*(400)$, embraces values from c. 0.007 to 0.28 $m^2\ g^{-1}$. The boundary values of $a_p^*(400)$, that is, for all SPM, are somewhat higher: we can state with a high degree of probability that they lie at c. 0.01 $m^2\ g^{-1}$ and c. 0.50 $m^2\ g^{-1}$. We do wish to stress, however, that these estimated values and ranges of natural variability of a_p^* and a_{NAP}^* may be encumbered with substantial errors.

[29] In the standard method, the filter is first weighed in "mint" condition, and then again with the particulate sediment, that is, after a water sample of given mass has been filtered through it.

TABLE 5.27. Ranges of variability of mass (SPM)-specific coefficients of light absorption at 400 nm for all SPM (a_p^* (400 nm) and for NAP (a_{NAP}^*(400 nm)) in various natural marine basins.

No	Region investigated	a_p^*(400 nm) [$m^2\ g^{-1}$]	a_{NAP}^*(400 nm) [$m^2\ g^{-1}$]	Ref.[a]
-1-	*-2-*	*-3-*	*-4-*	*-5-*
1	Various seas and lakes		0.05–0.28	1
2	Baltic Sea (Gulf of Gdańsk, Gdańsk Deep)	0.06–0.2	0.025–0.08	2
3	Pacific (coastal zone of southern California)	0.041–0.095		3
4	The seas around Europe		~0.007 – ~0.16	4

[a] 1. The data from various seas and lakes are cited in Bukata et al. (1995); 2. our own research data; 3. data gleaned from the paper presented by S..B. Woźniak et al. at the Ocean Sciences Meeting, Honolulu 2006 (S.B. Woźniak et al. 2006); 4. data for light (λ = 443 nm) assessed on the basis of Babin et al. (2003a; Figure 15 in this paper).

Light-Absorption Spectra of Nonalgal Particles (NAP): Shapes and Slopes

Apart from the absolute values of the mass-specific absorption coefficients a_p^* and a_{NAP}^*, which are characteristic of all SPM and of NAP from different natural waters, a second property of the spectra of $a_p^*(\lambda)$ and $a_{NAP}^*(\lambda)$ is their shape; this depends not on the concentration of SPM in the water, only on its chemical and physical properties. The shapes of the spectra of $a_p^*(\lambda)$ for all SPM in the sea are governed chiefly by the optical properties of phytoplankton pigments. As this question is addressed in great detail in Chapter 6, we concentrate in the remainder of this section on the shapes of the absorption spectra of $a_{NAP}^*(\lambda)$.

As we already mentioned in Section 5.3.1, most spectra of the VIS absorption coefficients of NAP, $a_{NAP}(\lambda)$ measured in Nature, and hence, too, the spectra of their mass-specific absorption coefficients $a_{NAP}^*(\lambda)$, exhibit a monotonic drop in value of the absorption coefficient with increasing light wavelength. Numerous authors (e.g., Yentsch (1962), Kirk (1980), Roesler et al. 1989, Bricaud et al. (1998), and Babin et al. (2003a)) have suggested that this drop is to a good approximation described by a one-termed expression, the simple exponential function:

$$a_{NAP}^* (\lambda) = a_{NAP}^* (\lambda_{ref}) e^{-S_{NAP} \cdot (\lambda - \lambda_{ref})}, \tag{5.37}$$

where the parameter S_{NAP}, the "spectral slope," is independent of the wavelength λ over the entire VIS spectral range; in other words, it is constant in the VIS region.

This assumption—the nonselectivity of S_{NAP} with respect to λ—may, however, give rise to some controversy. Inspection of the spectra of $a_{NAP}(\lambda)$ or $a_{NAP}^*(\lambda)$, plotted on a semi-logarithmic scale in Figures 5.37 and 5.38c,d shows that in many cases the spectral slope S_{NAP} in the longwave range of the VIS (especially above 600 nm) is decidedly smaller than in its shortwave range. The assumption of the nonselectivity of S_{NAP} remains in force, however, if it is further assumed that measurements of light absorption in the longwave range of VIS are encumbered with the "zero-absorption line" error. What this means is that they are distorted by the dark currents (electronic noise) of the recording apparatus, such that the absorption values determined from the measurements are higher than the real ones. Therefore, in order to take these effects into account, the value of S_{NAP} in approximations of the real spectra of $a^*_{NAP}(\lambda)$ (or $a_{NAP}(\lambda)$) using expression (5.37) should be determined solely on the basis of empirical absorption spectra measured in the shortwave range of the VIS (e.g., for $\lambda < 550$ nm).

Unfortunately, a further difficulty crops up as a result of the complexities of measuring the absorption of light by a layer of particles filtered out of a sample of sea water, the standard technique for measuring absorptions $a_p(\lambda)$ and $a_{NAP}(\lambda)$. This brings us to the problem of calibrating the apparatus and establishing the exact position of the zero absorption line; it has yet to be finally solved. An alternative approach is to assume that this overestimation of measured absorptions above the values given by Equation (5.37) in the longwave range of the VIS is not due to any fault of the apparatus but in fact represents the real state of affairs. Hence, some authors (e.g., Bowers and Binding (2006)) have stated that the real values of the absorption coefficients $a_{NAP}(\lambda)$ and $a^*_{NAP}(\lambda)$ are better described by a two-termed expression (rather than the one-termed Equation (5.37)) containing one exponential term and one constant term:

$$a^*_{NAP}(\lambda) = C_1 + C_2\, e^{-S_{NAP}(\lambda - \lambda_{ref})}, \tag{5.38a}$$

where C_1 and C_2 are the parameters of this relationship, which together yield the value of light absorption λ_{ref}:

$$a^*_{NAP}(\lambda_{ref}) = C_1 + C_2. \tag{5.38b}$$

Bowers and Binding (2006) suggest that in many cases, especially where suspensions of mineral particles are concerned (see Figure 5.39), expression (5.38) better approximates the real shapes of $a^*_{NAP}(\lambda)$ spectra than the one-termed expression (5.37).

It is not possible at this juncture to state definitively which of these two mathematical expressions—Equation (5.37) or Equation (5.38)—more accurately describes the real light absorption spectra of NAP. Both can be used for this purpose, but each one for different associations of suspended particles. The one-termed exponential expression (Equation (5.37)) may be more suited to cases where the SPM consists solely or mostly of organic particles; in contrast, the two-termed expression (Equation (5.38)) may better serve the cases where the SPM additionally contains some mineral components, whose absorption in the red and IR does not decrease, or when the SPM contains some other "black" particles (e.g., soot) with nonzero absorption coefficients that are nonselective with respect to light wavelength.

When analyzing the spectra of the absorption coefficients of NAP, it seems sensible to alternate between these two approximating expressions, depending

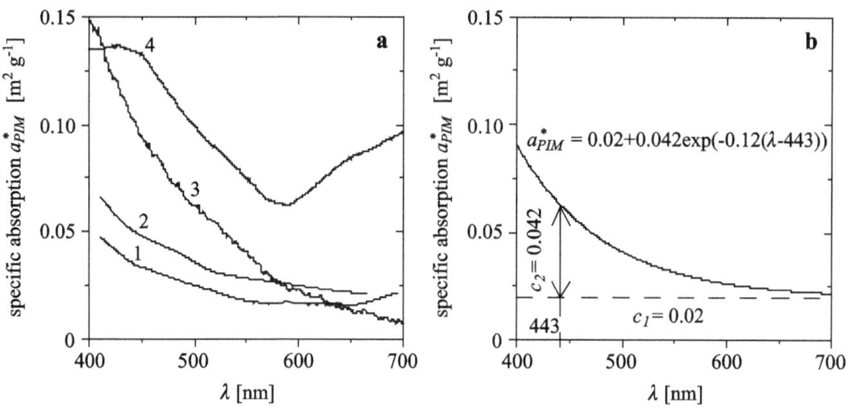

FIGURE 5.39. Spectra of mass-specific coefficients of light absorption by mineral (inorganic) particles $a^*_{PIM}(\lambda)$: (a) empirical examples for SPM from: 1. Chilko Lake, British Columbia (after Gallie and Murtha (1992)); 2. the Irish Sea (after Bowers and Binding (2006)); 3. the Menai Strait, Irish Sea (after Bowers et al. (1996)); 4. Lake Ontario (after Bukata et al. (1983)); (b) model spectrum calculated according to Equation (5.38a), suggested by Bowers and Binding (2006).

on our knowledge of the type of particles present in a given basin. Bearing in mind the possible errors due to the apparatus, however, it is best to determine the spectral slope S_{NAP} from absorptions measured in the shortwave range (e.g., $\lambda < 550$ nm). It should be noticed, too, that the values of S_{NAP} determined in this way are practically convergent for these two approximating expressions.

Tables 5.28 and 5.29 provide statistical data characterizing the average slopes S_{NAP} of the absorption spectra of NAP and the average mass-specific absorption coefficients at $\lambda_{ref} = 443$ nm for these associations of NAP in different basins. They were determined by various authors (including ourselves) from measurements made in different seas and lakes. The statistical data in Table 5.28 refer to light absorption by natural associations of NAP without any distinction being made between the proportions of organic and mineral particles. They were determined using the purely exponential approximating expression given by Equation (5.37). On the other hand, the statistical data in Table 5.29 refer only to suspended mineral particles (or to associations of NAP in which mineral particles are presumed to be in the majority). These data were determined with the aid of the two-termed approximating expression (Equation (5.38)). The table also gives (see item 9) the corresponding average values of S_{NAP} and $a^*_{NAP}(\lambda_{ref} = 443$ nm) for associations of mineral particles in natural water basins. In the opinion of Bowers and Binding (2006), these values can, to a first approximation, be regarded as a universal model for the totality of suspended mineral particles in natural water basins. In our view, however, this suggestion seems open to question.

TABLE 5.28. Statistically averaged parameters characterizing the spectra of the mass-specific absorption coefficient of non-algal particles in various marine basins.[a]

		a^*_{NAP}(443 nm) [m^2g^{-1}]				S_{NAP} [nm^{-1}]				
No.	Region investigated	Range of variability	Average	Standard deviation	Logarithmic average	Range of variability	Average	Standard deviation	N[b]	Authors of investigations
-1-	-2-	-3-	-4-	-5-	-6-	-7-	-8-	-9-	-10-	-11-
1	Various oceans (NW Pacific, Kiel Harbor, Sargasso Sea)					0.006-0.014	0.011	0.002		Various references in Roesler et al. 1989[d]
2	San Juan Islands						0.011	0.002	40	Roesler et al. 1989
3	Various oceanic Case 1 waters:									Bricaud et al. 1998
	Total:					0.008–0.016	0.011	0.0025	267	
	Ultra-oligotrophic[c]						0.012	0.002		
	Oligo/mesotrophic[c]						0.011	0.002		
	Eutrophic[c]						0.010	0.001		
4	Coastal waters around Europe:									Babin et al. 2003a
	Total:		0.041	0.023	0.031	0.0089–0.0178	0.0123	0.0013	348	
	Adriatic Sea		0.041	0.017		0.0114–0.0168	0.0128	0.0011	39	
	Atlantic		0.044	0.030		0.0089–0.0161	0.0124	0.0015	33	
	Baltic Sea		0.067	0.022		0.0114–0.0147	0.0130	0.0007	54	
	English Channel		0.035	0.018		0.0093–0.0155	0.0117	0.0011	82	
	Mediterranean Sea		0.036	0.019		0.0104–0.0178	0.0129	0.0016	52	
	North Sea		0.033	0.023		0.0089–0.0143	0.0116	0.0007	88	

5	Baltic Sea (coastal regions and bays):									Present authors' own studies
	Total:	0.016–0.055	0.030	0.011	0.028	0.0077–0.0120	0.0104	0.0006	33	
	$POM/SPM \geq 0.75$[e]	0.018–0.055	0.036	0.009	0.035	0.0077–0.0109	0.0099	0.0005	19	
	$POM/SPM < 0.75$[e]	0.016–0.035	0.021	0.005	0.020	0.0090–0.0120	0.0111	0.0005	14	

[a] $a^*_{NAP}(443\ \text{nm})$, mass-specific absorption coefficient of light ($\lambda_{ref} = 443$ nm); S_{NAP}, slope of the mass-specific absorption spectrum.

[b] Number of empirical spectra $a^*_{NAP}(443\ \text{nm})$.

[c] The basins were assigned to various trophic types, depending on the chlorophyll *a* concentration C_a [mg tot. chl *a* m^{-3}], as follows: ultra-oligotrophic ($C_a < 0.1$); oligo/mesotrophic ($0.1 \leq C_a \leq 1$); eutrophic ($C_a > 1$).

[d] Data inter alia from: Kishino et al. (1986), Maske and Haardt (1987), Iturriga and Siegel (1989), Morrow et al. (1989).

[e] The set of spectra of $a^*_{NAP}(\lambda)$ we analyzed was divided into subsets: 1. when the POM concentration was more than 75% of the total concentration of suspended particulate matter (SPM), and 2. when *POM* made up less than 75% *SPM*; these POM concentrations were estimated on the basis of measured particulate organic carbon (POC) contents, assuming that $POM \approx 2POC$.

TABLE 5.29. Parameters characterizing the spectra of mass-specific light absorption coefficients of mineral particles in various natural water basins.[a]

No.	Region investigated	a^*_{NAP} (443 nm) $[m^2g^{-1}]$	C_1 $[m^2g^{-1}]$	C_2 $[m^2g^{-1}]$	Spectral slope S_{NAP} $[nm^{-1}]$	Ref.[b]
-1-	-2-	-3-	-4-	-5-	-6-	-7-
	Lakes:					
1	Lake Ladoga	~0.21	—	—	—	1
2	Lake Ontario	~0.135	—	—	—	2
3	Chilko Lake (British Columbia)	0.035–0.047	0.019	0.025	0.016 ± 0.001	3
	Seas:					
4	Menai Strait (Irish Sea)	0.081	0.026	0.055	0.010	4
5	Menai Strait (Irish Sea)	0.059	0.021	0.038	0.006	4
6	Menai Strait (Irish Sea)	0.024	0	0.024	0.011	5
7	Irish Sea	0.070	0.016	0.054	0.011 ± 0.0003	6
8	Irish Sea (187 points)	0.041	0.016	0.035 ± 0.002	0.009 ± 0.0004	7
9	**Model description**					7
	(after Bowers and	Mean	0.020	0.042	0.012	
	Binding 2006)	St. deviations	0.004	0.012	0.02	

[a] a^*_{NAP}(443 nm), mass-specific absorption coefficient at λ_{ref} = 443 nm; C_1 and C_2, components of the coefficient a^*_{NAP}(443 nm) (according to Equation (5.38)); S_{NAP}, slope of the mass-specific absorption spectrum (adapted from Bowers and Binding (2006).

[b] 1. Bukata et al. (1991); 2. Bukata et al. 1983; 3. Gallie and Murtha (1992); 4. Harker (1997); 5. Bowers et al. (1996), as reworked by Babin et al. (2003a); 6. Binding et al. (2003); 7. Bowers and Binding (2006).

Determined by various authors, the slopes S_{NAP} of empirical absorption spectra of NAP in various marine basins (see column 7 in Table 5.28) lie within a wide range, from c. 0.006 nm^{-1} to c. 0.018 nm^{-1}; similar spectral slopes have also been recorded for suspensions of mineral particles (see column 6 in Table 5.29). Ignoring the extreme values, one can assume the range of variability of this parameter normally comes across in different seas to be much narrower. The average values of S_{NAP} given in column 8 of Table 5.28 testify to this: they all lie within the much narrower range from 0.010 to 0.013 nm^{-1} (the average for the World Ocean is S_{NAP} = 0.012). This result is thus wholly convergent with the analogous mean spectral slope of absorption by suspended mineral particles in various water basins (including lakes), postulated by Bowers and Binding (2006) as a model value for natural associations of suspended mineral particles (see item 9 in Table 5.29).

Do the Spectra of Mass-Specific Light Absorption Coefficients of Natural Associations of Organic Particles and Mineral Particles Differ? If So, How Do They Differ?

We answer these questions on the basis of the currently available empirical data.

The results of empirical studies of the spectral mass-specific absorption coefficients of NAP carried out so far are not reliable enough to give a precise response to the above questions. This is mainly because the standard measurement techniques determine the combined absorption of all SPM (NAP + phytoplankton pigments), or of NAP containing both organic non-pigment particles and mineral particles. To be sure, certain modifications to these techniques have been made in order to separate the effects due to mineral particles from those due to organic particles;[30] there are also various indirect methods, but none is sufficiently accurate and reliable (Bowers and Binding 2006). Hence, the empirical spectra of mass-specific absorption coefficients obtained separately for mineral particles and for other NAP are extremely small in number; in addition, they are very probably encumbered with considerable errors.[31] Practically all the relevant data available in the literature are presented in condensed form in Tables 5.27 through 5.29 in this section.

Leaving aside these possible errors, it is possible to draw one inference from the data in these tables. It is, in fact, the only possible inference, and a rather trivial one to boot. The spectra of mass-specific VIS absorption coefficients of both main types of NAP (mineral and organic) in natural associations of SPM display certain resemblances. This applies both to the values of these coefficients and to the shapes of their spectra. Evidence for this is provided by the similar values and range of variability of $a^*_{NAP}(443\ \text{nm})$ and by the spectral slopes S_{NAP} for all NAP (Table 5.28) and mineral particles (Table 5.29).

But because the range of possible values of this mass-specific absorption for mineral particles is very wide, their spectra are comparable only on a global scale. On the other hand, one cannot rule out situations where on a local scale, in a given basin and at a given time, the mass-specific absorption coefficients of these two types of particles differ in value, sometimes widely so. We recorded such a situation in the Gdańsk Basin (the Gulf of Gdańsk and the Gdańsk Deep) in the spring of 2004: the evidence is presented in Figure 5.40c,d (see also Figure 5.38 and the numerical data in Table 5.28, item 5). This shows the statistically averaged (including standard deviations) spectra of $a^*_{NAP}(\lambda)$ for the following sets of empirical spectra: all (33) the spectra of $a^*_{NAP}(\lambda)$ recorded in the relevant sea areas during the study period (red plots); 19 of these spectra for the case when more than 75% of the mass of NAP consisted of POM (i.e., when $POM/SPM > 0.75$) (green plots); the other 14 of these spectra, when *POM* made up less than 75% of the mass of *SPM* (black plots). The positions of these averaged spectra on

[30] The effects of nonmineral particles can be removed by extracting pigments in a solvent such as methanol (e.g., Kishino et al. (1984, 1986)), by bleaching them with sodium hypochlorite (Tassan and Ferrari (1995), or by combusting them in an oven (Bowers et al. 1996).

[31] With reference to the mass-specific absorption coefficients, these errors are exacerbated by the overlapping errors in estimated SPM concentrations.

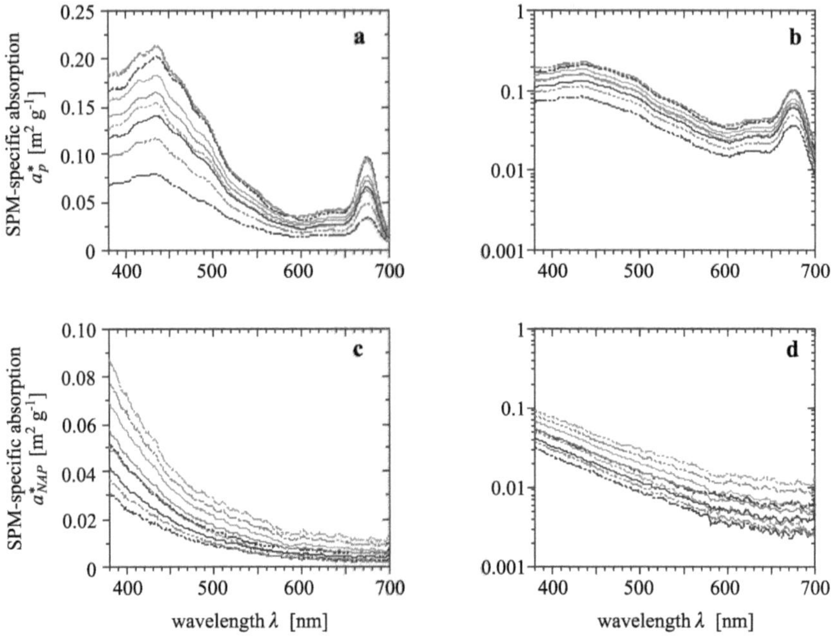

FIGURE 5.40. Averaged spectra (the middle solid lines of each color) and standard deviations (upper and lower dashed line of each color) of mass-specific light absorption coefficients of total (a,b) and nonalgal (c,d) particles in the Baltic Sea (the Gdansk Deep and the Gulf of Gdansk) in spring 2004: (a),(c) averaged in accordance with arithmetic statistics; (b),(d) averaged in accordance with logarithmic statistics; red shows averages of all the spectra, 33 data; green shows averages when $POM \geq 0.75$ SPM, 19 data; black shows averages when $POM < 0.75$ SPM, 14 data. (Based on the data bank IO PAS, Sopot.) (See Colour Plate 7)

the graph show that the values of the mass-specific absorption were much higher (nearly twice as high) when $POM \geq 75\%$ SPM than when $POM < 75\%$ SPM. All this simply means that the mass-specific absorption coefficients of organic NAP in the Gdańsk Basin in spring 2004 were much higher in value than the corresponding coefficients of the mineral suspensions present there.

Obviously, the above situation, when the mass-specific absorption coefficients of POM were much higher than the corresponding coefficients of inorganic (mineral) particles present at the same time and in the same sea, was a random one. Because these coefficients can take a wide range of possible values, one can expect a whole series of such situations in Nature, when the specific absorptions of mineral particles exceed those of organic particles, and vice versa, when the specific absorptions of organic particles are greater than those of mineral particles.

5.3.3 *Parameterization of the Particulate Matter Absorption Spectra for Oceanic Case 1 Waters*

The two previous sections (5.3.1 and 5.3.2) reviewed the empirical values of the VIS absorption coefficients of all SPM a_p and of the two principal components of these coefficients of all SPM: a_{pl}, light absorption by phytoplankton pigments and a_{NAP}, light absorption by nonalgal particles (NAP). They also characterized the most important spectral features of these absorptions. These data do not, however, form a systematized set, which would allow particular values of the spectral coefficients $a_p(\lambda)$, $a_{pl}(\lambda)$, and $a_{NAP}(\lambda)$ to be ascribed to given regions of seas and oceans. In view of the huge differentiation in the environmental factors governing the concentrations of various optically significant components (light absorbers and scatterers) in particular regions of seas and oceans, achieving such a systematized set of data is impossible, given the current state of knowledge. Nevertheless, an important step towards finding a solution to this problem was taken by Bricaud and his co-workers (Bricaud et al. 1998; see also Bricaud et al. (1995)).

They performed statistical analyses of more than 1000 empirical VIS absorption spectra (400–700 nm) of all SPM $a_p(\lambda)$ and of phytoplankton pigments $a_{pl}(\lambda)$ from a variety of oceanic Case 1 waters, where chlorophyll *a* concentrations C_a ranged from 0.02 to 25 mg tot. chl *a* m^{-3}. As a result of their analyses, these researchers developed two parallel statistical descriptions: a parameterization of chlorophyll-specific particulate absorption spectra (Bricaud et al. 1998) and a parameterization of chlorophyll-specific phytoplankton pigment absorption spectra[32] (Bricaud et al. 1995). Both parameterizations are founded upon statistically established relationships between the spectral chlorophyll-specific coefficient of light absorption by all SPM $a_p^{*chl}(\lambda)$ [m^2(mg tot. chl a)$^{-1}$] and the spectral chlorophyll-specific coefficient of light absorption by phytoplankton pigments $a_{pl}^{*chl}(\lambda)$ [m^2(mg tot. chl a)$^{-1}$] on the one hand, and the concentration of chlorophyll *a*, C_a, in the water. These relationships take the forms:

$$a_p^{*chl}(\lambda) = A_p(\lambda) C_a^{E_p(\lambda)-1}, \tag{5.39}$$

$$a_{pl}^{*chl}(\lambda) = A_{pl}(\lambda) C_a^{E_{pl}(\lambda)-1}, \tag{5.40}$$

where $A_p(\lambda)$, $E_p(\lambda)$, $A_{pl}(\lambda)$, $E_{pl}(\lambda)$ are wavelength-λ-dependent numerical coefficients determined by statistical analysis. The dependence of these coefficients on the light wavelength λ is illustrated in Figure 5.41.

---

[32] This parameterization of the chlorophyll-specific phytoplankton pigment absorption spectra is classified in this book among the single-component, homogeneous models describing light absorption by phytoplankton. It is presented, along with other models, in Chapter 6 (Section 6.7.3).

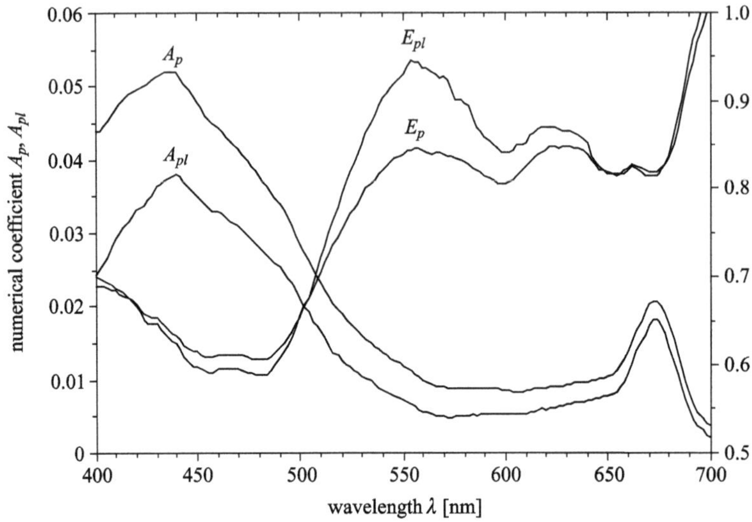

FIGURE 5.41. Spectral values of the numerical coefficients A_p, A_{pl}, E_p, E_{pl}, appearing in the expressions describing the dependence of the light absorption coefficients of all SPM and that of nonalgal particles on the chlorophyll *a* concentration in sea water. (Adapted from Bricaud et al. (1998).)

By these chlorophyll-specific coefficients $a_p^{*chl}(\lambda)$ and $a_{pl}^{*chl}(\lambda)$ are meant the ratios of the relevant absorption coefficients $a_p(\lambda)$ and $a_{pl}(\lambda)$ to the chlorophyll *a* concentration C_a in sea water; that is, $a_p^{*chl}(\lambda) = a_p/C_a$ and $a_{pl}^{*chl}(\lambda) = a_{pl}/C_a$. From these it is easy to work out the analogous chlorophyll-specific coefficient of light absorption by nonalgal particles $a_{NAP}^{*chl}(\lambda) = a_{NAP}(\lambda)/C_a$, which is equal to the difference:

$$a_{NAP}^{*chl}(\lambda) = a_p^{*chl}(\lambda) - a_{pl}^{*chl}(\lambda). \quad (5.41)$$

Spectra of the coefficients $a_p^{*chl}(\lambda)$, $a_{NAP}^{*chl}(\lambda)$ and $a_{pl}^{*chl}(\lambda)$, calculated using Equations (5.39) to (5.41), are shown in Figure 5.42.

Knowledge of the relationships between these three chlorophyll-specific coefficients of light absorption and the chlorophyll *a* concentration C_a in the sea also enables the relationships among the corresponding nonspecific absorption coefficients—$a_p(\lambda)$, $a_{pl}(\lambda)$, and $a_{NAP}(\lambda)$—and C_a to be calculated in a straightforward manner:

$$a_p(\lambda) = a_p^{*chl} \cdot C_a = A_p(\lambda) C_a^{E_p(\lambda)}, \quad (5.42)$$

$$a_{pl}(\lambda) = a_{pl}^{*chl} \cdot C_a = A_{pl}(\lambda) C_a^{E_{pl}(\lambda)}, \quad (5.43)$$

$$a_{NAP}(\lambda) = a_p(\lambda) - a_{pl}(\lambda). \quad (5.44)$$

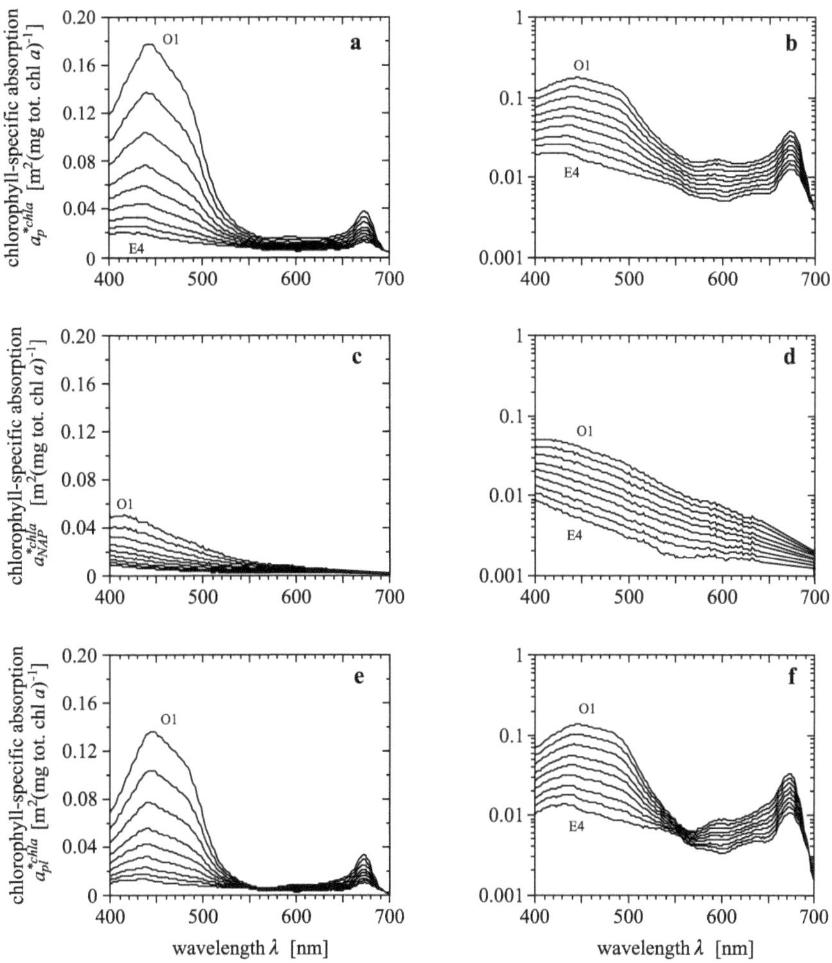

FIGURE 5.42. Spectra of chlorophyll-specific light absorption coefficients of: all SPM a_p^{*chl} (λ) (a),(b); nonalgal particles a_{NAP}^{*chl} (λ) (c),(d); phytoplankton pigments a_{pl}^{*chl} (λ) (e),(f); plotted on the basis of the parameterization of Bricaud et al. (1995, 1998) for oceanic Case 1 waters of different trophicity (the spectra from top to bottom correspond to increasing chlorophyll concentrations C_a [mg tot. chl *a* m^{-3}]: 0.035 (O1), 0.07 (O2). 0.15 (O3), 0.35 (M), 0.7 (I), 1.5 (E1), 3.4 (E2), 7.0 (E3), 15 (E4)).

These equations allow the spectra of light absorption coefficients to be determined for all SPM ($a_p(\lambda)$), and for its two principal constituents, phytoplankton pigments ($a_{pl}(\lambda)$) and nonalgal particles ($a_{NAP}(\lambda)$), in any trophic type of oceanic Case 1 waters. These spectra are illustrated in Figure 5.43.

This parameterization of the absorption spectra of SPM can be used to estimate these spectra from a knowledge of the chlorophyll *a* concentration

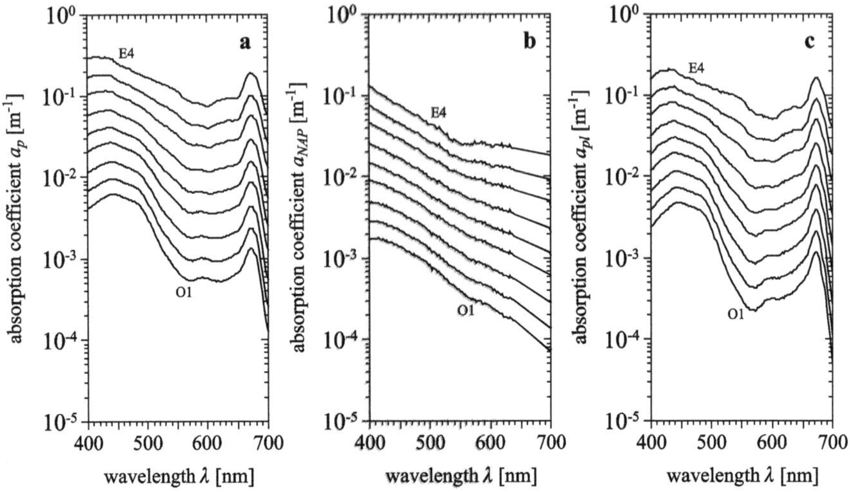

FIGURE 5.43. Absorption spectra of: all SPM (a), nonalgal particles (b), phytoplankton pigments (c), plotted on the basis of the parameterization of Bricaud et al. (1995, 1998) for oceanic Case 1 waters of different trophicity (the spectra from bottom to top correspond to increasing chlorophyll concentrations C_a [mg tot. chl a m^{-3}]: 0.035 (O1), 0.07 (O2). 0.15 (O3), 0.35 (M), 0.7 (I), 1.5 (E1), 3.4 (E2), 7.0 (E3), 15 (E4)).

C_a in Case 1 waters, that is, in most of the euphotic zone of the World Ocean. Unfortunately, it is not possible to perform such a simple parameterization for the remainder of the World Ocean, that is, for the coastal zones of oceans and for enclosed seas, which by and large contain Case 2 waters (see the definitions of water cases in Chapter 1). For such Case 2 waters it is impossible to establish general quantitative relationships among optical coefficients such as a_p, a_{NAP}, and a_{pl}, and the chlorophyll a concentration C_a in the sea. This is because these waters contain allogenous optical constituents that are unrelated to the concentration C_a. For these cases it is necessary to devise multicomponental optical models adapted to local conditions, in which the real optical properties of the sea water are functions of many variables, and not just of one independent variable, namely, the chlorophyll a concentration C_a.

6
Light Absorption by Phytoplankton in the Sea

In Chapter 5 we presented the theoretical (Section 5.1) and empirical (Section 5.2) principles of light absorption by suspended particulate matter (SPM) in the sea. We also described the current state of knowledge regarding the absorption properties of nonalgal particles present in sea waters (Section 5.3). In Chapter 6, we characterize light absorption by another group of SPM, the marine algae or phytoplankton: more precisely, by the pigments contained in their cells. The pigments contained in the cells of marine phytoplankton are the principal group of organic substances absorbing light in the ocean. These cells form a suspension, whose light absorption coefficient we denoted earlier by a_{pl} (Chapter 1, Equation (1.5)). As befits the importance and role of plant pigments in the absorption of solar radiation in the oceans and the supply of energy to marine ecosystems, we devote this separate and extensive chapter to a very detailed examination of light absorption by phytoplankton and the factors controlling this process in the sea. This is an important problem in marine optics and biooptics, if not the most important one; we now outline our reasons for classifying it as such.

First, the absorption of solar radiation by phytoplankton stimulates the photosynthesis of organic matter in the sea, which makes it the supreme phenomenon responsible for providing marine ecosystems with energy (Leith and Whittaker 1975, Steemann Nielsen 1975, Vinogradov and Shushkina 1987). But, governed as it is by the absorption of light by phytoplankton, photosynthesis in the sea is not merely a process supplying energy to marine ecosystems (the first link in the food chain of marine organisms). It is also one of the prime factors affecting the oxygen and carbon dioxide balance in the atmosphere, because these gases participate in marine photosynthesis and are exchanged through the sea surface. On a global scale, therefore, light absorption and the photosynthesis of organic matter by phytoplankton in the oceans determines the state of the greenhouse effect and climatic changes on the Earth (Trenberth 1992, Kożuchowski and Przybylak 1995).

Secondly, quantitative knowledge of the spectral light absorption coefficients of phytoplankton and their mathematical description are indispensable for modeling optical processes and photosynthesis in the sea (Sathyendranath et al. 1989,1994; Morel 1988,1991; Woźniak et al. 1992a,b; Dera 1995). Such models constitute the foundation of the optical methods—both remote sensing (from satellites) and in situ—for monitoring the state of marine ecosystems and their biological productivity, (see, e.g., Platt and Sathyendranath (1993a,b), Antoine and Morel (1996), Antoine et al. (1996), Woźniak et al. (2003), and the papers cited therein).

For these reasons, the theoretical and empirical study of the absorption properties of phytoplankton already has a long history (Yentsch (1962), Steemann Nielsen (1975), Koblentz-Mishke et al. (1985), and the references cited in these works), and the formulation of their mathematical descriptions is one of the chief tasks of contemporary marine biooptics (Kirk 1994; Bricaud et al. 1995,1998). In the present chapter, therefore, we discuss the current state of this research in as much detail as is feasible.

To begin with, we describe the most important abiotic factors governing the absorption capacities of marine algae (Section 6.1), the most important chemical and optical properties of phytoplankton pigments (Section 6.2), the resources and compositions of pigments in different phytoplankton species in different sea waters (Sections 6.3 and 6.4), and the effect of pigment packaging on the absorption properties of phytoplankton cells (Section 6.5). Having dealt with the multifarious conditions to which the absorption properties of phytoplankton in different basins of the World Ocean are subject, we move on to a detailed discussion of these properties based on the relevant empirical data (Section 6.6) and then of their mathematical models (Section 6.7). The questions addressed in this chapter, as well as some of the problems in marine optics described earlier, have been the objects of study of the authors of this book and their colleagues over a great many years. Hence, the arguments presented here are copiously illustrated with our research team's results, but always against the background and under consideration of the knowledge available in the world literature.

6.1 Abiotic Factors Governing Light Absorption by Phytoplankton in the Sea

The absorption properties of marine phytoplankton depend on the complex set of their own internal characteristics as well as numerous environmental factors. Owing to their great complexity, these relationships, both direct and indirect, are still not completely understood. Nevertheless, it is possible to indicate the principal direct underlying factors and a few of the more important indirect ones determining the light absorption spectra of marine algae.

According to the current state of knowledge (e.g., Morel and Bricaud (1981), Hoepffner and Sathyendranath (1991), and Woźniak et al. (1999, 2003), we can write the expression for the spectral light absorption coefficient of the monodispersant, uniform medium of a suspension of phytoplankton in water $a_{pl}(\lambda)$ as:

$$a_{pl}(\lambda) = a^*_{pl}(\lambda) C_a, \tag{6.1a}$$

$$a^*_{pl}(\lambda) = Q^*(\lambda) \sum_j \left[a^*_j(\lambda) \frac{C_j}{C_a} \right], \tag{6.1b}$$

where $a^*_{pl}(\lambda)$ is the chlorophyll-specific coefficient of light absorption of phytoplankton in vivo [m^2 (mg tot. chl. a)$^{-1}$], $Q^*(\lambda)$ is a dimensionless factor of the absorption change due to pigment packaging in the phytoplankton cells, the so-called packaging function (Hulst 1981, Morel and Bricaud 1981), C_a is the concentration of chlorophyll a in the water [mg tot. chl a m^{-3}], C_j is the concentration of the jth pigment (including chlorophyll a) [mg of the jth pigment m^{-3}], and $a^*_j(\lambda)$ is the mass-specific light absorption coefficient of the jth pigment [m^2 (mg of the jth pigment)$^{-1}$].

We see from Equation (6.1) that the resultant (overall) absorption properties of algae in the sea depend directly on the following factors, which we can divide into the following groups:

1. The absorption properties of the separate phytoplankton pigments, $a^*_j(\lambda)$.
2. The total chlorophyll a concentrations at given depths in the sea, $C_a(z)$.
3. The pigment composition in phytoplankton cells at particular depths in the sea $C_j(z)$, frequently described by the ratios of their concentrations $C_j(z)/C_a(z)$.
4. The structure (size and the intracellular concentration of pigments) of the light-absorbing phytoplankton cells, which governs the packaging effect of the pigments (defined by the packaging function $Q^*(\lambda,z)$).

Apart from these four "direct" factors, there are others in the marine environment indirectly affecting the value of the absorption coefficient. In the main these latter factors are abiotic ones, which we can divide into two groups (although with much simplification):

- The complex set of abiotic factors determining the trophicity of the waters in a basin, such as their nutrient content, temperature, or dynamic state. The trophicity directly affects the vertical distributions of chlorophyll a concentrations, C_a(z), and also governs the pigment packaging effect in cells $Q^*(\lambda,z)$ (e.g., Woźniak et al. (1999)).
- The spatial distributions and spectra of light fluxes determining the light field in the sea. Among other things, the light field influences the production of photosynthetic and photoprotecting pigments, in proportions depending on the intensity and spectrum of the incident light. This happens during photoadaptation of the phytoplankton cells to the light intensity and chromatic adaptation to the light spectrum (e.g., Steemann Nielsen (1975), Babin et al. (1996a,b,c), and Majchrowski and Ostrowska (2000)). At high intensities of shortwave radiation ($\lambda < 480$ nm), photooxidation of cell pigments can also occur.

These two factors, the trophicity of waters and the light field, are the most important indirect factors controlling light absorption by phytoplankton, and we cover them briefly in the present section. The four earlier mentioned principal direct factors we treat separately in the subsequent four sections (6.2–6.5).

6.1.1 The Trophicity of Marine Basins: A Factor Governing the Resources of Algae and Light Absorption

The concept of trophicity, or, to put it another way, the fertility of a basin, has a long tradition in limnology and in the ecology of lakes: it is the foundation of their typology. Such a classification is based on comparative studies of the biological productivity and fertility of waters and bottom sediments. We can thus distinguish three main types of lake: oligotrophic, mesotrophic, and eutrophic. Marine biologists use this concept in like manner to classify the waters in particular regions of oceans and seas, dividing them into oligotrophic basins (biologically poor seas, where productivity is low), mesotrophic basins (of intermediate productivity), and eutrophic basins (where productivity is high). The potential measure of a basin's productivity is usually taken to be the surface concentration of chlorophyll *a*, $C_a(0)$ (Vinogradov 1977). This quantity gives some indication of the potential productivity of the water and correlates with the total primary production of the basin (i.e. in the water column) (Koblentz-Mishke and Vedernikov 1977; Morel 1978, 1991; Antoine et al. 1996). That is why contemporary biooptics uses the chlorophyll *a* concentration $C_a(0)$ in the surface layer of the sea as the numerical *trophicity index* of a given sea (Woźniak and Ostrowska 1990a; Bricaud et al. 1995, 1998; Babin et al. 1996a). In the present work, we distinguish on the basis of the chlorophyll *a* concentration $C_a(0)$ several types of basin suggested earlier by Woźniak et al. (1992a,b); see Table 6.1. We would remind the reader, however, that this is a simplified and highly conventional division, dictated by the need to parameterize the optical properties of the sea.

As we have just stated, the trophicity of a basin is the result of the interaction of a complex set of abiotic environmental factors in the sea, although the usual definitions of trophicity (e.g., the class of biological productivity) or its indices, such as the surface chlorophyll *a* concentration $C_a(0)$, are, strictly speaking, biological concepts. This is because natural trophicity is in fact determined by multifarious combinations of the countless abiotic factors in the marine environment: nutrient concentrations, water temperature, irradiance conditions, the density stratification of water masses, and the dynamic states of waters, to mention just the most important ones. We discuss these particular questions in Section 6.3. Now how all these factors work together is not yet completely understood. What is known, though, is that the trophic type of water (the trophicity), represented by the index $C_a(0)$, is to a first approximation determined for most basins of the World Ocean by the

TABLE 6.1. The classification of marine basins (ecosystems) into trophic types.[a]

Trophic type of basin (and the symbol denoting it in this book)	Potential biological productivity class	Range of variability of surface chlorophyll *a* concentration $C_a(0)$ [mg tot. chl *a* m^{-3}]	Typical surface chlorophyll *a* concentration $C_a(0)$ [mg tot. chl *a* m^{-3}]
-1-	-2-	-3-	-4-
Oligotrophic basins O	Low productivity	<0.2	—
subtypes:			
O1		0.02–0.05	0.035
O2		0.05–0.10	0.075
O3		0.10–0.20	0.15
Mesotrophic basins M	Medium productivity	0.2–0.5	0.35
Transitional mesoeu-trophic basins (Intermediate) I	Transitional, medium/high productivity	0–1.0	0.75
Eutrophic basins E	High productivity	>1	—
subtypes:			
E1		1–2	1.5
E2		2–5	3.5
E3		5–10	7.5
E4		10–20	15
E5		20–50	35
E6		50–100	70

[a] After Woźniak et al. (1992a,b).

concentration of nitrogenous nutrients $N_{inorg}(0)$[1] and the water temperature *temp* in the euphotic zone (see, e.g., Koblentz-Mishke and Vedernikov (1977), Woźniak (1990a), Ficek (2001), and the papers cited therein). From the approximate relation between these two principal abiotic factors and the trophicity of waters, which we have worked out statistically, we can estimate to a first approximation the surface chlorophyll *a* concentration $C_a(0)$ on the basis of known concentrations of nutrients $N_{inorg}(0)$ and the water temperature *temp* in the sea. The following polynomial function describes this relation (after Woźniak et al. (2003)):

$$\log C_a(0) = \sum_{m=0}^{4} \left[\sum_{n=0}^{4} A_{m,n} \left(\log N_{inorg}(0)\right)^n \right] temp^m, \tag{6.2}$$

where the variables are expressed in the units: $C_a(0)$ [mg tot. chl *a* m^{-3}], $N_{inorg}(0)$ [μmol dm^{-3}], *temp* [°C], for the values of coefficients $A_{m,n}$; see Table 6.2.

[1] The total concentration of inorganic nitrogen: this includes the nitrogen contained in nitrates, nitrites, and ammonia dissolved in sea water ($N_{inorg} = N(NO_3) + N(NO_2) + N(NH_4)$), but not dissolved nitrogen gas N_2. This latter form (N_2) is not assimilated by most phytoplankton species in the sea.

TABLE 6.2. The values of coefficients $A_{m,n}$ in Equation (6.2).

n\m	0	1	2	3	4
0	–0.01662	0.3502	–0.04418	0.001785	-2.430×10^{-5}
1	–0.04148	–0.01815	0.001975	3.991×10^{-5}	-2.259×10^{-6}
2	–0.05814	–0.02717	–0.001333	0.0001978	-4.019×10^{-6}
3	0.005918	0.004394	-3.613×10^{-5}	-2.484×10^{-5}	6.079×10^{-7}
4	–0.02117	–0.0004761	0.0007483	-5.039×10^{-5}	8.540×10^{-7}

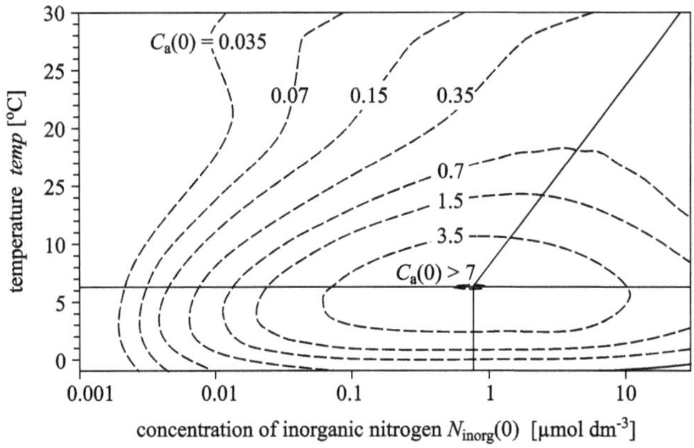

FIGURE 6.1. Mean surface concentration of chlorophyll *a* (isolines) $C_a(0)$ [mg tot. chl *a* m^{-3}] in different regions of the World Ocean with respect to water temperature *temp* [^{0}C] and the mean concentration of inorganic nitrogen N_{inorg} (0) [µmol dm^{-3}] in the surface layer of water (0–10m). (After Woźniak et al. (2003).)

The plot of the relationship described by Equation (6.2) is illustrated in Figure 6.1. It shows the surface chlorophyll *a* concentrations $C_a(0)$ commonly measured in the World Ocean, approximated by isolines ($C_a(0)$ = *const*) on the $N_{inorg}(0)$ versus *temp* graph, and takes the scale of variability of these parameters encountered in the World Ocean into account.

The relationships presented in Figure 6.1 endorse the hypothesis that the trophicity of basins, given by the value of $C_a(0)$, is a complex abiotic factor governing the growth and functional characteristics of algae, including the chlorophyll resources and the total light absorption coefficients of phytoplankton in the sea. We can draw further interesting and important conclusions from Figure 6.1. For instance, very warm, extremely nutrient-rich waters are by no means the richest in chlorophyll; for the most chlorophyll-rich waters we have to go to basins in which the water temperatures and nutrient concentrations take intermediate values ($N_{inorg}(0) \approx 1$ µmol dm^{-3}, $temp \approx 6°C$). It is in these latter types of water that we can also expect the largest absolute coefficients of light absorption by algae.

To conclude this description, we underscore the practical utility of the concept of trophicity as defined by the index $C_a(0)$. It plays a role of a parameter symbolizing the complex set of abiotic factors governing a variety of processes in the sea, including the absorption of light by phytoplankton. The weightiest arguments in favor of this concept are that:

1. In analyses and mathematical descriptions of a range of phenomena in the sea the trophicity index $C_a(0)$ can replace several variables (e.g., *temp* and $N_{inorg}(0)$; see Figure 6.1).
2. The trophicity of a basin, defined as the surface concentration of chlorophyll $C_a(0)$, can be used to classify the natural vertical distributions of chlorophyll and other pigments in the sea, and also the vertical distributions of the characteristics describing the optical effects of pigment packaging in phytoplankton cells. To this end we use the statistical relations between these magnitudes (see Sections 6.3 through 6.5).
3. $C_a(0)$ is not only the trophicity index of a basin, but also the index on which the biooptical classification of sea waters is based, that is, a parameterization of optical properties and models of seas, in particular, seas with Case 1 waters, of which more later in this section.
4. Whether as the index of trophicity or the biooptical water classification, $C_a(0)$ permits a suitable classification and systematization of the optical absorbing properties of algae (see Sections 6.6 and 6.7) living in the various basins of the World Ocean, given the relevant sets of abiotic factors prevailing there.

We therefore use the concept of the trophicity of a basin and apply it as being equivalent to the "surface concentration of chlorophyll *a*;" we denote this concentration $C_a(0)$. We call the value of $C_a(0)$ the "trophicity index" of waters, and we ascribe waters whose $C_a(0)$ values lie within certain ranges to the trophic types defined in Table 6.1. In some cases we also apply the trophicity concept to a water body at certain depth z denoting the index $C_a(z)$.

6.1.2 The Light Field: A Factor Governing the Composition of Light-Absorbing Pigments in Cells

The natural irradiance conditions in the waters of oceans and seas are highly diverse; this is a consequence of the variable content in these waters of optically active admixtures, primarily dissolved organic matter and all kinds of suspended particles. This leads to a corresponding diversity in the so-called apparent optical properties of these waters (see monographs: Jerlov (1976) and Dera (2003)), including the spectral coefficients of downward irradiance attenuation $K_d(\lambda)$. As a result, the range of spatial and spectral distributions of downward irradiance $E_d(\lambda, z)$ is very broad. Nevertheless, a number of regularities allow us to classify ocean and sea waters with optically similar properties. Indeed, numerous authors have developed various schemes for classifying or parameterizing the optical properties of sea waters. The very first of these was proposed by Jerlov (1976, 1978), and subsequent ones were devised and described by Pelevin and Rutkovskaya (1978), Smith and Baker

(1978), Baker and Smith (1982), Morel (1988), and Woźniak and Pelevin (1991). A discussion of the relative merits of these classifications can be found, for example, in Højerslev (1986).

Below we present a model description of the optical properties of various types of seas, which enables the most important spatial and spectral characteristics of the underwater irradiance to be defined on the basis of the biooptical classification designed by Woźniak and his co-workers (Table 6.3). The algorithm in Table 6.3 enables not only the determination of the fundamental quantitative and spectral characteristics of underwater light fields, but also of specific biooptical characteristics such as the potentially destructive radiation *PDR* and the spectral fitting function of photosynthetic

TABLE 6.3. Algorithm for determining selected characteristics describing underwater light fields.[a]

INPUT DATA

$C_a(z)$ [mg tot. chl *a* m^{-3}], concentration of total chlorophyll *a* at depth *z* $PAR(0)$ [Ein m^{-2} s^{-1}], surface downward irradiance in the photosynthetically available radiation spectral range (from c. 400 nm to 700 nm)

MODEL FORMULAS

1) Coefficient of downward spectral irradiance attenuation:

$$K_d(\lambda, z) = K_w(\lambda) + C_a(z)\,\{C_1(\lambda)\exp[-a_1(\lambda)\,C_a(z)] + K_{d,i}(\lambda)\} + \Delta K(\lambda)\,, \qquad \text{(T-1)}$$

where downward irradiance attenuation coefficient of the allochthonic admixture in the water:

$$\Delta K(\lambda) = \begin{cases} 0 & \text{for Case 1 Waters} \quad \text{(T-2)} \\ 0.068 \cdot \exp[-0.014(\lambda - 550)] & \text{for Baltic Case 2 Waters} \quad \text{(T-3)} \end{cases}$$

The constants $C_1(\lambda)$, $a_1(\lambda)$, $K_{d,i}(\lambda)$) and the attenuation of pure water $K_w(\lambda)$ for some light wavelengths:

λ [nm]	a_1[m^3(mg chl *a*)$^{-1}$]	C_1[m^3(mg chl *a*)$^{-1}$]	$K_{d,i}$[m^3(mg chl *a*)$^{-1}$]	K_w[m^{-1}]
400	0.441	0.141	0.0675	0.0209
450	0.550	0.107	0.0569	0.0181
500	0.610	0.0672	0.0389	0.0276
600	0.333	0.0171	0.0225	0.212
650	0.364	0.0164	0.0236	0.343
700	–	0	0.0125	0.626

The full set of these parameters for different wavelengths from 400 to 700 nm is given in Table A1.1 in Woźniak et al. (2003).

2) The downward irradiance relative spectral distribution functions:

$$f_E(\lambda, z) = f_E(\lambda, 0^+)\exp\left[-\int_0^z K_d(\lambda, z)\,dz\right], \qquad \text{(T-4)}$$

where the normalized typical spectral distributions of *PAR* irradiance entering the sea are:

$$f_E(\lambda, 0^+) = -1.3702 \times 10^{-12}\,\lambda^4 + 3.4125 \times 10^{-9}\lambda^3 - 3.1427 \times 10^{-6}\,\lambda^2 + 1.2647 \times 10^{-3}\,\lambda - 1.8381 \times 10^{-1}\,, \qquad \text{(T-5)}$$

where λ is expressed in [nm]

TABLE 6.3. Algorithm for determining selected characteristics describing underwater light fields.[a] — Cont'd.

3) The downward spectral irradiance

$$E_d(\lambda, z) = PAR(0)\, f_E(\lambda, z). \tag{T-6}$$

4) The overall irradiance in the PAR range:

$$PAR(z) = \int_{400nm}^{700nm} E_d(\lambda, z)\, d\lambda. \tag{T-7}$$

5) The mean values of the *chromatic acclimation factor* F_j and *photo-adaptation factor* PDR^* in a water layer $\Delta z = z_2 - z_1$:

$$<F_j>_{\mathbf{D}z} = \frac{1}{z_2 - z_1} \int_{z_1}^{z_2} F_j(z)\, dz, \tag{T-8a}$$

$$<PDR^*>_{\mathbf{D}z} = \frac{1}{z_2 - z_1} \int_{z_1}^{z_2} PDR(z)^*\, dz, \tag{T-9a}$$

where j is the pigment group index (chlorophylls *a, b, c* and photosynthetic carotenoids *PSC*), $z_1 = z - 30$ m if $z \geq 30$ m and $z_1 = 0$ if $z > 30$ m, $z_2 = z + 30$ m. The mean values in a water layer Δz have been taken in order to include the influence of water mixing.

The acclimation factors were defined as functions of the mass-specific absorption coefficients of various groups of phytoplankton pigments and of selected properties of the underwater light field (Woźniak et al. 1999):

- *Chromatic acclimation factor* for the jth group of pigments (so-called *Fitting Functions*):

$$F_j(z) = \frac{1}{a^*_{j,\,max}} \int_{400nm}^{700nm} f(\lambda, z)\, a^*_j(\lambda)\, d\lambda. \tag{T-8b}$$

- *Photo-acclimation factor* (known as the dayly mean Potentially Destructive Radiation):

$$PDR^*(z) = \int_{400nm}^{480nm} a^*_a(\lambda)\, \langle E_0(\lambda, z) \rangle_{day}\, d\lambda, \tag{T-9b}$$

where $f(\lambda, z) = f_E(\lambda, z)/T(z) = E_d(\lambda, z)/PAR(z)$ is the normalized spectral distribution of irradiance in the PAR spectral range at depth z; $a_j^*(\lambda)$ is the spectral mass specific absorption coefficient for the jth group of unpackaged pigments (i.e., in solvent), determined by Equation (6.3); values of the parameters of the equation are given in Table 6.8.

[a] Adapted from Woźniak et al. (2003).

pigments F_j. These two quantities are useful in analyses of the photoadaptation and chromatic adaptation of phytoplankton cells. This model, which we now present in an abbreviated and simplified form, was described in detail by Woźniak et al. (1992a, 2003). Here we use it to analyze and describe the adaptation of algae to light, as well as other factors governing the optical absorbing properties of phytoplankton. In addition, the results of the model calculations, presented on the relevant plots, are useful for understanding and illustrating the numerous phenomena discussed in this book.

We characterize the above-mentioned diversity of irradiance conditions in various seas and at different depths on the basis of the results of the model calculations and empirical data given in Figures 6.2 and 6.3. The first of these

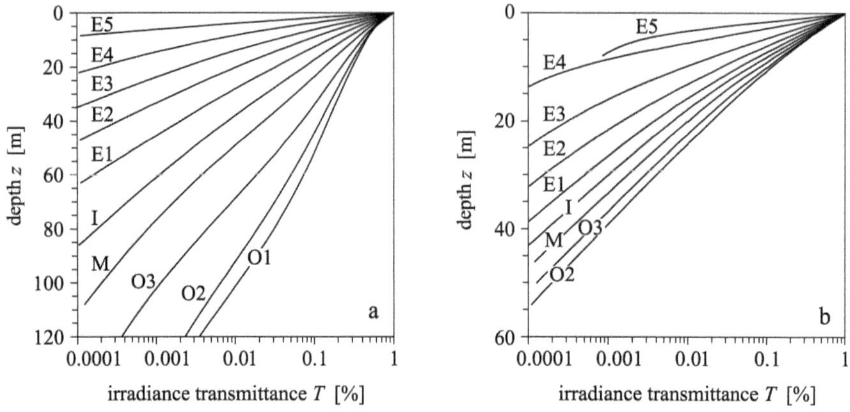

FIGURE 6.2. Modeled vertical profiles of the downward irradiance transmittance $T = PAR(z)/PAR(0)$ in basins of various trophicities: (a) Case 1 waters; (b) Case 2 waters from the Baltic Sea. The trophic types of basins and their symbols used on the graphs (O1, . . . , E5) are defined in Table 6.1. (After Woźniak et al. (2003).)

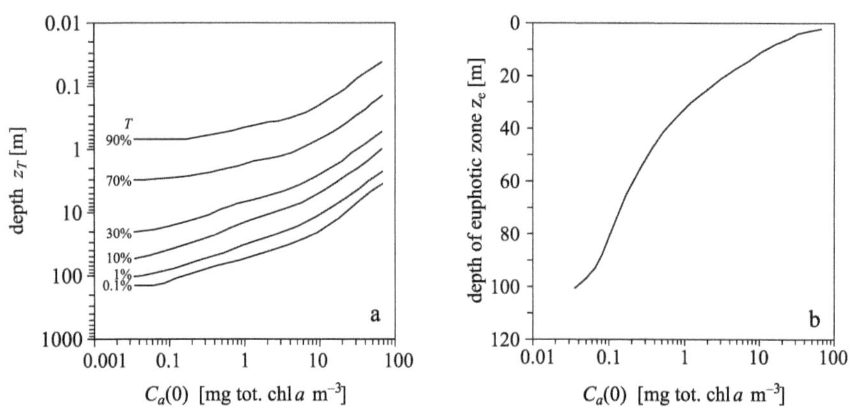

FIGURE 6.3. Dependence of the downward PAR penetration ranges $z_{T,}$ in Case 1 waters, on the type of basin, represented by the surface chlorophyll a concentration $C_a(0)$. (a) Isolines of the relative irradiances; that is, the transmittance $T = PAR(z)/PAR(z = 0^+)$. (b) Dependence of the depth of the euphotic zone $z_e = z(T = 1\%)$ on the trophicity of the basin $C_a(0)$.

figures (6.2) shows model depth profiles of the total downward irradiance transmittance in the PAR spectral range (400–700 nm): they express the reduction of this irradiance with depth z in absolute units. The separate profiles illustrate the drop in PAR irradiance in seas of different trophicity in two types of water (Case 1 waters: Figure 6.2a, and Case 2 waters: Figure 6.2b, distinguished by Morel and Prieur (1977); this division is explained in Chapter 1 of the present book). The trophicity index of the seas is the surface concentration of chlorophyll $C_a(0)$, in accordance with the convention discussed and accepted earlier (Table 6.1).

These profiles show that in all cases there is a substantial decrease in the level of irradiance with depth. Quantitatively, these reductions are varied and depend on the trophicity and optical type of the basin. This attenuation of irradiance with depth is weakest in oligotrophic seas, and becomes stronger as the value of $C_a(0)$ rises. It is also very much greater in Case 2 waters than in oceanic Case 1 waters.

An analysis of the depth range of penetration of this irradiance can give us some idea of the quantitative differences in *PAR* irradiance attenuation in the sea. By "depth range of penetration" we mean the depth to which a given fraction of the irradiance $PAR(z)$ penetrates with respect to the irradiance just below the surface $PAR(0)$. Some of these ranges of penetration—10 % for $PAR(z)$ = 10 % $PAR(0)$, 1% for $PAR(z)$ = 1% $PAR(0)$,[2] and 0.1% for $PAR(z)$ = 0.1% $PAR(0)$—obtained from model calculations done for Case 1 waters, are shown in Figure 6.3a. The analogous ranges of penetration for Case 2 waters can be c. 5 to 50% less. According to the data in Figure 6.3a, all these ranges of penetration have a similar scale of relative changes with trophicity variations in the basin. From optically very pure oligotrophic seas (e.g., type O1, where the concentration of chlorophyll $C_a(0) \approx 0.035$ mg tot. chl *a* m^{-3}) to supereutrophic seas (e.g., type E5, where $C_a(0) \approx 70$ mg tot. chl *a* m^{-3}) the range of penetration diminishes by a factor of 30–50. Let us take a 1% range by way of example: if the thickness of the eutrophic zone in superoligotrophic seas is around 110 m, the equivalent zone in biologically very rich waters will then be scarcely a few meters deep (Figure 6.3b).

The spectral distribution of underwater irradiances also depends on the basin type and the depth in the sea. This is illustrated in Figure 6.4 by empirical spectral distributions of all-day downward irradiance doses recorded in various seas (Figures 6.4a–d), and also by the model spectra determined for different water types with the same surface chlorophyll concentrations $C_a(0)$ (Figures 6.4e–h).

[2] The range of 1% irradiance penetration $z_{T=1\%}$ is often identified with the thickness of the *euphotic zone* z_e (Steemann Nielsen 1975, Dera 1995, Ficek 2001). This concept is understood to mean the surface layer of the sea delimited from below by the so-called *compensation depth*, that is, the depth at which the rates of metabolic and catabolic processes in phytoplankton are the same. Above this depth the quantity of oxygen released during photosynthesis exceeds its consumption during phytoplankton respiration; below this depth, this situation is reversed. Research has shown that this compensation depth coincides quite well with the depth to which c. 1% of the *PAR* entering the sea penetrates (hence $z_e \approx z_{T=1\%}$).

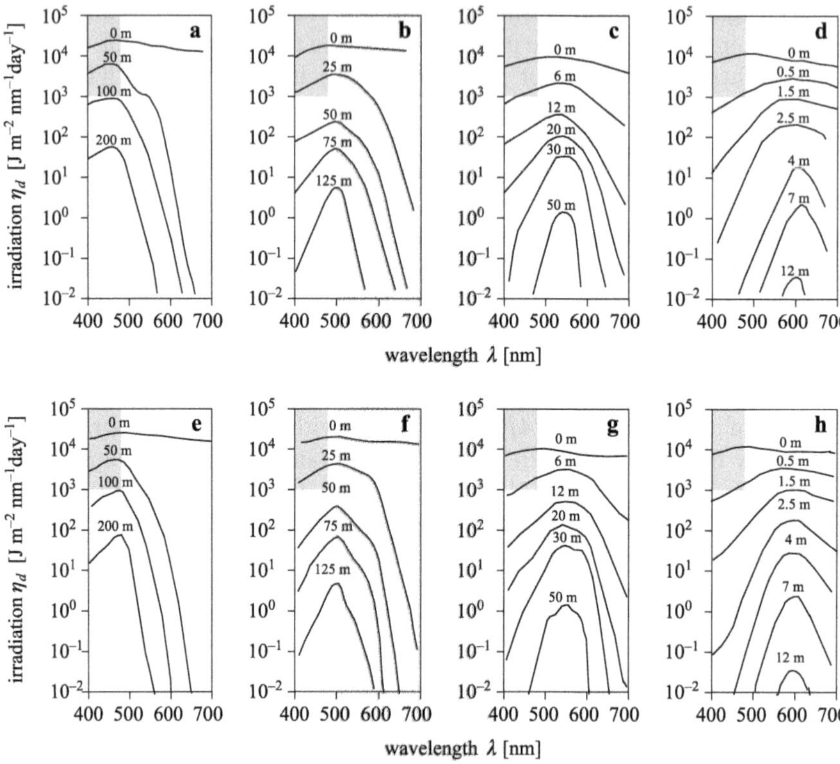

FIGURE 6.4. Spectral distributions of daily irradiance doses (irradiation) in Case 1 waters at different depths in the sea and for various surface concentrations of chlorophyll *a*, $C_a(0)$ [mg tot. chl *a* m^{-3}]: (a)–(d) measured in various seas: Indian Ocean, $C_a(0) = 0.035$ (a); Arabian Sea, $C_a(0) = 0.32$ (b); Gulf of Burgas (Black Sea), $C_a(0) = 3.7$ (c); Puck Bay (Baltic), $C_a(0) = 70$ (d); (e)–(h) modeled for different water types with the same surface irradiances assumed: O1 – $C_a(0) = 0.035$ (e); M – $C_a(0) = 0.35$ (f); E2 – $C_a(0) = 3.5$ (g); E6 – $C_a(0) = 70$ (h); the gray shading on the plots indicates the areas of *PDR* for the phytoplankton cells (explanation in the text). (Adapted from Woźniak et al. (2003).)

Figure 6.4 shows that, regardless of the fact that absolute values of light energy decrease with depth much more rapidly in biologically rich seas than in basins of lower productivity, all these cases possess one basic feature in common. This is the quantitative dominance, increasing with depth, of irradiance from the middle region of the visible light spectrum over irradiance from the other areas of this spectrum. In contrast, the main feature discriminating spectral distributions of irradiance in various types of seas is the position of the light energy peak with respect to the wavelength scale. Hence, in oligotrophic ocean basins, optically the purest ones (Figure 6.4a,e), the maximum energy reaching particular depths in the sea is due to blue-green light in the 470–480 nm range.

With increasing depth the position of the maximum does not change very much. The energy peak merely becomes more distinct; that is, the incident light spectrum becomes narrower as a result of its increasing dominance over other wavelengths in this spectral region. This situation changes, however, on moving into other types of basins. In all types of seas, the energy spectrum just below the surface is practically the same as in pure waters. Nevertheless, as the depth increases, the spectral peaks of the energy distributions shift towards the longer wavelengths. These shifts become more pronounced with increasing trophicity of the waters, that is, increasing chlorophyll *a* content $C_a(0)$. So, in supereutrophic waters, where $C_a(0) = 70$ mg tot. chl *a* m^{-3} (see Figure 6.4d,h), only yellow or yellow-red light in the range of wavelength $\lambda \approx 580$–600 nm and higher is present at suitably great depths. These characteristics of irradiance in different types of seas are due to the combined influence of suspended particles and dissolved organic substances (yellow substances) on the spectra of the light attenuation coefficients. Both of these components attenuate light more strongly in the shortwave region of the visible spectrum, and their concentrations in most cases increase with rising chlorophyll content. The upshot of this is that the irradiance transmittance maximum shifts towards the longwave end of the spectrum on going from optically pure oligotrophic waters to eutrophic seas.

It is these properties of underwater irradiance fields in seas of different trophicity and at different depths that determine the pigment composition in phytoplankton (Babin et al. 1996a; Majchrowski and Ostrowska 1999, 2000). They affect a variety of processes whereby marine organisms adapt to the ambient light conditions, as we show later. This influence applies not only to the level of irradiance, that is, to its absolute magnitudes, but also to the spectral composition of the ambient light in the water. We now discuss briefly these two aspects of the interaction of light with plant cells in the sea.

It is well known that sufficiently high levels of irradiance, especially light from the $\lambda < 480$ nm range, that is, the shortwave end of the visible spectrum and UV radiation, generally have an adverse effect on plant growth in that photoinhibition becomes irreversible (e.g., Grodziński (1978), Hall and Rao (1999), Ficek (2001), and others). One of the chief mechanisms of irreversible photoinhibition is the loss of chlorophyll as a result of its photooxidation. (We deal with the mechanism of this kind of photooxidation in Section 6.2.). Photooxidation is a process undergone by chlorophyll molecules excited to triplet states. These states are usually induced when a chlorophyll molecule has absorbed a quantum of sufficiently high-energy light (from the $\lambda < 480$ nm region): the molecule passes first into a higher, excited singlet state, then directly into the triplet state. As our analyses have shown, such a threshold value of natural irradiance from the $\lambda < 480$ nm region, above which there is a marked rise in the probability of finding chlorophyll molecules excited to triplet states (and which are oxidized in consequence) in the photosynthetic apparatus of algae, lies at an irradiance intensity with a spectral density of around $E_p(\lambda) \approx 0.2\ \mathrm{W \cdot m^{-2} \cdot nm^{-1}}$.

Situations such as these do occur in the surface layers of seas, particularly on sunny days; see Figure 6.4. The gray shading in the top left-hand corners of the plots indicates the areas of such destructive irradiances, following their conversion to daily doses; we have assumed a 12 h day. The figure shows that the penetration depths of such destructive light, causing photooxidation of chlorophyll molecules, depend on the basin's trophicity. In ultra-oligotrophic waters (O1), the irradiance level in this spectral region exceeds the above threshold value for the photooxidation of chlorophyll molecules in a very extensive layer of the sea surface; this may be as much as 100 m deep (Figure 6.4a,e). With increasing trophicity, the thickness of this photodestructive zone falls rapidly, so that in supereutrophic waters (E6), we need expect photooxidation of chlorophyll only in the top half-meter or so of the sea (Figure 6.4d,h).

As Woźniak et al. (1998) and Majchrowski and Ostrowska (1999, 2000) have demonstrated, we can also define a precise photometric characteristic of the underwater irradiance fields in the sea that is directly responsible for the excitation of higher energy states in chlorophyll molecules, that is, one linked indirectly to their photooxidation. This magnitude is the potentially destructive radiation (per unit of chlorophyll *a* mass) *PDR**, defined by expressions (T-9a,b) in Table 6.3. It is the mean daily energy dose of natural downward irradiance from the 400–480 nm spectral region absorbed by the unit mass of chlorophyll *a*. This value is usually averaged over depth in a layer of given thickness Δz (see the description below, and Equation (T-9a) in Table 6.3) to take account of the fact that the chlorophyll absorbing this energy is contained in algae which, as a result of the water mixing processes (e.g., Druet (1994), Massel (1999), and Ramming and Kowalik (1980)) migrate inertly in the water, and also in a vertical direction.

The vertical distributions of the potentially destructive radiation $PDR^*(z)$ have been modeled for different types of seas (Figure 6.5). In all cases and for obvious reasons, *PDR** is greatest just below the surface and decreases with depth. In general, however, it is always much higher in waters of low trophicity than in eutrophic seas. This has a significant effect on the content of the so-called photoprotecting pigments in phytoplankton cells, that is, the pigments protecting chlorophyll against photoinhibition. Therefore the *PDR** is also a photoadaptation factor. The pigment composition thus varies in basins of different trophicity and at different depths. These characteristic differences are due to the photoadaptation of cells, and are of crucial importance with regard to light absorption, of which more in the next sections.

Apart from the photoadaptation of phytoplankton stimulated by high absolute irradiance levels in the water, its pigment content also depends on the spectral composition of that irradiance. As we showed earlier, the spectra of this irradiance can vary a great deal. The adaptation of the photosynthetic apparatus of individual algae or of whole plant communities to the spectrally varied irradiance conditions is known as chromatic adaptation (or chromatic acclimation).

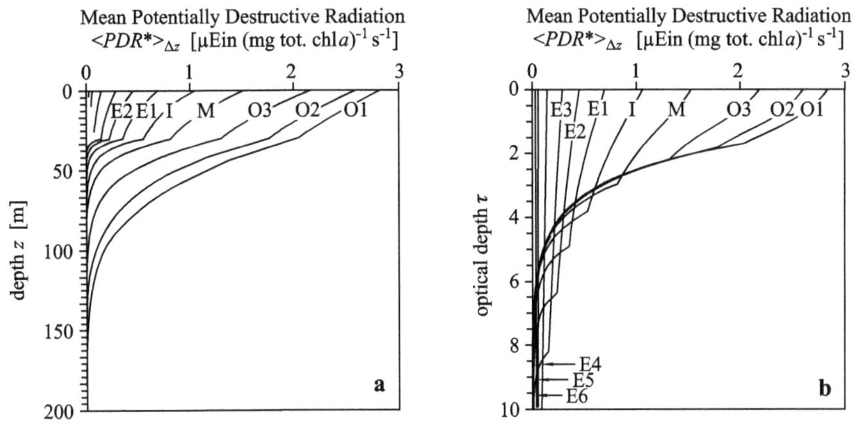

FIGURE 6.5. Modeled vertical profiles, the mean in water layers Δz, of the *PDR** in various trophic types of basins computed for a surface irradiance $PAR(0) = 695$ $\mu E \cdot m^{-2}\, s^{-1}$ for real depth (a) and for optical depth (b) the thicknesses of the water layers are defined as follow: $\Delta z = z_2 - z_1$, where $z_2 = z + 30$ m, $z_1 = 0$ if $z < 30$ m and $z_1 = z - 30$ m if $z \geq 30$. (After Woźniak et al. (2003).)

We have known for a long time that this process leads to the production (if we are talking about the transformation of a particular organism) or the arising (e.g., in the case of the succession of species) of such photosynthetic pigments as are capable of harvesting light from the environment[3] (Steeman Nielsen 1975). They must therefore possess characteristic absorption bands in those spectral regions from which photons in the external irradiance can be abstracted. In other words, the spectral distributions of light absorption for a given pigment must "fit", as it were, the spectral distributions of the natural irradiance in the plant's surroundings. The optical quantities that could be a measure (index) of such a fitting of individual pigments to the irradiance conditions are the so-called spectral fitting functions for the *j*th pigment, F_j, as suggested by our team (Woźniak et al. 1999, 2003; Majchrowski and Ostrowska 1999, 2000).

Known as chromatic adaptation (or acclimation) factors, they are defined by the expression (T-8a,b) given in Table 6.3. As can be seen from this expression, by "the spectral fitting function for the *j*th pigment," we mean the average value of its mass-specific absorption coefficient $a^*_j(\lambda)$ in the PAR spectral range (400–700 nm), weighted with the relative function $f(\lambda)$ of the spectral distribution of irradiance, and normalized with respect to the maximum in the spectrum of the mass specific absorption coefficient for that pigment $a^*_{j,max}$. As in the case of the photoadaptation factors (*PDR**), the chromatic adaptation

---

[3] That is, pigments whose role it is to acquire energy for photosynthetic purposes by the absorption of light. In the next section, we return to the types of pigment occurring in phytoplankton and the parts they play there.

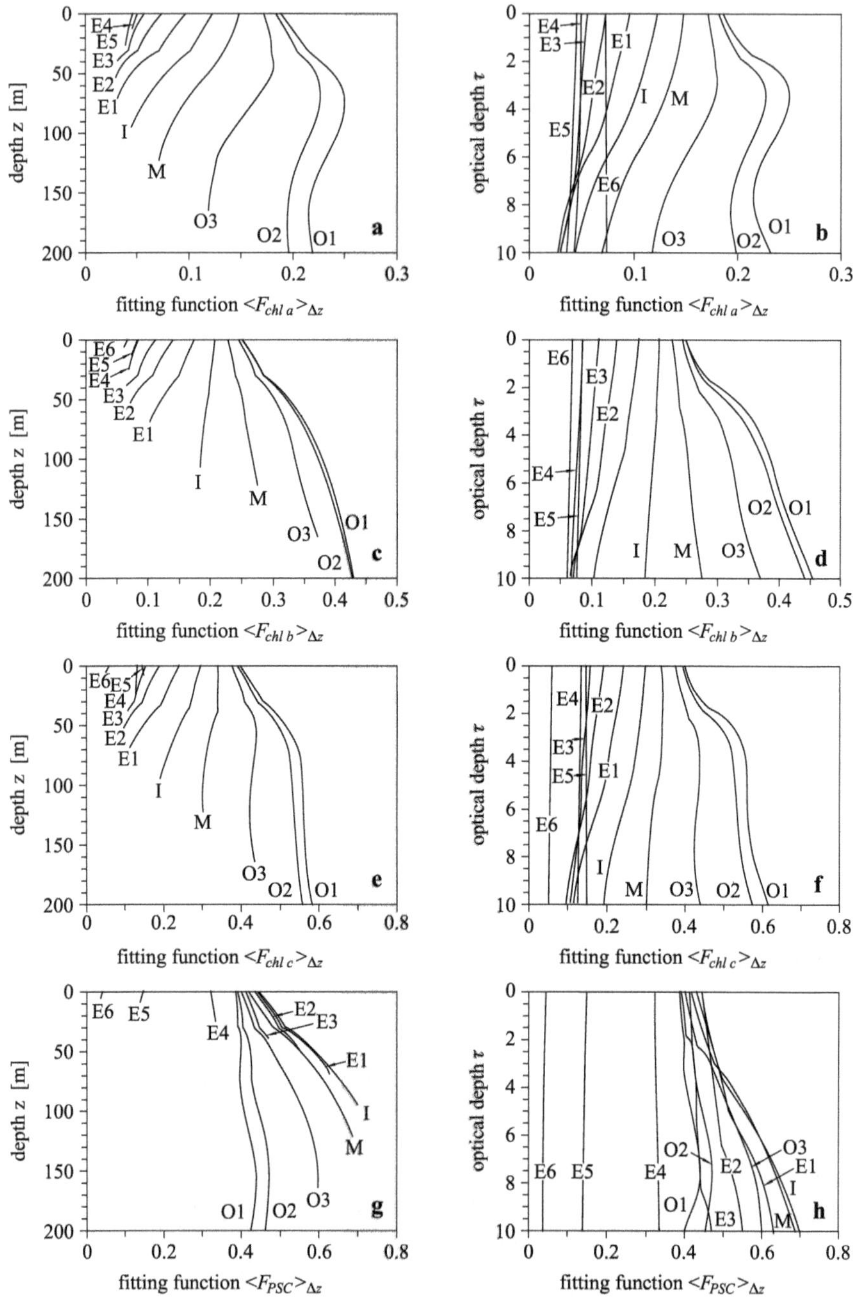

FIGURE 6.6. Modeled vertical profiles of mean fitting functions in water layer Δz for different groups of pigments (chl*a* (a),(b), chl*b* (c),(d), chl*c* (e),(f), *PSC* (g),(h)) for real (a),(c),(e),(g), and optical (b),(d),(f),(h) depths. The thicknesses of water layers are defined as follow: $\Delta z = z_2 - z_1$, where $z_2 = z + 30$ m, $z_1 = 0$ if $z < 30$ m and $z_1 = z - 30$ m if $z \geq 30$. (After Woźniak et al. (2003).)

factor of phytoplankton is usually the depth-averaged magnitude $\langle F_j\rangle_{\Delta z}$ rather than F_j, determined for separate fixed depths (see the comment in Table 6.3) in order to take account of the possible vertical migrations of algae resulting from the mixing of waters.

The vertical distributions of these fitting functions$\langle F_j\rangle_{\Delta z}$ in different types of Case 1 waters estimated with the aid of the model for the main group of photosynthetic pigments (apart from phycobilin) are exemplified in Figure 6.6. Notice that these functions F_j can take values from 0 to 1. A zero value is assigned when the spectrum of the pigment's absorption coefficient does not overlap anywhere with the spectrum of the underwater irradiance, that is, there is no fitting whatsoever, whereas a value of one is assigned when the wavelength of the ambient irradiance coincides with that of the maximum absorption coefficient of the pigment group. Thus, as we see in Figure 6.6, spectral fitting depends on trophicity, depth in the sea, and the type of pigment in question. As we go on to show, these regularities underlying the distributions of the vertical fitting functions influence in a fundamental way the vertical distributions of the concentrations of different groups of pigments in seas of different trophicity.

6.2 Phytoplankton Pigments and Their Electronic Absorption Spectra in the Visible Region

The organic matter in marine phytoplankton consists mainly of proteins (c. 50% by weight) and lipids (c. 30–50%) (Riley and Chester 1971, Romankevich 1977, Filipowicz and Więckowski 1979, Woźniak et al. 2005 a,b). The proportion of pigments in this composition is much smaller and usually varies from 0.1 to 10% (Koblentz-Mishke and Vedernikov 1977, Kirk 1994). Even so, pigments are the main and practically the only group of visible light absorbers in plant cells, so it is they that determine the absorption spectra of algae in the visible region (see Section 5.2.3). Any effects that the other organic components may have on the absorption of light by phytoplankton cells become detectable only in the UV region. This is because pigments contain very many more chromophores optically active in the VIS region than do the other organic components of phytoplankton. Hence, the absorption of light by phytoplankton pigments exerts a significant influence on the absorption properties of sea waters in the visible wavelength region.

6.2.1 The Role of Phytoplankton and the Main Types of Phytoplankton Pigments

The most important role of light absorption by phytoplankton pigments is to stimulate the photosynthesis of organic matter (Kirk 1994, Govindjee 1982a, Steemann-Nielsen 1975), which is essential for sustaining life in the ocean. The primary production of organic matter during photosynthesis is the factor determining the circulation of matter in ecosystems. Indirectly, this primary production also determines the nature and concentrations of the vast majority of suspended particles scattering and absorbing light in the sea (zooplankton, nekton, and organic detritus), as well as the content of light-absorbing organic substances dissolved in sea water.

Pigments are contained in the photosynthetic apparatus of plant cells; this in turn, is located within organelles known as chloroplasts. In addition, these organelles are filled with proteins and lipids, exist in a variety of shapes, and are usually a few micrometers in size. Without entering into the fine detail of the complex internal structure and location of the photosynthetic apparatus (this is more than adequately dealt with in a whole range of monographs, e.g., Grodziński (1978), Govindjee (1975, 1982a, b, 1987), Heldt (1997), and Hall and Rao (1999)), we should nonetheless note that from the point of view of light absorption, the most important aspect is the presence in the chloroplast of a number of centers absorbing light and trapping energy. Of two types, PS I and PS II (PS, photosystem), these centers have slightly different optical properties; they act independently when light is being absorbed, but in tandem during photosynthesis (Ort and Govindjee 1987, Okamura et al. 1982). Together they form a photosynthetic unit, a quantosome, that functions autonomously where the synthesis of the simplest organic compounds is concerned. Apart from other constituents, each quantosome contains from 300 to 5000 pigment molecules (Govindjee and Whitmarsh 1987, 1982).

Among the pigments that could be called photosynthetically active, there are three main types capable of absorbing and emitting visible light: the chlorophylls, the carotenoids, and the phycobilins. They occur in varying proportions depending on the species of cell and also on the abiotic conditions prevailing in the sea. Moreover, each group of pigments contains a number of chemical and optical (native) variants. Thanks to this diversity of pigments, phytoplankton can absorb light energy from different regions of the visible spectrum. By altering its chemical composition, a phytoplankton cell can adapt to the ambient light spectrum. This is one of the signs of the photoadaptation and chromatic adaptation of phytoplankton cells (Steemann-Nielsen 1975) that we mentioned in Section 6.1.

Nevertheless, the pigments in the photosynthetic apparatus play diverse parts. Table 6.4 shows how plant pigments are classified according to their functions in cells. As we can see, three groups of these pigments (chlorophylls, phycobilins, and carotenoids) act as "antennas": by means of photon absorption: they harvest solar radiation energy from the different spectral wavelengths incident on the cell. These are the photosynthetic pigments (PSP). The energy they absorb is then partially utilized to produce organic matter during photosynthesis in the quantosomes with the aid of the PS I and PS II centers. This happens independently in both photosystems, thanks to the chains of nonradiant migration of excitation energy between molecules of the different pigments or between molecules of the different native variants of the same pigment (Parson and Ke 1987, Govindjee 1982a, 1987).

TABLE 6.4. Classification of plant pigments according to their functions in the photosynthetic apparatus.

Pigments			
Photosynthetic (PSP[a])			
Principal pigment	"Antennas"	Photoprotecting (PPP[a])	Other functions
-1-	*-2-*	*-3-*	*-4-*
Chlorophyll *a* (Bacteriochlorophyll *a*)	Chlorophylls *a*, *b*, *c*, *and others*	Carotenoids PPC[a, b], including: *Antheraxanthin, Diadinoxanthin, Alloxanthin, Diatoxanthin, Dinoxanthin, Lutein, Violaxanthin, Neoxanthin, Zeaxanthin, β-Carotene*	Pheopigments
	Phycobilins, including: *Phycoerythrin, Phycocyanin, Allophycocyanin*		
	Carotenoids PSC[a, b], including: *Fucoxanthin, 19'Butfucoxanthin, 19'Hexfucoxanthin, Peridinin, Prasinoxanthin, α-Carotene*		

[a] PSP, photosynthetic pigments; PPP, photoprotecting pigments; PSC, photosynthetic carotenoids; PPC, photoprotecting carotenoids.
[b] The lists of PSC and PPC give the variants of carotenoids usually acting as photosynthesizing pigments (antennas) or photoprotecting pigments. However, in nature the reverse can occur, when a compound classified among the PSC fulfills a photoprotecting function and vice versa.

This energy migrates (some is dissipated) from molecules of the more "shortwave" pigments (those absorbing higher-energy photons, i.e., with higher levels of excitation) to the more "longwave" ones (with lower levels of excitation) and ends up in so-called "energy traps," which are directly involved in photosynthesis. Primary donors of electrons essential for initiating the cycle of photosynthesizing chemical reactions (Cramer and Crofts 1987), these traps are probably various longwave variants of chlorophyll *a* (in green plants) or bacteriochlorophyll *a* (in photosynthesizing bacteria). They are denoted by P_{700}– chlorophyll in PS I, P_{680}–chlorophyll in PS II, P_{890}–bacteriochlorophyll in PS I, and P_{870}– bacteriochlorophyll in PS II (the subscripts by the letters P denote the approximate wavelengths (in nm) of the photons, the energies of which correspond to the excitation energies of these pigment variants).

The differences in the properties of the PS I and PS II optical centers result from the fact that in green plants, for example, PS I centers harvest the energy absorbed by pigments in the $\lambda < 700$ nm spectral range, whereas energy uptake at PS II centers is from the $\lambda < 680$ nm range. It has also been shown that excited PS II centers fluoresce to a far greater extent than PS I centers do

(Grodziński 1972, 1978; Govindjee 1982a; Rubin 1995; Rubin et al. 1994; Hall and Rao 1999).

It emerges from the foregoing, then, that chlorophyll *a*, unlike the other pigments (the other variants of chlorophyll, carotenoids, and phycobilins), is directly involved in the photosynthesis of green plants and is, indeed, indispensable to this process. Bacteriochlorophyll *a* plays an equivalent role in photosynthesizing bacteria. Both of these pigments can be assumed to be absolutely fundamental. In contrast, the presence of the other pigments—the accessory pigments, which assist in the acquisition of energy—is not essential for photosynthesis to occur. They are part of a mechanism that allows plants to photosynthesize in less than optimum light conditions (Margalef 1967). Their composition and concentration therefore depend on external factors and also on the botanical association to which the algae in question belong.

Exceptional among these accessory pigments are some of the photoprotecting carotenoids (PPC), which in fact are always present together with chlorophyll *a* in all plants. In addition to playing a supporting role as absorption antennas, carotenoid pigments also protect chlorophyll against photooxidation. Experiments on algae have shown that mutants of algae deprived of carotenoids, although continuing to photosynthesize, perish rapidly as a result of the oxidation of chlorophyll *a* (Filipowicz and Więckowski 1979, 1983; Hall and Rao 1999). The mechanisms of the photooxidation and other deactivation processes of the chlorophyll molecule are probably the following:

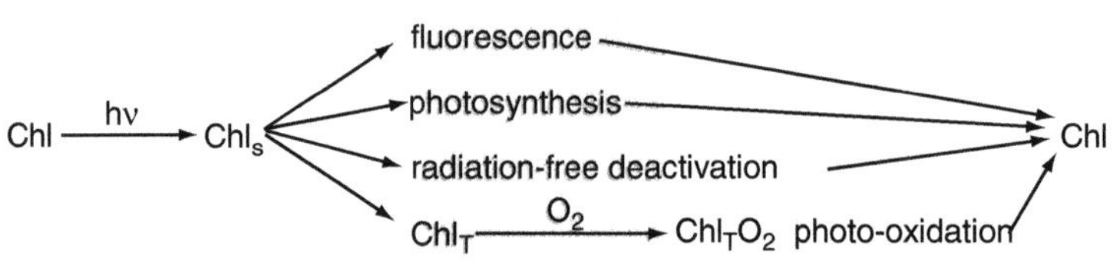

where the respective abbreviations Chl, Chl_S, and Chl_T stand for chlorophyll in the ground state, and in the singlet and the triplet excited states. As we can see, the alternative route to the deactivation of the excited singlet state of the chlorophyll molecule by fluorescence, photosynthesis, or radiation-free deactivation, is the transition to the triplet state and the oxidation of the molecule. This last process is prevented when the triplet states in chlorophyll are eliminated as a result of the radiation-free triplet–triplet transition of energy from chlorophyll to, say, β-carotene (see, e.g., Grodziński (1972) and Hall and Rao (1999)).

Apart from the three main groups of photosynthetically active pigments, there is another group of strong absorbers of light occurring in green plants, particularly in marine phytoplankton, at night or at great depths. These are the pheopigments (Yentsch 1965, Wolken 1975), and they contribute greatly to the overall absorption capabilities of phytoplankton. We know that the light energy absorbed by pheopigments is not used for photosynthesis: they play another role in this process, for example, by participating in intermediate systems, or in the transport of electrons in PS II centers (e.g., Shipman (1987), Ostrowska and Woźniak (1991a, b), Rubin (1995), and Hall and Rao (1999)).

6.2.2 *The Chemical Structure of Pigments*

As we have already said, the optical properties of pigments depend chiefly on the types of chromophores they contain, which in turn are determined by the chemical structures of the pigment molecules. In order to state which chromophores are typical of particular groups of phytoplankton pigments, let us briefly recall the structures of their molecules. These the reader can find in Figures 6.7–6.9 and in Table 6.5, which have been compiled on the basis of the following works: Grodziński (1972), Filipowicz and Więckowski (1983), Woźniak and Ostrowska (1991), and Frąckowiak et al. (1997).

Chlorophyll and related compounds. These are derivatives of porphyrin (Figure 6.7a). Although produced in plants, these compounds are synthesized according to mechanisms very similar to the ones producing haem. Various aspects of these problems are discussed in detail in Fradkin (1986a, b), Shlyk (1965, 1974), Govindjee (1982a, b, 1987), Rubin (1995), Rubin et al. 1994 and elsewhere. A glance at Figure 6.7 tells us that these compounds are all similar in structure. Because of their optical properties in the visible region, however, the most important aspect of these compounds is the porphyrin

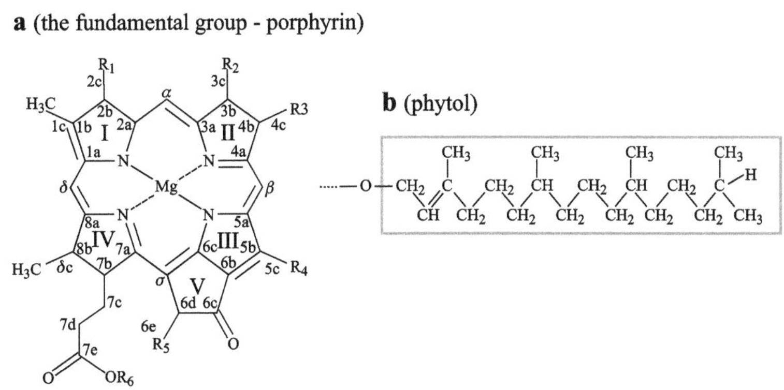

FIGURE 6.7. Structural aspects of the molecules of chlorophyll and related compounds: (a) the fundamental group, porphyrin; (b) the R_6 radical (phytol) in most chemical forms of chlorophyll. (The complete chemical structures of chlorophyll pigments can be obtained by taking into account the additional data given in Table 6.5.)

TABLE 6.5. Structures of selected chemical forms of chlorophylls and related compounds.[a]

	Bond[b]		Radical[b]					
Pigment	3b–4b	7b–8b	R_1	R_2	R_3	R_4	R_5	R_6
-1-	*-2-*	*-3-*	*-4-*	*-5-*	*-6-*	*-7-*	*-8-*	*-9-*
Chlorophyll *a*	Double	Single	$-CH{=}CH_2$	$-CH_3$	$-CH_2-CH_3$	$-CH_3$	$-COOCH_3$	Phytol
Chlorophyll *b*	Double	Single	$-CH{=}CH_2$	$-CHO$	$-CH_2-CH_3$	$-CH_3$	$-COOCH_3$	"
Chlorophyll c_1	Single	Double	$-CH{=}CH_2$	$-CH_3$	$-CH_2-CH_3$	$-CH_3$	$-COOCH_3$	–H
Chlorophyll c_2	Single	Double	$-CH{=}CH_2$	$-CH_3$	$-CH_2{=}CH_2$	$-CH_3$	$-COOCH_3$	"
Chlorophyll *d*	Double	Single	$-CHO$	$-CH_3$	$-CH_2-CH_3$	$-CH_3$	$-COOCH_3$	Phytol
Bacterio chlorophyll	Single	Single	$-COOCH_3$	$-CH_3$	$-CH_2-CH_3$	$-CH_3$	$-COOCH_3$	"
Chlorophylls *chlorobium* (bacterioviridine)	Double	Single	$-CHOHCH_3$	$-CH_3$	$-CH_2\ CH(CH_3)_2$ or $-CH_2CH_2CH_3$ or $-CH_2CH_3$	$-CH_2CH_3$ or $-CH_3$	–H	$-CH_2(CH{=}C_2H_3CH_2CH_2)_2CH{=}C(CH_3)_2$
Pheophorbide[c]	Double	Single	$-CH = CH_2$	$-CH_3$	$-CH_2CH_3$	$-CH_3$	$-COOCH_3$	–H
Chlorophyllide *a*	Double	Single	$-CH = CH_2$	$-CH_3$	$-CH_2CH_3$	$-CH_3$	$-COOCH_3$	–H
Pheophytin *a*[c]	Double	Single	$-CH = CH_2$	$-CH_3$	$-CH_2CH_3$	$-CH_3$	$-COOCH_3$	Phytol
Protochlorophyllide *a*	Double	Double	$-CH = CH_2$	$-CH_3$	$-CH_2CH_3$	$-CH_3$	$-COOCH_3$	Phytol

[a] Data in this table complement the structural aspects of these compounds shown in Figure 6.7; adapted from Grodziński (1972).
[b] See Figure 6.7.
[c] In molecules of pheophorbides and pheophytins the central magnesium atom Mg is replaced by two atoms of hydrogen H.

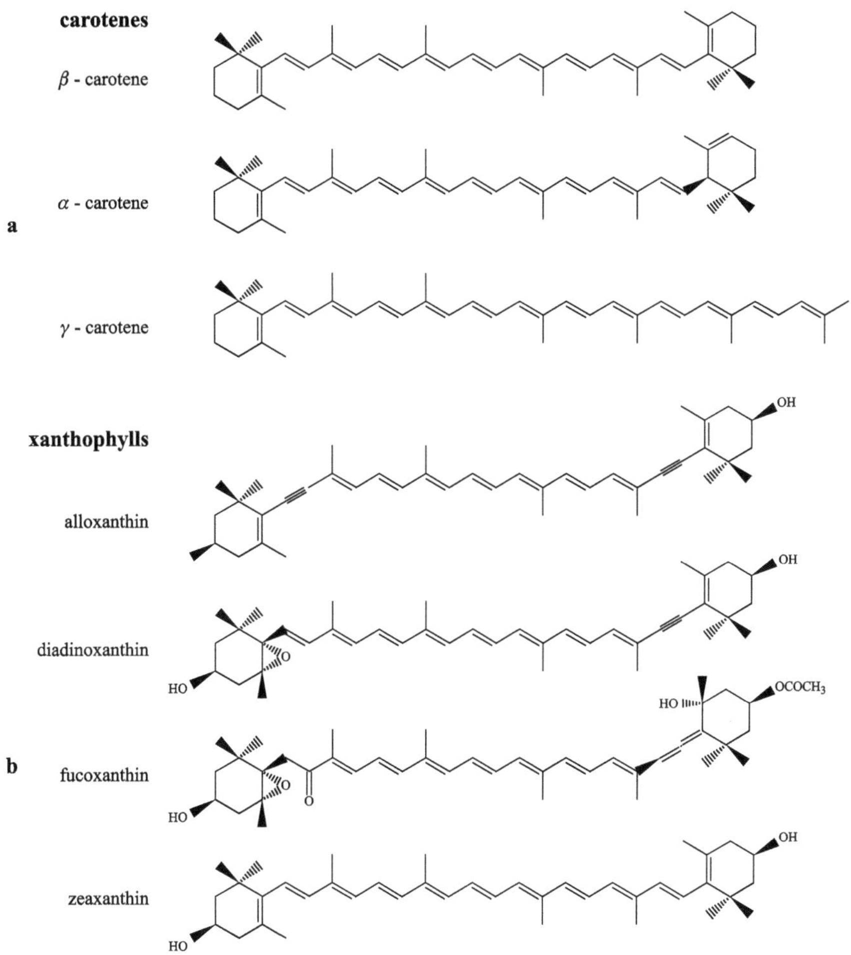

FIGURE 6.8. The chemical structures of selected carotenoids: (a) the principal carotenes; (b) xanthophylls.

arrangement, which is the principal chromophore in these compounds. This is a closed structure consisting of five pyrrole rings linked internally to a magnesium atom[4] (Figure 6.7a). The numbering of the rings here is given after Arnoff (1950). The differences between the pigments of this group

[4] If an atom of iron replaced the magnesium atom, we would have haem, a derivative of which is hemoglobin. Chlorophyll is thus the "brother" of hemoglobin, as it were. Note that the first of these "siblings" is essential to plant life; the second (in blood) determines the life of animals. If pyridine were combined with copper instead of magnesium or iron, we would obtain a blue color. Are there perhaps organisms out there with blue blood?

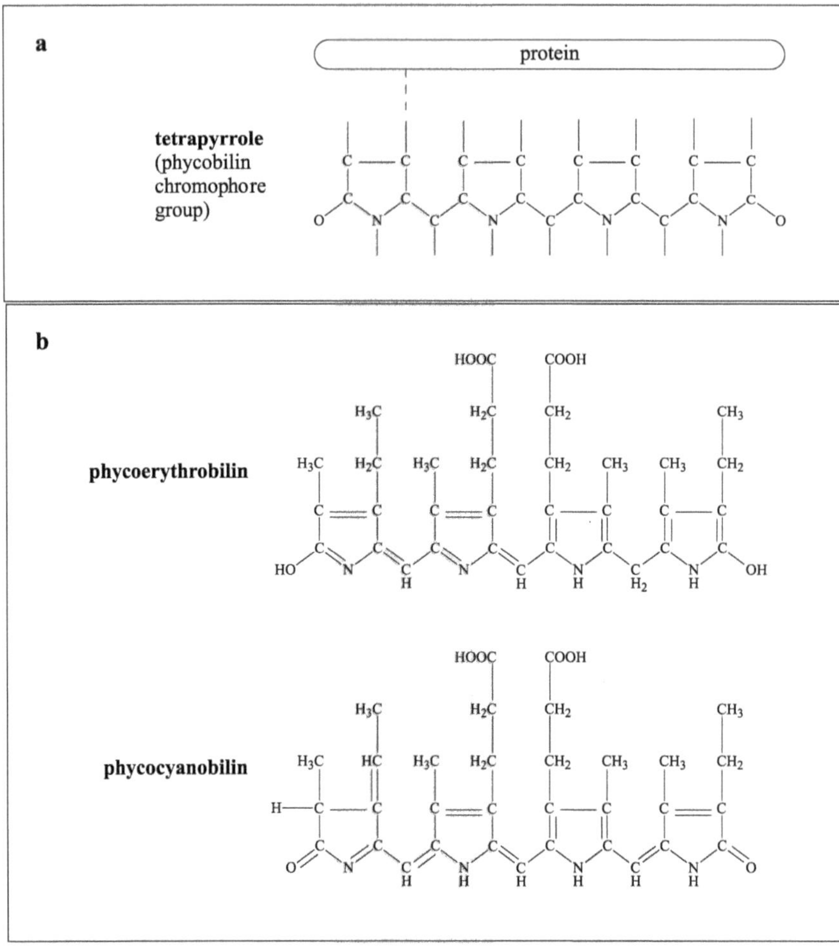

FIGURE 6.9. The chemical structures of phycobilin pigment chromophores: (a) the starting chromophore, tetrapyrrole; (b) the chromophores in phycoerythrin and phycocyanin.

arise out of the different multiplicities of some of the chemical bonds (bonds (3b–4b) and (7b–8b) in the figure) as well as the various possible substituents (R_1 to R_6 in the figure) on the pyrrole rings (Table 6.5). A feature common to most of the "plant versions" of these pigments is the fact that substituent R_6 of the carboxyl group attached to ring IV is the long-chain phytol (Figure 6.7b). However, this does not much affect the visible light absorption capabilities of chlorophylls, inasmuch as this chain remains unsaturated and contains hardly any π-electrons.

There are many chemical forms of these pigments in existence and a large number of their optical variations (native forms). For instance, in green plants at least ten chemical variants of chlorophyll can be distinguished, including chlorophylls *a, b, c, d, e,* bacteriochlorophylls, and also the so-called P_{700} and P_{830} pigments found in the photosynthetic centers; the latter may not be chemical variants of chlorophyll *a* or bacteriochlorophyll *a*, merely native forms of them. In addition, there is a whole range of chlorophyll decomposition products, such as the pheopigments (see below), the pheophorbides, and others which, although they absorb light, do not use its energy for photosynthesis. More than 50 photosynthetically active chemical and native forms of chlorophyll have been discovered in algae (Scheer 1991).

Pheopigments. This term covers primarily the group of substances known as pheophytins which, like the chlorophylls, are derivatives of pyridine. Pheophytins are conversion products of chlorophylls; that is, chlorophyll *a* is converted into pheophytin *a*, bacteriochlorophyll *a* is converted into bacteriopheophytin *a*, and so on. This conversion hardly disturbs the chemical structure of the pigments (compare chlorophyll *a* and pheophytin *a* in Table 6.5): it simply involves the replacement of the magnesium atom in the porphyrin arrangement by two hydrogen atoms. Under natural conditions, this conversion takes place when there is insufficient light energy in the water: at night or at great depths (Wolken 1975, Yentsch 1965, Fradkin 1986a, b). The chemical structures, and therefore the optical properties of pheophytins are similar to those of chlorophylls.

Carotenoids are derivatives of linear polyenes (the chromophoric core of these compounds) with an 11-fold π-electron conjugation. We touched upon their fundamental absorption properties in Section 3.2 (Figure 3.5 and Table 3.3A). Formally, carotenoids are a subgroup of the so-called isoprenoids, that is, compounds built up of isoprene units $(C_5H_8) \times n$; the carotenoids themselves are derivatives of octa-isoprenoids; that is, $n = 8$. The molecular formula $C_{40}H_{56}$ indicates the chemical composition of the carotenes, as this group of compounds, the most widespread of the carotenoids, is known. It comprises the three principal forms of these pigments—α-carotene, β-carotene and γ-carotene (Figure 6.8a)—as well as the numerous structural variants of these three basic forms. The α-carotene and β-carotene molecules differ only in the position of the double bond in one of the end of C_9H_{15} rings. In γ-carotene this ring is opened up, so that additional double bonds come into being.

Another large group of carotenoid compounds are the xanthophylls. The chemical structure of these pigments differs only marginally from that of the basic carotenes: they are oxygen derivatives of the latter in that they contain carboxyl groups (see the examples in Figure 6.8b). Despite these very slight differences, more than 600 chemical variants of these compounds have been identified. Their detailed structures and properties, as well as the complicated mechanisms of their formation and conversion in plants are discussed by Feofilova (1974), Mantoura and Llewellyn (1983), and Jeffrey et al. (1997), among others.

Phycobilins are groups of chromophoric phycobiliproteids (i.e., pigment–protein complexes), and are components of the antenna complexes (phycobilisomes) in cyanobacteria and some plants, including certain species of algae (e.g., Grabowski (1984)). Formally, phycobilins are members of the large family of compounds known as chromoproteins. They are protein substances containing prosthetic chromophore groups (Zvalinsky 1986, Filipowicz and Więckowski 1983). In addition to playing a part in photosynthesis, these compounds perform numerous other biological functions, and are present in both plants (for instance, the phytochromes in plants are regulators of phototaxis and photomorphogenesis; see, e.g., Frąckowiak et al. (1997)) and animals (e.g., bile pigments).

As in the chlorophylls, the chromophoric groups of many chromoproteins contain four pyrrole rings, the so-called tetrapyrrole group; however, the chain remains open. This kind of starting chromophore of many chromoproteins is illustrated in Figure 6.9a. The three characteristic main groups of phycobilins have been identified in the phycobilisomes of cyanobacteria and certain other algae: phycoerythrin (PE), phycocyanin (PC), and allophycocyanin (APC); see the examples in Figure 6.9b. However, the individual structures of these three phycobilins may vary slightly, depending on the species of alga or cyanobacteria in which they occur. Hence, a letter denoting the origin of these pigments is added to their abbreviated designation: C refers to cyanobacteria, R to rhodophytes, and so on. Thus CPE stands for C-phycoerythrin, RPC for R-phycocyanin, and so forth. The molecular mass of these phycobilins can also vary, because the number of chromophores (usually three) can differ, depending on the pigment's origin. The reader will find much more information on this subject in Grodziński (1978), Grabowski (1984), and Frąckowiak et al. (1997) among others.

6.2.3 The Individual Absorption Properties of Pigment Extracts

In practice it is not possible to measure directly the light absorption spectra of the separate pigments in vivo. At most, what can be done is to measure the overall spectra with respect to the complete photosynthetic apparatus, which contains natural mixtures of these various pigments. Only on the basis of these overall "summary" spectra and taking the dominance of certain groups of pigments into account is it possible to define indirectly and approximately the positions of the peaks and the magnitudes of the spectral absorption coefficients of individual pigments. A more accurate determination of these features involves the laborious method of decomposing the summary spectra, for example, by applying differential spectroscopy (of which more later in this section). Now the light absorption spectra of the separate pigments extracted in various solvents are well known,[5] so we deal with them first.

Examples of such spectra for extracts of the more common phytoplankton pigments are given in Table 6.6 and illustrated in Figure 6.10.

Figure 6.10 shows that plant pigments absorb light mainly in two regions of the visible spectrum: in the blue band, known as the Soret band, and in the red band. Practically all pigments except phycobilin absorb in the Soret band; hence, chlorophyll *a* and the accessory pigments, that is, the other chlorophylls and the carotenoids, absorb violet light. Red light, on the other hand, is absorbed as a rule only by the chlorophylls, chiefly by chlorophyll *a*. Light from the middle regions of the visible spectrum is absorbed to only a small extent by the carotenoids, especially those classified among the photosynthetic pigments. The absorption bands of the carotenoids extend farther in the direction of the long wavelengths than do the shortwave bands of the chlorophylls. The phycobilin pigments, however, are more intensive absorbers of light from these middle regions, that is, of yellow, orange, and orange-red light.

The absorption properties of pigment extracts we have just outlined usually deviate quite a lot from the actual light absorption spectra of these pigments in the living plant. This problem is illustrated by the data in Table 6.7. Not only does this give the exact positions of the peaks of the principal

TABLE 6.6. Spectra of mass-specific light absorption coefficients $a^*_{pig,extr}$ of selected phytoplankton pigments (chlorophylls and carotenoids) in 90% acetone.[a]

	$a^*_{pig.extr.}$ $[10^{-3}\ m^2\ mg^{-1}]$ for Wavelength λ [nm]:								
Pigment	400	420	450	480	510	630	645	655	685
-1-	*-2-*	*-3-*	*-4-*	*-5-*	*-6-*	*-7-*	*-8-*	*-9-*	*-10-*
Chlorophyll *a*	13.4	16.3	2.0	0.4	0.6	2.7	3.7	15.4	0.8
Chlorophyll *b*	2.5	6.2	12.4	3.1	0.8	2.9	10.5	1.5	0
Chlorophyll *c*	5.5	8.6	18.1	1.2	0.5	2.4	1.0	0.3	0
β-Carotene	18.5	34.1	51.5	55	10.4	0	0	0	0
Fucoxanthin	24.5	38.9	57.3	46.7	13.0	0	0	0	0

[a] Calculated from extinction data in Richards (1952) and Richards and Thompson (1952).

[5] In the case of the chlorophylls and carotenoids, these are various solvents, for example, acetone (see Richards (1952) and Richards and Thompson (1952)). Knowledge of the individual absorption spectra of the various pigments in 90% acetone is made use of, for example, in the spectrophotometric methods of determining their concentrations routinely used in oceanography (Strickland and Parsons 1968). On the other hand, the phycobilins are insoluble in organic media (but are water soluble), and their extraction from chloroplasts requires the application of other, more sophisticated techniques (see, e.g., Gantt and Lipschultz (1972), Yakamura et al. (1980), Gagliano et al. (1985), Zoha et al. (1998)).

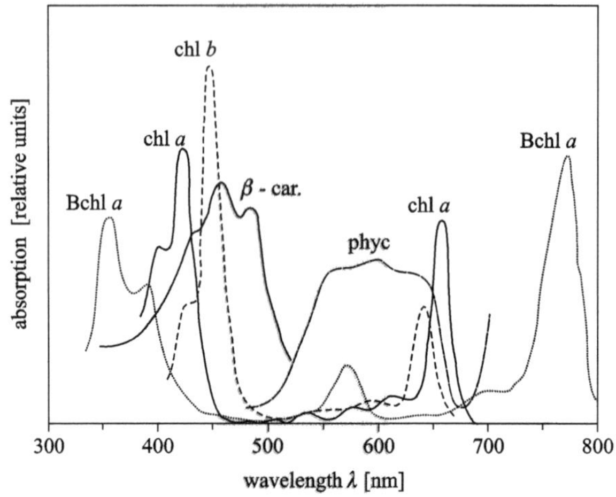

FIGURE 6.10. Light absorption spectra of selected pigments dissolved in 90% acetone: chl *a*, chlorophyll *a*; chl *b*, chlorophyll *b*; chl *c*, chlorophyll *c*; Bchl *a*, bacteriochlorophyll *a*; *β*-car, β carotene, and natural forms of phycobilins (phyc) extracted from algae. (After various authors, collected in Woźniak and Ostrowska (1991).)

TABLE 6.7. Positions of the maxima in the absorption spectra of selected phytoplankton pigments λ_{max} [nm].[a,b]

Pigment	In acetone	In vivo
-1-	*-2-*	*-3-*
Chlorophyll *a*		
Short wave max.	420	<u>435</u>
Long wave max.	663	661–663, 669, 677–680, 684, 687, 695–699, (in PS-I), 650, 663, 669, <u>672–674</u>, <u>677–680</u> (in PS-II)
Chlorophyll *b*		
Short wave max.	453	470–480
Long wave max.	645	650–653
Chlorophyll *c*		
Short wave max.	445	460
Long wave max.	631	633
α-Carotene	<u>450–460</u>, 480	445, <u>470</u>, 500
Fucoxanthin	448–449	<u>490</u>, 540

[a] The most intensive maxima are underlined.
[b] Data after various authors, collected in Woźniak and Ostrowska (1991).

absorption bands of 90% acetone extracts of chlorophylls and carotenoids (column 2 in Table 6.7), it also gives the positions of these peaks ascribed to the various pigments in vivo (column 3). The latter values have been visually defined in an approximate way by numerous authors from reviews of the summary light absorption spectra in different living plants. On comparing the two sets of values, we notice that the absorption bands in the acetone extracts are characteristically shifted, to a greater or smaller extent, in relation to the absorption in living plants, usually towards the shorter wavelengths. More or less the same applies to the absorption spectra of pigments dissolved in other solvents. Nevertheless, these absorption peak shifts do depend on the concentration and the type of solvent, and can move in either direction, towards the longer or the shorter wavelengths.

The main absorption bands in the in vivo spectra are broader than those obtained from the extracts, often exhibiting structures composed of a number of narrower absorption bands. These correspond to the various natural "optical" forms of pigments, the so-called "native" forms. The occurrence of these forms in the PS I and PS II photosystems is variable; it also depends on the properties of the plant, both as a species and as an individual organism. That is why the natural light absorption spectra due to the same pigment but from different plants can vary. It is only by extracting the pigments that we can put some semblance of order into their absorption spectra. This happens when the component absorption bands characteristic of a pigment under a given set of natural conditions are replaced by one or a small number of well-defined absorption bands characteristic of that pigment in a particular solvent. That is why the absorption spectra of plant extracts are highly suitable for identifying plant pigments and are used in spectrophotometric techniques for determining their concentrations (e.g., Strickland and Parsons 1968).

As far as the native "optical" forms of the pigments are concerned, we return to them in Section 6.2.5.

6.2.4 The Individual Absorption Properties of Pigments in Vivo

Measuring the optical absorption properties of individual phytoplankton pigments in vivo is an exceedingly complex undertaking. Besides the spectrophotometric measurement of summary light absorption spectra of plants, this process also requires the application of complex statistical procedures. These are based on the spectral decomposition of the summary spectra (see, e.g., Woźniak et al. (1999, 2000) into the component absorption spectra contributed by the separate pigments, followed by their tabulation or quantitative description in the form of mathematical formulas.

All the while it has to be borne in mind that the summary absorption spectra consist of the absorption coefficients of the separate pigments in their "packed" state in plant cells (see Chapter 5 and Section 6.5). These values are thus lower in relation to the absorption coefficients of the various pigments when "unpacked" in the solvent. This packing effect depends, among other things, on the dimensions of the phytoplankton cells and on the intracellular concentration of pigments, and differs for different wavelengths (e.g., Morel and Bricaud (1981)). In view of these methodological difficulties, the individual spectra of the "unpacked" coefficients of light absorption by particular plant pigments in vivo, expressed in absolute units, remained undefined for a long time. At present, however, we have the three following quantitative descriptions of the chlorophyll-specific coefficients of light absorption by algal pigments in vivo:

- The tabular description by Bidigare et al. (1990) of oceanic phytoplankton from basins where the packing effect can be neglected.
- The description in the form of 13 Gaussian bands by Hoepffner and Sathyendranath (1991) obtained from analyses of the absorption bands of selected cultures of algae.
- The description in the form of 21 Gaussian bands by the authors of the present monograph (Woźniak et al. 1999) referring generally to the pigments in marine phytoplankton (also to those in which the packing effect is significant), but only in Case 1 waters.[*)]

Notice that these descriptions do not refer to all the chemical variants of these pigments, but only to the following five groups of pigments: chl *a*, that is, the combined total of all variants of chlorophyll *a* and its decomposition products; chl *b*, that is, the combined total of all variants of chlorophyll *b*; chl *c*, that is, the combined total of all variants of chlorophyll *c*; PSC, that is, the combined total of all photosynthetic carotenoids; and PPC, that is, the combined total of all photoprotecting carotenoids. Phycobilins, on the other hand, are taken into account only by the first of the above descriptions (Bidigare et al. 1990), and that in a preliminary form which refers solely to two separate variants of phycoerythrin: 8103PE and DC2PE.

Without going into the details of the methods for determining the spectra of the light absorption coefficients of particular groups of "unpacked" pigments from the measured summary coefficients for all the "packed" pigments in phytoplankton in vivo (see Woźniak et al. (1999, 2000); also Sections 6.5 and 6.7), we move straight on to a presentation and discussion of these spectra. The absorption coefficients of "unpacked" pigments, according to the concepts of the three research teams mentioned above, are illustrated in Figure 6.11. Moreover, the resolutions of these coefficients into Gaussian band components, which are the foundation of the quantitative description by Woźniak et al. (1999, 2000), are given in Table 6.8. The parameters given in this table enable the determination of the spectra of the mass-specific light absorption coefficients of selected groups of "unpacked" pigments (and also of all these pigments jointly) with the aid of the following sum of Gaussian bands:

$$a_j^*(\lambda) = \sum_i a_{max,\,i}^* \cdot \exp\left[-\frac{1}{2}\left(\frac{\lambda - \lambda_{max,\,i}}{\sigma_i}\right)^2\right], \tag{6.3}$$

where $\lambda_{max,i}$ [nm] is the middle of the spectral band peak, σ_i [nm] is the band dispersion, $a^*_{max,i}$ [m^2(mg pigment)$^{-1}$] is the mass-specific absorption coefficient at the spectral band peak, and i is the number of the Gaussian band for the

[*)] In our subsequent research we derived analogous description of light absorption by pigments for phytoplankton in the Baltic Case 2 waters (see Ficek et al. 2004).

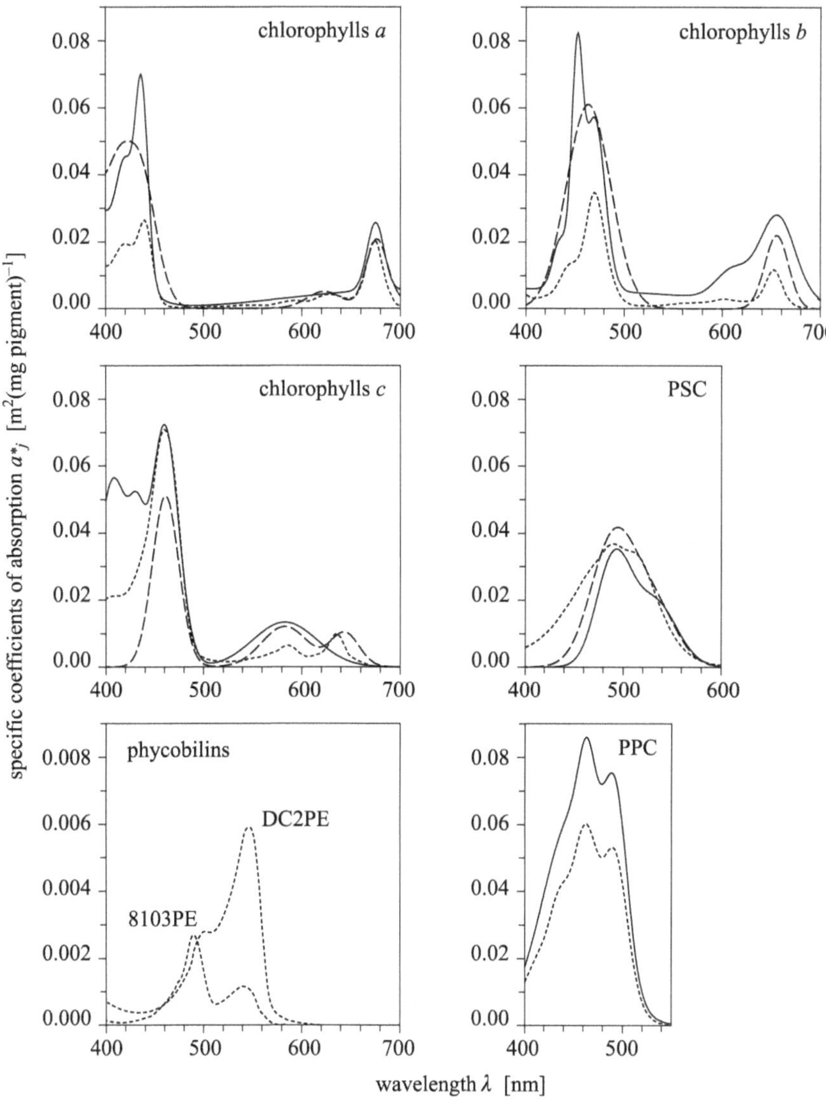

FIGURE 6.11. Spectra of the mass-specific absorption coefficients in vivo for the main groups of "unpacked" pigments, calculated in accordance with the model of Woźniak et al. (1999, 2000) (continuous line), and those proposed by other authors: Hoepffner and Sathyendranath (1991) (dashed line); Bidigare et al. (1990) (dotted line).

principal groups of phytoplankton pigments (i.e., chlorophylls *a*, chlorophylls *b*, chlorophylls *c*, photosynthetic carotenoids, and photoprotecting carotenoids).

As we see in Figure 6.11, the spectra of the mass-specific light absorption of the separate groups of pigments, obtained by all three of the above research teams, are similar in shape, although quantitatively somewhat different. These

TABLE 6.8. Parameters of model Gaussian bands describing the spectrum of the mass-specific absorption coefficients for five groups of phytoplankton pigments.[a,b]

Chlorophylls *a* (A_1–A_6)

	Number of gaussian band					
Parameter	A-1	A-2	A-3	A-4	A-5	A-6
-1-	*-2-*	*-3-*	*-4-*	*-5-*	*-6-*	*-7-*
$\lambda_{max,i}$	381	420	437	630	675	700
σ_i	33.8	8.25	6.50	89.8	8.55	101
$a^*_{max,i}$	0.0333	0.0268	0.0580	0.0005	0.0204	0.005

Chlorophylls *b* (B_1–B_6)

	Number of gaussian band					
Parameter	B-1	B-2	B-3	B-4	B-5	B-6
-1-	*-2-*	*-3-*	*-4-*	*-5-*	*-6-*	*-7-*
$\lambda_{max,i}$	380	442	452	470	609	655
σ_i	194	7.45	5.6	10.5	16.0	18.5
$a^*_{max,i}$	0.0059	0.0145	0.0631	0.0514	0.0083	0.0257

Chlorophylls *c* (C_1–C_4)

	Number of gaussian band			
Parameter	C-1	C-2	C-3	C-4
-1-	*-2-*	*-3-*	*-4*	*-5-*
$\lambda_{max,i}$	408	432	460	583
σ_i	16.1	7.93	14.2	32.2
$a^*_{max,i}$	0.0561	0.0234	0.072	0.0133

Photosynthetic Carotenoids (PSC-1, PSC-2) and Photoprotecting Carotenoids (PPC-1–PPC-3)

	Number of gaussian band				
Parameter	PSC-1	PSC-2	PPC-1	PPC-2	PPC-3
-1-	*-2-*	*-3-*	*-4-*	*-5-*	*-6-*
$\lambda_{max,i}$	490	532	451	464	493
σ_i	17.1	22.8	32.0	8.60	12.0
$a^*_{max,i}$	0.0313	0.0194	0.0632	0.0253	0.0464

[a] See Equation (6.3).
[b] After the authors' model (Woźniak et al. 1999, 2003).

differences may be the result of errors in the method of determining the absolute values of the absorption coefficients; they could also be the consequence of the diverse optical properties of natural pigment associations (e.g., a different content of the native forms of the same pigments) on which those descriptions were based. It is hard to determine which of these three descriptions of the optical properties of phytoplankton pigments most closely reflects reality.

The light absorption spectra of the separate groups of pigments in vivo, as shown in Figure 6.11, are in agreement with the aforementioned regularities ensuing from the chemical structures of these pigments, or, to put it more precisely, from the chromophores they contain. The chromophores are those parts of the pigment molecules where, over short distances, there are a large number of double bonds involving π-electrons and, possibly, additional n-type electrons (we discussed this question in detail in Section 3.2). Such situations give rise to conjugations of π-electrons or a mixture of π and n-electrons; these conjugations, in turn, yield relatively low energy π-electrons that can interact with quanta of visible light. In consequence, photons are absorbed ($\pi \rightarrow \pi^*$ or $n \rightarrow \pi^*$ transitions) or emitted ($\pi^* \rightarrow \pi$ or $\pi^* \rightarrow n$ transitions). Hence, the more double bonds there are over a short distance, the greater is the multiplicity of the π-electron conjugations, and the absorption bands are shifted towards the longer wavelengths. This is confirmed by the spectra of absorption coefficients shown in Figure 6.11 (also Figure 6.10).

The highest degree of conjugation in the plant pigments we are discussing here is found in the chromophoric group of chlorophyll *a*: this is the porphyrin arrangement and its immediate surroundings (Figure 6.7). Here we have some 16 conjugated π and n electrons. This is why chlorophyll *a* also absorbs red light, with a peak around 675 nm. This corresponds to an electron transition in the molecule from the ground singlet state to the first excited state. Phycobilins have 14 or 15 π and n electrons (Figure 6.9) and absorb light from the middle of the visible spectrum, peaking mostly in the 500–600 nm region, and in extreme cases from c. 490 nm (some native forms of phycoerythrin) to c. 660 nm (some native forms of allophycocyanin). In carotenoid molecules (Figure 6.8) from 11 or 12 π electrons (carotenes) to 13 or 14 π and n electrons (xanthophylls) can be involved in conjugation. Of all the plant pigments, the carotenoids have the lowest multiplicity of electron conjugations and absorb light from the shorter-wave end of the visible spectrum, with peaks usually in the 480–500 nm region (less often 530 nm), and some also in the 450–460 nm region.

One more thing has to be said about the chlorophylls: in addition to their ability to absorb red light, which is due to the high degree of π-electron conjugation in the molecules, they can also absorb blue light in the Soret band (c. 440 nm). This is the effect of the transition of the molecules from the ground singlet state to the second principal excited singlet state (i.e., at a level higher than the first excited state). In actual fact, there are more such bands for chlorophylls; this was demonstrated in Section 3.2 (Figures 3.11–3.13). The results of the quantum-mechanical calculations set out there show that the photosynthetic pigments from the chlorophyll group are characterized by a substantial number of excited singlet and triplet states. As a result, they are capable of absorbing visible light in some seven to nine bands and not just in two. The results of the Gaussian analysis we performed (Table 6.8) also indicate quite a large number of these bands. Even so, light absorption by chlorophylls in these other bands is usually much weaker than that in the two main bands, that is, the red (675 nm) and the Soret band (440 nm).

Having said that, however, the existence of these other bands, associated with the large number of possible singlet states of the pigment molecules, enhances the ability of plants to absorb energy from different regions of the visible spectrum and also facilitates the migration of absorbed energy from the pigment antennas to the photosynthetic centers. A phenomenon even more significant than this large number of singlet states, which also augments the ability of plants to absorb visible light from different spectral regions and the migration of excitation energy to the photosynthetic centers, is the existence of native forms of pigments, a subject we have already mentioned. The existence of these forms makes it possible, as a result of an electron transition between the same two singlet states, for chemically indistinguishable molecules of the same pigment to absorb or emit light of slightly different wavelengths. What we have here is the splitting of the electronic absorption or emission bands into their elementary, spectrally differentiated, component substructures. We now proceed to analyze this phenomenon using the example of the bands due to the absorption and emission of red light by chlorophylls.

6.2.5 The Native Forms of Chlorophyll Pigments

The structural complexity of the absorption and fluorescence spectra of the photosynthetic apparatus of plants is due, among other things, to the existence of so-called native forms, that is, various natural forms of chemically identical pigment molecules. These forms differ slightly in their optical properties (Shlyk 1965, 1974; Godnev 1963). They come into being as a result of intermolecular interactions in the chloroplasts, principally between the various pigment molecules, but also between these molecules and proteins. These interactions cause the molecules to be bonded into supramolecular structures. The intermolecular forces of these bonds, and hence their effects on the chromophores, vary in different parts of the chloroplast, in which a variety of molecular complexes—aggregates, dimers, oligomers, or pigment polymers—are formed. This is exemplified in Figure 6.12, which shows empirical spectra of fluorescence and its excitation[6] for *Chlorella* algae at a low tem-

[6] As in the case of absorption, the fluorescence of pigments in the natural state has a complicated structure owing to the various native forms of the pigments. Hence, studies of the emission phenomena of various substances are very helpful for assessing their absorption capabilities (see, e.g., Mashke and Haardt (1987), Ostrowska and Woźniak (1991a,b), Woźniak and Ostrowska (1991), and Ostrowska (2001)). Thus, the fluorescence excitation spectra of the various pigments and also of the entire photosynthetic apparatus in vivo, are, as in the case of all other substances that fluoresce, practically identical with the radiation absorption spectra. On the other hand, the fluorescence spectra of the separate pigments, for example, in solution, are shifted, usually by about 4–8 nm towards the longer wavelengths in comparison with the absorption bands corresponding to the transitions between these same energy states of the molecule, only in the opposite direction (e.g., absorption $S_0 \rightarrow S_1^*$ an emission $S_1^* \rightarrow S_0$). Moreover, in accordance with Levshin's rule, the structure of the fluorescence bands is usually a mirror image of that of the absorption bands.

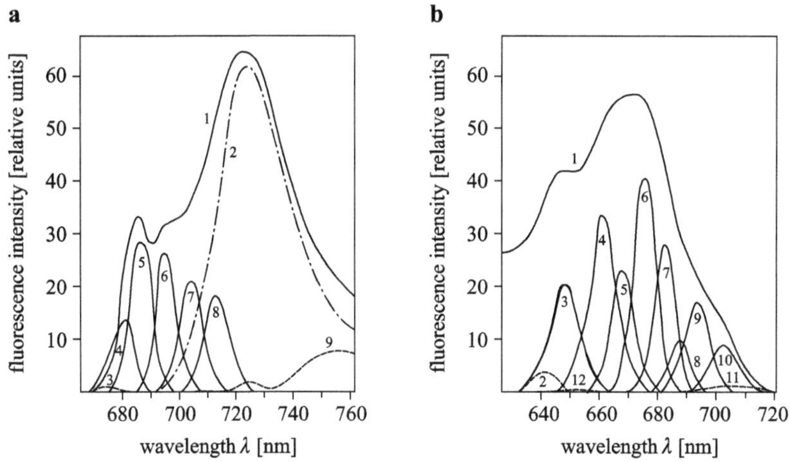

FIGURE 6.12. Fluorescence spectra (a) and fluorescence excitation spectra (b) of *Chlorella* at –196°C. Analysis of the spectra into a number of Gaussian curves. (a): 1, fluorescence spectrum, λ_{exc} = 640 nm; 2, broad band, λ_{exc} = 705 nm; 3–8, narrow bands; 9, curve 1 minus curves 2–8; (b): 1, excitation spectrum λ_m = 713 nm; 2–11, Gaussian curves; 12, curve 1 minus curves 2–11. (Reproduced from Litvin et al. (1974), *Systems of the native forms of chlorophyll, its role in primary processes of photosynthesis and development in process of the plant leafs greening* in: *Chlorophyll*, A. A. Shlyk (ed.) (in Russian), with the kind permission of Belaruskaya Nauka, Minsk.)

perature (–196°C), as well as its resolution into the component bands derived from the separate native forms of the pigments (after Litvin et al. (1974)). The two summary spectra shown in Figure 6.12 (fluorescence and its excitation) refer to the red region of the light spectrum. The elementary component bands therefore refer to the various native forms of chlorophyll that are optically active in this region of the spectrum. Research has shown (Litvin et al. 1974, Zvalinsky 1986) that other photosynthetic pigments, too, can occur in various native forms with different optical properties. Nonetheless, there are usually fewer such natural optical variants of these pigments than of the chlorophylls.

Litvin and his team discovered other important optical properties of these native forms of chlorophylls (Litvin et al. 1974, Litvin and Sineshchekov 1975). They analyzed the elementary components of the light absorption spectra of more than 30 plant species and also the optical spectra of isolated chlorophyll in different media (solutions, monolayers, emulsions, etc.). They found that the optical properties of the native forms of chlorophyll are universal in character: the positions of the elementary absorption and fluorescence band maxima are not arbitrary; rather, they are the same for all the plants, and are also the same as the peak positions for chlorophyll in different media. Likewise, the half-widths of the repeating elementary bands are the same.

The observed differences in the summary absorption and fluorescence spectra of the photosynthetic apparatuses in different plants are due to the varying participation of the particular elementary bands. This, in turn, is dependent on the plant species in question and the conditions in which it is growing.

The native forms of chlorophylls *a* and *b* identified by Litvin and his co-workers, together with the positions of their characteristic absorption and fluorescence bands, are listed in Table 6.9. As we can see, there are two native forms of chlorophyll *b* found in nature. These are represented in the spectra by narrow absorption and fluorescence bands, which at low temperatures (–196°C) achieve half-widths of c. 8 nm. The Stokes shift between the absorption and fluorescence peaks is c. 5 nm.

TABLE 6.9. Spectral characteristics of the native forms of chlorophyll and its aggregates.

	In a cell		In a model system		
Chl form[a]	Half bandwidth [nm] (at –196°C)	Relative area of absorption band (for *phaseolus*)	Chl form[a]	Half bandwidth [nm] (at –196°C)	Conditions (at –196°C)
-1-	*-2-*	*-3-*	*-4-*	*-5-*	*-6-*
		Chlorophyll *b*: Narrow Bands			
$Chl_{(640)}^{(645)}$	9	–	–	–	–
$Chl_{648}^{(653)}$	8	0.19	–	–	–
		Chlorophyll *a*: Narrow Bands			
Chl_{662}^{668}	10	0.18	Chl_{663}^{669}	10	Hexane $10^{-6\text{-}3}$M
Chl_{668}^{674}	8	0.15	Chl_{670}^{676}	8–9	Hexane film
Chl_{676}^{680}	8	0.29	Chl_{677}^{681}	9	Hexane 10^{-5}M
Chl_{682}^{686}	8	0.10	Chl_{682}^{687}	14	Hexane 10^{-5}M (20°C)
Chl_{687}^{692}	7.5	0.04	$Chl_{(686)}^{691}$	9	Hexane 10^{-4} M film
Chl_{693}^{697}	7.5	0.02	$Chl_{690}^{695-696}$	9–11	Hexane cyclohexane
Chl_{698}^{702}	7	–	Chl_{698}^{702}	7–8	Hexane 10^{-5}M
$Chl_{703}^{706-707}$	8–10	0.02	Chl_{702}^{706}	Shoulder	Hexane 10^{-3}M 10^{-2}M
Chl_{707}^{712}	10	0.02	Chl^{712}	Shoulder	Hexane film
$Chl_{(712)}^{718}$	10–12	0.01	$Chl_{715}^{(720)}$	16	Hexane 10^{-2}M (20°C)
$Chl_{(738)}^{743}$	9–11	0.01	Chl^{740}	11–13	Hexane 10^{-5} M
$Chl_{(755)}^{760}$	11–12	0.01	$Chl_{(755)}^{760}$	10–12	Octane 10^{-2} M
		Chlorophyll *a*: Wide Bands			
$Chl_{(702)}^{722}$	29	?	$Chl_{(700)}^{720}$	22–24	Hexane 10^{-5}M
Chl_{704}^{726}	30	?	$Chl_{700-705}^{726}$	27–30	Octane 10^{-3} M
$Chl_{(706)}^{732}$	32	?	Chl_{705}^{730}	30	Hexane 10^{-3} M
Chl_{710}^{738}	36	?	Chl_{710}^{740}	38–35	Film on glass

Reproduced from Litvin et al. (1974), *Systems of the native forms of chlorophyll, its role in primary processes of photosynthesis and development in process of the plant leafs greening*) in: *Chlorophyll*, A. A. Shlyk (ed.), (in Russian), with the kind permission of Belaruskaya Nauka, Minsk.

[a] Positions of maximum fluorescence (λ[nm] – the upper number) and maximum absorbance (λ[nm] – the lower number).

In contrast, the structure of chlorophyll *a* is much more complicated. In nature, it can occur in many forms, sixteen of which are detailed in Table 6.9. Twelve of these are characterized by narrow absorption and fluorescence bands with half-widths (at low temperatures) of the order of 7–12 nm and Stokes shifts of c. 4–6 nm. In addition, there are four other, wideband forms of this pigment: their absorption and fluorescence bands have half-widths of c. 30 nm and more, and Stokes shifts between 16 and 26 nm.

Another important regularity of the native forms of chlorophyll *a* (and also of bacteriochlorophyll *a*) discovered by Litvin and his team (Litvin et al. 1974, Litvin and Sineshchekov 1975) is the rule describing the positions of the absorption peaks of its narrow-band forms. There is, as it turns out, a linear dependence between the wave numbers of these peaks, which takes the form:

$$\nu^*(N) = \nu_0^* - (\mathrm{N} - 1)\Delta\nu^*, \tag{6.4}$$

where N is the form number arranged in the order of increasing wavelength (i.e., for "662" Chl – N = 1; for "668" Chl – N = 2; for "676" Chl – N = 3; etc.); $\nu_0^* = 15079\ \mathrm{cm}^{-1}$ is the absorption maximum of the chlorophyll form with the longest wavelength maximum; and $\Delta\nu^* = 137\ \mathrm{cm}^{-1}$ is the constant difference between the wave numbers of the successive basic bands.

The existence of these native forms of chlorophyll and the associated rich energy structure of the pigments strongly influences the transfer of excitation energy at the photosynthetic centers, and hence, the functioning of the entire photosynthetic apparatus. This question is discussed in detail in Litvin et al. (1974), Litvin and Sineshchekov (1975), and Ort and Govindjee (1987); see also Woźniak and Ostrowska (1991).

6.3 Phytoplankton Resources and Chlorophyll *a* Concentrations in Oceans and Seas

Apart from the individual absorption properties of plant pigments, the factor determining the absolute light absorption coefficients of marine algae are the resources of phytoplankton. A useful measure of these resources is the concentration in the water of the principal plant pigment: chlorophyll *a* (together with the phaeophytins). This concentration is denoted by C_a, and its value in the surface layer of the sea, denoted by $C_a(0)$, is treated as the trophic index (or trophicity) of a sea. We discussed the significance of this index in Section 6.1.

The chlorophyll *a* concentrations recorded in biologically active, surface water layers in different oceanic regions (Figure 6.13) and seas (e.g., the Baltic) can vary over c. four orders of magnitude. This becomes clear from the maps and graphs in this section. Chlorophyll concentrations C_a range from c. 0.02 mg tot. chl *a* m^{-3} and less in the central oceanic regions to c. 100 mg tot. chl *a* m^{-3} and more in the bays of enclosed seas. These concentrations

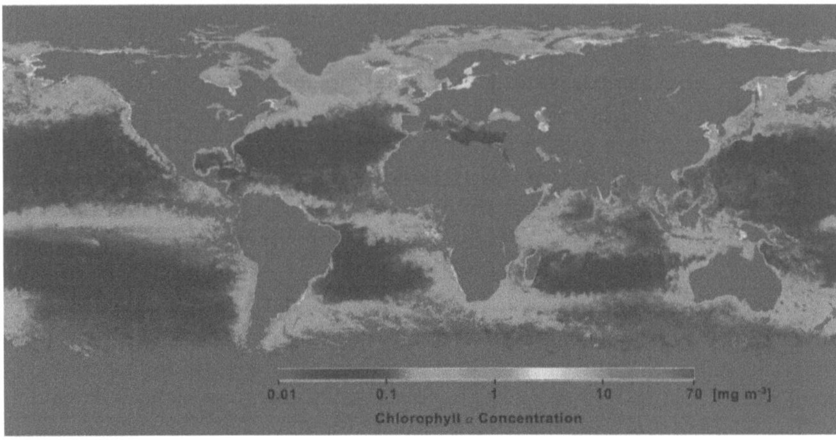

FIGURE 6.13. Map of the monthly average chlorophyll *a* concentration for August 2005 calculated from MODIS Aqua data. (Data supplied by NASA; image courtesy of M. Darecki, IO PAS, Sopot.) (See Colour Plate 8)

also vary strongly with depth in the sea. Because a detailed analysis of this diversity in seas and oceans would go beyond the scope of the present volume, we refer the interested reader to the following publications: Krey and Babenerd (1976), Mordasova (1976), Karabashev (1987), Lewis et al. (1983), Morel and Berthon (1989), Woźniak and Ostrowska (1990b), and Vinogradov et al. (1992). Below we present only the most important natural factors governing the magnitude of marine phytoplankton resources, as well as some of the regularities in the occurrence of chlorophyll in different regions of the World Ocean and at different depths. We also outline the analytical description of the vertical distributions of chlorophyll concentrations that are essential to modeling the spatial distributions of the light absorption coefficients of algae in seas of different trophicity.

6.3.1 The Principal Natural Factors Governing Phytoplankton Resources in the World Ocean

The major factors governing the resources of marine phytoplankton (including chlorophyll *a* and other pigments absorbing visible light) are the irradiance conditions, concentrations of nutrients (organic compounds of nitrogen, phosphorus, and other elements), the water temperature, and the density stratification of the water masses resulting from their thermal and dynamic states (Steeman Nielsen 1975, Bougis 1976, Woźniak and Ostrowska 1990b, Woźniak 1990a, Dera 1995). The diversity of these factors in the vast spaces of the ocean and their variability in time strongly influences the temporal and spatial diversity of the natural resources of phytoplankton in the various parts of the World Ocean. The most productive areas, that is, where the largest resources of phytoplankton are available, are the ones where "good" light conditions combine with a suitably high level of nutrients.

In the ocean, the upper illuminated layer of water—the euphotic zone—receives its supply of nutrients mostly from deeper waters. Given the right conditions, such as the right wind action, divergence of ocean currents, and other hydrodynamic factors, currents rising to the surface are formed; this phenomenon is known as "upwelling" (Druet 1994, 2000; Massel 1999; Ramming and Kowalik 1980; Summerhayes et al. 1995; Tomczak and Stuart 1995; Mordasova 1976). These rising currents carry nutrient-rich waters up to the surface. The nutrients are formed in deep waters when organic matter derived from the metabolism and the decay of marine organisms sinks towards the bottom. Wherever the hydrodynamics of the waters promote upwelling, the associated euphotic zone is extremely fertile. So where upwelling is a permanent feature, as it is off the west coasts of the continents, the waters are eutrophic. In contrast, where upwelling does not occur, the waters are practically unproductive (oligotrophic) and consequently contain very little chlorophyll (Table 6.1).

The influence of the stratification of waters on primary production and the chlorophyll concentration varies according to climatic zone and time of year. So, in cool temperate waters, for example, where seasonal temperature changes are usual, both the appearance and the disappearance of stratification always exert a positive influence on phytoplankton growth (Semina 1957, 1985). In summer, when the water is subjected to intense heating, the resulting clear-cut stratification to some extent prevents phytoplankton cells, both living and dead, from sinking beyond the limits of the euphotic zone. Hence, the living phytoplankton and the nutrients (decay products of dead cells) remain within the euphotic zone. And with stratification promoting phytoplankton growth in this way, pigment concentrations will be high. In winter, the water cools, and stratification is reduced or disappears altogether. But now conditions favor convection and wind-mixing, and hence upwelling, which carries nutrients up to the surface from the deeper water layers.

The reverse situation obtains in the tropical zones of the oceans: here the temperature hardly varies from one season to the next, so the stratification of the water masses remains relatively stable. This prevents the vertical mixing of waters, and hence the formation of upwelling currents transporting nutrients up from abyssal waters. So, even though they are well illuminated, the tropical regions of the oceans are for the most part the least productive areas of the World Ocean; highly oligotrophic, they are sometimes referred to as "marine deserts."

The conditions regulating the productivity of sea waters apply in principle to the open waters of oceans. In semi-enclosed oceans (bays) and in enclosed seas such as the Baltic or the Black Sea, it is usually the local conditions, highly differentiated in time and space, that exert the major influence on phytoplankton growth and chlorophyll concentration. It is therefore not possible to describe the conditions governing their productivity with the aid of a single universal model. In general, however, these seas are mostly eutrophic, principally because they receive vast loads of nutrients flowing in with river waters. In shallow seas, moreover, it is often the case that the strong mixing of surface waters with nutrient-rich, near-bottom waters also contributes to the overall productivity of their waters.

The conditions controlling the growth of phytoplankton and the concentrations of its pigments in the water are embraced quantitatively (Section 6.1) by the relationship among the chlorophyll *a* concentration, the nutrient content, and the water temperature in the surface layer of the sea (Figure 6.1 and Equation (6.2)). It is these conditions that determine the horizontal and vertical distributions of the chlorophyll *a* concentration in the World Ocean. Because the chlorophyll *a* concentration C_a is used as an index in the optical classification of sea waters, which appears in the formulas describing their optical properties, we now proceed to outline these conditions.

6.3.2 The Distribution of Chlorophyll in the World Ocean

For many decades now, attempts have been made to estimate the resources of phytoplankton in the World Ocean on the basis of measurements of chlorophyll *a* concentrations in the surface waters of the various ocean basins. The in situ studies of these concentrations undertaken by a number of authors have yielded chlorophyll distribution maps for selected seasons and ocean basins, separately for the Atlantic, Pacific, and Indian Oceans (e.g., Mordasova (1974a,b), Krey (1971, 1973), and Krey and Babenerd (1976)). Mordasova (1976) undertook to summarize these results for almost the entire World Ocean. However, despite the thousands of empirical data employed in the analyses, these maps are only very approximate, because of the randomness and sparse distribution of the measurement points in time and space.

Unfortunately, they do not meet present-day demands to define the phytoplankton biomass resources and the flux balances in the oceanic carbon cycle for the purposes, say, of modeling and forecasting climatic changes. Significant, although still insufficient, progress towards a solution of this problem has been made in the last 30 years as a result of the advances in optical remote-sensing methods and the systematic recording of chlorophyll concentrations in oceans and seas by satellite imagery (Barale and Schlittenhardt 1993, McClain et al. 2004, Kręźel et al. 2005, Woźniak et al. 2004). The first research mission in this direction was completed with the aid of the satellite-mounted Coastal Zone Color Scanner (CZCS) in 1978–1986 (Barale and Schlittenhardt, 1993). The second such mission with common accessible data has been in progress since September 1997 (and is on-going as we write this book in 2005) with the Sea-viewing Wide Field-of-View Sensor (SeaWiFS; see McClain et al. (2004)).

Thanks to these missions and their successors, maps are now available illustrating the instantaneous distribution of surface chlorophyll *a* concentrations over the entire area of the World Ocean or of the distributions of these concentrations averaged over longer periods of time, for example, a month, a year, or the last few years. Unfortunately, the accuracy of this particular remote-sensing technique for determining chlorophyll concentrations in the sea is still unsatisfactory. As far as the open ocean is concerned, the statistical error is in excess of 30%, and with respect to bay regions, coastal and enclosed seas (mostly basins with Case 2 waters), this error is usually much greater.[7] Figure 6.13 shows such a chlorophyll concentration map, based on SeaWiFS data, and averaged over the whole of the year 2003. From this map we can infer the following regularities, characterizing the chlorophyll distribution in the surface layers of the ocean:

1. Vast expanses of the oceans consist of waters very poor in phytoplankton (i.e., they are highly oligotrophic); as already mentioned, these areas are sometimes called marine deserts. They occur in tropical and subtropical regions in the areas of downwelling at the centers of the northern and southern anticyclonic gyres. There, concentrations of chlorophyll *a* drop to values of $C_a(0) < 0.25$ mg tot. chl *a* m^{-3} in the Atlantic and Pacific; in the Indian Ocean they can be lower still ($C_a(0) < 0.05$ mg tot. chl *a* m^{-3}).
2. It is rare to find open ocean waters rich in chlorophyll, for example, where $C_a(0) > 1$ mg tot. chl *a* m^{-3} (i.e., eutrophic and intermediate between meso- and eutrophic); where they do occur, they cover only small areas. In the low latitudes and temperate zones, these more fertile ocean waters occur only in continental shelf regions. These are richer in nutrients as a result of river run-off or upwelling, the latter especially off the western coasts of continents. These more fertile ocean basins, with surface chlorophyll concentrations $C_a(0) > 1$ mg tot. chl *a* m^{-3}, occur more often in the cold climatic zones ($\varphi \sim 50°–70°$), particularly in the Arctic and Antarctic, where in summer, the insolation is intense and the local hydrodynamics promote the supply of nutrients to the euphotic zone.
3. The fertility of ocean waters in the temperate latitudes of the northern hemisphere is intermediate between meso- and eutrophic, with chlorophyll contents of $C_a(0) \sim 0.5–1$ mg tot. chl *a* $m^{-3.}$
4. The relatively small circumequatorial regions are usually mesotrophic, with values of $C_a(0)$ around 0.25 mg tot. chl *a* m^{-3}.
5. In shelf seas, especially enclosed and semi-enclosed ones, chlorophyll concentrations display considerable temporal and spatial variation. Usually much richer in phytoplankton than the oceans, most of these basins can be classified as intermediate between meso- and eutrophic: for instance, in the central Baltic $C_a(0)$ ranges from 0.5 to 1.2 mg tot. chl *a* m^{-3}, and in much of the Black Sea from 0.5 to 0.8 mg tot. chl *a* m^{-3}. In littoral areas, the waters of these seas are usually eutrophic with chlorophyll concentrations

[7] The problem of the accuracy of satellite algorithms for determining chlorophyll concentrations in Case 1 and Case 2 waters is analyzed in detail by Sathyendranath (2000), Ruddick et al. (2000), and Darecki et al. (2003), among others.

reaching, in extreme cases, c. 10 and even 20–50 mg tot. chl *a* m^{-3}. Occasionally, even higher concentrations have been recorded.[8]

The regularities outlined above regarding the occurrence of chlorophyll in the World Ocean are based on analyses of the mean annual concentrations of this pigment. They therefore take no account of these large seasonal amplitudes during the year and other temporal variabilities of chlorophyll concentration. As much of the earlier in situ research (e.g., Mordasova (1974a, b, 1976); Bogorov (1974); see also Woźniak and Ostrowska (1990b)) and also the present-day satellite data have shown, the changes in chlorophyll concentration with time also display a number of regularities.

First and foremost, they are dependent on the geographical zone. It has now become possible to analyze these regularities in minute detail on the basis of this very satellite data. Because, however, such an analysis would exceed the scope of this book, we just illustrate the point with a simple sketch diagram of the seasonal variations in phytoplankton content (and hence of chlorophyll) in the oceans in different climatic zones (Figure 6.14). This depicts the most important trends in basins in different latitudes. The time axis covers two consecutive years and the scale is divided into months. However, the phytoplankton (or chlorophyll) concentration axes, that is, the vertical distances on the individual plots, merely express the trends and the place of the maxima in time: they are uncalibrated, approximate, and are expressed in units that are not comparable. Nevertheless, they do demonstrate a number of interesting regularities.

During the year there are two phytoplankton blooms (i.e., increases in chlorophyll concentration): in autumn (August, September) and in spring (March). The spring maximum is usually higher than the autumn one. Between these two blooms there is an interval of about half a year. Towards higher latitudes, the spring bloom occurs later and the autumn one earlier. These bloom maxima in the polar zone thus overlap in time and a single bloom is observed during the polar summer, peaking in June. This effect is a natural consequence of the astronomical factors (the inclination of the Earth's axis and its orbit around the Sun) that govern the daily cycles and the seasons.

Moving towards the Equator from the temperate zones, however, these trends are reversed: the spring bloom happens earlier, and the autumn one later. Thus, in the tropical zone, the bloom peaks also coincide to give a single bloom, but this time in winter. The apparent paradox that the numbers of plankton are very much larger in winter than in summer can probably be explained by changes in the activity of the entire ecosystem, which are manifested, for example, in the faster rate of phytoplankton grazing by zooplankton in summer than in winter, and also the quicker exhaustion of nutrient supplies.

[8] For example, during the international SOZOPOL-1986 experiment held in the Gulf of Burgas (Black Sea) in spring–summer, the chlorophyll *a* concentration reached 0.5 and more kg tot. chl *a* m^{-3} (unpublished data of an international research team, which included the authors of the present monograph).

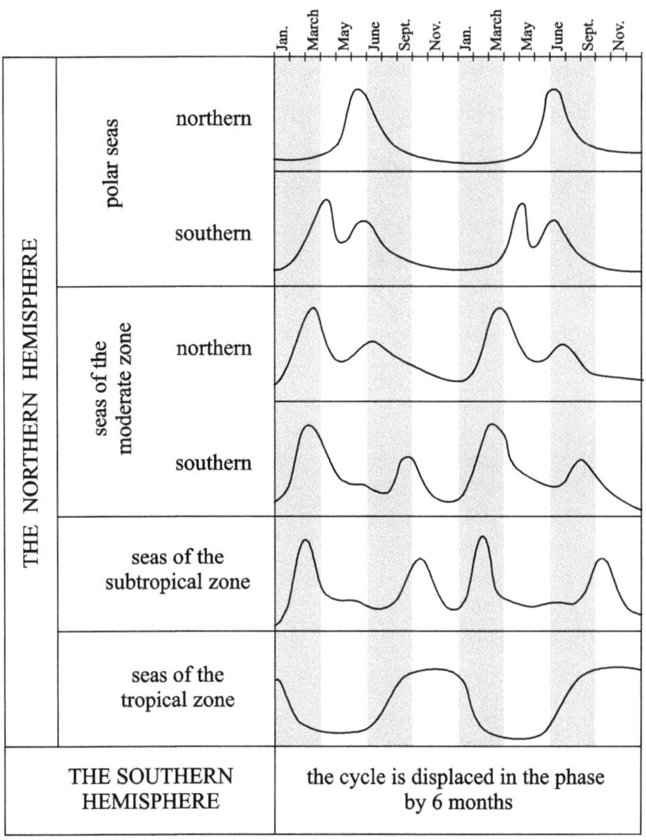

FIGURE 6.14. Outline of seasonal changes in phytoplankton content for different climatic zones of the World Ocean. (Adapted from Bogorov (1974).)

Moreover, the seasonal amplitudes in the quantities of phytoplankton differ markedly in various latitudes (Figure 6.14 does not show this). At the Equator and in the tropical zone, these amplitudes are minimal, but rise steeply with increasing latitude. This can be explained primarily by the annual cycle of temperature and insolation of the sea. During the year, these values are practically constant at the Equator, but highly diverse in higher latitudes.

6.3.3 *Vertical Distributions of Chlorophyll* a *in the Sea*

The horizontal differentiation in chlorophyll *a* concentration in the World Ocean is further complicated by its vertical distribution, inasmuch as the concentrations of chlorophyll and all the other phytoplankton pigments vary greatly with depth in the sea. This can be seen in Figure 6.15a, which illustrates empirical profiles of the vertical chlorophyll *a* concentrations $C_a(z)$ recorded in seas of different trophicity. In actual fact, the $C_a(z)$ distributions

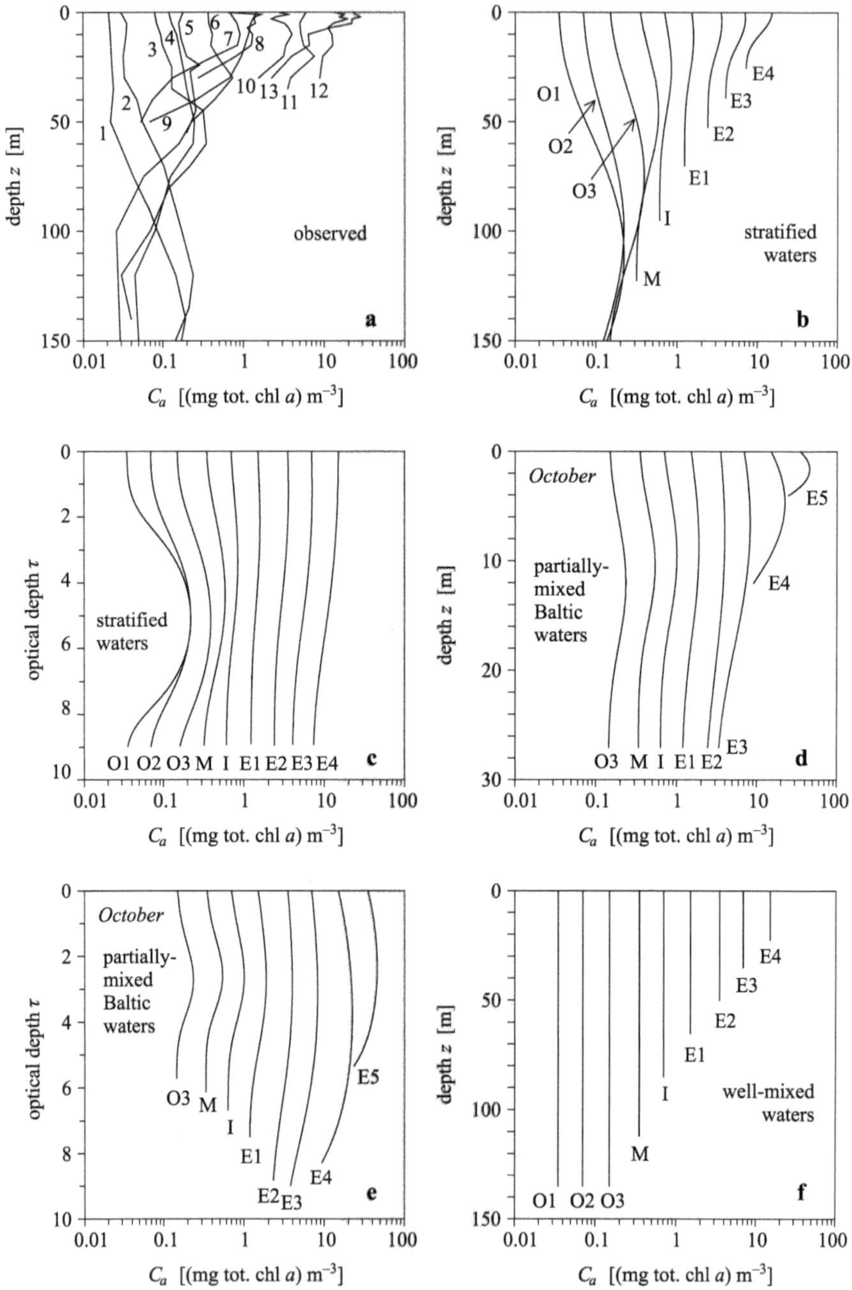

FIGURE 6.15. Vertical distributions of chlorophyll *a* concentrations C_a in different oceans and seas: (a) empirical profiles: 1–3, Indian Ocean; 4–6, Atlantic Ocean; 7–9, Black Sea; 10–13, Baltic Sea; (b)–(f) model profiles of $C_a(z)$ in sea waters of different trophicity (according to formula (6.5) and Table 6.10): stratified waters (b),(c); partially mixed Baltic waters (d),(e); well-mixed waters (f). The trophicities of basins and their symbols shown on the plots (O1, . . . , E4) are defined in Table 6.1. (After Woźniak et al. (2003).)

are even finer in structure, something that is not shown in the figure. This is due to the microstructure of the water masses, which can be recorded only with continuous techniques of detecting chlorophyll in the vertical, such as fluorimetry (see, e.g., Karabashev (1987), Kolber and Falkowski (1993), Ostrowska (1990, 2001), and Ostrowska et al. (2000a,b)).

The diversity of the vertical distributions of chlorophyll concentrations $C_a(z)$ is the consequence of the multifarious environmental factors governing the phytoplankton growth in different regions and at different depths in seas and oceans. The influence of these factors on vertical chlorophyll distributions is extremely complex and has been examined by numerous authors. We now present a number of features characteristic of these vertical $C_a(z)$ profiles and indicate how they are governed by environmental factors.

1. The general outlines of vertical chlorophyll concentration profiles $C_a(z)$ (we ignore their microstructure) can vary a great deal. Having analyzed several thousand such profiles, recorded in situ in uncalibrated units with the aid of an immersion fluorimeter, Karabashev (1987) distinguished five main types (Figure 6.16). They may be quasi-homogeneous (plot 1), contain one single peak (plot 2), or have a more intricate structure (plots 3 and 4). The distributions are generally unimodal, with a single main chlorophyll concentration peak at a certain depth.

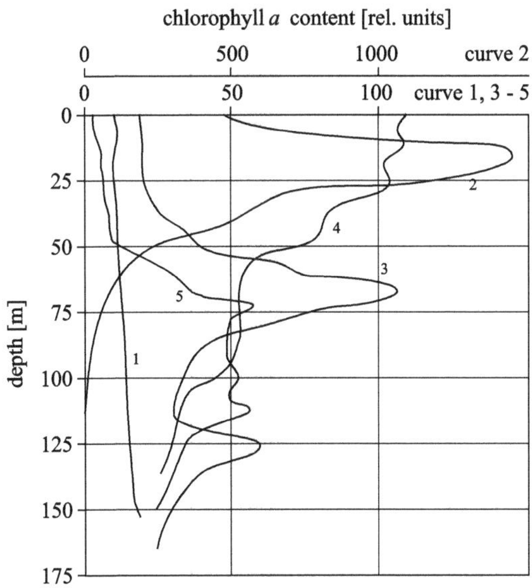

FIGURE 6.16. Types of vertical distribution of chlorophyll *a* concentrations in the World Ocean: 1, quasi-homogeneous; 2, unimodal, with one principal peak (maximum); 3, polymodal, with several well-developed peaks (maxima) of comparable amplitude; 4, with the maximum at the water surface; 5, complex. (Based on fluorescence measurements; adapted from Karabashev (1987).)

2. How clearly delineated this peak is at a certain depth depends largely on the seasonal variations in the environmental conditions and the stability of the water masses. A very intense maximum can appear during the spring phytoplankton bloom at the depth where the rate of photosynthesis is highest. Below this peak, the fall in chlorophyll *a* concentration may resemble the drop in the rate of photosynthesis with depth (Steele 1964). Nevertheless, such distributions are rarely observed, and then only at the height of the bloom. At other times, the maximum chlorophyll *a* concentration is less sharply defined or is absent altogether (Renk 1973, Woźniak and Ostrowska 1990b, Sathyendranath et al. 1989). The most distinct maxima are recorded in regions where or in seasons when the water masses are extremely stable (Parsons 1963, Yentsch 1965, Morel and Berthon 1989).
3. The depth at which the main peak lies depends on the transparency of the water and is the resultant of the rate of primary production, the rate of sinking of phytoplankton and its grazing by zooplankton, and the movements of the water masses. The main chlorophyll concentration peak usually lies below the maximum rate of photosynthesis, and like the latter, is recorded the deeper, the greater the transparency of the water.
4. In shallow seas such as the Baltic, the differences between the positions of the chlorophyll maxima and the fastest rate of photosynthesis are not great: from zero to a few meters. This is probably due to the phytoplankton cells sinking deeper into the water (Renk 1973). On the other hand, in oceans these distances can be much larger, from c. 10 m to 40–50 m. This suggests that other causes, in addition to the gravitational sinking of phytoplankton, are at work here.
5. The effect of the dynamics and thermal regimes of water masses on vertical chlorophyll profiles in the sea is a complicated one. In oceans, maximum chlorophyll concentrations often occur at depths between the seasonal pycnocline and the permanent halocline. However, in subtropical regions, where the halocline is replaced by a permanent thermocline (at depths of 140–160 m), convectional mixing of waters takes place throughout the year. Under such conditions, a fairly high, almost evenly distributed concentration of chlorophyll is maintained down to about 100 m (Mordasova 1976). But in the tropical zone, where the thermocline lies very deep, most of the phytoplankton is distributed fairly evenly throughout the surface layer of waters subject to wind-mixing (Semina 1957), that is, in the entire mixed layer (Niiler and Kraus 1977).
6. In addition to the main chlorophyll *a* concentration peak, it is possible in seas with a complex water mass structure such as the Baltic that there are more or less distinct second- and higher-order maxima, which will include the microstructure. They are usually strongly correlated with the vertical differentiation of temperature or salinity (Karabashev 1987).
7. The maximum thickness of layers in which chlorophyll *a* is detectable ranges in the oceans from c. 100 m in the Antarctic and sub-Antarctic zones to 200 m in the tropics (Mordasova 1976). In other seas, chlorophyll is usually detectable only to the limits of the thermocline (Renk 1973).

Summarizing the description of these factors, we can state that the overall trend in most naturally occurring vertical $C_a(z)$ profiles is the presence of one main concentration peak. The clarity, and the breadth and depth of occurrence in the sea of this maximum, depend on the trophicity of the basin $C_a(0)$ and on the degree of vertical stability of the water masses. This tendency is strongest in the most distinctly stratified waters, but it is much less strong, if it occurs at all, in mixed waters, and even less so in well-mixed waters. Stratified waters, mostly oligotrophic and mesotrophic, are found in the vast central regions of oceans, and coincide to a large extent with the basins containing the Case 1 waters distinguished by Morel and Prieur (1977; also see Chapter 1).

Mixed and well-mixed waters, in contrast, are found in dynamically active regions of oceans, that is, in the divergence and convergence zones of ocean currents, and in shelf and inland seas. The latter consist mostly of Case 2 waters. One should, however, always bear in mind that both these water classifications, the optical and the dynamic, are a matter of convention, only an approximation of the real state of affairs, and do not entirely overlap. So, for example, in regions of divergence and convergence the waters are mixed; even so, they are placed as a rule in the first optical category. And conversely, in eutrophic, semi-enclosed seas, a clearly defined stratification may develop during the phytoplankton blooms, and yet these basins are usually classified from the optical point of view as containing Case 2 waters: the Baltic Sea, for instance (Ficek 2001).

6.3.4 Statistical Formulas Describing the Vertical Distributions of Chlorophyll Concentration

In order to take account of the aforementioned regularities in the distributions $C_a(z)$, a number of authors worked out statistical formulas describing these distributions in different types of sea, for example, Lewis et al. (1983), Platt et al. (1988), Morel and Berthon (1989), and Sathyendranath et al. (1989), mainly for the needs of satellite remote-sensing techniques. The formulas express the concentration of chlorophyll *a* (including the phaeophytins) at any depth z in the sea $C_a(z)$ as a function of the surface concentration $C_a(0)$, which can be determined efficiently and for large areas by remote sensing. Below we present model formulas of this kind, the ones that we postulated in our earlier publications (Woźniak et al. 1992a,b, 1995b). They describe the relationships among the vertical distributions of concentration $C_a(z)$ at different depths in the sea and the surface concentration $C_a(0)$, expressed as the sum of the constant, depth-independent component C_{const} and the variable component described by the Gaussian function:

$$C_a(z) = C_a(0)\frac{C_{const} + C_m \exp\left\{-\left[(z - z_{max})\sigma_z\right]^2\right\}}{C_{const} + C_m \exp\left\{-(z_{max}\sigma_z)^2\right\}}. \qquad (6.5)$$

TABLE 6.10. Model descriptions of the vertical distributions of chlorophyll *a* concentrations in the sea.

A. Stratified Ocean Basins[a]

Input data $C_a(0)$ [mg tot. chl*a* m^{-3}], Surface concentration of chlorophyll *a* (chl a + pheo)
Basic expression for the vertical distribution of the chlorophyll *a* concentration $C_a(z)$ [mg·tot. *chl a* m^{-3}] in the water:

$$C_a(z) = C_a(0)\frac{C_{const} + C_m \exp\left\{-\left[(z - z_{max})\sigma_z\right]^2\right\}}{C_{const} + C_m \exp\left\{-(z_{max}\sigma_z)^2\right\}},$$

where

$C_{const} = 10^{[-0.0437 + 0.8644 \log C_a(0) - 0.0883 (\log C_a(0))^2]}$
$C_m = 0.269 + 0.245 \log C_a(0) + 1.51(\log C_a(0))^2 + 2.13(\log C_a(0))^3 + 0.814(\log C_a(0))^4$
$z_{max} = 17.9 - 44.6 \log C_a(0) + 38.1(\log C_a(0))^2 + 1.32(\log C_a(0))^3 - 10.7(\log C_a(0))^4$
$\sigma_z = 0.0408 + 0.0217\, C_a(0) + 0.00239(\log C_a(0))^2 + 0.00562(\log C_a(0))^3 + 0.00514(\log C_a(0))^4$.

B. Partially Mixed Baltic Basins[b]

Input data $C_a(0)$ [mg tot. chl*a* m^{-3}], Surface concentration of chlorophyll *a* (chl a + pheo), n_d – consecutive day number in the year
Basic expression for the vertical distribution of the chlorophyll *a* concentration $C_a(z)$ [mg tot. chl*a* m^{-3}] in the water:

$$C_a(z) = C_a(0)\frac{C_{const} + C_m \exp\left\{-\left[(z - z_{max})\sigma_z\right]^2\right\}}{C_{const} + C_m \exp\left\{-(z_{max}\sigma_z)^2\right\}},$$

where

$$C_{const} = \left[0.77 - 0.13\cos\left(2\pi\frac{n_d - 74}{365}\right)\right]^{C_a(0)}$$

$$C_m = \frac{1}{2M}\left[(0.36)^{C_a(0)} + 1\right] \times \left[M + 1 + (M - 1)\cos\left(2\pi\frac{n_d - 120}{365}\right)\right]$$

$M = 2.25\,(0.765)^{C_a(0)} + 1$
$z_{max} = 9.18 - 2.43 \log C_a(0) + 0.213(\log C_a(0))^2 - 1.18(\log C_a(0))^3$
$\sigma_z = 0.118 - 0.113 \log C_a(0) - 0.0139\,(\log C_a(0))^2 + 0.112(\log C_a(0))^3$.

[a] After Woźniak et al. (1992a,b).
[b] After Woźniak et al. (1995).

The various constants appearing in this expression are functions of the surface concentration $C_a(0)$. Table 6.10 sets out the forms of these functions, determined on the basis of the relevant statistical analyses for stratified oceanic waters (Table 6.10A) and for partially mixed Baltic Sea waters (Table 6.10B). In the case of the latter, the model description of $C_a(z)$ also takes account of the seasonally dependent degree of mixing of the waters (where the season is defined in the formulas by the consecutive day number in the year n_d).

We presented some plots of these model relationships earlier in Figures 6.15.b–e. The statistical errors of these models are depth-dependent. In the case of stratified oceanic waters, the error of our model (Formula 6.5) is small in the surface layer, but increases with depth: for example, it is about 13% at optical depth $\tau = 1$, c. 33.7% at $\tau = 4.6$, that is, a depth extending to the bottom of the euphotic zone, and 56.8% at $\tau = 6.9$. In the case of Baltic waters, the statistical error is slightly greater. The respective values are: c. 14.5 % at $\tau = 1$, c. 49% at $\tau = 2.3$, and c. 68% at $\tau = 4.6$.

Mixed waters are more complex, so it has not yet been possible to find general analytical expressions to define the distributions $C_a(z)$ in them. As they are depth-dependent to a much lesser degree than stratified waters, the constant value

$$C_a(z) = C_a(0) = \text{const} \tag{6.6}$$

is usually applied at the present stage of their modeling, which implies homogeneous distributions (Figure 6.15f). In further research this assumption will necessarily require adjustment.

6.4 The Composition of Chlorophyll *a* and Accessory Pigments in Marine Algae

As with the concentration of chlorophyll *a*, so too the concentrations of the accessory pigments in phytoplankton, which absorb light of various wavelengths, vary strongly in time and over the vast spaces of the oceans. The scale of this diversity is also similar, ranging as it does over some four orders of magnitude. Furthermore, the differentiation in the relative contents of accessory pigments (referred, e.g., to chlorophyll *a*) depends, among other things, on the trophicity of the water and depth in the sea. This is illustrated by the observed vertical distributions of the relative concentrations of the four main groups of accessory pigments: the photoprotecting carotenoids (PPC), the photosynthetic chlorophylls *b* and *c*, and the photosynthetic carotenoids (PSC; see further in this section Figure 6.22a,d,g,j). Highlighted in the figure, this diversity in the relative concentrations of accessory pigments at various depths and in waters of different trophicity points to their dependence on the irradiance of the plant cells. The mechanism of this dependence involves the photoadaptation and chromatic adaptation of the cells, as well as acclimation at the plant community level. This section covers the most important aspects of this question.

6.4.1 *Pigment Compositions Characteristic of Various Classes of Phytoplankton*

The pigment composition in different photosynthesizing plants varies a great deal. Nonetheless, certain sets of pigments are characteristic of particular plant subassemblages. Qualitatively, these sets have been studied for a good many years with respect to various taxonomic groups of marine phytoplankton (e.g., Lorentzen (1968), Goodwin and Mercer (1972), Parsons et al. (1977), Jeffrey (1980a, b), Lorentzen and Jeffrey (1980), Gieskes and Kraay (1983), Guillard et al. (1985), Goericke and Repeta (1992), and Claustre (1994)). A summary of much of this work can be found in Jeffrey et al. (1997). The overall results of this research are set out in Table 6.11, in which a three-point scale is used to identify the intensity of occurrence of pigments in the principal classes of marine phytoplankton.

TABLE 6.11. Pigments in the various classes of algae and cyanobacteria.

	Class										
Pigments	*Cyanobacteria*	*Rhodophyceae*	*Cryptophyceae*	*Chlorophyceae*	*Prasinophyceae*	*Euglenophyceae*	*Eustigmatophyceae*	*Bacillariophyceae*	*Dionophyceae*	*Prymnesiophyceae*	*Chrysophyceae*
-1-	*-2-*	*-3-*	*-4-*	*-5-*	*-6-*	*-7-*	*-8-*	*-9-*	*-10-*	*-11-*	*-12-*
Chlorophylls: *a*	1	1	1	1	1	1	1	1	1	1	1
b				1	1	1					
c1								1		1	1
c2			1					1	1	1	1
c3										1	
MgDVP					1						
Carotenes: α		2	2	3	3					3	
β	2			2	2	2	2	3	3	3	3
γ				3							
Lycopene			3								
Phycobilins: Allophycocyanin	1	1									
Phycocyanin	1	3	1								
Phycoerythrin	1	1	1								
Xanthophylls: Alloxanthin			1								
Anteraxanthin				3	3	3					
19′-Butanyloxy fucoxanthin										1	1
Diadinoxanthin						1		1	1	1	1
Diatoxanthin						3		3	3	3	3
Dinoxanthin									2		
Fucoxanthin								1		1	1
19′-Hexanyloxy-fucoxanthin										1	1
Lutein				1	2						
Monadoxanthin			3								
Neoxanthin				1	1	3					
Peridinin									1		
Prasinoxanthin					1						
Violaxanthin				1	1		1				
Zeaxanthin	1	1		2			3				

Adapted from Jeffrey and Vesk (1997).

A three-point scale is used to identify the intensity of occurrence of pigments:

1. The principal pigment, making up over 10% of the total carotenoid or chlorophyll content in a group.
2. Pigment present in smaller amounts (1 – 10%).
3. Pigment present in trace amounts (<1%).

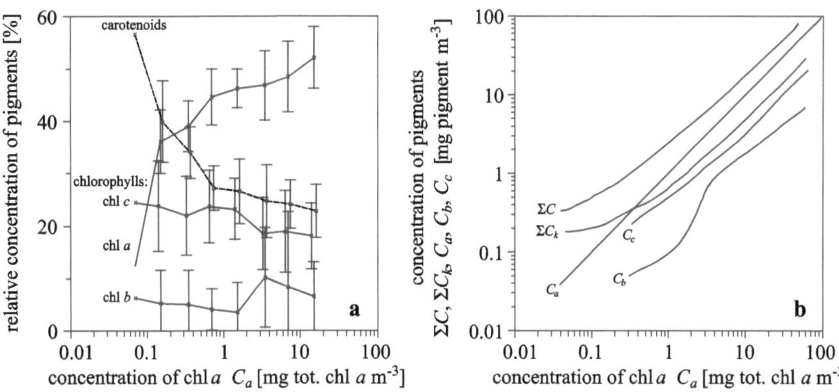

FIGURE 6.17. Occurrence of the various groups of pigments in natural marine plant communities. (a) Statistical profiles of the relative concentrations of the individual phytoplankton pigments in the total pigment concentration with respect to the chlorophyll *a* concentration, C_a. (b) The absolute concentration of pigments with respect to the chlorophyll *a* concentration: C_a, chlorophyll *a* concentration; C_b, chlorophyll *b* concentration; C_c, chlorophyll *c* concentration; ΣC_k, total concentration of all PSC and PPC carotenoids; ΣC, total concentration of all pigments except phycobilin. (After Woźniak and Ostrowska 1990b).)

In all the classes, chlorophyll *a* is the principal photosynthetic pigment; accessory pigments are present in various combinations. Thus, chlorophyll *a* always occurs together with larger or smaller amounts of PPC, both carotenes and xanthophylls, which are there to protect the cellular chlorophyll against photooxidation. However, the presence or absence of the other accessory antenna pigments—chlorophylls *b* and *c*, phycobilins and PSC—depends on the class of plants in question. Three types of photosynthetic apparatus can be distinguished in phytoplankton, not to mention higher plants (Parsons et al. 1977). In addition to PPC and chlorophyll *a*, the following pigments are present in these types as light energy antennas:

Type I: chlorophyll *c* and selected PSC.
Type II: chlorophyll *b*.
Type III: phycobilins.

The first type is the most common one in natural phytoplankton populations. Type II, on the other hand, is more typical of higher plants, although in marine phytoplankton, chlorophyll *b* does occur in three classes: the *Prasinophyceae*, the *Chlorophyceae*, and the *Euglenophyceae*. Type III is the least frequent, because cyanobacteria and two other classes in which phycobilins are present, the *Rhodophyceae* and the *Cryptophyceae*, are only occasional components of natural marine plant communities. Nevertheless, phycobilins are found much more often in the littoral zones of oceans and in semi-enclosed seas, especially in the lower layers of the euphotic zone, as our recent observations have shown.

The pigment composition in the photosynthetic apparatus can vary at the species level (limited adaptation) and at the community level (succession). In poor illumination, a species "goes for survival" by reducing its numbers, and its place in the ecological niche is taken by another species, whose existing pigment composition is optimal under the given light conditions.

6.4.2 Pigment Compositions in Natural Plant Communities, in Different Types of Sea and at Different Depths

Because Table 6.11 provides only a qualitative assessment of the pigment compositions of the various classes of phytoplankton, the information it contains cannot be used to make a quantitative prediction of the optical properties of natural phytoplankton populations in the sea, especially as these are represented by many different classes and species. Moreover, at the present time there are too few generalizations emerging from the relevant research data that could be used to formulate a universally applicable description of the occurrence of algal taxa in different regions of the sea, at different seasons, and at different depths. Attempts are being made, however, to find overall quantitative descriptions of the composition and concentrations of the various pigments (or at least to establish certain statistical regularities in such descriptions) in the diverse natural populations of algae in different oceanic regions and depths (e.g., Lazzara et al. (1996)). We now present the results of these attempts on the basis of our own research and that of our colleagues in the late 1980s and early 1990s (Woźniak 1990b, Woźniak and Ostrowska 1990b), as well as the work of Babin et al. (1996a).

Woźniak and Ostrowska (1990b) provide a statistical analysis of the relative and absolute relationships between the concentrations of the various groups of pigments (except phycobilin), found in the oceans and seas, as a function of chlorophyll *a*, that is, as a function of the trophicity of waters C_a (Section 6.1, Table 6.1). The pigment concentrations were determined using the spectrophotometric methods standard in oceanography at that time. The results of this work are shown in Figure 6.17: this shows clearly how the pigment composition changes with the trophicity. The proportion of pigments accessory to chlorophyll *a* rises on going from the biologically rich eutrophic waters to the poor waters of oligotrophic basins. Thus, in highly productive eutrophic basins the chlorophyll *a* concentration makes up almost 60% of the total pigment content, and the combined content of all the other pigments is slightly over 40%. In unproductive oligotrophic basins, these proportions are reversed: in highly oligotrophic waters the chlorophyll *a* concentration drops to almost 10%; in such waters, the accessory pigments are dominant, with an overall concentration as high as 90%.

Of the three groups of accessory pigments under discussion here (chl *b*, chl *c*, carotenoids), chlorophyll *b* is present in the smallest concentrations, usually <10%. Hence, the dominant accessory pigments are principally carotenoids and, to a lesser extent, chlorophylls *c*. The concentration of the former in oligotrophic seas is on average more than twice that of chlorophyll *a*, and in extreme cases can be more than four times as high. As Figure 6.17 shows, it is in eutrophic seas, that is, in conditions most favorable to photosynthesis, that the optimum pigment composition is established: c. 55% chl *a* and c. 45% accessory pigments. In less propitious conditions for phytoplankton growth, for example, in unproductive oligotrophic waters, the algae bring defensive mechanisms into play. According to a number of authors (Margalef, 1967, Koblentz-Mishke, 1971, Woźniak and Ostrowska, 1990b), the processes enabling algae to adapt to inferior conditions are reflected in the faster production of additional quantities of accessory pigments. This would explain the dominance of these over chlorophyll.

For a long time now, the pigment index $P_{i,extr}$ of acetone extracts of pigments has been used as a measure of the diversity of the pigment content of phytoplankton; it, too, was analyzed by Woźniak and Ostrowska (1990b). It is given by

$$P_{i,extr} = A_{433}/A_{661}, \quad (6.7)$$

which is the ratio of the extinction of light A_{433} at wavelength $\lambda = 433$ nm to the extinction of light A_{661} at wavelength $\lambda = 661$ nm, the extinctions being measured in acetone extracts of the phytoplankton. Practically all pigments except phycobilin absorb in the first of these bands (433 nm) to a greater or lesser extent. But in the 661 nm band, chlorophyll *a* is the principal absorber. Hence, the ratio of these two extinctions provides a certain "optical" measure of the ratio of the total pigment concentration to the chlorophyll *a* content. Figure 6.18 shows the statistical relationships between the pigment

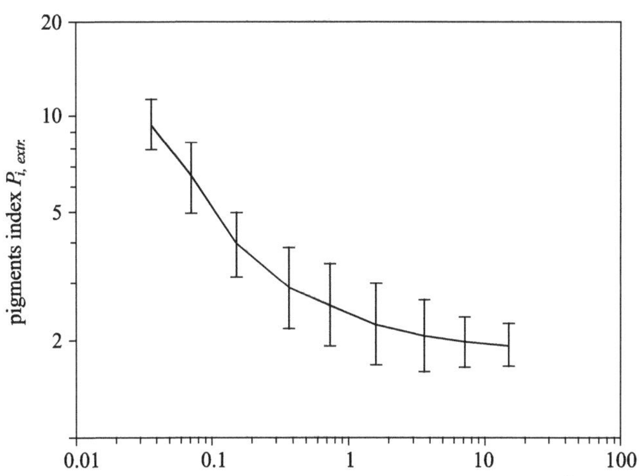

FIGURE 6.18. Statistical relationship between the pigment index of marine phytoplankton pigments $P_{i,extr}$, and the chlorophyll *a* concentration, C_a. The vertical intervals on the profiles correspond to the standard deviations of $P_{i,extr}$. (After Woźniak et al. (1992a).)

index $P_{i,extr}$ and the chlorophyll *a* concentration, obtained at different depths in various seas and oceans. The value of $P_{i,extr}$ declines markedly from unproductive oligotrophic seas to highly productive eutrophic waters. This is because of the very much larger proportions of accessory pigments in the photosynthetic apparatus in oligotrophic than in eutrophic communities. This confirms the results of the research into pigment compositions in various seas that we discussed earlier.

The regularities regarding pigment occurrence exhibit a wide scatter of experimental points in comparison to the averaged profiles (see the standard deviation intervals in Figure 6.18). Notice, however, that these profiles have been averaged from data sets obtained from various, unspecified depths in the sea. On the other hand, the dependences of the pigment index $P_{i,extr}$ on depth in these various seas (Figures 6.19 and 6.20, after Woźniak and Ostrowska (1990b)) reflect the diversity of pigment compositions at different depths. Certain statistical regularities of these profiles are discernible in Figures 6.19 and 6.20. It is usually the case that at a certain depth, the pigment index falls to a more or less distinct minimum. In clearest oligotrophic basins, this minimum of $P_{i,extr}(z)$ becomes broadened. As we go on to show in this section, the increase in the pigment index in the water layer beyond the minimum of $P_{i,extr}(z)$ is due to photoadaptive processes taking place in the algal cells and involving the production of accessory photoprotecting pigments (PPP). In very deep waters, however, the rise in value of the pigment index $P_{i,extr}(z)$ is due to the chromatic adaptation of cells, as a result of which the relative quantity of photosynthetic antenna pigments increases.

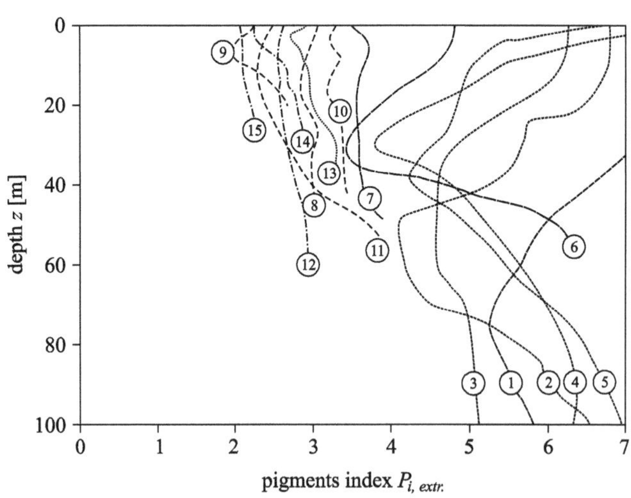

FIGURE 6.19. Experimental depth profiles of the pigment index $P_{i,extr}$: 1, Central Indian Ocean; 2–5, Central Atlantic; 6–7, Atlantic: Ezcurra Inlet; 8–11,13,Baltic and Gulf of Gdańsk; 12,14,15, Black Sea and Gulf of Burgas. (After Woźniak and Ostrowska (1990b).)

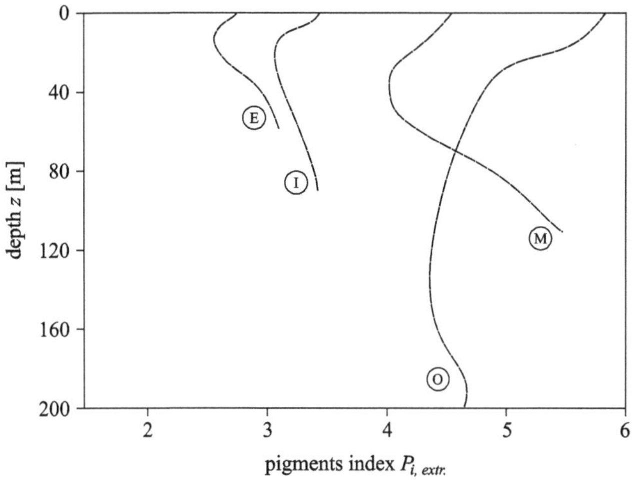

FIGURE 6.20. Vertical profiles of the pigment index $P_{i,extr}$ of acetone extracts of phytoplankton pigments averaged for the main trophic types of sea (see chlorophyll concentrations in Table. 6.1): E, eutrophic basins; I, intermediate (transitional) meso-eutrophic basins; M, mesotrophic basins; and O, oligotrophic basins. (After Woźniak and Ostrowska (1990b).)

Nevertheless, it is not enough merely to state these facts: in particular, the problem of distinguishing between PPP and photosynthetic pigments (PSP) demands closer scrutiny. The results of the analyses by Woźniak and Ostrowska (1990b) arose out of the joint treatment of photosynthetic (PSC) and photoprotecting carotenoids (PPC), an approach imposed by the limitations of the traditional spectrophotometric method of determining pigment concentrations which these authors used. The division into PSP and PPP was earlier applied by Babin et al. (1996a), who measured phytoplankton pigment concentrations by means of HPLC in different parts of the Central Atlantic. On the basis of their findings, they determined the mean depth profiles of what they called the *nonphotosynthetic pigments index (NPP)* for basins of three trophicity types (Figure 6.21). They took *NPP* to be the ratio of the concentration of photoprotecting carotenoids to that of all pigments in the phytoplankton: $NPP = C_{PPC}/(C_{PPP} + C_{PSP})$.

As can be seen in Figure 6.21, the relative PPP content is fairly small in eutrophic basins, but is much greater in oligotrophic waters. The drop in this content with depth, especially in transparent oligotrophic waters, is also characteristic. It is thus quite easy to perceive the correlation between *NPP* and the absolute quantities of radiant energy entering the water. In clean seas and at shallow depths more of this energy is available, so at the same time, as a result of the appropriate adaptive mechanisms, the phytoplankton there produce more PPP in order to prevent the photodestruction of chlorophyll. We now analyze this question in detail.

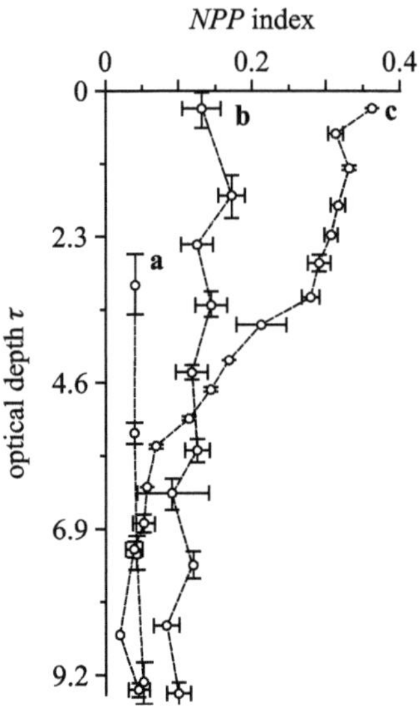

FIGURE 6.21. Mean vertical profiles of the non-photosynthetic pigments index $NPP(\tau)$ for waters of different trophicity (where τ is the optical depth). Plot a: eutrophic waters, plot b: mesotrophic waters, plot c:oligotrophic waters. The error columns indicate the standard deviations. (Based on Babin et al. (1996a).)

6.4.3 Photoadaptation and Chromatic Adaptation; Model Descriptions of Pigment Concentrations in Different Seas

As a result of the adaptation of natural plant communities to the light conditions obtaining in the sea, the pigment combinations found in marine algae differ according to the water's trophicity and depth in the sea. This is illustrated by the vertical distribution profiles (Figure 6.22a,d,g,j) of the relative concentrations of these pigments, as recorded in various regions of the World Ocean. These adaptive processes involve a change in the pigment composition in a single phytoplankton cell, or at the species level (limited adaptation), or even at the community level as a result of succession.

As we have already mentioned, there are two types of such adaptations, which we have named "photoadaptation" and "chromatic adaptation." Photoadaptation controls the content of PPC. As Figure 6.22a shows, the relative concentration of these carotenoids decreases with depth in all cases. This is because at small depths, particularly in oligotrophic waters, we record high values of the absolute irradiance, and this includes light from the blue region of the spectrum, that is, light which can bring about the photooxidation of chlorophyll *a*. In these circumstances, the plant produces large amounts of protective pigments, chiefly carotenoids, that absorb this undesirable light. Chromatic adaptation, on the other hand, controls the content of photosynthetic pigments in plant cells. The dependence of their concentration on depth is usually the reverse of that of PPP (Figure 6.22d,g,j). The relative content of these accessory antenna pigments, which harvest the available light for photosynthesis, generally increases at depths where there is no more light directly absorbable by chlorophyll. However, this is by no means a

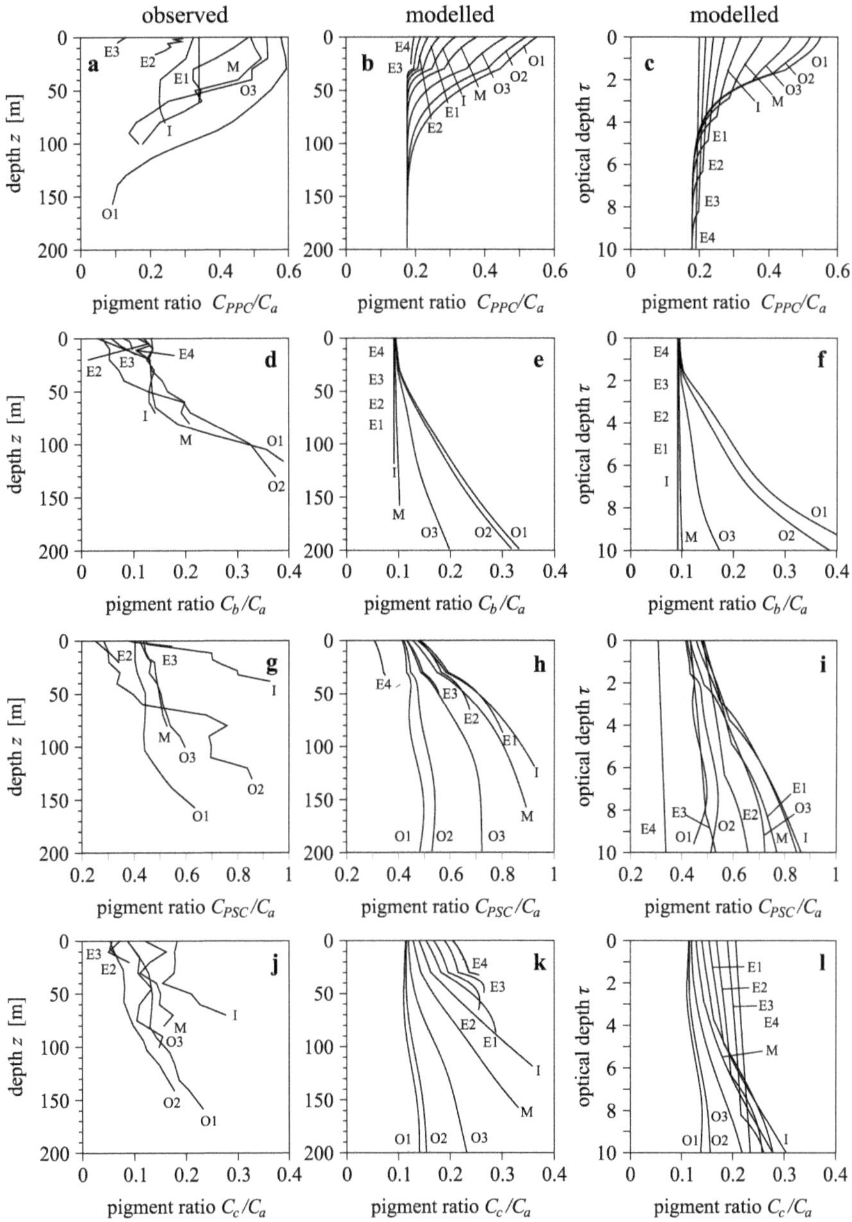

FIGURE 6.22. Typical depth profiles of the relative concentrations of accessory pigments: measured (a,d,g,j), modeled for real depth z [m] (b,e,h,k) and modeled for optical depth τ (c,f,i,l): photoprotecting carotenoids C_{PPC}/C_a (a,b,c); chlorophyll b, C_b/C_a (d,e,f); photosynthetic carotenoids C_{PSC}/C_a (g,h,i); chlorophyll c, C_c/C_a (j,k,l); these profiles were determined for various trophic types of waters defined on the basis of the surface concentration of chlorophyll a. The trophicities of basins and their symbols used on the plots (O1, . . . , E4) are defined in Table 6.1. (After Woźniak et al. (2003).)

universal feature of this process. In some cases, from a certain depth downwards, the PSP content can fall off. Even more complex is the nature of these relationships with regard to the water's trophicity as given by the chlorophyll *a* concentration. This is where the spectral properties of the underwater light field come in, an aspect which we now consider.

The adaptation of the photosynthetic apparatus in phytoplankton cells to the ambient underwater irradiance is an extremely complex problem, one that has been studied by numerous authors, for example, Steemann Nielsen (1975 and the papers quoted therein), Zvalinsky (1986), and Babin et al. (1996a,c). Although this research has demonstrated the existence of links between the concentrations of the individual accessory pigments in cells and the various optical characteristics of natural light fields in the sea, these links have not been subjected to detailed quantitative scrutiny. The first quantitative descriptions of these links have only recently been compiled by our research team on the basis of analyses of appropriate empirical data from some 400 vertical profiles in various regions of the World Ocean[9] (Woźniak et al. (1999), Majchrowski and Ostrowska (1999, 2000), Majchrowski (2001), and the review in Woźniak et al. (2003)). We now briefly discuss the results of these analyses.

With respect to photoadaptation we have discovered that the factor regulating the PPC content in cells is radiation in the shortwave (blue) region of the PAR spectrum. Statistical analyses indicate that the relative concentrations of these pigments correlate well with the values of the *Potentially Destructive Radiation* (PDR^*), defined in Section 6.1 (see expressions (T-9a) and (T-9b) in Table 6.3). We have derived a mathematical formula to describe the dependence of the relative PPC concentration (expressed in relation to the concentration of chlorophyll *a*) as a function of the PDR^*, averaged in the layer Δz in order to take mixing into account (thickness of the layer Δz is defined under the Table 6.12). This dependence is given in Table 6.12, Equation (T-1).

Examination of the chromatic adaptation process has revealed links among the concentrations of the individual Accessory Photosynthetic Pigments, that is, chlorophylls *b*, chlorophylls *c*, and photosynthetic carotenoids (PSC), and the properties of the underwater irradiance. It turns out that the concentrations of these pigments depend strongly on the relative distribution of the spectral irradiance $f(\lambda,z) = E_0(\lambda,z)/PAR_0(\lambda,z)$, but only to a slight extent on the absolute irradiance $E_d(\lambda)$. The relevant statistical formulas describing the dependences of the relative concentrations of these accessory photosynthetic pigments on the spectral fitting functions F_j, averaged in the layer Δz, are given in Table 6.12, Equations (T-2) to (T-4). (For a definition of F_j, the chromatic acclimation factor or spectral fitting function, see Section 6.1 (Table 6.3, expression (T-8b)). Statistical

[9] These empirical data comprised the vertical profiles of pigment concentrations and the vertical spectral distributions of the downward irradiance. A total of more than 4500 such datasets from different depths in the sea was analyzed.

TABLE 6.12. Model statistical formulas permitting the determination of accessory pigment concentrations from a known chlorophyll *a* concentration C_a and the characteristics of the underwater irradiance.[a]

Pigment	Formula	Equation No.
-1-	*-2-*	*-3-*
Photoprotecting carotenoids	$C_{PPC}/C_a = 0.1758 \cdot \langle PDR^* \rangle_{\Delta z} + 0.1760$	(T-1)
Photosynthetic carotenoids	$C_{PSC}/C_a = 1.348 . < F_{PSC} >_{\Delta z} - 0.093$	(T-2)
Chlorophyll *b*	$C_b/C_a = 54.068 . < F_b >_{\Delta z}^{5.157} + 0.091$	(T-3)
Chlorophyll *c*	$C_c/C_a = < F_c >_{\Delta z} \cdot 0.0424 . < F_a >_{\Delta z}^{-1.197}$	(T-4)

$<PDR^*>_{\Delta z}$ (potentially destructive radiation) and $<F_i>_{\Delta z}$ (spectral fitting functions). The thicknesses of the water layers are defined as follow: $\Delta z = z_2 - z_1$, where $z_2 = z + 30$ m, $z_1 = 0$ if $z < 30$ m and $z_1 = z - 30$ m if $z \geq 30$ (After Woźniak et al. (2003)).

analyses have shown that the relative concentration of these pigments correlate well with the spectral fitting functions F_j. On the basis of these model formulas it is possible to predict the probable vertical distribution profiles of the relative concentrations of the main accessory pigments in different trophic types of sea water, from oligotrophic to eutrophic. The results of these computations are illustrated by profiles b,c,e,f,h,i,k,l in Figure 6.22. The model depth concentration profiles of these pigments clearly resemble the experimental distributions, and so give a good idea of their real nature (cf. profiles a,d,g,j in the same figure). Easily recognized on the profiles, the main trends and characteristics of the vertical distributions of pigments in basins of different trophicity are readily explained on the basis of the characteristics of the light fields in these waters. These we discussed earlier in Section 6.1.2.

Thus, Figures 6.22a–c present the depth profiles of the relative PPC concentrations C_{PPC}/C_a in different trophic types of sea. For obvious reasons, these profiles are similar to the vertical $PDR^*(z)$ profiles (Figure 6.5). In eutrophic seas, C_{PPC}/C_a diminishes rapidly with depth. Here there are relatively few PPP. Shortwave (blue) light is already absorbed at small depths, so only red light penetrates into deeper waters. As this red light does not cause photooxidation of chlorophyll, it does not affect the photosynthetic apparatus (e.g., Figures 6.4d,h). The presence of PPP in phytoplankton under these conditions would therefore be superfluous. The situation in oligotrophic basins, however, is quite different. The waters here being highly transparent, blue light can pass through them to great depths, which puts the photosynthetic apparatus of phytoplankton cells in danger of photooxidation (e.g., Figs. 6.4a,e). The cells counteract this threat by producing PPP; this effect is clearly visible in Figs. 6.22a–c. In oligotrophic waters PPC concentrations $C_{PPC}(z)$ are much higher in relation to that of chlorophyll $C_a(z)$ and decrease with depth much more slowly than in eutrophic waters.

Model depth profiles of the relative concentrations of accessory PSP, that is, PSC, chlorophyll *b*, and chlorophyll *c*, are shown in profiles e,f,h,i,k, and l in Figure 6.22. They are similar in nature to the vertical distributions of the corresponding fitting functions $<F_j>_{\Delta z}(z)$ (Figure 6.6). We now examine this, using the photosynthetic carotenoids *PSC* as an example. The relative concentration C_{PSC}/C_a generally rises with depth. But if we compare these PSC concentrations in seas of different trophicity $C_a(0)$, then, as Figures 6.22g–i show, their relative concentration reaches a maximum in mesotrophic and intermediate waters. In oligotrophic waters, shortwave (blue) light is dominant in very deep waters. The PSC absorption maximum is around 490 nm and does not overlap the spectrum of the light entering these waters (Figure 6.4). Here, the chlorophylls, in particular chlorophyll *a*, are the principal absorbers.

The relative concentration C_{PSC}/C_a hardly changes with depth, and can even become smaller at great depths. In mesotrophic waters, the spectral light maximum in the water shifts with increasing depth towards the long waves, tending towards the PSC absorption peak. C_{PSC}/C_a thus increases with $C_a(0)$ and depth, reaching a maximum in mesotrophic and intermediate waters ($C_a(0) = 0.7$ mg tot. chl *a* m^{-3}), where in deeper waters the spectral irradiance maximum (see Figure 6.4 and Section 1.2) coincides with the maximum PSC absorption coefficient. In these circumstances, PSC play the dominant role, acting as antennas in promoting photosynthesis. Any further increase in the surface concentration of chlorophyll *a* (i.e., in the basin's trophicity) causes the spectral maximum to shift toward the red end of the spectrum, so that the irradiance maximum recedes more and more from the PSC absorption maximum. The upshot is a drop in the relative PSC concentration.

The role of photosynthetic antennas is then taken over by phycobilins. However, the regularities governing the occurrence of these compounds have not yet been thoroughly elucidated and so are not discussed here. Nonetheless, it is to be expected that these regularities are similar in nature to those of PSC, and that they too are determined by the relevant spectral fitting functions.

6.5 The Packaging Effect of Pigments in Marine Phytoplankton Cells

A further important factor affecting the magnitude of the light absorption coefficients of phytoplankton in sea water, in addition to the absolute contents and compositions of phytoplankton pigments, is the packaging of these pigments in phytoplankton cells, and hence their discontinuous distribution in the water. The function taking account of this effect, the packaging function Q^*, was already introduced in Equation (6.1), which describes the dependence of the light absorption coefficient of phytoplankton in water $a_{pl}(\lambda)$ and of the chlorophyll-specific light absorption coefficient $a^*_{pl}(\lambda)$ of the phytoplankton on a number of factors.

The point about the packaging effect of pigments in phytoplankton cells on light absorption by these pigments in the sea is that the molecules of these absorbers are not dissolved or distributed evenly throughout the water but occur in clusters, that is, only in the phytoplankton cells, which are a form of suspended particles. Water containing phytoplankton (and/or other complex aggregations of suspended particles) is thus an optically discontinuous medium; it is, in fact, a polydispersing medium, containing as it does optical inhomogeneities in the form of diverse suspended phytoplankton cells, whose refractive indices differ from (are usually larger than) the refractive index of the surrounding water together with the substances dissolved in it. In this section we give a formal description of the packaging effect in marine phytoplankton, and in Section 6.5.2 we present a preliminary statistical description of the product $C_{chl}D$ for phytoplankton in different types of sea water (where C_{chl} is the intracellular concentration of chlorophyll *a*, and *D* is the equivalent spherical diameter of the phytoplankton cell). The product $C_{chl}D$ is a kind of parameter determining the quantitative changes in light absorption by phytoplankton cells brought about by pigment packaging.

6.5.1 An Approximate Formal Description of the Packaging Effect for Marine Phytoplankton

We discussed the electrodynamic description of the interaction of light with dispersing (mono- and polydispersing) media in the previous chapter (Section 5.1). So let us then just recall that the optical properties of dispersing media diverge from the optical properties of homogeneous media (solutions) with the same molecular composition (Mie 1908). This is because the former are subject to the packaging effect, which usually means that:

- Light-scattering coefficients (forward scattering directions are especially privileged) are larger than in a homogeneous medium.
- Light absorption coefficients a_p in a dispersing medium are smaller than absorption coefficients a_{sol} in a homogeneous medium containing the same quantity of dissolved absorbers (i.e., in solution).

The ratio of these two absorption coefficients can be given by the dimensionless factor Q^*, known as the packaging function Q^*:

$$Q^* = \frac{a_p}{a_{sol}} = \frac{a_p^*}{a_{sol}^*}, \tag{6.8}$$

where a_{sol}, a_{sol}^* are the coefficient and specific coefficient, respectively, of light absorption for a given absorber in the solvent or dissolved state (i.e., in a homogeneous medium); a_p, a_p^* are the coefficient and specific coefficient, respectively, of light absorption for a given absorber in the coagulated state (i.e., in a dispersing medium) in the form of suspended particles.

The relative difference between the values of these coefficients is expressed by the dimensionless factor Δ_a, the absorption defect due to the packaging effect:

$$\Delta_a = \frac{a_p - a_{sol}}{a_p} = \frac{a_p^* - a_{sol}^*}{a_p^*} = 1 - Q^*. \tag{6.9}$$

In order, therefore, to predict the extent to which the absorption properties of dispersing media deviate from those of homogeneous media with the same molecular composition, one needs to know the packaging function Q^* (or the absorption defect Δ_a) characteristic of the given medium, that is, dependent on the sizes, shapes, and optical constants of the particles present in the medium. This can be calculated from known absorption efficiency factors Q_a, which are determined, for example, with the aid of Mie's theory, described in Section 5.1 (see the algorithm given in Table 5.1), because the existence of the following simple relationship between functions Q^* and Q_a, emerging inter alia from the definitions of these functions,[10] is readily demonstrated:

$$Q^*(\rho') = \frac{3Q_a(\rho')}{2\rho'}, \tag{6.10}$$

where ρ' is the optical particle size parameter, in this particular case, of a phytoplankton cell. *This size parameter* ρ' is in turn related to the equivalent spherical diameter D of the suspended phytoplankton cell with an intracellular chlorophyll *a* concentration C_{chl}, and to the specific coefficient of light absorption by all the pigments in the cell in the scattered state a^*_{sol} or to the imaginary part of refractive index n' of its intracellular matter. These relations are given by the expression:

$$\rho'(\lambda) = a^*_{sol}(\lambda) \cdot C_{chl} \cdot D = 4\pi \cdot D \cdot n'/\lambda. \tag{6.11}$$

Now, defining the packaging function Q^* on the basis of Equation (6.10) for all the optical inhomogeneities in a dispersing medium is a highly complex theoretical problem in the field of electrodynamics (see Section 5.1). Determining the optical efficiency factor Q_a for every single kind of optical inhomogeneity (e.g., every kind of phytoplankton cell present in a given type of water) requires for each one a separate solution of the Maxwell equations accounting for its internal structure and the complex boundary conditions describing its shape (e.g., cell shape). In practice this is impossible. In marine biooptics, therefore, a series of simplifying assumptions is made in order to take account of the effect of packaging on the light absorption coefficients of marine phytoplankton. These assumptions are, first, that a phytoplankton cell is an "optically soft" particle (this means that its real part of refractive

[10] Equations (T-10c) (Table 5.1) and (5.7c), the definition of function Q_a, yield the following expression for the absorption a_p: $a_p = Q_a N_v \pi D^2/4$ (where N_v is the number of particles in unit volume and D is the particle diameter). Application, on the other hand, of the relationship between the chlorophyll concentrations in the medium C_a and in the cell C_{chl} ($C_a = C_{chl} N_v \pi D^3/6$) enables the expression for the absorption $a_{sol} = C_a a^*_{sol}$ to be reduced to the form $a_{sol} = a^*_{sol} C_{chl} N_v \pi D^3/6$. Substitution of these expressions for a_p and a_{sol} in Equation (6.8), the definition of Q^*, gives the relationship between Q^* and Q_a described by Equation (6.10).

index relative to that of water approaches unity, and that its imaginary part of refractive index is small); second, that such a particle is spherical; and third, that the pigments it contains are distributed evenly, that is, that the particle is internally homogeneous. Application of these assumptions and the relevant solutions of the Maxwell equations for such a spherical symmetry yields the following approximate expression, alternative to the full Mie theory algorithm (after Duysens (1956), Hulst (1981), and Morel and Bricaud (1981)), describing the relation between the efficiency factor Q_a and the optical size parameter ρ':

$$Q_a(\rho') = 1 + \frac{2e^{-\rho'}}{\rho'} + \frac{2e^{-\rho'} - 1}{\rho'^2}. \tag{6.12}$$

Hence, from Equations (6.12) and (6.10), approximate formulas are obtained describing the dependence of the packaging function Q^* and the absorption defect Δ_a on the optical particle size parameter ρ':

$$Q_a{}^* = 3\left(\frac{1}{2\rho'} + \frac{e^{-\rho'}}{\rho'^2} + \frac{(e^{-\rho'} - 1)}{\rho'^3}\right) \tag{6.13}$$

$$\Delta_a = 1 - 3\left(\frac{1}{2\rho'} + \frac{e^{-\rho'}}{\rho'^2} + \frac{(e^{-\rho'} - 1)}{\rho'^3}\right). \tag{6.14}$$

Some results of the effect of pigment packaging on light absorption by phytoplankton are illustrated in Figure 6.23. This shows the approximate theoretical dependence of the absorption defect Δ_a, the packaging function Q^*, and the absorption efficiency Q_a on the optical particle size parameter ρ', determined from Equations (6.12) through (6.14). It is obvious that the larger the optical size parameter ρ' of the sphere, the greater the absorption defect.

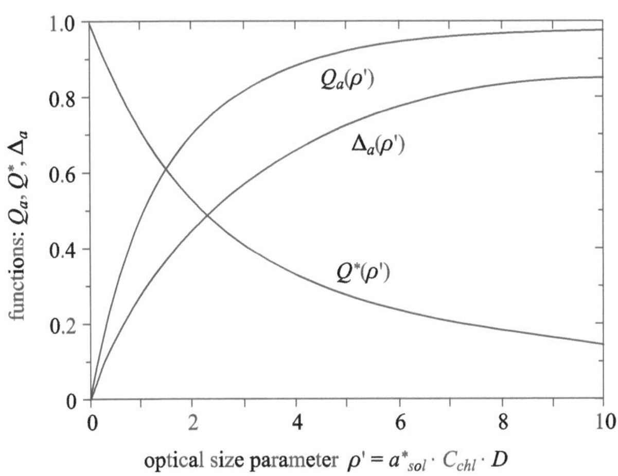

FIGURE 6.23. Approximate theoretical dependence of the packaging function Q^*, absorption defect Δ_a, and absorption efficiency Q_a on the optical particle size parameter ρ', determined on the basis of Equations (6.12) through (6.14).

In reality, therefore, light absorption by such a sphere is less than that due to its pigments in the scattered state (in solution). This absorption diminishes with an increase in any of the three factors in the expression for ρ' (Equation (6.11)): the specific absorption of the matter contained in the sphere a^*_{sol}, the size of the sphere D, or the intracellular chlorophyll concentration C_{chl} of the sphere.

To describe the spectral distribution of the packaging function one can also apply the set of equations (6.11)–(6.14), because the coefficient a^*_{sol} depends in the general case on the light wavelength λ. This is illustrated by the plots in Figures 6.24 and 6.25. They exemplify the results of our simulated calculations of the spectral chlorophyll-specific coefficients of light absorption by phytoplankton cells of various dimensions for a fixed value of the intracellular chlorophyll concentration (see the figure captions).

It is evident from Figures 6.24a and 6.25 that the values of all spectral coefficients $a^*_{pl}(\lambda)$ fall with increasing phytoplankton cell size. The extent of this

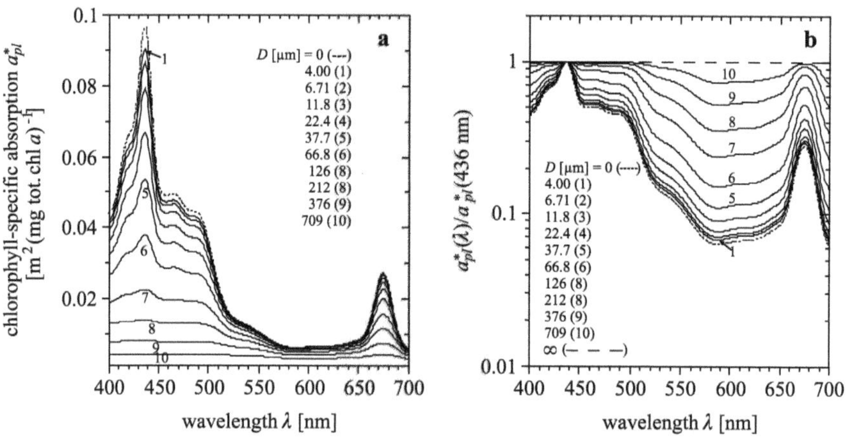

FIGURE 6.24. Effect of cell size of hypothetical phytoplankton on its chlorophyll-specific coefficient of light absorption $a^*_p(\lambda) = a^*_{pl}(\lambda)$. The simulation was done on the basis of Equation (6.1) and the set of Equations (6.11)–(6.14), on the assumption of a known spectrum of the pigments of this phytoplankton (pigment composition typical of algae in the surface layer of Case 1 waters, with an average trophicity $C_a \approx 0.7$ mg tot. chl a m^{-3} and proportions C_a: C_b: C_c: C_{PSC}: C_{PPC} = 1:0.107:0.137:0.478:0.316) in the insolvent state, $a^*_{sol}(\lambda) = a^*_{pl,sol}(\lambda)$. The spectra for different cell diameters D were calculated assuming an intracellular chlorophyll a concentration, constant in all cases, of $C_{chl} = 5$ kg m^{-3}. (a) Absolute absorption $a^*_{pl}(\lambda)$ on a linear scale.(b) Relative absorption $a^*_{pl}(\lambda)/a^*_{pl}$ (436 nm) (normalized to the peak at $\lambda = 436$ nm) on a logarithmic scale.

Dotted curve is the absorption of pigments $a^*_{pl,sol}$ in the insolvent state, equivalent to absorption by phytoplankton cells $a^*_{pl}(D = 0)$ with diameters $D \to 0$; unbroken curves from (1) to (10) are the absorptions a^*_{pl} for cells with various diameters (detailed on the figure); dashed curve ((b) only) are the absorption $a^*_{pl}(D \to \infty)$ for phytoplankton cells with infinitely large diameters (the "black-body" state).

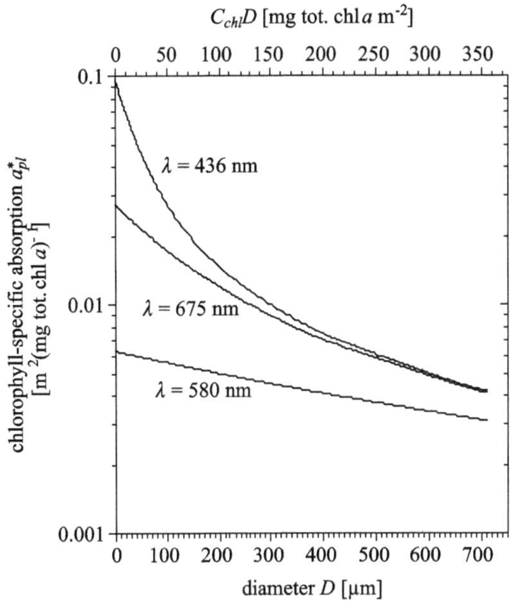

FIGURE 6.25. Model dependences of the chlorophyll-specific coefficient of light absorption for hypothetical phytoplankton cells (pigment concentrations and intracellular chlorophyll a concentrations as in Figure 6.24) at selected light wavelengths on the cell diameter D (lower scale), and on the product $C_{chl}D$ (upper scale), where C_{chl} is the intracellular chlorophyll a concentration. Calculations performed in accordance with the equations given in the caption to Figure 6.24.

decrease differs, however, depending on the light wavelength. It is the greatest in those spectral regions where the total absorption of light by phytoplankton pigments $a^*_{pl,sol}(\lambda)$ in the insolvent state is large (e.g., in the bands of absorption peaks at 436 and 675 nm; see the relevant plots in Figure 6.25), but is much weaker in those regions where the absorption $a^*_{pl,sol}(\lambda)$ takes low values (e.g., at λ = 580 nm; see the relevant plot in Figure 6.25).

A second feature of the effect of packaging on the absorption spectra of phytoplankton is the "flattening" of the spectra with increasing cell diameter; this is clearly visible in the plots in Figure 6.24b. So for sufficiently large algal cell diameters a state can be attained in which the specific absorption becomes practically independent of the wavelength (see Figure 6.24b, plot $a^*_{pl}(D \to \infty)$ for phytoplankton cells with infinitely large diameters). This is the so-called "black-body" state (Morel and Bricaud 1981). Note that these two features of the packaging effect—the reduced absorption and the flattened spectra—are due not only to increasing cell size, but also to rising intracellular chlorophyll concentration C_{chl}. To be exact, they are related to the increase in the product of both parameters, that is, $C_{chl}D$.

To conclude this description, we draw attention to the practical utility of the approximate theoretical formulas describing the effect of packaging on light absorption by phytoplankton. On the basis of these formulas and known specific coefficients of light absorption by all phytoplankton pigments $a^*_{pl,sol}(\lambda)$ in the solvent state, it is possible to determine approximate coefficients of light absorption by phytoplankton in the packaged state, characteristic of different seas and of different depths in them. To this end, knowledge is also needed of $C_{chl}D$ (the product of the intracellular chlorophyll *a* concentration and the equivalent diameter of the phytoplankton cells). Even though quite a lot is known about the intracellular concentrations C_{chl} (see, e.g., the references in the caption to Figure 5.19) and cell sizes D (see, e.g., Table 5.23 and the references cited therein) of various marine phytoplankton species, this problem has not yet been studied in sufficient depth. For some time, it has been known only that values of $C_{chl}D$ tend to fall with increasing chlorophyll concentrations C_a in the water; this was suggested by Bricaud et al. (1995), among others, on the basis of observed relationships of Q^* ($\lambda = 675$ nm) versus C_a. On the other hand, the first statistical generalizations of this significant product $C_{chl}D$, enabling rough estimates to be made of its characteristic values at different depths in different seas, were worked out in late 1998 and early 1999 by the authors of this book and their co-workers (see Woźniak et al. (1998, 1999)). We describe these generalizations in the following section.

6.5.2 *The Product $C_{chl}D$ for Phytoplankton in Different Types of Seas: A Preliminary Statistical Description*

Starting out from a Gaussian analysis of more than 400 empirical spectra of $a^*_{pl}(\lambda)$ defined for oceanic Case 1 waters and applying the relevant statistical fitting to this set of empirical spectra,[11] we were able to establish their characteristic "pigment packaging states," as defined by the products $C_{chl}D$ (Woźniak et al. 1998,1999). It turned out that these products $C_{chl}D$, determined for samples of phytoplankton from different seas and different depths, are correlated quite well with the chlorophyll *a* concentration C_a in the water. This is shown by the plots in Figure 6.26. Comparison of Figures 6.26a and 6.26.b shows that the vertical distributions of $C_{chl}D = f(z)$ tend to vary with depth in much the same way as the vertical distributions of the chlorophyll *a* concentration in the sea $C_a = f(z)$. Likewise, the absolute values of $C_{chl}D$ increase with the trophicity of the water (Figure 6.26c). One can therefore assume that the product $C_{chl}D$ for phytoplankton in oceanic Case 1 waters depends, to a first approximation, only on the concentration of chlorophyll *a*, C_a. We described this relationship using the statistical formula:

$$C_{chl}D = 24.65C_a^{0.75} \tag{6.15}$$

in which the equivalent diameter of the cell D is expressed in [m], and the intracellular concentration of chlorophyll *a*, C_{chl}, and the total chlorophyll *a* concentration C_a in the sea water (i.e., its mass per unit volume of sea water) in [mg tot. chl $a \cdot m^{-3}$].

The formula is applied to determine $C_{chl}D$ in the relevant equations of the phytoplankton absorption model presented later in this book (Chapter 6.7,

[11] Some details of these analyses are given in Section 6.7.5.

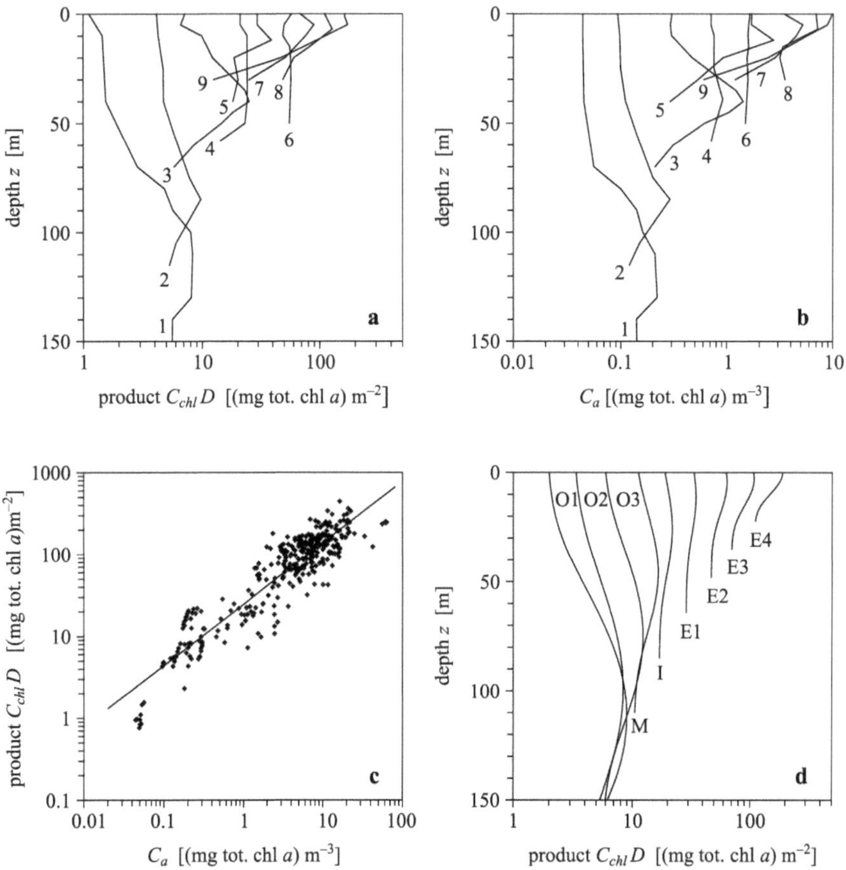

FIGURE 6.26. Relationship of the product $C_{chl}D$ with the total chlorophyll *a* concentration C_a and depth in the sea. (a) Empirical vertical profiles of the product $C_{chl}D$: plots 1–3, Atlantic; plots 4–9, Baltic. (b) Empirical total chlorophyll *a* concentration profiles $C_a(z)$ for the same stations as in (a). (c) Relationship between the product $C_{chl}D$ and the chlorophyll concentration C_a; observed (points) and approximated by Equation (6.15) (line). (d) Modeled vertical profiles of $C_{chl}D$ in various trophic types of stratified case 1 waters (curves O1–E4 correspond to various water trophicities as defined in Section 6.1, Table 6.1). In (d) the $C_a = f(C_a(0),z)$ model from Woźniak et. al. (1992a,b) was applied. (After Woźniak et al. (1999).)

Sections 6.7.5 and 6.7.6). The model dependence of the chlorophyll concentration C_a on depth and the surface chlorophyll concentration, given earlier in the equations in Table 6.10A, can be applied together with formula (6.15) to determine the distribution of the products $C_{chl}D$ in various types of seas (see the examples in Figure 6.26d).

As this figure shows, the typical values of $C_{chl}D$ vary over a range of almost three orders of magnitude and increase as the trophic index $C_a(0)$

does so. They are also depth-dependent: the nature of these changes is similar to that displayed by the vertical variability in chlorophyll concentration in oceanic Case 1 waters of different trophicity (see, e.g., Figures 6.26d and 6.15b).

In our subsequent research (e.g., Ficek et al. (2004)) we derived a formula, analogous to Equation (6.15), describing the relationship of $C_{chl}D$ for phytoplankton in Baltic waters of different trophicity. It takes the form:

$$C_{chl}D = 10.77C_a^{0.3767}, \tag{6.16}$$

where the units of the several magnitudes in this equation are identical to those in formula (6.15).

To end with, we emphasize that both these formulas for quantitatively characterizing the product $C_{chl}D$ in oceanic Case 1 waters (6.15) and in Baltic Case 2 waters (6.16) are approximate and of a preliminary nature. A more exact mathematical description of the dependence of $C_{chl}D$ on C_a, and possibly on other state parameters of the marine environment, requires further research.

6.6 Total Light Absorption by Marine Algae: Results of Empirical Studies

The overall light absorption coefficients of phytoplankton $a_{pl}(\lambda)$ in vivo is the sums of absorption coefficients of the component pigments modified (mostly reduced) by the pigment packaging effect in the cells. This reduction in absorption due to the pigment packaging effect is given by the packaging function $Q^*(\lambda)$ (see Equation (6.1) in Section 6.1), which is dependent on the size of the cells and the pigment concentrations in them (covered in the previous section). The great diversity in the composition and absolute concentrations of pigments in cells, also in the cell sizes of marine algae, is mirrored by a similar diversity in the absorption properties of phytoplankton. In the present section we review the empirical data concerning these properties, taking into account our own copious data from various seas, as well as the results of other authors available in the literature. These data form the basis of our description of the spectral characteristics and absolute light absorption coefficients of phytoplankton and their relationships with the chlorophyll *a* concentration at different depths in various regions of the World Ocean.

6.6.1 *Methodological Problems*

Determining the spectra of the light absorption coefficient of phytoplankton $a_{pl}(\lambda)$ is a complex empirical problem. The various methods that have been applied to its solution have often yielded very different results. For this reason, we begin by outlining the most commonly applied groups of methods.

1. *Hydrooptical methods* are based on the direct determination of phytoplankton absorption spectra. Directly measured are the total apparent and inherent optical properties of the sea, that is, the properties resulting from the sum of all the optically active components of sea water. Only then, on the basis of these overall properties, are the absorption properties of selected sea water components, such as the phytoplankton absorption spectra, obtained by calculations, usually approximate ones. Numerous versions of these methods have been applied by different authors, for example, Morel and Prieur (1977) and Morel (1978). However, they are encumbered with considerable error and are usually not accurate enough to permit detailed analyses of the spectral properties of phytoplankton.
2. *Fluorimetry* is used to measure the spectra of light-inducing fluorescence. The assumption here is that a similarity exists between the spectra of light inducing the fluorescence of chlorophyll in phytoplankton and the phytoplankton absorption spectra (Ostrowska and Woźniak 1991b). For measurements of this type, the fluorimeter needs to be calibrated in such a way that the relevant units for expressing the absorption coefficients [m^{-1}] can be obtained on the basis of the units for measuring fluorescence. In oceanology, however, these methods are seldom applied (see, e.g., Mashke and Haardt (1987)). Various aspects of their application have been discussed, for example, by Mitchell and Kiefer (1984, 1988), SooHoo et al. (1986), and Lazzara et al. (1996).
3. *Extraction spectrophotometry*, in which light absorption spectra in phytoplankton extracts (usually in acetone) are measured. This is the standard method in oceanology for identifying pigments in phytoplankton (Jeffrey and Humphrey 1975, Lorentzen 1968, Strickland and Parsons 1968). Nevertheless, the results obtained by this technique do not reflect the real absorption properties of phytoplankton in vivo, among other things, because the absorption properties of the individual pigments in the solvent are different from those they display in vivo. This question was analyzed in Section 6.2 (Tables 6.6 and 6.8). Moreover, phycobilins are not present in the absorption spectra of organic phytoplankton extracts.[12]
4. *Nonextraction spectrophotometry*, the most complex methods, involve the direct measurement of light absorption spectra in samples of isolated phytoplankton in vivo. These samples may take the form of concentrates, or, more commonly, filtered samples of natural phytoplankton taken from the sea or from cultures. In order to eliminate errors due to light scattering during the measurements, the spectrophotometer is fitted with integrating spheres or milk-glass diffusors (see, e.g., Dera (2003)). Even so, the so-called the path-length amplification effect has to be taken into account.

[12] Phycobilins are the only group of phytoplankton pigments that do not dissolve in organic solvents (acetone); however, they are readily water soluble (Filipowicz and Więckowski 1979,1983).

This effect arises as a result of light rays being obscured by the dense packaging of the particles (algae and other suspended matter) deposited on the filter or in the concentrated suspension. That is why a correction needs to be made to the measurements, the so-called β-factor (Butler 1962), which is determined experimentally. In order to obtain the absorption spectrum of phytoplankton pigments, the absorption of the entire sediment deposited on the filter is first measured. The pigments are then decolorized (so that they are no longer optically active in visible light), after which the absorption spectrum of the sediment on the filter is measured once again. The difference between the absorption coefficients from these two measurements is the absorption of the pigments. Various methods of decolorizing the experimental material have been tried, such as UV irradiation, or purging with solvent vapors. Recently, good results have been obtained by treating samples with $Ca(OCl)_2$ for a few minutes. The methodology of these nonextraction spectrophotometric techniques is continually being refined, and there are many publications on the subject, such as Morrow et al. (1989), Bricaud and Stramski (1990), Konovalov (1992), Babin et al. (1993), Tassan and Ferrari (1995), and Koblentz-Mishke et al. (1995). A paper by Allali et al. (1997) describes a novel version of this method, which to a large extent eliminates the need to determine the β-factor.

Of all these techniques, the nonextraction ones are the most suitable for analyzing the optical properties of live phytoplankton, as they permit the absorption spectra of phytoplankton to be measured directly. But as we have already mentioned, this involves serious experimental difficulties. The upshot is that the number of light absorption spectra of live phytoplankton published in the world literature is relatively small. In contrast, there is abundant empirical material on the absorption properties of phytoplankton extracts. In experimental practice, these spectra are obtained, as it were, automatically during the standard research procedure, because they usually serve to determine pigment concentrations. It would thus be worth analyzing the possibility of estimating the natural absorption properties of live phytoplankton from the absorption spectra of their extracts. This problem was investigated experimentally by, for example, Konovalov (1979), Woźniak (1990b), and Woźniak and Ostrowska (1990a). It was found that the overall absorption, that is, the mean absorption coefficients in the 400–700 nm range of visible light, determined from the extract spectra and in vivo pigment spectra are similar (Figure 6.27). It was also found that there is a statistically significant correlation between the pigment indices of these two types of spectra[13] (Figure 6.28), which can be approximately described by the relationship (after Woźniak and Ostrowska (1990a)):

[13] The pigment index is taken to mean the relationship of the extinction or absorption at the two main light absorption peaks of phytoplankton. Thus, in the case of phytoplankton extracts, the pigment index $P_{i,extr} = a_{pl,e}(\lambda = 430\ \text{nm})/a_{pl,e}(\lambda = 663\ \text{nm})$, and for phytoplankton in vivo, $P_i = a_{pl}(\lambda = 441\ \text{nm})/a_{pl}(\lambda = 675\ \text{nm})$.

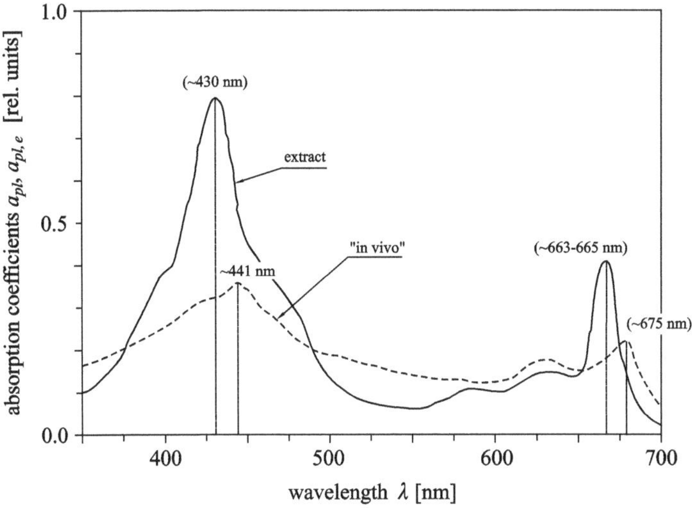

FIGURE 6.27. Comparison of the light absorption spectra of phytoplankton in vivo and its acetone extract.

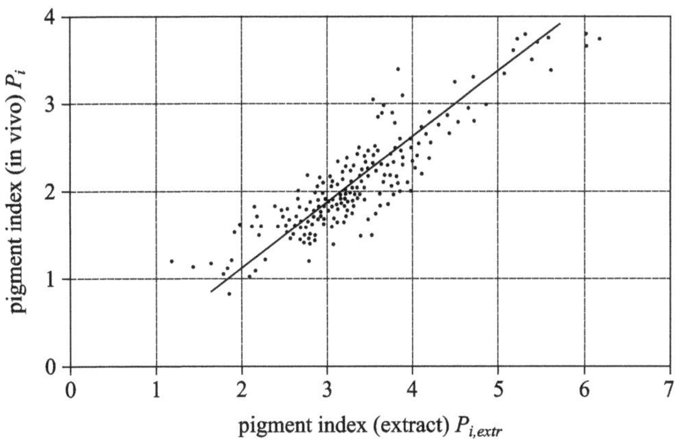

FIGURE 6.28. The observed relationship between the pigment indices: $P_i = a_{pl}(\lambda = 441\ \text{nm})/a_{pl}(\lambda = 675\ \text{nm})$ for the light absorption spectra of phytoplankton in vivo; $P_{i,extr} = a_{pl}(\lambda = 430\ \text{nm})/a_{pl}(\lambda = 663\ \text{nm})$ for the light absorption spectra of acetone extracts of phytoplankton. Points are empirical data from various seas, line is the linear approximation according to Equation (6.17). (After Woźniak and Ostrowska (1990a).)

$$P_i = 0.74P_{i,extr} - 0.38, \tag{6.17}$$

where P_i are in vivo pigment indices and $P_{i,extr}$ are pigment indices of the acetone extracts. By making use of these regularities and taking into account the peak shifts in the two main absorption bands (Figure 6.27), which we discussed earlier, we can predict the approximate profiles of the in vivo light absorption spectra of phytoplankton from the absorption properties of its acetone extract.

6.6.2 Light Absorption Spectra of Phytoplankton: A General Outline

A few experimental light absorption spectra of $a_{pl}(\lambda)$ for natural phytoplankton populations in vivo, as measured by various authors, are illustrated in Figure 6.29. They exemplify the wide range of absorption properties of phytoplankton in different environments, from the most productive, highly eutrophic basins such as lakes, through productive and fairly productive eutrophic and mesotrophic basins such as the Baltic Sea, to the unproductive oligotrophic waters of oceans. The general shape of the absorption spectra of $a_{pl}(\lambda)$ is similar in seas of all trophicities. The same applies to the spectra of the chlorophyll-specific absorption $a_{pl}(\lambda)/C_a$, both by natural phytoplankton communities and algal cultures (Figure 6.34). In the great majority of cases, these spectra are characterized by two broad absorption bands. One of them, the stronger (taller) and broader Soret band, lies in the blue region with a peak usually at c. 435–445 nm (mean 441 nm). Its half-width exceeds 100 nm, sometimes by a considerable margin. Situated in the red region with a maximum absorption of c. 675 nm, the other band is weaker and narrower, with a half-width of 20–30 nm. The occurrence of these two peaks in the phytoplankton light absorption spectrum is governed by the absorption properties of the pigments it contains (e.g., Figure 6.11). Thus, the absorption band in the red region is due almost entirely to chlorophyll *a*, the dominant pigment, and only to a small extent to chlorophyll *b*, inasmuch as there is little of it in phytoplankton. All the chlorophylls and the carotenoids contribute to the extensive absorption band in the blue and blue-green regions. Yellow light, on the other hand, is absorbed by the phycobilins, although only to a small degree, owing to the relatively low concentrations of these pigments, a property common to most natural populations of phytoplankton (Table 6.11).

Apart from these similarities, a feature distinguishing this two-banded structure of the light absorption spectra for algae living in different environments is the ratio between the band maxima. The corresponding pigment index P_i, equal to the ratio of the absorption coefficients at these maxima, $P_i = a_{pl}(441 \text{ nm})/a_{pl}(675 \text{ nm})$, defined for natural phytoplankton communities, is usually very much higher for phytoplankton in oligotrophic basins (i.e., for low concentrations of chlorophyll *a*) than is the case for chlorophyll-rich eutrophic waters (Figure 6.30). As we have already mentioned, this emerges from the greater accessory pigment content in the photosynthetic apparatus of phytoplankton (absorbing in the blue region of the spectrum) in unproductive than in highly productive waters (Figures 6.17 and 6.18).

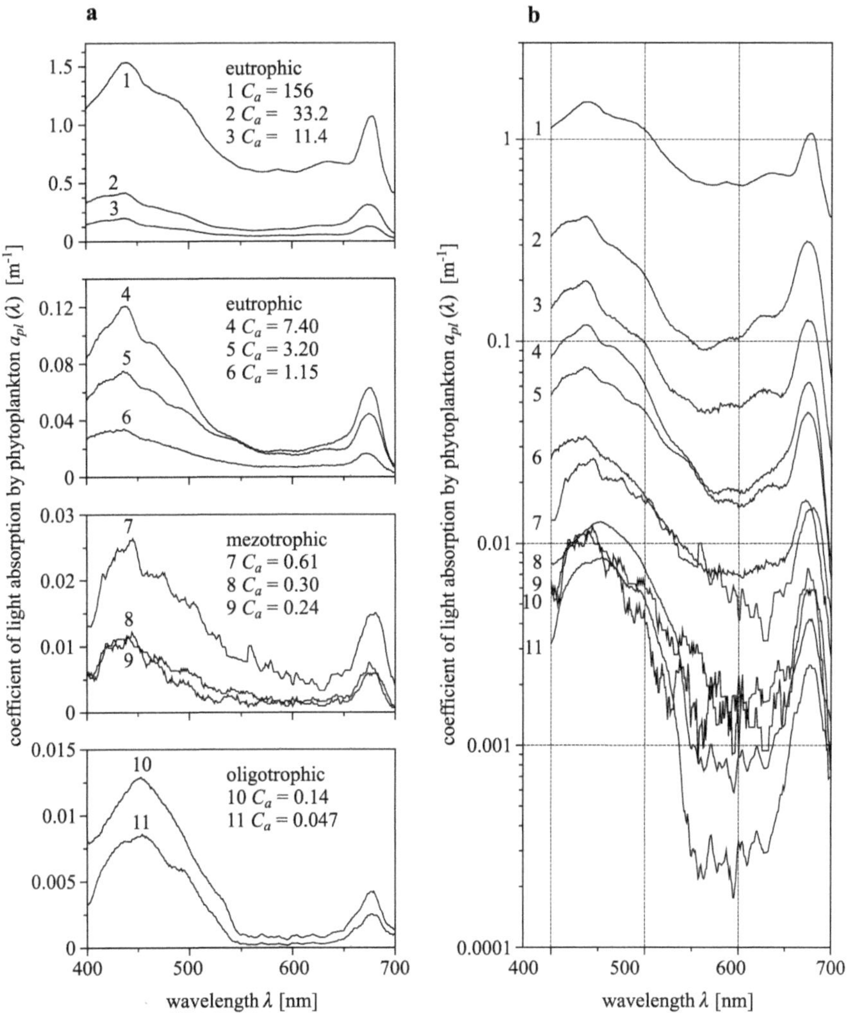

FIGURE 6.29. Empirical spectra of light absorption coefficients of phytoplankton a_{pl} (λ) in various seas: (a) coefficients $a_{pl}(\lambda)$ drawn on a linear scale; (b) coefficients $a_{pl}(\lambda)$ drawn on a logarithmic scale. The numerals on the plots denote the respective concentrations of chlorophyll a C_a [mg tot. chl a m^{-3}]: 1. $C_a = 156$; 2. $C_a = 33.2$; 3. $C_a = 11.4$; 4. $C_a = 7.4$; 5. $C_a = 3.2$; 6. $C_a = 1.15$; 7. $C_a = 0.61$; 8. $C_a = 0.30$; 9. $C_a = 0.24$; 10. $C_a = 0.14$; 11. $C_a = 0.047$. Measurement sites: 1, Lake Fukami-ike, Japan (after Takematsu et al. (1981)); 2–5, Baltic Sea 1994 ULISSE Experiment (Ooms 1996); 6–8, Baltic Sea 1990 results from the cruise of r/v "Professor Shtockman"; 9–11, Atlantic Ocean 1973 results from the cruise of r/v "Mendeleev"; 10, Pacific Ocean (Kishino et al. 1986); plots 6–11 from the data bank of IO PAS Sopot. (Adapted from Woźniak et al. (2003).)

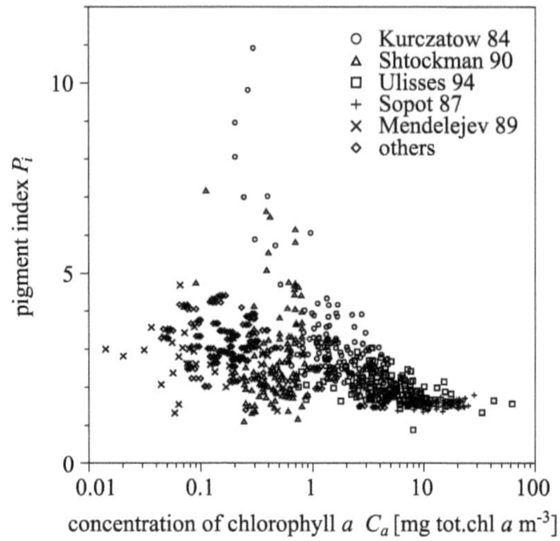

FIGURE 6.30. The empirical relationship between the pigment index of phytoplankton absorption spectra $P_i = a_{pl}(441 \text{ nm})/a_{pl}(675 \text{ nm})$ and concentrations of chlorophyll a in the sea, where: Kurchatov 94 is data from the cruise of r/v "Kurchatov" in the Baltic in 1994; Shtockman 90 is data from the cruise of r/v "Professor Shtockman" in the Baltic in 1990; Ulisses 94 is the ULISSE experiment (1994) (Ooms 1996) in the southern Baltic; Sopot 87 is the shore experiment (1987) with data from the Gulf of Gdańsk; Mendeleev 89 is data from the cruise of r/v "Mendeleev" in the Atlantic in 1989; Others are data from the Atlantic, Pacific, Mediterranean, and Black Sea, various experiments and cruises. (Data bank of IO PAS Sopot.)

6.6.3 Light Absorption Spectra of Phytoplankton: Fine Structure

In spite of the fact that the light absorption spectra of $a_{pl}(\lambda)$ for various taxonomic groups of phytoplankton living in widely different environments are generally similar in outline, the great natural diversity of pigment compositions is responsible for the corresponding diversity of spectral fine structures and also for their great complexity.

The most commonly occurring fine-structural features of $a_{pl}(\lambda)$ spectra in the 375–750 nm range are listed in Table 6.13 and illustrated in Figure 6.31. These data show that the spectra of the light absorption coefficients of phytoplankton $a_{pl}(\lambda)$ can consist of some 15 absorption bands, and in extreme cases, as many as 21 (after Woźniak et al. (1999)); they can also display up to five absorption minima (light transmittance maxima). These absorption bands can be classified into three magnitude classes. The first of these classes contains the two main absorption bands (439 and 675 nm), already mentioned; they are of the first magnitude. These bands are always distinctive as tall peaks in $a_{pl}(\lambda)$ spectra and are present in practically all cases. Somewhat less intense signs of absorption are classified as second magnitude bands: from 0 to 4 such bands can occur simultaneously. When they occur, they always form local peaks in $a_{pl}(\lambda)$ spectra.

TABLE 6.13. The possible characteristic peaks in the visible light absorption spectra of phytoplankton.

Type of peak	Number of peaks in the spectrum	Position of peaks [nm]		Peak in the spectrum is expressed as a:
Band maximum: 1st magnitude absorption	2	439 ± 5	675 ± 3	Maximum, local maximum
Band maximum: 2nd magnitude absorption	0–4	465 ± 5 493 ± 5	583 ± 4 630 ± 4	Local maximum
Band maximum: 3rd magnitude absorption	0–9	381 ± 2 408 ± 2 420 ± 2 451 ± 3	532 ± 2 609 ± 2 655 ± 2 700 ± 3 712 ± 3	Shoulder
Transmittance maximum	1–5	450 ± 2 480 ± 3	573 ± 4 605 ± 2 650 ± 2	(main) Local minimum

Adapted from Woźniak and Ostrowska (1990a).

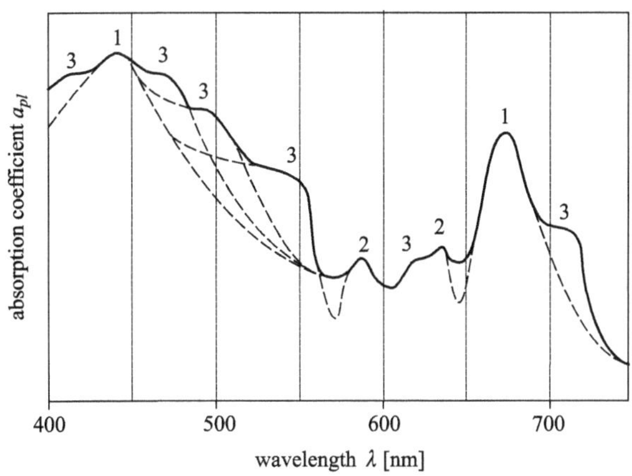

FIGURE 6.31. Positions of the first, second, and third magnitude class maxima in light absorption spectra of phytoplankton. (Adapted from Konovalov (1979).)

Finally, the absorption bands making the smallest contribution to the overall phytoplankton absorption spectrum are classified as being of the third magnitude. These bands are always to be found on the relatively steep slopes of the spectral bands in the other two magnitude classes. As a result, they never form separate peaks, but manifest themselves only as shoulders in the overall spectrum. There may be from 0 to 9 such bands present at the same time.

There can be from one to five transmittance maxima, that is, minima of the absorption coefficients $a_{pl}(\lambda)$. By way of example, three such maxima are shown in Figure 6.31. They are always formed in the middle part of the spectrum, between the first and second magnitude bands, or between two second magnitude bands.

The occurrence of these features in the absorption spectra of different phytoplankton samples varies considerably and depends on the accessory pigment composition. Individual spectra differ not only in the number and presence of the above-mentioned absorption maxima and minima, but also in the mutual relations of their intensities. This is due to the diversity in pigment composition, mentioned earlier, with respect both to individual phytoplankton species and to their communities flourishing in different seas and at different depths. The fine structure of absorption spectra is usually at its most intricate in strongly oligotrophic seas, where the relative concentrations of accessory pigments in the photosynthetic apparatus are the highest (Figure 6.17a). In this phytoplankton absorption spectrum we see the largest number of higher-order absorption bands and transmittance bands. But the number of these bands drops with rising trophicity. This diversity also applies to depth in the sea. The fine structure of $a_{pl}(\lambda)$ spectra is usually more elaborate in the surface layers of the sea (because of the greater accumulation of photoprotecting pigments) and in the deeper zones of phytoplankton occurrence (because of the presence there of accessory photosynthetic pigments). In contrast, the least complex absorption spectra are obtained from intermediate depths of the euphotic zone, where the plants enjoy optimal growing conditions and do not need to produce accessory pigments.

6.6.4 *Absolute Values of Total and Specific Absorption Coefficients*

The overall light absorption coefficients of phytoplankton $a_{pl}(\lambda)$ recorded in the World Ocean and in other natural basins can vary over a range of more than four orders of magnitude (Figure 6.29), from $a_{pl}(\lambda) \approx 10^{-4}$ m^{-1} in the ultra-oligotrophic centers of oceans for light from the mid-regions of the visible spectrum, to $a_{pl}(\lambda) \approx 1$ m^{-1} and more in the highly eutrophic coastal zones of seas, and in eutrophic lakes at particular times of the year. In addition to this trophicity-linked variation in the values of $a_{pl}(\lambda)$, these coefficients are also depth-dependent (see later Figure 6.35). We characterize in detail these two relationships of $a_{pl}(\lambda)$ with trophicity and depth a little later on the basis of the descriptions in Sections 6.1 and 6.5 of the main factors governing the absorption properties of algae in the sea.

The principal factor determining absolute light absorption coefficients of oceanic phytoplankton $a_{pl}(\lambda)$ is the concentration of chlorophyll *a* in the water C_a. With certain exceptions, the higher the chlorophyll *a* concentration in the sea, the greater the values of $a_{pl}(\lambda)$ for all wavelengths of visible and near-UV light (Figure 6.29). Examples of these relationships of $a_{pl}(\lambda)$ for six different wavelengths are illustrated in Figure 6.32.

Disregarding the scatter of experimental points, we can see from the plots (Figure 6.32) that the absorption coefficients $a_{pl}(\lambda)$ rise with increasing chlorophyll *a* concentration in all cases. However, these relationships are not linear, as the lines and formulas in the plots show. The absorption coefficients $a_{pl}(\lambda)$ generally increase in value more slowly than the chlorophyll concentrations,

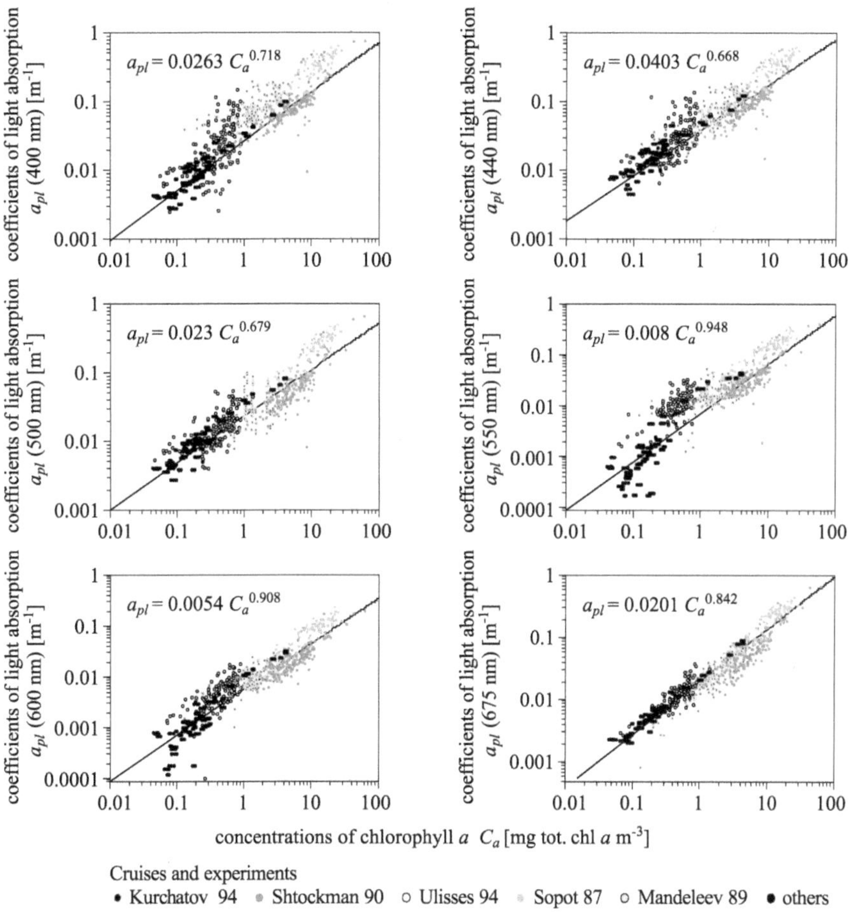

FIGURE 6.32. Dependences of empirical light absorption coefficients of phytoplankton in vivo $a_{pl}(\lambda)$ on the chlorophyll a concentration C_a in sea water for selected wavelengths λ (400 nm, 440 nm, 500 nm, 550 nm, 600 nm, 675 nm) in different basins, where: Kurchatov 94 – data from the cruise of r/v 'Kurchatov' in the Baltic in 1994, Shtockman 90 – data from the cruise of r/v 'Professor Shtockman' in the Baltic in 1990, Ulisses 94 – the ULISSE experiment ULISSE of 1994 (Olaizola 1996) – southern Baltic, Sopot 87 – shore experiment 1987 r. – data from the Gulf of Gdańsk, Mendeleev 89 - data from the cruise of r/v 'Mendeleev' in the Atlantic in 1989, Others – data from the Atlantic and Pacific Oceans, and from the Mediterranean and Black Seas – various experiments and cruises. The continuous lines and formulas on the plots illustrate the empirical (Bricaud et al. 1995) and statistically averaged relationships between coefficients $a_{pl}(\lambda)$ and chlorophyll concentrations C_a in the sea (adaptation from Majchrowski 2001).

and any deviations from this linearity depend on wavelength. Confirmation of this can be found in the statistical generalizations obtained by Bricaud et al. (1995); see the formulas and the continuous lines in the plots. The deviations from a simple proportionality between C_a and $a_{pl}(\lambda)$ are usually greater for light from the shorter-wave end of the visible spectrum than for longer wavelengths. This is borne out by the values of the power exponent B for variable C_a in the formula describing this relationship:

$$a_{pl} = A\ C_a^{\ B}, \tag{6.18}$$

obtained by Bricaud et al. (1995) on the basis of empirical studies. The value of this exponent is very much less than unity for blue light (e.g., for $\lambda = 440$ nm, $B = 0.67$) than for red light (e.g., for $\lambda = 675$ nm, $B = 0.84$). This is also illustrated by the empirical dependences of the chlorophyll-specific absorption coefficients $a^*_{pl} = a_{pl}/C_a$ on the chlorophyll concentration C_a (Figure 6.33) and by the spectra of these specific coefficients $a^*_{pl}(\lambda)$ (Figure 6.34a) recorded in seas of different trophicity. These figures show that for blue light ($\lambda = 440$ nm) these chlorophyll-specific absorption coefficients a^*_{pl} fall abruptly with increasing chlorophyll concentration, on average by 15 times (and c. 40 times in extreme cases) in supereutrophic basins in comparison with ultra-oligotrophic ones. For red light (675 nm), this drop in value is much less: only about three times. Hence, the corresponding values $C = -(B-1)$, of the power exponent (see the equations on the plots in Figure 6.33) after Bricaud et al. (1995):

$$a^*_{pl} = A\ C_a^{\ -C} \tag{6.19}$$

are relatively high for blue light (for $\lambda = 440$ nm, $C = 0.33$) but much lower for red light (e.g., for $\lambda = 675$ nm, $C = 0.16$). The reader will notice that if the proportionality were simple, C would be equal to 0.

We now attempt to explain the typical features of these coefficients of light absorption by algae in vivo under different conditions.

The coincidence in value of the chlorophyll-specific coefficients of red-light absorption is the consequence of the decisive influence of chlorophyll *a* on the overall absorption of light in this region of the spectrum, because the other pigments (apart from some of the phycobilins) practically do not absorb light from this region (Figure 6.11). The pigment composition in phytoplankton does not therefore seriously affect its longwave absorption band. Variations in the value of a^*_{pl} in this band are possible only as a result of the different effects of pigment packaging in cells. For light of 675 nm, that is, in the maximum absorption of this band, $a^*_{pl}(675\ \text{nm})$ usually takes values ranging from 0.01 to 0.03 [m^2 (mg tot. chl $a)^{-1}$] (Figure 6.34).

In the mid-region of the spectrum (550–650 nm), that is, in its broad absorption minimum, the diversity of $a_{pl}(\lambda)$ profiles is relatively greater than in the red. One reason for this could be the concentrations of phycobilins, which vary greatly, depending on the conditions. Moreover, this is an area where absorption coefficients take very low values, that is, an area where there is thus a high risk of experimental error.

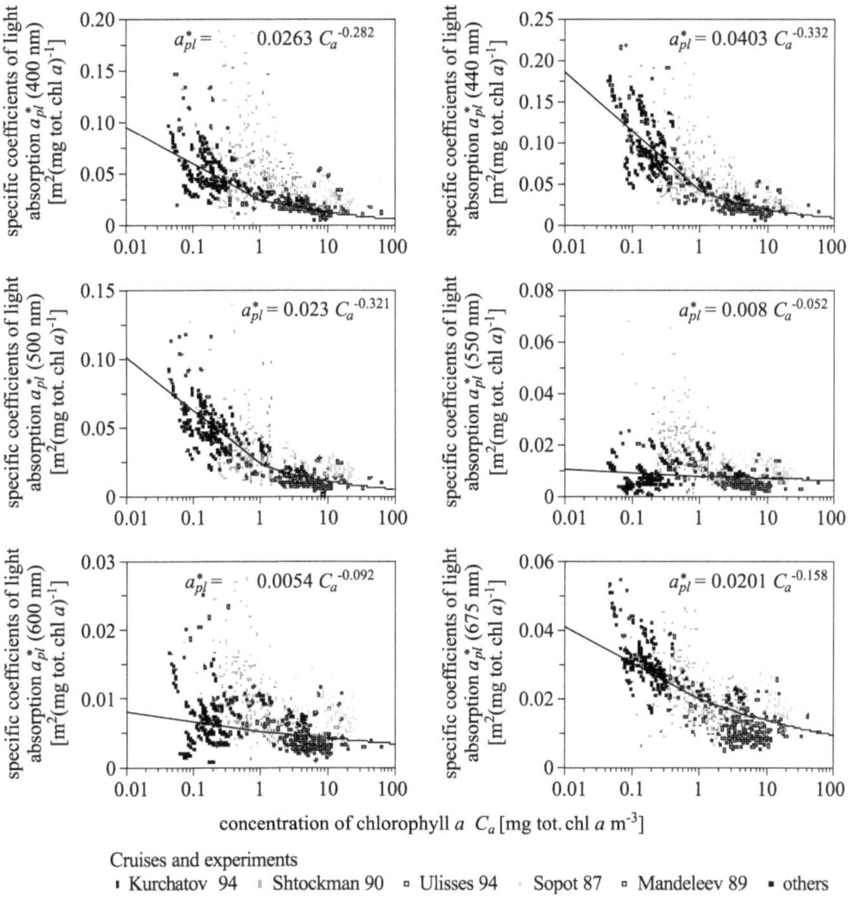

FIGURE 6.33. Dependences of empirical chlorophyll-specific light absorption coefficients of phytoplankton $a^*_{pl}(\lambda)$ on the chlorophyll *a* concentration C_a in sea water for selected wavelengths λ (400 nm, 440 nm, 500 nm, 550 nm, 600 nm, 675 nm) in different basins, where: Kurchatov 94 – data from the cruise of r/v 'Kurchatov' in the Baltic in 1994, Shtockman 90 – data from the cruise of r/v 'Professor Shtockman' in the Baltic in 1990, Ulisses 94 – the ULISSE experiment ULISSE of 1994 (Olaizola 1996) – southern Baltic, Sopot 87 – shore experiment 1987 r. – data from the Gulf of Gdańsk, Mendeleev 89 - data from the cruise of r/v 'Mendeleev' in the Atlantic in 1989, Others – data from the Atlantic and Pacific Oceans, and from the Mediterranean and Black Seas – various experiments and cruises (Data Bank of IO PAS Sopot). – The continuous lines and formulas on the plots illustrate the empirical (Bricaud et al. 1995) and statistically averaged relationships between coefficients $a^*_{pl}(\lambda)$ and chlorophyll concentrations C_a in the sea (adaptation from Majchrowski 2001).

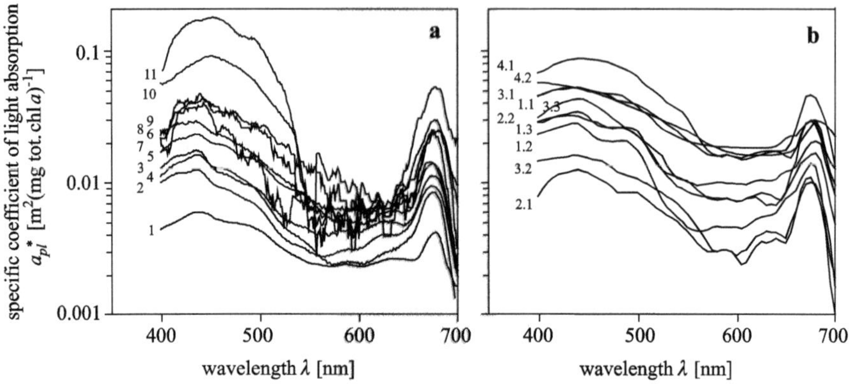

FIGURE 6.34. Spectra of chlorophyll-specific light absorption coefficients for: (a) natural plant communities (the numerals indicate the trophic type of sea and the measurement location, as in Figure 6.29); (b) selected species from cultures of four groups of phytoplankton (measurements by various authors; gleaned from Smekot-Wensierski et al. (1992)): (1) *Prasinophyceae* and *Chlorophyceae*: 1.1. *Platymonas* sp., 1.2. *Platymonas suecica*, 1.3. *Platymonas suecica*; (2) *Haptophyceae*: 2.1. *Cricosphaera carterae*, 2.2. *Hymenonas elongata*; (3) *Bacillariophyceae*: 3.1. *Coccolithus huxleyi*, 3.2. *Chaetoceros protuberans*, 3.3. *Phaeodactylum tricornutum*; (4) *Dinophyceae*: 4.1. *Gymnodinium kowalewski*, 4.2. *Gymnodinium* sp.

Finally, the greatest diversity of chlorophyll-specific absorption coefficients $a_{pl}^*(\lambda)$ that has been found in natural waters is due mainly to the absorption band with the maximum in the blue region. So, for example, a_{pl}^*(440 nm) usually ranges from c. 0.005 [m^2 (mg tot. chl a)$^{-1}$] in eutrophic basins to c. 0.20 [m^2 (mg tot. chl a)$^{-1}$] and more in oligotrophic waters. This also affects the pigment index, which in unproductive oligotrophic waters reaches values of around P_i = 3.5, and occasionally more, whereas in eutrophic basins it can drop to around 1 (after Woźniak and Ostrowska (1990b)). This differentiation in coefficients a_{pl}^* in the blue region of the spectrum, observed in natural populations of phytoplankton, is due both to the diversity of pigment compositions in the plant communities of different seas and to the pigment packaging effect. The influence of the latter on the absorption properties of cells is minimal in oligotrophic waters, but increases with rising chlorophyll concentrations in the sea (see Section 6.5).

This diversity in the absolute values of the overall and chlorophyll-specific light absorption coefficients of algae in the sea applies to waters of different trophicity, regardless of the depth at which the phytoplankton occur. As we have already shown in Sections 6.2 and 6.5, the concentration and composition of phytoplankton pigments and also the quantitative characteristics of the packaging effect depend not only on a basin's trophicity, but also vary with depth in the sea (Figures 6.19 through 6.22). Consequently, the absorption properties of phytoplankton change with depth in the sea; this is exemplified for light of wavelength 440 nm in Figure 6.35. In all cases the depth

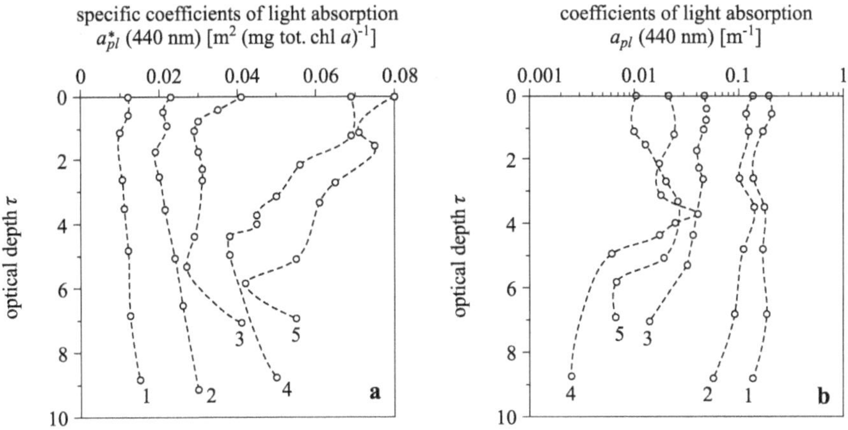

FIGURE 6.35. Examples of changes in optical depth τ in the sea of chlorophyll-specific absorption coefficients a^*_{pl} (a) and absorption coefficients a_{pl} (b) for blue light (λ = 440 nm) recorded in basins of different trophicity ($C_a(0)$ [mg tot. chl a m^{-3}]: Baltic Sea: 1. $C_a(0) \approx 16$; 2. $C_a(0) \approx 5.9$; 3. $C_a(0) \approx 1.1$; Black Sea: 4. $C_a(0) \approx 0.31$; 5. $C_a(0) \approx 0.13$. The graphs are based on empirical data from the cruise of r/v Siedlecki in the Baltic (1981) and the cruise of r/v Vitiaź in the Black Sea (1978) contained in the work by Koblentz-Mishke et al. (1985).

profiles of the chlorophyll-specific coefficient of absorption $a^*_{pl}(\lambda)$ (Figure 6.35a) are characterized by a minimum at certain depths. This is quite strongly marked in oligotrophic waters, but less so in eutrophic seas. The value of a^*_{pl} increases above and below the depths of these minima.

This nature of the $a^*_{pl}(\lambda)$ profiles is due mainly to changes in the pigment composition. The highest values of a^*_{pl} in the surface layer are thus due to the accumulation of large quantities of photoprotecting carotenoids PPC in the algae living there (Figures 6.19 to 6.22a,b,c). But the increase in value of a^*_{pl} in the layer below the minimum is the consequence of the rising proportions of accessory PSCs in the photosynthetic apparatus of algae (e.g., chl b, chl c, PSC, phycobilins), usually in very deep waters (Figures 6.19, 6.21, 6.22d–l). By comparison, the depth profiles of the overall absorption coefficients $a_{pl}(\lambda)$ display tendencies opposite to those of the vertical distributions of $a^*_{pl}(\lambda)$ (Figure 6.35b). As with the depth profiles of chlorophyll a concentrations (Figure 6.15a), so with absorption a_{pl}: maxima are recorded at certain intermediate depths. Absorption diminishes above and below these depths, a fact directly correlated with the drop in chlorophyll a content.

To end this section on the in situ absorption properties of natural populations of marine phytoplankton, which consist of a mixture of various species of algae, it is worth comparing these properties with those of the separate species of algae cultivated artificially. A few spectra of the chlorophyll-specific coefficients of light absorption by different cultures are illustrated in

Figure 6.34b. In this case too, we see that the absolute coefficients a^*_{pl} display considerable variation, and on a scale similar to the one obtaining in natural populations. This is obvious, and emerges from the fact that the absorption properties of natural communities of marine algae are a linear combination of the absorption properties of mixtures of different species of algae. (These problems are discussed in greater detail, e.g., in the papers by Smekot-Wensierski et al. (1992) and Stramski et al. (2001)).

6.7 Model Descriptions of the Absorption Properties of Marine Phytoplankton: A Review

In the previous section we presented the absorption properties of phytoplankton as elucidated by the results of empirical studies. The most important spectral features and regularities pertaining to light absorption by phytoplankton in the World Ocean were characterized. This was done mostly in a qualitative manner because the empirical data concerning such absorption have not been classified in any systematic way. They do not by any means present a full quantitative picture of the wide range of light absorption coefficients of phytoplankton in different marine environments, in particular, in waters of different trophicity and at different depths in the sea; neither do they take into account the diverse conditions under which natural irradiance occurs. Finding such regularities has been the primary cognitive objective of the statistical analyses of the empirical material and of the mathematical modeling undertaken by numerous authors.

An important practical aim of this research has been to create mathematical algorithms for estimating the absorption properties of marine algae which could be utilized in the remote sensing of the primary production of organic matter in seas and of various parameters relating to the state of marine ecosystems (e.g., Antoine and Morel (1996), Antoine et al. (1996), and Woźniak et al. (2004)). The assumption underlying these nontrivial algorithms is that the spectral light absorption coefficients of phytoplankton pigments (separate values for photosynthetic and photoprotecting pigments) at different depths in the sea can be estimated from a small number of parameters characterizing the surface layer of water that can be measured from a satellite (Majchrowski 2001, Ficek 2001, Woźniak et al. 2003). The most important of these parameters are the surface chlorophyll *a* concentration $C_a(0)$ and the irradiance of the sea surface (e.g., Krężel (1997) and Sathyendranath et al. (2000)).

The last 20 years have witnessed the development of a good number of statistical generalizations or semi-empirical models of the absorption properties of marine algae, all of which satisfy these requirements to a greater or lesser extent. However, most of them are of a local nature; that is, they usually apply only to the optical properties of algae from certain relatively small areas of the sea (e.g., Mitchell and Kieffer (1988), Morrow et al. (1989), Konovalov et al. (1990), Carder et al. (1991), Sosik and Mitchell (1995), and

Cleveland (1995)) or to selected species of algae (e.g., Konovalov (1985), Hoepffner and Sathyendranath (1991), Smekot-Wensierski et al. (1992), and the papers cited therein). Only a small number of these papers (cited later) can aspire to be general relationships or models, that is, are applicable to the wide range of variation in the absorption characteristics of algae living in a wide diversity of environmental conditions in the World Ocean. In this section we present the main types of model descriptions, including examples of highly original models, as well as those that have led to a breakthrough in the development of a semi-empirical mathematical model of the absorption properties of marine phytoplankton.

6.7.1 The Principal Model Descriptions of Light Absorption by Phytoplankton

A feature common to all semi-empirical models or statistically generalized descriptions of the light absorption coefficients of phytoplankton a_{pl} is the assumption that $a_{pl}(\lambda)$ is equal to the product of the chlorophyll-specific coefficient of absorption $a^*_{pl}(\lambda)$ and the chlorophyll *a* concentration C_a in the sea:

$$a_{pl}(\lambda) = C_a \, a^*_{pl}(\lambda), \tag{6.20a}$$

where the chlorophyll-specific coefficient is a function of environmental factors:

$$a^*_{pl} = f\,(environmental\ conditions). \tag{6.20b}$$

In view of this assumption, and depending on the degree of complexity of Equation (6.20b), variously constructed by different authors, we can divide the existing model descriptions into the following types:

1. *Classical descriptions* (single-component models, homogeneous). In these Equation (6.20a) for the absorption coefficient takes the form:

$$a_{pl}\,(\lambda) = C_a \, a^*_{pl}\,(\lambda) \quad \text{and} \quad a^*_{pl}\,(\lambda) = const\,(\lambda). \tag{6.21}$$

   In other words, these descriptions rely on one single, independent variable describing light absorption $a_{pl} = f\,(C_a)$, namely, the chlorophyll *a* concentration C_a in sea water (hence single-component models). It is further assumed in these models that the spectral chlorophyll-specific coefficients of light absorption $a^*_{pl}(\lambda)$ are identical under all environmental conditions (hence homogeneous) (e.g., Kopelevitch (1983) and Morel (1988)).
2. *Single-component, nonhomogeneous models*. These, too, rely on only one independent variable (C_a), but assume that the chlorophyll-specific absorption $a^*_{pl}(\lambda)$ varies with trophicity, which is the concentration of chlorophyll *a* in the sea (hence nonhomogeneous), that is:

$$a_{pl}\,(\lambda) = C_a \, a^*_{pl}\,(\lambda) \quad \text{and} \quad a^*_{pl} = f(C_a). \tag{6.22}$$

   Examples of such models are the two discussed later on in this section, by Woźniak and Ostrowska (1990a) and Bricaud et al. (1995, 1998).

3. *Multicomponent, homogeneous models.* These take into consideration the effect on absorption $a_{pl}(\lambda)$ of the concentrations of all groups of phytoplankton pigments $\{C_j\}$, not just that of chlorophyll *a* (hence multicomponent models). They can be symbolically represented as:

$$a_{pl}(\lambda) = C_a\, a^*_{pl}(\lambda) \quad \text{and} \quad a^*_{pl}(\lambda) = f(\{C_j/C_a\}). \qquad (6.23)$$

   This type of model includes the algorithm proposed by Bidigare et al. (1990) for determining light absorption coefficients of phytoplankton in oligotrophic seas. We discuss this algorithm later in greater detail. Formally, however, it is a single-component model because, as we show in due course, the chlorophyll-specific absorption is identical for fixed compositions of phytoplankton pigments ($\{C_j/C_a\} = \{const\}$) and does not depend on their absolute concentrations $\{C_j\}$.
4. *Multicomponent, nonhomogeneous models.* Here, as in the multicomponent, homogeneous models, the absorption $a_{pl}(\lambda)$ depends on the concentrations of all the groups of phytoplankton pigments and is described by the relationships:

$$a_{pl}(\lambda) = C_a\, a^*_{pl}(\lambda) \quad \text{and} \quad a^*_{pl}(\lambda) = f(Q^*(\lambda)\{C_j/C_a\}), \qquad (6.24)$$

   where $Q^*(\lambda)$ is the packaging effect function. Such models include the one worked out by our team, which we discuss later in detail (Woźniak et al. 1998,1999). This model is nonhomogeneous, because it also assumes the dependence of the chlorophyll-specific coefficients of absorption on the pigment packaging effect $Q^*(\lambda)$ in seas of different trophicity and at various depths, where the coefficients a^*_{pl} can take different values even when the pigment composition is identical.

In addition to these simple models, there are also complex utilitarian models, which we also discuss later on the basis of the multicomponent complex model MCM (Woźniak et al. 2003).

6.7.2 Classical Models

It was in the 1980s that the first classical descriptions of light absorption by phytoplankton appeared: they had been developed for application in analyses of this process in different seas. Those models were founded on the assumption that the spectral characteristics of the chlorophyll-specific light absorption coefficients of phytoplankton a^*_{pl} were constant, regardless of trophicity and depth in the sea. These characteristics were usually established on the basis of a statistical analysis of empirical absorption spectra $a_{pl}(\lambda)$ in natural phytoplankton communities or in cultures. One such model is that developed by Kopelevitch (1983). Here, the chlorophyll-specific light absorption coefficients of phytoplankton are presented in tabular form and correspond to the means of the coefficient $a^*_{pl}(\lambda)$ calculated from several hundred measurements in different regions of the World Ocean. (Figure 6.36, plot 1)

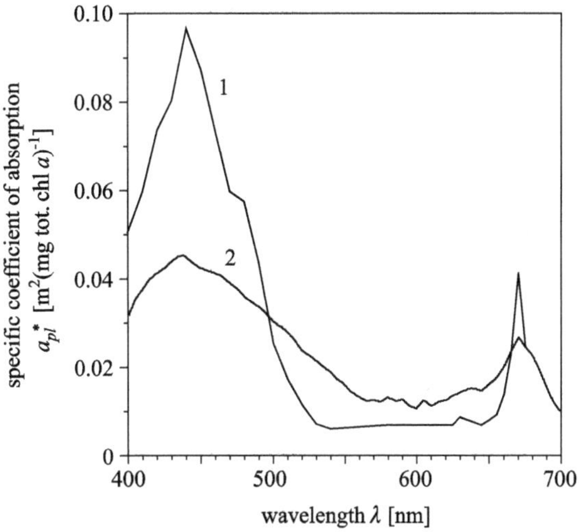

FIGURE 6.36. Spectral distribution of the chlorophyll-specific light absorption coefficient of phytoplankton a_{pl}^*: plot 1, after Kopelevitch (1983), plot 2, after Morel (1988).

illustrates the spectral distribution of this coefficient. Another example of this type of modeling is the description obtained by Morel (1988) on the basis of light absorption measurements in cultures of selected species of algae (Figure 6.36, plot 2). In principle, it differs from the Kopelevitch model in the different values established for the chlorophyll-specific light absorption coefficient of phytoplankton.

6.7.3 *Single-Component, Nonhomogeneous Models*

Empirical studies have shown that the absolute chlorophyll-specific light absorption coefficients of phytoplankton $a_{pl}^*(\lambda)$ depend strongly on the trophicity of the water (Figure 6.34a), a dependence that is particularly strong with respect to the blue region of the light spectrum. The highest values of $a_{pl}^*(\lambda)$ are recorded in oligotrophic seas, the lowest in eutrophic waters. The first analytical description to take account of this dependence is the model by Woźniak and Ostrowska, described in Woźniak (1990b) and developed further in Woźniak and Ostrowska (1990a). It is also the first nonhomogeneous single-component model of a general nature, describing as it does the absorption properties of phytoplankton from different seas and oceans. Basing their analysis on around 300 spectra of $a_{pl}^*(\lambda)$ [m^2 (mg tot. chl a)$^{-1}$], Woźniak and Ostrowska were able to approximate the spectral distribution of this coefficient by summing three Gaussian bands and making it dependent on the wavelength λ [nm] and the chlorophyll a concentration C_a [mg tot. chl a m^{-3}] by invoking the pigment index $P_{i,extr}$:

$$a^*_{pl}(\lambda) = (0.0187P_{i,extr} - 0.011)e^{-0.00012(\lambda-441)^2} + 0.00645e^{-0.00035(\lambda-608)^2} + 0.0233e^{-0.0014(\lambda-675)^2} \quad (6.25)$$

where $P_{i,extr}$ is the pigment index of acetone extracts of phytoplankton, that is, the ratio of the coefficient $a^*_{pl}(\lambda)$ for the two principal peaks of light absorption by phytoplankton dissolved in acetone (430 nm and 663 nm). It is linked to the chlorophyll concentration by the formula:

$$P_{i,extr} = 10^{[0.516 - 0.161x + 0.0422x^2 - 0.0584x^3 + 0.0360x^4]} \quad (6.26)$$

where $x = \log C_a(0)$

Two of these three Gaussian bands (441 nm and 675 nm) reflect physical reality, because they coincide with the maxima of light absorption by chlorophyll *a*. The third band (max. $\lambda = 608$ nm) was introduced as an adaptive correction.

Figure 6.37 compares the spectra of the chlorophyll-specific coefficients of light absorption $a^*_{pl}(\lambda)$ determined using this model (Fig. 6.37b) with the corresponding empirical spectra (Figure 6.37a). In comparison with the empirical data, the model overestimates the values of $a^*_{pl}(\lambda)$, a discrepancy that is conspicuous with regard to eutrophic waters (plots 1,2,3, . . . in Figure 6.37b). This overestimation is due to the fact that, in the development of this model, empirical data were employed which ignored the effect of light scattering in the phytoplankton samples during the measurement of the absorption coefficients (the so-called β-factor; see Section 6.6.1).

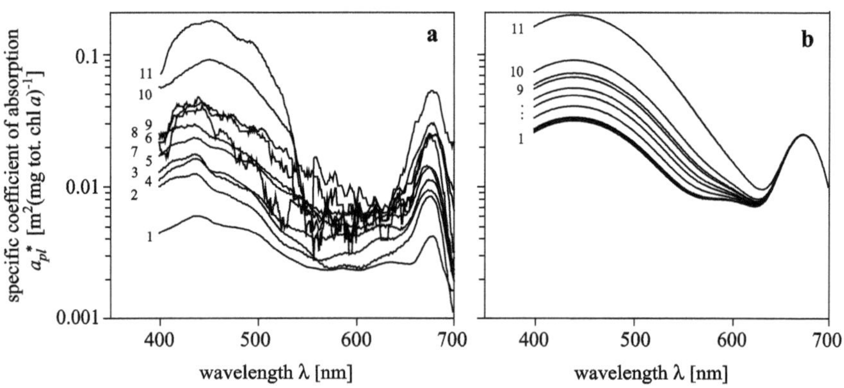

FIGURE 6.37. Comparison of chlorophyll-specific light absorption coefficients $a_{pl}^*(\lambda)$ of phytoplankton: (a) measured in situ; (b) determined using the model by Woźniak and Ostrowska (1990a). The numerals on the figures indicate the trophicity of the basin and the measurement station, as on Figure 6.29 in Section 6.6.2 (from highly eutrophic, 1, to extremely oligotrophic, 11).

A single-component model of greater precision than the one just described is that developed later by Bricaud et al. (1995,1998);[14] it is frequently cited in the literature. Here, the expression for the chlorophyll-specific light absorption coefficient of phytoplankton takes the form (as the Equation (5.40)):

$$a^*_{pl}(\lambda) = A_{pl}(\lambda) C_a^{E_{pl}(\lambda)-1}, \tag{6.27}$$

where $A_{pl}(\lambda)$, $E_{pl}(\lambda)$ are wavelength-dependent parameters defined on the basis of statistical analyses of c. 800 empirical spectra $a_{pl}(\lambda)$ and tabulated in the 400–700 nm spectral range with a 2 nm step (see Table 2 in Bricaud et al. (1995); see also Figure 5.41).

Figure 6.38 compares the chlorophyll-specific coefficients of absorption $a^*_{pl}(\lambda)$ determined from this model with experimental spectra. The Bricaud model yields a value of $a^*_{pl}(\lambda)$ smaller than that obtained by the Woźniak and Ostrowska model and more closely approaches real magnitudes of absorption. It is also encumbered by far smaller errors than the latter.

The single-component models by Woźniak and Ostrowska (1990a) and Bricaud et al. (1995,1998) in principle claim only to describe in general terms the diversity of the light absorption coefficients of phytoplankton resulting from the differences in trophicity of natural waters. They do not allow for the depth-related variations in $a^*_{pl}(\lambda)$ caused by adaptive processes in phytoplankton cells. Neither is it possible on their basis to separate absorption by photosynthetic pigments and by photoprotecting pigments, crucial factors

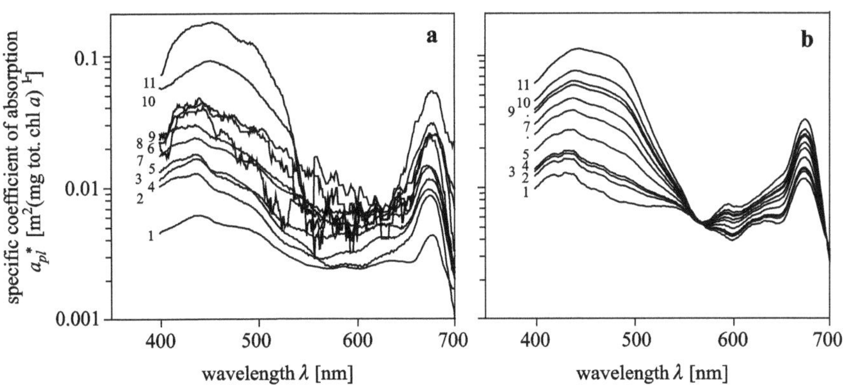

FIGURE 6.38. Comparison of chlorophyll-specific light absorption coefficients $a^*_{pl}(\lambda)$ of phytoplankton: (a) measured in situ; (b) determined using the model of Bricaud et al. (1995). The numerals on the figures indicate the trophicity of the basin and the measurement station, as in Figure 6.29 in Section 6.6.2 (from highly eutrophic, 1 to extremely oligotrophic, 11).

[14] In these papers, Bricaud et al. presented not only the statistical dependences of the specific light absorption coefficients of phytoplankton on the chlorophyll a concentration, but also similar statistical relationships for the absorption of light by other marine suspensions (see Section 5.3.3).

that must be taken into account in analyses and modeling of the photosynthetic process. Fulfilling these requirements only becomes possible with multicomponent models that take into account the effect on light absorption not only of chlorophyll *a,* but also of the other groups of phytoplankton pigments. It is to these models that we now turn our attention.

6.7.4 The Multicomponent, Homogeneous Model

In order to formulate a multicomponent model of the absorption properties of algae, we need to know the individual spectra of the mass-specific light absorption coefficients $a_j^*(\lambda)$ of all the main groups of phytoplankton pigments in the unpackaged dispersed state. As we have already mentioned (Sections 6.2 and 6.6) these spectra long remained unknown because they could not be defined by direct spectrophotometric measurements. It was probably Bidigare et al. (1990) who were the first to define the mean spectra of absorption $a_j^*(\lambda)$ for each of the main groups of phytoplankton pigments (chlorophyll *a* and its derivatives, chlorophyll *b* and its derivatives, the various forms of chlorophyll *c*, the natural associations of phycobilins, photosynthetic carotenoids, and photoprotecting carotenoids). They did so indirectly, by means of "adaptive" statistical analyses of c. 180 empirical spectra of the overall absorption of light by phytoplankton from various depths in the clean waters of the Central Atlantic (the Sargasso Sea area), and taking into account known concentrations of these pigments (determined by HPLC). These spectra were defined in the form of spectral tables (in the 400–700 nm range with a 2 nm step). They can be found in Section 6.2 alongside the other profiles in Figure 6.11.

Utilizing these spectra $a_j^*(\lambda)$ and assuming additionally that the packaging effect in the case of oceanic phytoplankton does not significantly affect its absorption properties, Bidigare's group put forward the following spectral reconstruction algorithm, which would enable the spectra of the overall light absorption coefficients of phytoplankton to be determined from known concentrations of pigments (C_j):

$$a_{pl}(\lambda) = \sum_j a_j^*(\lambda) C_j. \tag{6.28}$$

This formula is a special case of Equation (6.1), and assumes that the value of the packaging function is equal to unity for all wavelengths ($Q^*(\lambda) = 1$).

In accordance with the criteria we have chosen for classifying these models, the algorithm proposed by Bidigare et al. is formally a "multicomponent" model, in which the chlorophyll-specific coefficient of absorption is both a function of the trophicity (the chlorophyll *a* concentration C_a) and depends on the composition of accessory pigments:

$$a_{pl}^*(\lambda) = \frac{a_{pl}(\lambda)}{C_a} = a_a^*(\lambda) + \frac{C_b}{C_a} a_b^*(\lambda) + \frac{C_c}{C_a} a_c^*(\lambda) + \frac{C_{PSC}}{C_a} a_{PSC}^*(\lambda) + \frac{C_{PPC}}{C_a} a_{PPC}^*(\lambda) = const(C_j/C_a). \tag{6.29}$$

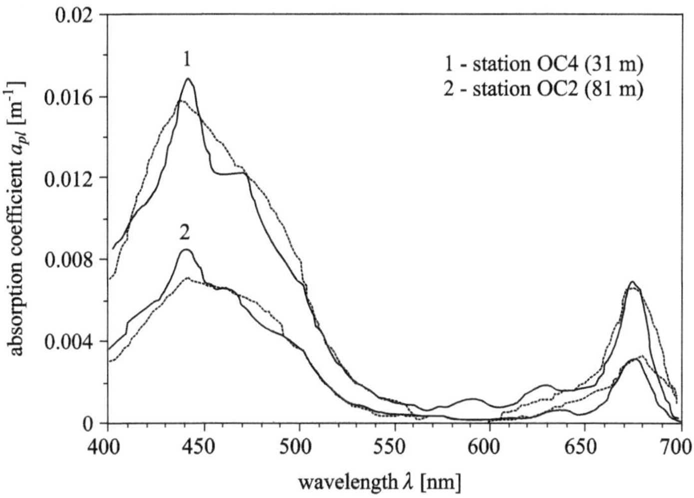

FIGURE 6.39. Spectra of light absorption coefficients of phytoplankton in vivo: calculated from the model (continuous line) and measured in phytoplankton samples taken from various depths in the Sargasso Sea. (Data taken from Bidigare et al. (1990).)

However, it is clear from this equation that Bidigare's model is homogeneous, because the spectral chlorophyll-specific coefficients of absorption remain constant for fixed pigment compositions $\{C_j/C_a\}$ and are independent of their absolute concentrations $\{C_j\}$. In fact, these coefficients $a^*_{pl}(\lambda)$ can vary, even if the pigment compositions are the same: this is primarily due to the packaging effect.

Bidigare's model of the absorption properties of algae is thus applicable solely to oligotrophic basins (the central regions of oceans), because the effect of packaging on these properties has been neglected. In such waters, the cells of algae are relatively small and the packaging functions will be close to unity ($Q^* \approx 1$). The spectra of $a_{pl}(\lambda)$ measured in such waters are indeed convergent with the values obtained from this model (Figure 6.39). Our calculations have shown, however, that if it is applied to phytoplankton in different seas, especially where the trophicity $C_a \geq 0.5$ mg tot. chl $a \cdot \mathrm{m}^{-3}$, without taking account of the packaging function, the results are encumbered with serious errors, and is therefore not to be recommended.

6.7.5 *The Multicomponent, Nonhomogeneous Model*

The first multicomponent model of the absorption properties of algae, which is nonhomogeneous because of the necessity to take account of the pigment packaging effect, is the one gradually being developed by our own research team and described in successive papers: Woźniak et al. (1998, 1999, 2000) and Majchrowski et al. (2000). It emerged from thorough analyses of empirical material with the implementation of differential spectroscopy and a battery of statistical methods. Without going into the details of the various stages of construction of the model (these were described in the above-mentioned papers),[15] we now outline the most important aspects.

- (*Stage I*) Differential spectroscopy was used to analyze more than 1400 empirical spectra of phytoplankton absorption $a_{pl}(\lambda)$ in vivo and the spectra of their fourth derivative. It was thereby demonstrated that the spectra of $a_{pl}(\lambda)$ in the general case can be stated with very considerable precision as the sum of 21 elementary absorption bands described by Gaussian functions.
- (*Stage II*) The relevant "adaptive" statistical analyses were performed on a set of more than 400 empirical spectra of the chlorophyll-specific absorption a^*_{pl}, in vivo, for which the pigment compositions were defined very precisely by chromatography. As a result of these analyses, the elementary light absorption bands were ascribed to particular groups of pigments and the pigment packaging states were established for these spectra $a^*_{pl}(\lambda)$. These states are well defined by the products $C_{chl}D$ (where C_{chl} is the intracellular concentration of chlorophyll a and D is the cell diameter; see Equation (6.15)). It turns out that the products $C_{chl}D$ calculated for the various phytoplankton samples are quite well correlated with the chlorophyll a concentration C_a in the sea (see Figure 6.26c.).

  For modeling purposes, it is thus assumed that the product $C_{chl} \cdot D$ depends to a first approximation only on the chlorophyll a concentration, and can be described as given by Equation (6.15):

  $$C_{chl}D = 24.65 C_a^{0.75}, \tag{6.30}$$

  where the relevant magnitudes are expressed in the following units: intracellular concentration of chlorophyll a, C_{chl} [mg tot. chl *a* m^{-3}]; cell diameter D [m]; chlorophyll a concentration in sea water C_a [mg tot. chl *a* m^{-3}].

- (*Stage III*) The spectra of the chlorophyll-specific absorption of light by phytoplankton with pigments in the "unpackaged" (in solvent) state $a^*_{pl,S}(\lambda)$ were reproduced for the set of 400 empirical spectra of the chlorophyll-specific absorption of light by phytoplankton $a^*_{pl}(\lambda)$ in vivo. Following the Gaussian analysis of this set of "unpackaged" spectra $a^*_{pl,S}(\lambda)$, Gaussian decomposition of these spectra was performed and averaged formulas were found for the spectral mass-specific coefficients of light absorption by the various groups of pigments $a^*_j(\lambda)$ (for chlorophyll a, $a^*_a(\lambda)$; for chlorophyll b, $a^*_b(\lambda)$; for chlorophyll c, $a^*_c(\lambda)$; for photosynthetic carotenoids, $a^*_{PSC}(\lambda)$; for photoprotecting carotenoids, $a^*_{PSC}(\lambda)$). These formulas refer to light absorption by pigments in the unpackaged (insolvent, dispersed) state and take the form of the sum of Gaussian bands (see Equation (6.3) and the parameters in Table 6.8).

---

[15] The complete set of mathematical techniques used in the Gaussian and statistical analyses is discussed in Woźniak (2000), Woźniak et al. (1999) .

TABLE 6.14. Algorithm for determining the absorption properties of phytoplankton.

INPUT DATA

C_j [mg pigment m^{-3}] are the concentrations of the separate (principal) groups of pigments (chl *a*; chl *b*; chl *c*; PSC; PPC)

MODEL FORMULAS

1. Spectra of mass-specific coefficients of light absorption by the separate groups of pigments in the unpackaged, dispersed state:

$$a_j^*(\lambda) = \Sigma a_{max,i}^* e^{-\frac{1}{2}\left(\frac{\lambda - \lambda_{max,i}}{\sigma_i}\right)^2}, \qquad \text{(T-1)}$$

where $\lambda_{max,i}$ [nm] is the center of the *i*th spectral band maximum; σ_i is the band dispersion; $a_{max,i}^*$ [m^2 (mg pigment)$^{-1}$] is the mass-specific coefficient of absorption for the wavelength $\lambda_{max,i}$ corresponding to the center of the spectral band maximum; *i* is the number of the Gaussian band for the principal groups of phytoplankton pigments.
The values of these parameters are given in Table 6.8 in Section 6.2.

2. Chlorophyll-specific "unpackaged" coefficients of light absorption by photosynthetic pigments a_{PSP}^*, photoprotecting pigments a_{PPP}^*, and all pigments $a_{pl,S}^*$:

$$a_{PSP}^*(\lambda) = \frac{1}{C_a}\left[C_a a_a^*(\lambda) + C_b a_b^*(\lambda) + C_c a_c^*(\lambda) + C_{PSC} a_{PSC}^*(\lambda)\right] \qquad \text{(T-2)}$$

$$a_{PPP}^*(\lambda) = \frac{1}{C_a}\left[C_{PPC} a_{PPC}^*(\lambda)\right] \qquad \text{(T-3)}$$

$$a_{pl,S}^*(\lambda) = a_{PSP}^*(\lambda) + a_{PPP}^*(\lambda). \qquad \text{(T-4)}$$

3. Coefficients of absorption in vivo and chlorophyll-specific coefficients of absorption in vivo for plankton and the main groups of pigments:
-For plankton (all pigments):

$$\left.\begin{aligned} a_{pl}(\lambda) &= C_a \cdot a_{pl}^*(\lambda) \\ a_{pl}^*(\lambda) &= Q^*(\lambda) \cdot a_{pl,S}^*(\lambda) \end{aligned}\right\} \qquad \text{(T-5)}$$

-For photosynthetic pigments *PSP*:

$$\left.\begin{aligned} a_{pl,PSP}(\lambda) &= C_a\, a_{pl,PSP}^*(\lambda) \\ a_{pl,PSP}^*(\lambda) &= Q^*(\lambda) a_{PSP}^*(\lambda) \end{aligned}\right\} \qquad \text{(T-6)}$$

-For photoprotecting pigments *PPP*:

$$\left.\begin{aligned} a_{pl,PPP}(\lambda) &= C_a\, a_{pl,PPP}^*(\lambda) \\ a_{pl,PPP}^*(\lambda) &= Q^*(\lambda) a_{PSP}^*(\lambda) \end{aligned}\right\}, \qquad \text{(T-7)}$$

where $Q^*(\lambda)$ is the packaging function, equal to:

$$Q^*(\lambda) = \frac{3}{2\rho'(\lambda)}\left[1 + \frac{2e^{-\rho'(\lambda)}}{\rho'(\lambda)} + 2\frac{e^{-\rho'(\lambda)} - 1}{\rho'^2(\lambda)}\right] \qquad \text{(T-8)}$$

$$\rho'(\lambda) = a_{pl,S}^*(\lambda) C_{chl} D,$$

where $C_{chl}D$ is the product of the intracellular concentration of chlorophyll C_{chl} [mg tot. chl *a* m^{-3}] and the cell diameter *D* [m] given by the equation:

$$C_{chl}D = 24.65 C_a^{0.75}. \qquad \text{(T-9)}$$

In accordance with the model by Woźniak et al. (1999, 2000, 2003).

The results of these spectroscopic and statistical analyses thus yield a full set of analytical formulas, from which we can formulate the algorithm this model is based on. Table 6.14 gives the mathematical methods of this algorithm.

As the table shows, the model provides for the estimation of overall (a_{pl}) and chlorophyll-specific (a^*_{pl}) absorptions, jointly for all phytoplankton pigments, and separately for the individual groups of pigments. It thus suffices to establish the quantitative composition of the pigments in the given phytoplankton. In particular, it is important that the absorption in vivo by photosynthetic pigments ($a_{pl,PSP}$, $a^*_{pl,PSP}$) and by photoprotecting pigments ($a_{pl,PPP}$, $a^*_{pl,PPP}$) are investigated separately. This is one of this model's great advantages, as it enables, for example, the utilization of light energy during photosynthesis to be analyzed and this process to be modeled under diverse conditions in the sea. A further virtue of the model is the quite good estimation of the light absorption spectra of algae (distinctly better than with the earlier models), together with aspects of their fine structure. Evidence for this is the comparison of the spectra of absorption $a^*_{pl}(\lambda)$ calculated using this model with the corresponding empirical spectra, shown in Figure 6.40. The errors in the estimates of the absorption coefficients using the model are relatively low. As Woźniak et al. (2003) showed, the statistical error as regards the mean chlorophyll-specific absorption in the PAR spectral interval (400–700 nm) for this model is c. 33%. It is thus comparable with the methodological error inherent in state-of-the-art empirical methods of determining the chlorophyll-specific light absorption coefficients of algae in vivo. At the same time, this error is smaller than in the case of the earlier single-component methods: c. 43% of the error in the model by Bricaud et al. (1995, 1998) and c. 59% of the error in the model by Woźniak and Ostrowska (1990a).

To be fair, we should add that despite its merits, this model still has numerous shortcomings and requires further refinement. The first problem that

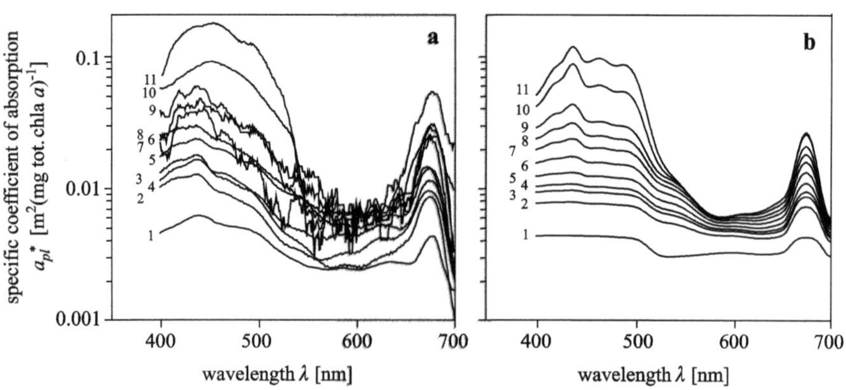

FIGURE 6.40. Comparison of the spectral chlorophyll-specific light absorption coefficients of phytoplankton: (a) measured; (b) determined from the multicomponent nonhomogeneous model developed by Woźniak et al. (2000). The numerals on the figures indicate the trophic type of sea and the measurement station, as in Figure 6.29 in Section 6.6 (from highly eutrophic, 1 to extremely oligotrophic, 11).

needs addressing is the absorption of light by phycobilins, which the model in its present form has neglected. On-going analyses are showing that this model is less effective when applied to the algae of coastal waters, that is, in Case 2 waters such as those of the Baltic, which may well contain these particular pigments more frequently and in significant quantities.[16] Moreover, by making the pigment packaging effect in the current version of the model dependent only on the chlorophyll concentration (see, e.g., the set of equations (T-8) and (T-9) in Table 6.14), we have only a first approximation of this description; here too, further study and refinement are needed.

6.7.6 *Complex Utilitarian Models*

Apart from these main types of simple semi-empirical models, with which we can determine the absorption properties of algae from known concentrations of chlorophyll *a* (C_a), or known concentrations of all the main types of phytoplankton pigments ($\{C_j\}$), there is also a class of complex utilitarian models. Constructed from a number of more or less complicated modules, they are usually a synthesis of one of the above-mentioned simple models of light absorption by algae with models of other phenomena taking place in natural waters. Able to fulfill various functions in the research, monitoring, and forecasting of processes in the marine environment, they generally involve the indirect determination of the absorption properties of phytoplankton at different depths on the basis of "more distant" environmental parameters, and not directly from known concentrations of pigments. A particularly important group of these models are the satellite algorithms for determining the absorption properties of algae.

A good example is the algorithm for determining absorption coefficients devised by Antoine and Morel (Antoine and Morel 1996, Antoine et al. 1996) to estimate distributions of primary production in the World Ocean on the basis of satellite data obtained from the CZCS scanner. Another example is our own Multicomponent Complex Model (MCM; Woźniak et al. 2003, Ficek et al. 2003), a brief description of which follows.

The algorithms in both these models enable the light absorption coefficients of algae to be estimated at any depth especially in Case 1 waters in the sea on the basis of only one (in the Antoine and Morel model) or at most two (in our MCM) parameters describing the state of the surface layer of the sea. The first of these is the surface concentration of chlorophyll *a*, $C_a(z \cong 0) \equiv C_a(0)$, which, as we have already stated several times, is the trophic index or trophicity of a basin. The second parameter is the irradiance of the sea surface by natural daylight. Adapting complex utilitarian models so that they can make use of these two parameters is of real practical importance. The parameters

[16] Up to the present moment we and our team have managed to develop a preliminary version of such a model, which also includes phycobilins among the photosynthetic pigments in phytoplankton (see Ficek et al. (2004)). This model describes the basins of the southern Baltic, but further research will be required before it can be extended to other basins.

can be determined efficiently and over the vast expanses of the oceans by remote-sensing techniques; on the other hand, the models allow us to study the energy inflow to marine ecosystems and also to monitor the primary production of organic matter and the natural carbon cycle. We shortly reveal the essence of these complex utilitarian models using our MCM as an example (see Woźniak et al. (2003) and the papers cited therein).

One of the principal components of the MCM is the multicomponent homogeneous model of the absorption properties of algae by Woźniak et al. (2000), which we discussed earlier. The MCM also includes a number of other blocks consisting of models, developed by our research team, of various phenomena in the sea. These are shown in the block diagram of that part of the MCM algorithm which applies to the absorption properties of algae in oceanic Case 1 waters (Figure 6.41). The originators of the component models are cited there. This simplified block diagram consists of three sectors: (A) *input data*, indispensable for performing the computations; (B) *model formulas,* which define the methods of calculation; they emerge from the set of partial models of different phenomena in the sea. The MCM is an attempt at a synthesis of this set of partial models; and (C) *calculations*, that is, a detailing of the calculated abiotic and biotic characteristics of the environment, from the vertical profiles of chlorophyll *a* concentration $C_a(z)$ (block 9 in Figure 6.41) to the coefficients of light absorption by phytoplankton, together with—and this is of crucial significance for the study of photosynthesis—the separate coefficients of absorption by photosynthetic and photoprotecting pigments (block 18 in Figure 6.41).

A detailed description and the computational algorithm of the MCM would go beyond the scope of this monograph. For these, the reader is directed to the papers, cited earlier, by the members of our research group. Nevertheless, the relevant calculations can be performed with the information given in this book: all that is needed is to apply the diagram in Figure 6.41 and a few of the partial algorithms, all of which are given in the present chapter[17] (Tables 6.10, 6.3, 6.12, and 6.14).

[17] Formally the principle of these computations is as follows. The input data are the two magnitudes that can be determined by remote sensing: surface chlorophyll $C_a(0)$ (block 1 in Figure 6.41), surface irradiance $PAR_0(0)$ (block 3); the third input datum is an independent variable: depth in the sea z (block 2). On this basis the following can be determined in succession: (block 9) depth profiles of chlorophyll $C_a(z)$ using their modeled profiles (block 4) based on the algorithm given in Table 6.10; (blocks 10–12) various optical characteristics of the irradiance fields in the sea, the spectral fitting functions of the pigments $F_i(z)$, the photodestructive radiation $PDR(z)$ on the basis of biooptical models of the sea (block 5), and the definitions of the adaptation factors (block 6), the algorithms for which are given in Table 6.3; (blocks 13 and 14) the relative concentrations of different phytoplankton pigments, based on the relevant relationships between the adaptation factors and these concentrations, given in Table 6.12; (blocks 15–18) factor $C_{chl}D$, the packaging function $Q^*(\lambda,z)$ and the various unpackaged (in solvent) and in vivo coefficients of light absorption by algae and the groups of their pigments on the basis of the model of light absorption by algae, the algorithm for which is given in Table 6.14.

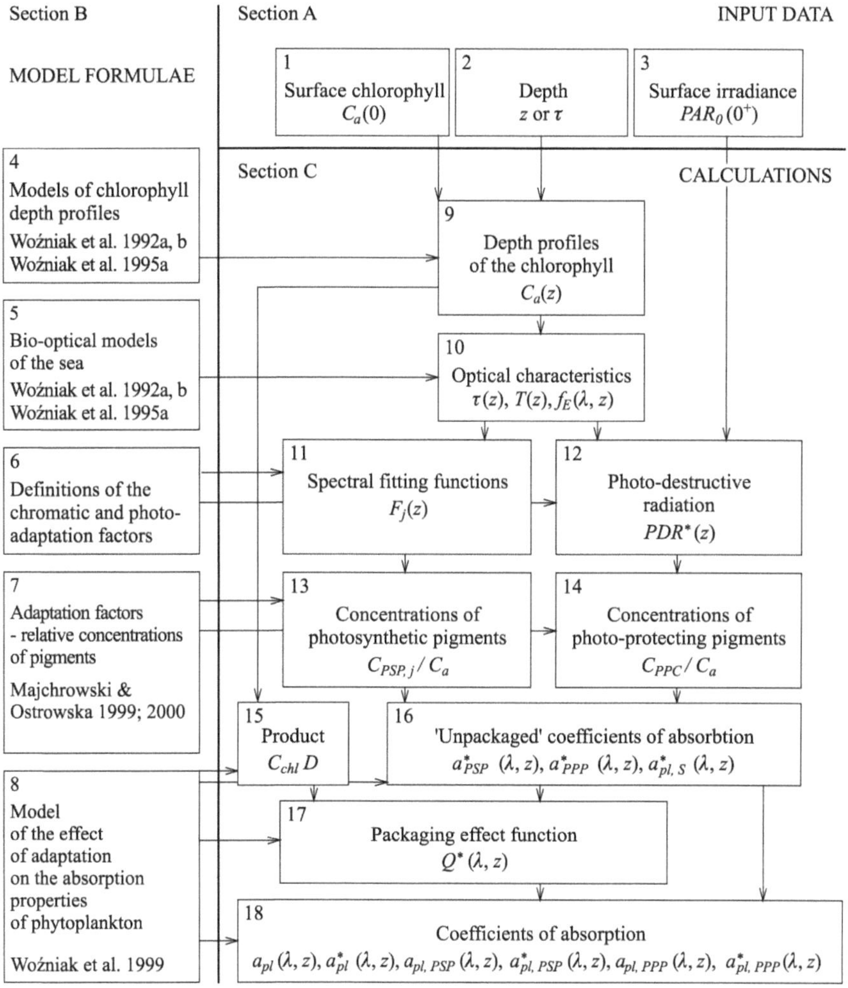

FIGURE 6.41. Block diagram of the model of the spectral absorption properties of phytoplankton pigments in vivo (i.e., that part of the MCM algorithm referring to the absorption properties of algae in oceanic Case 1 waters).

Apart from its applications in satellite algorithms, the MCM makes it possible to systematize and to quantify more precisely our knowledge of the regularities and the scale of diversity in the absorption properties of marine algae. This application of the model is what we now look at.

6.7.7 Modeled Absorption Properties of Algae in Different Types of Sea

Thus far in this chapter, we have described, on the basis of empirical results, the nature of light absorption by phytoplankton and the main environmental factors governing this process. In doing so, we have laid stress on the principal regularities and the scale of diversity of the absorption properties of phytoplankton in different trophic types of sea and at different depths. These descriptions are primarily analytical, not synthetic, and the optical characteristics of phytoplankton we have presented are mostly only qualitative examples.

Now we attempt to systematize and synthesize a general quantitative description of the absorption properties of phytoplankton in the World Ocean. This description is based on the calculations of these properties using the MCM algorithm (Figure 6.41). The input data are the trophicity of the basin $C_a(0)$, the zonal mean daily irradiance of the sea surface $<PAR(0)_{day}>$, and, as an independent variable, depth in the sea.

The natural diversity in the light absorption coefficients of phytoplankton covers four orders of magnitude and is correlated mainly with the trophicity of waters $C_a(0)$, or more precisely, with the in situ chlorophyll *a* concentration $C_a(z)$. This is shown by the profiles calculated using the model (Figure 6.42). These show the depth profiles of the mean coefficients of light absorption (in the 400–700 nm spectral range) in waters of different trophicity by all phytoplankton pigments $\bar{a}_{pl}$, and similar distributions of these coefficients for photosynthetic pigments (*PSP*) $\bar{a}_{pl,\,PSP}$. These coefficients were determined by averaging the spectra $a_{pl}(\lambda)$ and $a_{pl,PSP}(\lambda)$ estimated with the aid of the model:

$$\bar{a}_{pl} = \frac{1}{300}\int_{400nm}^{700nm} a_{pl}(\lambda)\,d\lambda \quad \text{and} \quad \bar{a}_{pl,\,PSP} = \frac{1}{300}\int_{400nm}^{700nm} a_{pl,\,PSP}(\lambda)\,d\lambda \qquad (6.31)$$

As Figure 6.42 shows, the plots of these depth profiles of absorption coefficients resemble the vertical profiles of chlorophyll *a* concentrations C_a in waters of different trophicity (Figure 6.15). However, on moving from basins of lower trophicity to more fertile waters, the absolute values of absorption rise more slowly than do the chlorophyll concentrations. This is demonstrated by the chlorophyll-specific light absorption coefficients of phytoplankton ($\bar{a}^*_{pl}$ *and* $\bar{a}^*_{pl,\,PSP}$), which are at once the coefficients of the proportionality between the absorption coefficients and the chlorophyll concentrations; that is, $\bar{a}_{pl} = \bar{a}^*_{pl} C_a$ and $\bar{a}_{pl,\,PSP} = \bar{a}^*_{pl,\,PSP} C_a$).

The model vertical profiles of these chlorophyll-specific coefficients of absorption (which can take values differing by a factor of several tens) are illustrated in Figure 6.43. This shows that the diversity of these coefficients is correlated with the trophicity of the basin and with depth. Nevertheless, these are correlations emerging from indirect relationships, because directly, the chlorophyll-specific absorption coefficients are governed by the packaging effect and the pigment composition in the phytoplankton. Here, the packaging effect is the major influence on the value of the chlorophyll-specific absorption coefficient. This influence is manifested by a c. 20-fold differentiation in the coefficients $\bar{a}^*_{pl}$ *and* $\bar{a}^*_{pl,PSP}$, and is further dependent on the trophicity. These coefficients are highest in oligotrophic seas and decrease in value with rising trophicity, from c. $\bar{a}^*_{pl}$ = 0.032 [m^2 (mg tot. chl *a*)$^{-1}$] (for

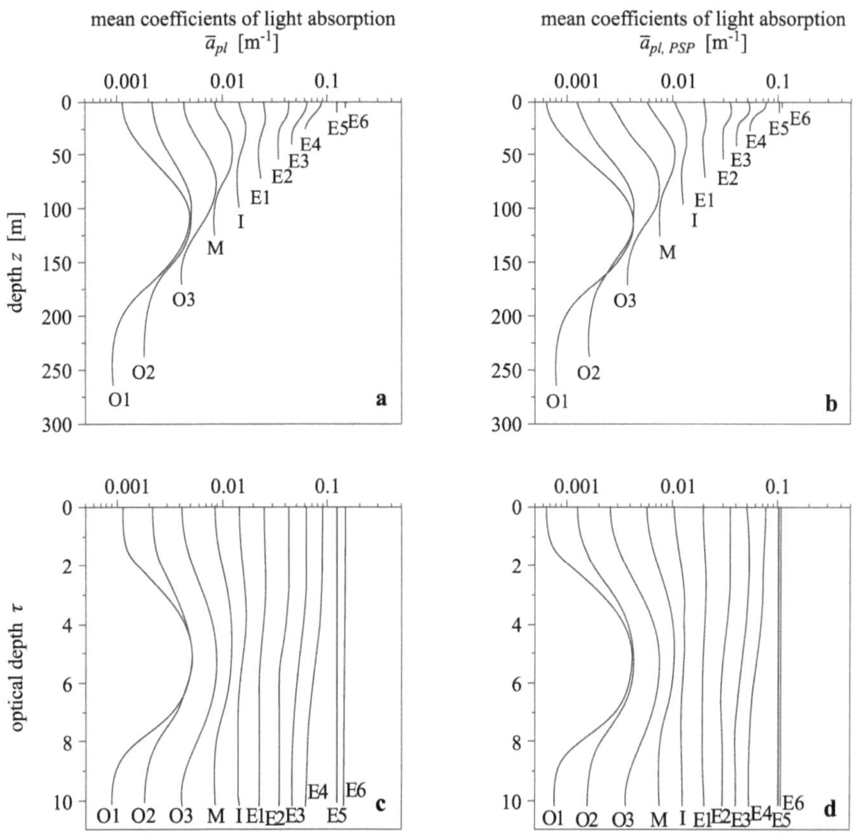

FIGURE 6.42. Depth profiles of mean coefficients of light absorption by phytoplankton in oceanic Case 1 waters, estimated using the model: (a) by all phytoplankton pigments $\bar{a}_{pl}$ depending on real depth in the sea z; (b) by photosynthetic pigments in phytoplankton $\bar{a}_{pl,\,PSP}$ with respect to depth z; (c) by all phytoplankton pigments $\bar{a}_{pl}$ with respect to optical depth τ; (d) by photosynthetic pigments in phytoplankton $\bar{a}_{pl,\,PSP}$ with respect to optical depth τ. For a surface irradiance of $PAR(0^+) = 695$ µEin m^{-2} s^{-1} and waters of different trophicities, from highly eutrophic to extremely oligotrophic O1. The trophicity symbols are explained in Table 6.1.

type O1, $C_a(0) \approx 0.035$ [mg tot. chl a m^{-3}]) to c. $\bar{a}^*_{pl} = 0.0016$ [m^2 (mg tot. chl a)$^{-1}$] (for type E6, $C_a(0) \approx 70$ [mg tot. chl a m^{-3}]).

Directly, this is due to the diversity in phytoplankton cell size, which in oligotrophic waters are smallest (diameters D are small, hence the product $C_{chl}D$ takes small values; see earlier, Equation (6.15) and Figures 6.26). Thus, the influence of the pigment packaging effect on light absorption by phytoplankton in oligotrophic waters is small, frequently negligible. On the other hand, as chlorophyll concentrations rise, so do the cell dimensions, and the effect of pigment packaging on light absorption by these cells makes itself

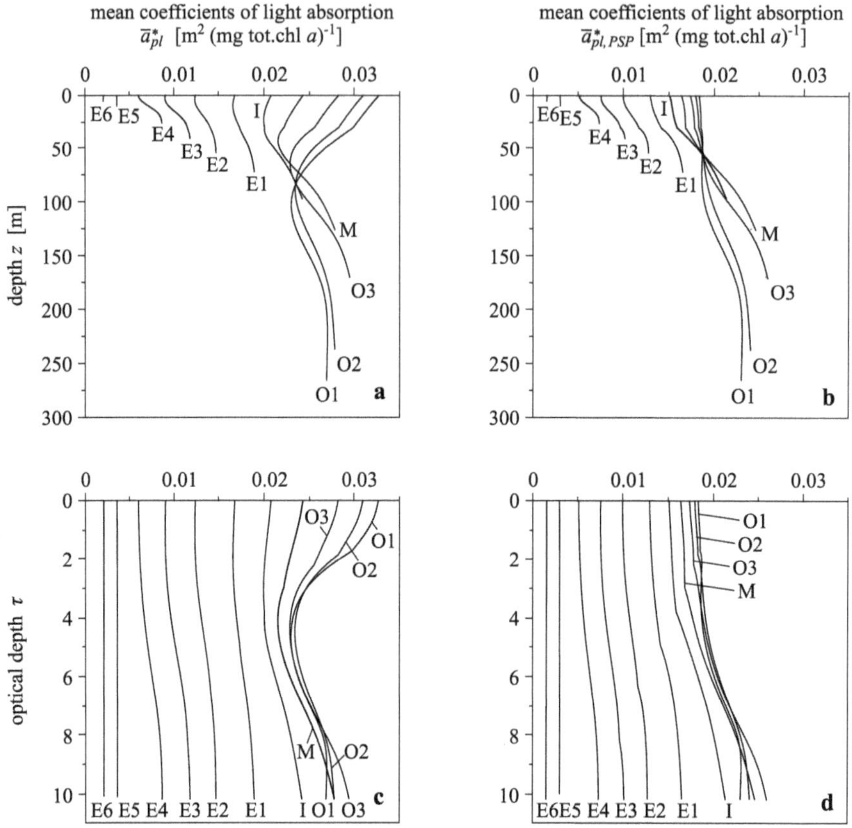

FIGURE 6.43. Depth profiles of chlorophyll-specific coefficients of light absorption by phytoplankton in oceanic Case 1 waters, estimated using the model: (a) by all phytoplankton pigments $\bar{a}^*_{pl}$ depending on real depth in the sea z; (b) by photosynthetic pigments in phytoplankton $\bar{a}^*_{pl, PSP}$ with respect to depth z; (c) by all phytoplankton pigments $\bar{a}^*_{pl}$ with respect to optical depth τ; (d) by photosynthetic pigments in phytoplankton $\bar{a}^*_{pl, PSP}$ with respect to optical depth τ. For a surface irradiance of $PAR(0^+) = 695$ μEin m^{-2} s^{-1} and waters of different trophicities, from highly eutrophic to extremely oligotrophic O1. The trophicity symbols are explained in Table 6.1.

ever more strongly felt, which leads to a fall in the value of the chlorophyll-specific absorption coefficients. Thus, as a result of the pigment packaging effect, chlorophyll-specific light absorption coefficients of phytoplankton in eutrophic basins are much lower than in oligotrophic basins, and, in the particular case of blue light, may be smaller by a factor of several tens. Of course, this diversity of chlorophyll-specific absorption with trophicity is also affected by other factors, but their influence is very much less than that of the pigment packaging effect. A certain influence is exerted by the different pigment compositions in different phytoplankton communities. The pigment composition has a considerable influence on the depth-related diversity of chlorophyll-specific absorption coefficients. The shapes of the profiles $\bar{a}^*_{pl}(z)$ and $\bar{a}^*_{pl,\,PSP}(z)$ are sufficient to illustrate this, not to mention a comparison of these shapes (Figure 6.43). In the case of PSP, the mean absorption coefficient $\bar{a}^*_{pl,\,PSP}$ increases in value with depth, which is due to the vertical increase in the relative concentration of these pigments. On the other hand, in the case of the mean chlorophyll-specific coefficient of absorption $\bar{a}^*_{pl}$ we see for all phytoplankton pigments a depth-related minimum (the lowest values of this coefficient in the profile). From oligotrophic to eutrophic waters this minimum shifts towards the surface. This increase in $\bar{a}^*_{pl}$from the minimum towards the surface is caused by a rise in the relative concentrations of PPC, and the increase below the minimum is due to a rise in the relative concentrations of PSP. These effects, more pronounced in oligotrophic than in eutrophic waters, are governed by the processes of photoadaptation and chromatic adaptation, described in Section 6.4.

Examples of the similar influence of photo- and chromatic adaptation, not on the mean chlorophyll-specific coefficients of absorption, but on their spectra, are shown in Figure 6.44. These provide firm evidence for the originality of the MCM and its usefulness, and also for describing the adaptation of phytoplankton to ambient light conditions in the sea.

Figure 6.44a shows spectra of chlorophyll-specific coefficients of absorption by phytoplankton $a^*_{pl}(\lambda)$ determined for mesotrophic waters using the model for different optical depths (from $\tau = 0$ to twice the depth of the euphotic zone; i.e., $\tau = 9.2$). The values of this coefficient are highest at the sea surface, but decrease with depth to reach a minimum at around the limit of the euphotic

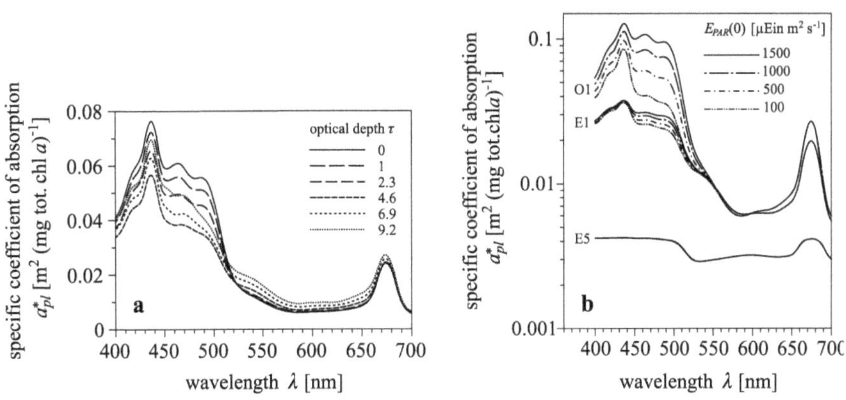

FIGURE 6.44. Spectra of chlorophyll-specific absorption coefficients of phytoplankton in oceanic Case 1 waters, determined using the MCM: (a) for different optical depths, irradiance $PAR(0^+)$ = 1400 μEin m^{-2} s^{-1} and a surface concentration of chlorophyll a, $C_a(0)$ = 0.35 mg tot. chl a m^{-3}; (b) for phytoplankton in oligotrophic O1 ($C_a(0)$ = 0.035 mg tot. chl a m^{-3}), eutrophic E1 ($C_a(0)$ = 1.5 mg tot. chl a m^{-3}), and supereutrophic waters E5 ($C_a(0)$ = 35 mg tot. chl a m^{-3}) for an optical depth τ. = 1 and for various values of the surface irradiance $PAR(0^+)$ from 100 to 1500 μEin · m^{-2} s^{-1}. (After Majchrowski et al. (2000).)

zone ($\tau = 4.6$). They then begin to increase again, for the reasons we explained earlier.

Figure 6.44b illustrates spectra of the chlorophyll-specific light absorption coefficient of phytoplankton in oligotrophic and eutrophic waters for an optical depth of $\tau = 1$ and for various values of the surface irradiance. It is known that an increase in the surface irradiance can raise the amount of photodestructive radiation *PDR** (defined in Section 6.1: Table 6.3, expressions (T-9a) and (T-9b)). This in turn brings about an increase in the concentrations of photoprotecting carotenoids. According to the figure, the chlorophyll-specific coefficient of absorption by phytoplankton $a^*_{pl}(\lambda)$ is the highest in value when the surface irradiance is the greatest, but as the irradiance decreases, $a^*_{pl}(\lambda)$ drops to the level of the chlorophyll-specific coefficient of absorption by PSP. It is characteristic that values of $a^*_{pl}(\lambda)$ are highest in oligotrophic waters: in this case, their large variations in relation to the irradiance are clearly visible. In the case of eutrophic waters, these values fall and reactions to changes in the irradiance are not as pronounced.

References

Aas, E., 1996, Refractive index of phytoplankton derived from its metabolite composition, *J. Plankton Res.* **18**:2223–2249.
Ackelson, S.G., Robins, D.B., and Stephens, J.A., 1988, Distributions in phytoplankton refractive index and size within the North Sea, *SPIE, Vol. 925, Ocean Optics, IX,* 317–325.
Adamczewski, I., 1965, *Ionization and Conductivity of Liquid Dielectric*, PWN, Warszawa, p. 445 (in Polish).
Adedokun, J.A., Emofurieta, W.O., and Adedeji, O.A., 1989, Physical, mineralogical and chemical properties of Harmattan dust at Ile-Ife, Nigeria, *Theor. Appl. Climatol.* **41**:161–169.
Aglintzev, K.K., 1961, *Ionizing Radiation Dosimetry*, PWN, Warszawa (in Polish, translated from Russian: Gostekhizdat, Moskva, 1957).
Ahn, Y.H., Bricaud, A., and Morel, A., 1992, Light backscattering efficiency and related roperties of some phytoplankters, *Deep-Sea Res.* **39**:1835–1855.
Aiken, G.R., McKnight, D.M., Wershaw, R.L., and McCarthy, P. (Eds.), 1985, *Humic Substances in Soil, Sediments, and Water: Geochemistry, Isolation and Characterization*, 1st ed., Wiley, New York, p. 532.
Alekin, O.A., 1970, *Hydrochemistry*, Gidrometeoizdat, Leningrad, p. 444 (in Russian).
Allali, K., Bricaud, A., and Claustre, H., 1997, Spatial variations in the chlorophyll-specific absorption coefficients of phytoplankton and photosynthetically active pigments in the equatorial Pacific, *J. Geophys. Res.* **102**:12413–12423.
Alpert, N.L., Keiser, W.E., and Szymański, H.A., 1974, *Infra-Red Spectroscopy*, PWN, Warszawa, p. 357 (in Polish).
Aly, K.M. and Esmail, E., 1993, Refractive index of salt water: Effect of temperature, *Opt. Mater.* **2**:195–199.
Anderson, L.G., 2002, DOC in the Arctic Ocean, in: D.A. Hansell and C.A. Carlson (Eds.), 2002 (op. cit.), pp. 665–683.
Antia, N.J., McAllister, C.D., Parsons, T.R., Stephens, K., and Strickland J.H.D.,1963, Further measurements of primary production using a large-volume plastic sphere, *Limnol. Oceanogr.* **8**:166–183.
Antoine, D. and Morel, A., 1996, Oceanic primary production: 1. Adaptation of spectral light-photosynthesis model in view of application to satellite chlorophyll observations, *Global Biogeochem. Cy.* **10**(1):42–55.
Antoine, D., André, J.M., and Morel, A., 1996, Oceanic primary production: 2. Estimation at global scale from satellite (Coastal Zone Color Scanner) chlorophyll, *Global Biogeochem. Cy.* **10**(1):56–69.
Armstrong, F.A.J. and Boalch, G.T., 1961, Ultraviolet absorption of sea water and its volatile components. A Symposium on Radiant Energy in the Sea (Helsinki), *IGGU Monogr.* **10**:63–66.

Armstrong, S.H., Jr, Budka, M.J.E., Morrison, K.C., and Hasson, M., 1947, Preparation and properties of serum and plasma proteins. XII. The refractive properties of the proteins of human plasma and certain purified fractions, *J. Am. Chem. Soc.* **69**:1747–1753.

Arnoff, S., 1950, Chlorophyll. *Bot. Rev.* **16**:525–588.

Arst, H., 2003, *Properties and Remote Sensing of Multicomponental Water Bodies*, Springer, Praxis, New York, p. 231.

Asano, S., and Yamamoto, G., 1975, Light scattering by a spheroidal particle. *Appl. Opt.* **14**:29–49.

Asano, S., 1979, Light scattering properties of spheroidal particles, *Appl. Opt.* **18**:712–722.

Babin, M., Therriault, J.-C., Legendre, L., and Condal, A., 1993, Variations in the specific absorption coefficient for natural phytoplankton assemblages: Impact on estimates of primary production, *Limnol. Oceanogr.* **38**(1):154–177.

Babin, M., Morel, A., Claustre, H., Bricaud, A., Kolber, Z., and Falkowski, P.G., 1996a, Nitrogen- and irradiance-dependent variations of the maximum quantum yield of carbon fixation in eutrophic, mesotrophic and oligotrophic marine systems, *Deep-Sea Res.* **43**(8):1241–1272.

Babin, M., Morel, A., and Gentili, B., 1996b, Remote sensing of sea surface sun-induced chlorophyll fluorescence: Consequences of natural variations in the optical characteristics of phytoplankton and the quantum yield of chlorophyll *a* fluorescence, *J. Remote Sens.* **17**(1):2417–2448.

Babin, M., Sadoudi, N., Lazzara, L., Gostan, J., Partensky, F., Bricaud, A., Veldhuis, M., Morel, A., and Flkowski, P.G., 1996c, Photoacclimation strategy of *Prochlorococcus sp.* and consequences on large scale variations of photosynthetic parameters, *Ocean Optics 13*, *Proc. SPIE*, **2963**:314–319.

Babin, M., Stramski, D., Ferrari, G. M., Claustre, H., Bricaud, A., Obolensky, G., and Hoepffner, N., 2003a, Variations in the light absorption coefficients of phytoplankton, nonalgal particles, and dissolved organic matter in coastal waters around Europe, *J. Geophys. Res.* **108**(C7):3211, doi:10.1029/2001JC000882.

Babin, M., Morel, A., Fournier-Sicre, V., Fell, F., and Stramski, D., 2003b, Light scattering properties of marine particles in coastal and open ocean waters as related to the particle mass concentration, *Limnol. Oceanogr.* **48**:843–859.

Babin, M. and Stramski, D., 2004, Variations in the mass-specific absorption coefficient of mineral particles suspended in water, *Limnol. Oceanogr.* **49**:756–767.

Babko, A.K. and Pilipienko, A.T., 1972, *Photometric Analysis* (translated from Russian), Wydaw. Nauk.-Tech., Warszawa (in Polish).

Bader, H., 1970, The hyperbolic distribution of particle size, *J. Geophys. Res.* **75**:2822–2830.

Baker, K.S. and Smith, R.C., 1982, Bio-optical classification and model of natural waters 2, *Limnol. Oceanogr.* **27**(3):500–509.

Bandaranayake, W.M., 1998, Mycosporines: Are they nature's sunscreens?, *Nat. Prod. Rep.* **15**(2):159–172.

Banna, M.S., McQuaide, B.H., Malutzki, R., and Schmidt, V., 1986, The photoelectron spectrum of water in the 30 to 140 eV photon energy range, *J. Chem. Phys.* **84**(9):4739–4744.

Banwell, C.N., 1997, *Fundamentals of Molecular Spectroscopy*, Hill McGraw, London.

Banwell, C.N., 1985, *Fundamentals of Molecular Spectroscopy* (translated to Russian), Mir, Moskva, p. 384.

Barale, V. and Schlittenhardt, P.M. (Eds.), 1993, *Ocean Colour: Theory and Applications in a Decade of CZCS Experience*, Kluwer, Dordrecht, p. 367.

Barber, J. (Ed.), 1976, *The Intact Chloroplast*, Elsevier, Amsterdam.

Barber, J. (Ed.), 1977, *Primary Processes of Photosynthesis*, Elsevier, Amsterdam, p. 516.

Barer R., 1966, Phase contrast and interference microscopy in cytology, in: *Physical Techniques in Biological Research. Vol. 3, Part A. Cells Ilnd Tissues,* A.W. Pollister (Ed.), 2nd ed. Academic Press, New York, pp. 1–56.
Barret, J. and Mansell, A.L., 1960, Ultra-violet absorption spectra of the molecules H_2O, HDO and D_2O, *Nature* **187**(4732):138–138.
Barrow, G.M., 1969, *Introduction to Molecular Spectroscopy*, PWN, Warszawa, p. 356 (in Polish).
Baryła, J., 1983, *Regulation of Cell Metabolism,* PWN, Warszawa, p. 596 (in Polish).
Battin, T., 1998, Dissolved organic matter and its optical properties in a blackwater tributary of the upper Orinoco River, Venezuela, *Org. Geochem.* **28**:561–569.
Bauer, J. E., 2002, Carbon isotopic composition of DOM, D.A. Hansell and C.A. Carlson (Eds.), 2002 (op. cit.), pp. 405–453.
Becquerel, J. and Rossignol, J., 1929, *Int. Crit. Tab.* **5**, p. 268.
Benedict, W.S., Gailar, N., and Olyer, E.K., 1956, *J. Chem, Phys.* **21**:1301.
Benner, R., 2002, Chemical composition and reactivity, D.A. Hansell and C.A. Carlson (Eds.), 2002 (op. cit.), pp. 59–90.
Bergs, J.B., 1975, *Organic Molecular Photosynthesis*, Vols. 1 and 2, Springer Verlag, Berlin etc.
Berlman, I.B., 1971, *Handbook of Fluorescence Spectra of Aromatic Molecules*, 2nd ed., Academic Press, New York.
Bernard, S., Probyn, T.A., and Barlow, R.G., 2001, Measured and modelled optical properties of particulate matter in the southern Benguela, *South African J.Sci.* **97**:410–420.
Bernath, P.F., 2002, The spectroscopy of water vapour: Experiment, theory and applications, *Phys. Chem. Chem. Phys.* **4**:1501–1509.
Berry, L.G. and Mason, B., 1959, *Mineralogy: Concepts, Descriptions, Determinations,* Freeman, San Francisco.
Bidigare, R., Ondrusek, M.E., Morrow, J.H., and Kiefer, D.A., 1990, 'In vivo' absorption properties of algal pigments, *Ocean Optics 10, Proc. SPIE* **1302**:290–302.
Bigelow, C.A., Bowman, D.C., and Cassel, D.A., 2004, Physical properties of three sand size classes amended with inorganic materials of sphagnum peat moss for putting green rootzones, *Crop. Sci.* **44**:900–907.
Binding, C.E., Bowers, D.G., and Mitchelson-Jacob, E.G., 2003, An algorithm for the retrieval of suspended sediment concentrations in the Irish Sea from SeaWiFS ocean colour satellite imagery. *Int. J. Remote Sensing.* **24**:3791–3806.
Bird, R.E. and Riordan, C.J., 1986, Simple solar spectral model for direct and diffuse irradiance on horizontal and tilted planes at the earth's surface for cloudless atmospheres, *J. Clim. Appl. Meteorol.* **25**(1):87–97.
Blough, N.V. and Zafiron, O.C., 1985, *Inorg. Chem.* **24**:3502–3504.
Blough, N.V., Zafiriou, O.C., Bonilla, J., 1993, Optical absorption spectra of water from the Orinoco River outflow: Terrestial input of colored organic matter to the Caribbean, *J. Geophys. Res.* **98**:2271–2278.
Blough, N.V. and Del Vecchio, R., 2002, Chromophoric DOM in the coastal environment, in: D.A. Hansell and C.A. Carlson (Eds.), 2002 (op. cit.), pp. 509–546.
Bogdanov, Y.A. and Lisitzin, A.P., 1968, Distribution and composition of suspended organic matter in Pacific waters, *Okeanologicheskie Issledowania*, No. 18, Nauka, Moskva (in Russian).
Bogorov, V.G., 1968, The productivity of the ocean, primary production and its utilisation in food chains, in: *Fundamental Problems in Oceanology*, Nauka, Moskva (in Russian).
Bogorov, V.G., 1974, *Plankton of the World Ocean,* Nauka, Moskva, p. 320 (in Russian).
Bohren, C.F. and Huffman, D.R., 1983, *Absorption and Scattering of Light by Small Particles*, Wiley, New York, p. 530.

Boivin, L.P., Davidson, W.F., Storey, R.S., Sinclair, D., and Earle, E.D, 1986, Determination of the attenuation coefficients of visible and ultraviolet radiation in heavy water, *Appl. Optics* **25**:877–882.
Born, M. and Wolf, E., 1968, *Principles of Optics*, Pergamon Press, Oxford, p. 719 (Russian edn. by Nauka, Moskva, 1973).
Bougis, P., 1976, *Marine Plankton Ecology*, North-Holland, Amsterdam, p. 355.
Bowers, D.G., Harker, G.E.L., and Stephan, B., 1996, Absorption spectra of inorganic particles in the Irish Sea and their relevance to remote sensing of chlorophyll, *Int. J. Remote Sens.* **17**:2449–2460.
Bowers, D.G. and Binding, C.E., 2006, The optical properties of marine suspended particles: A review and synthesis, *Estuarine, Coastal and Shelf Sci.* **67**:219–230.
Brandt, K. and Raben, E., 1920, Zur Kenntnis del' ehemischen Zusarnmensetzung des Planktons und einiger Bodenorganismen, *Wiss. Meeresuntersuch. Abt. Kiel, N.F.* **19**:175–210.
Brehm, J.J. and Mullin, W.J., 1989, *Introduction to the Structure of Matter: A Course in Modern Physics*, Wiley, New York.
Bricaud, A., Morel, A., and Prieur, L., 1981, Absorption by dissolved organic matter of the sea (yellow substance) in the UV and visible domains, *Limnol. Oceanogr.* **26**:43–53.
Bricaud, A., Morel, A., and Prieur, L., 1983, Optical efficiency factors of some phytoplacters, *Limnol. Oceanogr.* **28**(5):816–832.
Bricaud, A. and Morel, A., 1986, Light attenuation and scattering by phytoplanctonic cells: A theoretical modeling, *Appl. Optics* **25**:571–580.
Bricaud, A., Bedhomme, A.L., and Morel, A., 1988, Optical properties of diverse phytoplanktonic species: Experimental results and theoretical interpretation, *J. Plankton Res.* **10**:851–873.
Bricaud, A. and Stramski, D., 1990, Spectral absorption coefficient of living phytoplankton and nonalgal biogenous matter. A comparison between the Peru upwelling area and the Sargasso Sea, *Limnol. Oceanogr.* **35** (3):562–582.
Bricaud, A., Babin, M., Morel, A., and Claustre, H., 1995, Variability in the chlorophyll-specific absorption coefficients of natural phytoplankton:analysis and parameterisation, *J. Geophys. Res.* **100**(C7):13321–13332.
Bricaud, A., Morel, A., Babin, M., Allali, K., and Claustre, H., 1998, Variations of light absorption by suspended particles with chlorophyll *a* concentration in oceanic (case 1) waters: Analysis and implications for bio-optical models, *J. Geophys. Res.* **103**(C13):31033–31044.
Briegleb, G., 1961, *Elektronen-Donator-Komplexe*, Springer Verlag, Berlin.
Bristow, M.F. and Nielsen, D., 1981, Remote monitoring of organic carbon in surface waters, *Proj. Rep. EPA-600/4-81-001* (U.S. Environ. Protect. Agency, Las Vegas, Nevada), p. 88.
Bronk, D.A., 2002, *Dynamics of DON*, D.A. Hansell, and C.A. Carlson (Eds.), 2002 (op. cit.), pp. 153–247.
Browell, E.V. and Anderson, R.C., 1975, Ultraviolet optical – constants of water and ammonia ices, *J. Opt. Soc. Am.* **65**(8):919–926.
Brown, M., 1974, Transmission spectroscopy investigations of natural waters, *Rep. Inst. Fys. Oceanogr. Københavns Univ.* **28**:1–26.
Brown, M., 1980, Standarization of natural water fluorescence intensity by Raman emission, *Rep. Inst. Fys. Oceanogr. Københavns Univ.* **42**:29–38.
Brown, J., Colling, A., Park, D., Phillips, J., Rothery, D., and Wright, J., 1989, *Seawater: Its Composition, Properties and Behaviour*, Pergamon Press, Oxford, p. 165.
Brubach, J.B, Mermet, A., Filabozzi, A., Gerschel, A., and Roy, P., 2005, Signatures of the hydrogen bonding in the infrared bands of water, *J. Chem. Phys.* **122**(18):184509.
Bruge, F., Bernasconi, M., and Parrinello, M., 1999, Ab initio simulation of rotational dynamics of solvated ammonium ion in water, *J. Am. Chem. Soc.* **121**:10883–10888.

Brun-Cottan, J.C., 1971, Etude de la granulométrie des particules marines: mésures effectuées avc un compteur Coulter, *Cahiers Océanographiques* **23**:193–205.

Buiteveld, H., Hakvoort, J.M.H., and Donze, M., 1994, The optical properties of pure water, *Ocean Optics XII, Proc. SPIE*, J.S. Jaffe (Ed.), **2258**:174–183.

Bukata, R.P., Bruton, J.E., Jerome, J.H., Jain, S.C., and Zwick, H.H., 1981a, Optical water quality model of lake Ontario: 2 Determination of chlorophyll – a and suspended mineral concentration of natural waters from submersible and low altitude optical sensors, *Appl. Opt.*, **20**(9):1704–1714.

Bukata, R.P., Jerome, J.H., Bruton, J.E., Jain, S.C., and Zwick, H.H., 1981b, Optical water quality model of lake Ontario: 1 Determination of the optical cross sections of organic and inorganic particles in lake Ontario, *Appl. Opt.* **20**(9):1696–1703.

Bukata, R.P., Bruton, J.E., and Jerome, J.H., 1983, Use of chromaticity in remote measurements of water quality, *Remote Sens. Environ.* **13**(2):161–178.

Bukata, R.P., Jerome, J.H., Kondratyev, K.Y., and Pozdnyakov, D.V., 1991, Estimation of organic and inorganic matter in inland waters: Optical cross sections of lakes Ontario and Ladoga. *J. Great Lakes Res.* **17**(4):461–469.

Bukata, R.P., Jerome, J.H., Kondratev, K. J., and Pozdynakov, D.V., 1995, *Optical Properties and Remote Sensing of Inland and Coastal Waters*, CRC Press, New York, p. 362.

Burdige, D.J., 2002, *Sediment Pore Waters*, in: D.A. Hansell and C.A. Carlson (Eds.), 2002 (op. cit.), p. 611–663.

Butcher, R.W., 1959, *An Introductory Account of the Smaller Algae of British Coastal Waters. Part I. Introduction and Chlorophyceae*. Fishery Investigations, London, series 4, p. 74.

Butler, W.L., 1962, Absorption of light by turbid materials, *J. Opt. Soc. Am.* **52**(3):292–299.

Campbell, I.D. and Dwek, R.A., 1984, *Biological Spectroscopy*, Benjamin/Cummings, Menlo Park, CA, p. 404.

Campbell, M.J., Liesegang, J., Riley, J.D., Leckey, R.C.G., Jenkin, J.G., and Poole, R.T., 1979, The electronic structure of the valence bands of solid NH_3 and H_2O studied by ultraviolet photoelectron spectroscopy, *J. Electron Spectrosc. Relat. Phenom.* **15**:83.

Carder, K.L., 1970, Particles in the eastern Pacific Ocean: Their distribution and effect upon optical parameters. *Ph.D. thesis, Oregon State University, Corvallis*, pp. 833–839.

Carder, K.L., Beardsley G.F., Jr., and Pak, H., 1971, Particle size distribution in the Eastern Equatorial Pacific, *J. Geophys. Res.* **76**:5070–5077.

Carder, K.L. and Schlemmer, F.C., 1973, Distribution of particles in the surface waters of the Eastern Gulf of Mexico: an indicator of circulation, *J. Geophys, Res.*, **78**:6286–6299.

Carder, K.L. and Meyers, D.J., 1979, New optical techniques for particle studies in the bottom boundary layer, *Ocean Optics VI, Proc. SPIE* **208**:151–158.

Carder, K.L. and Steward, R.G., 1985, A remote-sensing reflectance model of a red-tide dinoflagellate off west Florida, *Limnol. Oceanogr.* **30**:286–298.

Carder, K.L., Steward, R.G., Harvey, G.R., and Ortner, P.B., 1989, Marine humic and fulvic acids: Their effects on remote sensing of ocean chlorophyll, *Limnol. Oceanogr.* **34**(1):68–81.

Carder, K.L., Hawes, S.K., Baker, K.A., Smith, R.C., Steward, R.G., and Mitchell, B.G., 1991, Reflectance model for quantifying chlorophyll *a* in the presence of productivity degradation products, *J. Geophys. Res.* **96**:20599–20611.

Carlson, C.A., 2002, *Production and Removal Processes*, in: Hansell, D.A., and Carlson, C.A. (Eds.), 2002 (op. cit.), pp. 91–151.

Carlson, T.N. and Benjamin, S.G., 1980, Radiative heating rates for Saharan dust, *J. Atmos. Sci.* **37**:193–213.

Carreto, J.I., Carignan, M.O., Daleo, G., and DeMarco, S.G., 1990a, Occurrence of ycosporine-like amino acids in the red-tide dinoflagellate Alexandrium excavatum: UV-photoprotective compounds? *J. Plankton Res.* **12**:909–921.

Carreto, J.I., Lutz, V.A., DeMarco, S.G., and Carignan, M.O., 1990b, *Influence and wavelength dependence of mycosporine-like aminoacid synthesis in the dinoflagellate Alexandrium excavatum*, in: *Toxic Marine Photoplankton*, Graneli, E., Edler, L., Sundstrom, B., and Anderson, D.M. (Eds.), Elsevier, New York, pp. 275–298.

Cauwet, G., 2002, *DOM in the Coastal Zone*, in: D.A. Hansell, and C. A. Carlson (Eds.), 2002 (op. cit.), pp. 579–609.

Chamot, E.M. and Mason, C.W., 1944, *Handbook of Chemical Microscopy*, Vol. 1. Wiley, New York.

Chandrasekhar, S., 1960, *Radiative Transfer*, Dover, New York, p. 393.

Chaplin, M., 2006, *Water Structure and Behavior; Molecular Vibration and Absorption*, London South Bank Univ. (http://www.lsbu.ac.uk/water/vibrat.html).

Christian, J.R. and Anderson, T.R., 2002, *Modeling DOC Biogeochemistry*, in: D.A. Hansell and C.A. Carlson (Eds.), 2002 (op. cit.), pp. 717–755.

Ciborowski, S., 1962, *Radiation Chemistry of Non Organic Compounds*, PWN, Warszawa, p. 290 (in Polish).

Clark, R.N., 1983, Spectral properties of mixtures of montmorillonite and dark carbon grains: implication for remote sensing minerals containing chemically and physically adsorbed water, *J. Geophys. Res.* **88**(B12):10,635–10,644.

Clarke, A.D., Shinozuka, Y., Kapustin, V.N., Howell, S., Huebert, B., Doherty, S., Anderson, T., Covert, D., Anderson, J., Hua, X., Moore II, K.G., McNaughton, C., Carmichael, G., and Weber, R., 2004, Size distribution and mixtures of dust and balckcarbon aerosol in Asian outflow: Physiochemistry and optical properties, *J. Geophys. Res.* **109**: D15S09, doi:10.1029/2003JD004378, p 20.

Clarke, G.L. and James, H.R., 1939, Laboratory analysis of the selective absorption of light by sea water, *J. Opt. Soc. Am.* **29**:43–55.

Claustre, H., 1994, The trophic states of various oceanic provinces as revealed by phytoplankton pigment signatures, *Limnol. Oceanogr.* **39**:1207–1211.

Claustre, H., Morel, A., Hooker, S. B., Babin, M., Antoine, D., Oubelkheir, K., Bricaud, A., Leblanc, K., Queguiner, B., and Maritorena, S., 2002, Is desert dust making oligotrophic waters greener? *Geophys. Res. Lett.*, **29**:10.1029 2001GL014056.

Cleveland, J.S., 1995, Regional models for phytoplankton absorption as a function of chlorophyll *a* concentration, *J. Geophys. Res.* **100**:13333–13344.

Cohen-Tannoudji, C., Diu B., and Laloë, F., 1977, *Quantum Mechanics*, Vols. 1 and 2, John Wiley, New York.

Colarco, P.R., Toon, O.B., Torres, O., and Rasch, P.J., 2002, Determining the UV imaginary index of refraction of Saharan dust particles from total ozone mapping spectrometer data using a three-dimensional model of dust transport, *J. Geophys. Res.* **107**(D16):4289, 10,1029/2001 JD000903.

Collins, J.R., 1925, A new infra-red absorption band of liquid water at 2.52 μ, *Phys. Rev.* **55**:470–472.

Collyer, D.M. and Fogg, G.E., 1955, Studies on fat accumulation by algae. *J. Exp. Bot.* **6**:256–275.

Copin-Montegut, G., Ivanoff, A., and Saliot, A., 1971, Coefficient d'attention des eaux de mer dans l'ultraviolet, *C. R. Acad. Sci. Paris* **272**(25):B-1459–B-1466.

Coulter, W.H., 1973, Theory of the Coulter Counter Instruction Manual for Coulter Counter Model ZBI (Biological), (2nd edn.) Coulter Electron. Ltd., U.K.

Cramer, W.A. and Crofts, A.R., 1982, Electron and proton transport, in: Govindjee (Ed.), 1982 (op. cit.), pp. 387–467.

Cramer, W.A. and Crofts, A.R., 1987, Transport of electrons and protons, in: *Photosynthesis*, Govindjee (Ed.), Vol. 1, Mir, Moskva, pp. 540–632 (in Russian).

Curcio, J. and Petty, C.C., 1951, The near infrared absorption spectrum of liquid water, *J. Opt. Soc. Am.* **41**:302–304.

Daniels, J., 1971, Bestimmung der Optischen Konstanten von Eis aus Energieverlustmessungen von Schnellen Elektronen, *Optics Comm.* **3**(4):240–243.

Darecki, M., Weeks, A., Sagan, S., Kowalczuk, P., and Kaczmarek, S., 2003, Optical characteristics of two contrasting Case 2 waters and their influence on remote sensing algorithms, *Cont. Shelf Res.* **23**(3)–(4):237–250.

Davies, H.G., Wilkins, M.H.F., Chayen, J., and La Cour, L.F., 1954, The use of the interference microscope to determine dry nlass in living cells and as a quantitative cytochemical method. *Q. J. Microsc. Sci.* **95**:271–304.

Davis-Colley, R.J. and Vant, W.N., 1987, Absorption of light by yellow substance in freshwater lakes, *Limnol. Oceanogr.* **32**(2):416–425.

Degens, E.T., 1970, Molecular nature of nitrogenous compounds in sea water and recent marine sediments, in: *Organic Matter in Natural Waters*, D.S. Hood (Ed.), Inst. Mar. Sci. Occas. Publ. (University of Alaska), **1**:77–106.

Deirmendijan, D., 1964, Scattering and polarization properties of water clouds and hazes in the visible and infrared, *Appl. Optics* **3(2)**:187–196.

Deirmendijan, D., 1969, *Electromagnetic Scattering on Spherical Polidispersions*, Elsevier, Amsterdam, p. 290.

Delany, A.C., Parkin, D.W., Griffin, J.J., Goldberg, E.D., and Reimann, B.E.F., 1967, Airborne dust collected at Barbados, *Geoch. Cosmochim. Acta* **31**:885–909.

Del Grosso, V.A., 1978, Optical transfer function measurements in the Sargasso Sea, *SPIE* vol.160. *Ocean Optics* V, 74–101.

Dera, J., Gohs, L., Hapter, R., Kaiser, W., Prandke, H., Rting, W., and Woźniak, B., 1974, Untersuchungen zur Wechselwirkung zwischen den optischen, physikalischen, biologischen und chemischen Umweltfaktoren in der Ostsee, *Geodatische und Geophysikalische Veröffentlichungen* **4**(13):5–100.

Dera, J., Gohs, L., and Woźniak, B., 1978, Experimental study of the composite parts of the light-beam attenuation process in waters of the Gulf of Gdańsk, *Oceanologia*, **10**:5–26.

Dera, J., 1980, Oceanographical investigation of the Ezcurra Inlet during the 2nd Antarctic Expedition of the Polish Academy of Sciences, *Oceanologia*, **12**:5–26.

Dera, J., 1992, *Marine Physics*, Elsevier-PWN, Amsterdam, p. 516.

Dera, J., 1995, *Underwater Irradiance as a Factor Affecting Primary Production*, Diss. and monogr., No **7**, Inst. Oceanol. PAS, Sopot, p. 110.

Dera, J., 2003, *Marine Physics*, 2nd ed., PWN, Warszawa, p. 541 (in Polish).

Diebel-Langohr, D., Doerffer, R., Reuter, R., Dörre, F., and Hengstermann, T., 1986, Long term stability of Gelbstoff concerning its optical properties, in:. Grassl et al. (Eds.), 1986b (op. cit.), Vol.2, Appendix 6, p. 24.

Diercksen, G.H.F., Kraemer, W.P., Rescigno, T.N., Bender, C.F., McKoy, B.V., Langhoff, S.R., and Langhoff, P.W., 1982, Theoretical studies of photoexcitation and ionization in H_2O, *J. Chem. Phys.* **76**:1043.

Doronin, J.P. and Chieysin, H.J., 1975, *Marine Ice* (in Russian: *Morskoj Led*), Gidromietizdat, Leningrad.

Dorsey, N.E., 1940, *Properties of the Ordinary Water Substance in All Its Phases*, Am. Chem. Soc. Monogr. Ser., No **8**, Reinhold, New York.

Druet, C., 2000, *The Dynamics of the Sea*, Wydaw. Uniw. Gdańsk.-Gdańsk. Tow. Nauk., Gdańsk, p. 288 (in Polish).

Druet, C., 1994, *Dynamics of a Stratified Ocean*, PWN, Warszawa, p. 225 (in Polish).

Du, H., Fuh, R., Li, A., Corkan, J., and Lindsey, J.S., 1998, Photochem CAD: A computer-aided design and research tool in photochemistry, *Photochem. Photobiol.* **68**: 141–142.

Dunn, M.E., Evans, T.M., Kirschner, K.N., and Shields, G.C., 2006, Prediction of accurate anharmonic experimental vibrational frequencies for water clusters $(H_2O)_n$, n = 2 – 5, *J. Phys. Chem. A* **110**:303–309.

DuRand, M.D., and Olson, R.J., 1998, Diel patterns in optical properties of the chlorophyte *Nannochloris* sp.: Relating individual-cell to bulk measurements. *Limnol. Oceanogr.,* **43**(6):1107–1118.

DuRand, M.D., Green, R.E., Sosik, H.M., and Olson, R.J., 2002, Diel variations in optical properties of *Micromonas Pusilla* (Prasinophyceae), *J. Phycol.* **38**:1132–1142.

Duursma, E.K. and Dawson, R. (Eds.), 1981, *Marine Organic Chemistry. Evolution, Composition, Interactions and Chemistry of Organic Matter in Seawater*, Elsevier, Amsterdam, p. 521.

Duursma, N.E., 1965, The dissolved organic constituents of sea water, in: *Chemical Oceanography*, Vol. 1, Academic Press, London, pp. 435–475.

Duysens, L.M., 1956, The flattening of the absorption spectra of suspension as compared to that of solution, *Biochim. Biophys. Acta* **19**:1–12.

Edler, L. (Ed.), 1979, *Recommendations on methods for marine biological studies in the Baltic Sea -Phytoplankton and chlorophyll.* The Baltic Marine Biologists. Publ. 5– Working Group 9, p. 160.

Egan, W.G., and Hilgeman, T.W., 1979, *Optical Properties of Inhomogeneous Materials; Application to Geology, Astronomy, Chemistry, and Geology,* Academic Press, New York.

Eisenberg, D. and Kauzmann, W., 1969, *The Structure and Properties of Water*, Oxford University Press, London.

Eisma, D., Schuhmacher, T., Boekel, H., van Heerwaarden, J., Franken, H., Laan, M., Vaars, A., Eijgenraam, F., and Kalf, J., 1990, A camera and image-analysis system for in situ observation of flocs in natural waters, *Neth. J. Sea Res.* **27**:43–56.

Ejchart, W., Jędrzejewski, J., and Siegoczyński, R.M., 1978, Some optical properties of polyacenaphthylene with benzanthrone admixture". *J. Molecular Structure* **47**:417–421.

Euler, B.V. and Jansson, B.,1931, Beziehungen zwischen Ergosterin und Carotin. *Ark. Kem. Miner. Geol.* **10B(**17):6.

Favre-Bonvin, J., Arpin, N., and Brevard, C., 1976, Structure de la mycosporine (P310), *Can. J. Chem.* **54**:1105–1113.

Feofilova, E.P., 1974, *Pigments of Micro-Organisms*, Nauka, Moskva, p. 216 (in Russian).

Ficek, D., 2001, *Modeling the Quantum Yield of Photosynthesis in Different Marine Basins*, Rozpr. i monogr., Nr **14**, Inst. Oceanol. PAN, Sopot, p. 224 (in Polish).

Ficek, D., Kaczmarek, S., Stoń-Egiert, J., Woźniak, B., Majchrowski, R., and Dera, J., 2004, Spectra of light absorption by phytoplankton pigments in the Baltic; conclusions to be drawn from a Gaussian analysis of empirical data, *Oceanologia* **46**(4):533–555.

Filipowicz, B. and Więckowski, W., 1979, *Biochemistry,* Vol. 1, PWN Warszawa, p. 482 (in Polish).

Filipowicz, B. and Więckowski, W., 1983, *Biochemistry,* Vol. 2, PWN Warszawa, p. 624 (in Polish).

Fischer, J., Doerffer, R., and Russo, M., 1986, The influence of yellow substance on remote monitoring of water substances in coastal waters, in: Grassl et al. (Eds.), 1986b (op. cit.), Vol.2, Appendix 2, p. 32.

Flaig, W., Bentelspacher, H., and Reitz, E., 1975, Chemical composition and physical properties of humic substances, in: *Soil Components,* Vol. 1, *Organic Components*, J.E. Gieseking (Ed.), Springer Verlag, New York, pp. 1–212.

Forsythe, W.E., 1954, *Smithsonian Physical Tables,* 9th ed. Smithsonian Institute, Washington.

Fox, D.F., Oppenheimer, C.H., and Kitterge, J.S., 1953, Microfiltration in oceanographic research, *J. Mar. Res.* **12(2).**

Frąckowiak, D., Dudkowiak, A., and Bryl, K., 1997, Biophysical basis of photobiology, in: *Biophysics for Bbiologist* (*Biofizyka dla Biologów*, in Polish), M. Bryszewska and W. Leyko (Eds.), PWN, Warszawa, pp. 376–417.

Fradkin, L.I., 1986a, Formation of the photosynthetic pigment system, *Zh. Vsekhsoyuz. Khim. Obshch.* **31**(6):577–582 (in Russian).

Fradkin, L.I., 1986b, Formation of the pigment system of photosynthesis (Formirowanie pigmentnoy sistemy fotosinteza), *Zh. Vsekhsoyuz. Khim. Obshch.* **21**(6):97–102 (in Russian).
Frey, A.,1926, Die submikroskopische Struktur del' Zellmembranen. *Jahrb. Wiss. Bot.* **65**:195–223.
Friedman, D., 1969, Infrared characteristics of ocean water, *Appl. Optics* **8**(10):2073–2078.
Fry, E.S., Kattawar, G.W., and Pope R.M., 1992, Integrating cavity absorption meter, *Appl. Optics* **31**:2055–2065.
Gagliano, A.G., Hoarau, J., Breton, J., and Geacintov, N.E., 1985, Orientation of pigments in phycobilisomes of Porphyridium sp. Lewin. A linear dichroism study utilizing electric and gel orientation methods. *Biochim. Biophys. Acta* **808**:455–463.
Gaiduk, V.I. and Vij, J.K. 2001, The concept of two stochastic processes in liquid water and analytical theory of the complex permittivity in the range 0–1000 cm^{-1}, *Phys. Chem.-Chem. Phys.* **3**:5173–5181.
Gallegos, C.L., Correll, D.L., and Pierce, J.W., 1990, Modeling spectral diffuse attenuation, absorption, and scattering in a turbid estuary, *Limnol. Oceanogr.* **35**(7): 1486–1502.
Gallie, E.A. and Murtha, P.A., 1992, Specific absorption and backscattering spectra for suspended minerals and chlorophyll-a in Chilko Lake, British Columbia, *Remote Sens. Environ.* **39**:103–118.
Ganor, E. and Foner, H., 1996, The mineralogical and chemical proprties and the behavirof of aeolian Saharan dust over Israel, in: *The Impact of Desert Dust Across the Mediterranean*, S. Guerzoni and R. Chester (Eds.), pp. 163–171, Kluwer, Dordrecht.
Gantt, E. and Lipschultz, C.A., 1972, Phycobilisomes of Porphyridium cruentem, *J. Cell Biol.* **54**:313–324.
Garaj, J., Bercik, J., Bustin, D., Cernak, J., Stefanec, J., and Traiter, M., 1981, *Physical and Physicochemical Methods of Analysis*, Wydaw. Nauk.-Tech., Warszawa, p. 503 (in Polish).
Garcia-Pichel, F., 1994, A model for internal self-shading in planktonic organisms and its implications for the usefulness of ultraviolet sunscreens, *Limnol. Oceanogr.* **39**(7):1704–1717.
Gershanovich, D.E. and Muromtsev, A.M., 1982, *Ocenological Principles of the Biological Productivity of the World Ocean*, Gidrometeoizdat, Leningrad, pp. 220 (in Russian).
Gershun, A., 1958, *Selected Papers in Photometry and Lighting Engineering*, Fizmatgiz, Moskva (in Russian).
Gibbs, R.J., 1977, Transport phases of transition metals in the Amazon and Yukon Rivers, *Geol. Soc. Am. Bull.* **88**:829–843.
Gibbs, T.R.P., Jr., 1942, *Optical Methods of Chemical Analysis*, McGraw-Hill, New York.
Gieskes, W.W.C. and Kraay, G.W., 1983, Unknown chlorophyll *a* derivatives in the North Sea and the tropical Atlantic Ocean revealed by HPLC analysis, *Limnol. Oceanogr.* **28**:757–766.
Godnev, T.N., 1963, *Chlorophyll. Its Structure and Production in Plants* (in Russian: *Hlorofill. Ego Sztroenie i Obrazovanie v Rasztenii*), Akad. Nauk Belorus. SSR, Minsk.
Goericke, R. and Repeta, D., 1992, The pigments of *Prochlorococcusmarinus*: The presence of divinyl-chlorophyll *a* and *b* in a marine procaryote, *Limnol. Oceanogr.* **37**:425–433.
Gohs, L., Dera, J., Gędziorowska, D., Hapter, R., Jonasz, M., Prandke, H., Schenkel, G., Siegel, H., Olszewski, J., and Woźniak, B., 1978, Untersuchungen zur Wechselwirkung zwischen den optischen, physikalischen, biologischen und chemischen Umweltfaktoren in der Ostsee, *Geod. Geophys. Veröff.* (AdW, DDR), R. IV, H. **25**.

Gołębiewski, A., 1969, *Quantum Chemistry of Inorganic Compounds,* PWN, Warszawa, p. 370 (in Polish).

Gołębiewski, A., 1973, *Quantum Chemistry of Organic Compounds,* PWN, Warszawa, p. 389 (in Polish).

Gołębiewski, A., 1982, *Elements of Mechanics in Quantum Chemistry,* PWN, Warszawa, p. 479 (in Polish).

Goodwin, T.W. and Mercer, E.I., 1972, *Introduction to Plant Biochemistry*, Pergamon Press, New York, p. 359.

Gordon, H.R. and Brown, O.B., 1972, A theoretical model of light scattering by Sargasso Sea particulates, *Limnol. Oceanogr.* **17**:826–830.

Gordon, H.R., Brown, O.B., and Jacobs, M.M., 1975, Computed relationships between inherent and apparent optical properties of a flat, homogeneous ocean, *Appl. Opt.* **14**:417–427.

Gordon, H.R., Brown, O.B., Evans, R.H., Brown, J.W., Smith, R.C., Baker, K.S., and Clark, D.K., 1988, A semi-analytical radiance model of ocean color, *J. Geophys. Res.* **93**:10909–10924.

Gordon, H.R., 2002, Inverse methods in hydrologic optics, *Oceanologia* **44**(1):9–58.

Gorshkova, T., 1972, Organic matter in Baltic Sea sediments and its biological significance, *Tr. VNIRO* **75**:191–219 (in Russian).

Govindjee (Ed.), 1975, *Bioenergetics of Photosynthesis*, Academic Press, New York, p. 698.

Govindjee (Ed.), 1982a, *Photosynthesis*, Vol. 1, *Energy Conversion by Plants and Bacteria*, Academic Press, New York, p. 799.

Govindjee (Ed.), 1982b, *Photosynthesis*, Vol. 2, *Development, Carbon Metabolism, and Plant Productivity*, Academic Press, New York, p. 580.

Govindjee (Ed.), 1987, *Photosynthesis*, Vol. 1, pp. 727–Vol. 2, p. 470, Mir, Moskva (in Russian).

Govindjee and Whitmarsh, J., 1982, Introduction to photosynthesis: Energy conversion by plants and bacteria, in: *Photosynthesis*, Govindjee (Ed.), Vol. 1, Academic Press, New York, pp. 1–16.

Govindjee and Whitmarsh, J., 1987, Introduction to photosynthesis: Energy conversion by plants and bacteria, in: *Photosynthesis*, Govindjee (Ed.), Vol. 1, Mir, Moskva (in Russian), pp. 90–107.

Grabowski, J., 1984, Phycobiliproteids and their natural complexes—the phycobilisomes (Structure and migration of electronic excitation energy), in: *Current Problems in Phytobiology*, Zesz. Probl. Post. Nauk Roln., Nr **271,** pp. 171–178 (in Polish).

Grassl, H., Doerfer, R., and Jager, W., 1986a, *The Use of Chlorophyll Fluorescence Measurements from Space for Separating Constituents of Sea-water,* Vols. 1 and 2, GKSS Research Centre Geesthacht [ESA Contract No. RFQ 3-5059/84/NL/MD].

Grassl, H., Doerfer, R., and Jager, W., 1986b, *The Influence of Yellow Substances on Remote Sensing of Sea-water Constituents from Space,* Vols. 1 and 2, GKSS Research. Centre Geesthacht [ESA Contract No. RFQ 3-5060/84/NL/MD].

Green A.E.S., Deepak, A., and Lipofsky, B.J., 1971, Interpretation of the San's Aureole based on Atmospheric Aerosol Models. *Appl. Optics* **10**:1263–1279.

Green, R.C., Sosik, H.M., Olson, R.J., and DuRand, M.D., 2003, Flow cytometric determination of size and complex refractive index for marine particles: Comparison with independent and bulk estimates, *Appl. Optics* **42**(3):526–541.

Green, S.A. and Blough, N.V., 1994, Optical absorption and fluorescence properties of chromophoric dissolved organic matter in sea water, *Limnol. Oceanogr.* **39**:1903–1916.

Greenwood, N.N. and Earnshaw, A., 1997, *Chemistry of the Elements*, 2nd ed., Butterworth-Heinemann, Oxford, pp. 1600.

Grenfell, T.C. and Maykut, G.A., 1977, The optical properties of ice and snow in the Arctic Basin, *J. Glaciol.* **18**(80):445–463.

Grenfell, T.C. and Perovich, D.K., 1981, Radiation absorption coefficients of polly-crstalline ice from 400–1400 nm, *J. Geophys. Res.* **86**(C8):7447–7450.
Grenfell, T.C., 1983, A theoretical model of the optical properties of ice in the visible and near infrared, *J. Geophys. Res.* **88**(C14):9723– 9735.
Grenfell, T.C. and Hedrick, D., 1983, Scattering of visible and near infrared radiation by NaCl ice and glacier ice, *Cold Reg. Sci. Technol.* **8**:119–127.
Grenfell, T.C. and Perovich, D.K., 1984, Spectral albedos of sea ice and incident solar irradiance in the southern Beaufort Sea, *J. Geophys. Res.*, **89**:3573–3580.
Grenfell, T.C. and Perovich, D.K., 1986, Optical properties of ice and snow in polar oceans, II: Theoretical calculations, *Ocean Optics VIII, Proc. SPIE* **637**:242–251.
Grenfell, T.C., 1991, A radiative transfer model for sea ice with vertical structure variations, *J. Geophys. Res.* **96**(C9):16991–17001.
Grodziński, D.M., 1972, *Plant Biophysics*, Nauk. Dumka, Kiyev, p. 403 (in Russian).
Grodziński, D.M., 1978, *Plant Biophysics*, PWR & L, Warszawa, p. 405 (in Polish).
Grzybowski, W. and Pempkowiak, J., 2003, Preliminary results on low molecular weight organic substances dissolved in the waters of the Gulf of Gdańsk, *Oceanologia* **45**(4):693–704.
Guillard, R.R.L., Murphy, L.S., Foss, P., and Liaaen-Jensen, S., 1985, *Synechococcus spp.* as likely zeaxanthin-dominant ultraphytoplankton in the North Atlantic, *Limnol. Oceanogr.* **30**(2):412–414.
Guo, J.H., Luo, Y., Augustsson, A., Rubensson, J.E, Såthe, C., Ågren, C., Siegbahn, H., and Nordgren, J., 2002, X-ray emission spectroscopy of hydrogen bonding and electronic structure of liquid water, *Phys. Rev. Lett.* **89**(13):137402-1 to 137402-4.
Haken, H. and Wolf, H.C., 1995 [1998], *Molekülphysic und Quantenchemie*, Springer; (Polish translation, PWN, Warszawa, p. 521).
Haken, H. and Wolf, H.C., 1996 [2002]; *Atom-und Quantenphysik*, Springer (Polish translation, PWN, Warszawa, p. 569).
Hale, G.M. and Querry, M.R., 1973, Optical constants of water in the 200 nm to 200 μm wavelength region, *Appl. Optics* **12**:557–562.
Hall, D.O. and Rao, K.K., 1999, *Photosynthesis*, Wydaw. Nauk.-Tech., Warszawa, p. 289.
Hanesiak, J.M., Barber, D.G., De Abreu, R.A., and Yackel, J.J., 2001, Local and regional albedo observations of Arctic first-year sea ice during melt ponding, *J. Geophys. Res.* **106**:1005–1016.
Hansell, D.A., 2002, *DOC in the global ocean carbon cycle*, in: D.A. Hansell, and C.A. Carlson, 2002 (op. cit.), 685–715.
Hansell, D.A. and Carlson, C.A. (Eds.), 2002, *Biogeochemistry of Marine Dissolved Organic Matter*, Acad. Press-Elsevier Sci. Imprint, Amsterdam, p. 774.
Harker, G.E.L., 1997, A comparison between optical measurements made in the field and in the laboratory, and the development of an optical model. *PhD thesis, University of Wales, Bangor, U.K.*
Harris, J.E., 1977, Characterization of suspended matter in the Gulf of Mexico II: Particle and lysis of suspended matter from deep water, *Deep-Sea Res.* **24**:1055–1061.
Harrison, E.F., Minnis, P., Barkstrom, B.R., and Gibson, G.G., 1993, Radiation budget at the top of the atmosphere, in: *Atlas of Satellite Observations Related to Global Change*, R.J. Gurney, J.L. Foster, and C.L. Parkinson (Eds.), Cambridge University Press, Cambridge, pp. 5–18.
Harvey, G.R., Boran, D.A., Chesal, L.A., and Tokar, M., 1983, The structure of marine fulvic and humic acids, *Mar. Chem.* **12**:119–132.
Hasler, P. and Nussbaumer, T., 1998, Particle size disrtribution of the fly ash from biomass combustion, *Biomass for Energy and Industry, 10 European Conference and Technology Exhibition,* Würzburg (Germany), p. 4.
Hayase, K. and Tsubota, K., 1985, Sedimentary humic acid and fulvic acid as fluorescent organic materials, *Geochim. Cosmochim. Acta* **49**:159–163.

Hayashi, H., Watanabe, N., Udagawa, Y., and Kao, C.C., 2000, The complete optical spectrum of liquid water measured by inelastic x-ray scattering, *Proc. Nat. Acad. Sci. USA* **97**(12):6264–6266.

Hedges, J.I., 1980, Chemical indicators of organic river sources in rivers and estuaries, *Workshop Rep. 'Flux of organic carbon by rivers to the ocean,'* Woods Hole, MA, Sept. 21–25, pp. 109–141.

Hedges, J.I., 2002, Why dissolved organic matter? in: D.A. Hansell and C.A. Carlson (Eds.), 2002 (op. cit.), pp. 1–33.

Heldt, H.W., 1997, *Plant Biochemistry and Molecular Biology*, Oxford University Press, Oxford, p. 522.

Henderson, M.A., 2002, The interaction of water with solid surfaces: Fundamental aspects revisited, *Surf. Sci. Rep.* **46**:5–308.

Herzberg, G., 1950, *Molecular Spectra and Molecular Structure*, I: *Spectra of Diatomic Molecules*, 2nd ed., Van Nostrand, New York (reprint ed., 1989, Krieger, Melbourne, FL, p. 678).

Herzberg, G., 1963–1967, *Molecular Spectra and Molecular Structure,* Vol. 1, p. 658–Vol. 2, p. 632-Vol. 3, p. 745, Van Nostrand, New York.

Herzberg, G., 1992, *Molecular Spectra and Molecular Structure,* reprint ed., Krieger, Melbourne, FL.

Hobbs, P.V., 1974, *Ice Physics,* Clarendon, Oxford.

Hobson, L.A., 1967, The seasonal and vertical distribution of suspended particlulate matter in an area of the northeast Pacific Ocean, *Limnol. Oceanogr.* **12**:642–649.

Hodgson, R.T. and Newkirk, D.D., 1975, Pyridine immersion: A technique for measuring the refractive index of marine particles, in: *Ocean Optics* L.F. Drummeter and L.E. Mertens (Eds), *Proc. SPIE* **64**:62–64, San Diego, CA.

Hoepffner, N. and Sathyendranath, S., 1991, Effect of pigment composition on absorption properties of phytoplankton, *Mar. Ecol. Prog. Ser.* **73**(1):11–23.

Hoidale, G.B. and Blanco, A.J., 1969, Infra-red absorption spectra of atmospheric dust over an interior desert basin, *Pure Appl. Geophys.* **74**:151–164.

Hollas, J.M., 1992, *High Resolution Spectroscopy*, 2nd ed., Wiley, New York.

Hoppel, W.A., Fitzgerald, J.W., Frick, G.M., Larson, R.E., and Mack, E.J., 1990, Aerosol size distributions & optical properties found in the marine boundary layer over the Atlantic Ocean, *J. Geophy. Res.* **95** (D4):3659–3686.

Horne, R.A., 1969, *Marine Chemistry*, Wiley, New York, p. 568.

Højerslev, N.K., 1980a, Water color and its relation to primary production, *Baund.-Layer Meteorol.* **18(**203).

Højerslev, N.K., 1980b, On the origin of yellow substance in marine environment, *Rep. Inst. Fys. Oceanogr. Københavns Univ.* **42**:57–81.

Højerslev, N.K., 1981, Optical water mass classification in Skagerrak and the Eastern North Sea, *Proc. The Norwegian Coastal Current Symp. (Voss)*, pp. 331–339.

Højerslev, N.K., 1986, Optical properties of sea water, in: *Landolt-Bornstein, Numerical Data and Functional Relationships in Science and Technology*, New series, Vol. 3, Oceanography, Springer Verlag, Berlin, pp. 386–462.

Hubert, K.P. and Herzbeg, G., 1979, *Molecular Spectra and Molecular Structure* IV; *Constants of Diatomic Molecules*, Reinhold, NewYork.

Hulburt, E.O., 1928, The penetration of ultrafiolet light into pure water and sea water, *J. Opt. Soc. Am.* **17**(1):15–22.

Hulburt, E.O., 1934, The polarization of light in sea, *J. Opt. Soc. Am.* **24**:35–42.

Hulburt, E.O., 1945, Optics of distilled and natural water, *J. Opt. Soc. Am.* **35**(11): 698–705.

Hulst Van de, H.C., 1958 (1961), *Light Scattering by Small Particles*, Wiley, New York (Russian translation: 1961, Mir, Moskva).

Hulst Van de, H.C., 1981, *Light Scattering by Small Particles*, Wiley, New York, p. 470.

Hwang, S.H., Lee, K.P., Lee, D.S., and Powers S.E., 2002, Models of estimating soil particle-size distribution, *Soil. Sci. Soc. Am. J.* **66**:1143–1150.

Irvine, W.M. and Pollack, J.B., 1968, Infrared optical properties of water and ice spheres, *Icarus* **8**:324–360.

Ishiwatari, R., 1973, Chemical characterization of fractionated humic acids from lake and marine sediments, *Chem. Geol.* **12**:113–126.

Ittekkot, V., 1981, Verteilung von gelösten organischem Kohlenstoff, gelösten Zuckern und Aminosäuren im Fladengrund, nördliche Nordsee (Flex 1976), *Mitt. Geol. Paläontol. Inst. Univ. Hamburg* **51**:115–187.

Iturriaga, R. and Marra, J., 1988, Temporal and spatial variability of chroococcoid cyanobacteria *Synechococcus* spp. specific growth rates and thair contribution to primary production in Sargasso Sea, *Marine Ecology Progress Series* **44**:175–181.

Iturriaga, R. and Siegel, D.A., 1988, Microphotometric distinction of phytoplankton and detrital absorption properties, *Ocean Optics 9, Proc. SPIE*, **925**:277–287.

Iturriaga, R. and Siegel, D.A., 1989, Microphotometric characterization of phytoplankton and detrital absorption in the Sargasso Sea, *Limnol.Oceanogr.* **34**(8): 1706–1726.

Ivanenkov, V.N., 1979, *General knowledge of nitrogen, phosphorus and silicon,* in: *Marine Chemistry*, O.K. Bordovskiy and V.N. Ivanenkov (Eds.), Vol. 1, Nauka, Moskva, pp. 176–183 (in Russian).

Ivanoff, A., 1975, *Introduction à L'Océanographie*, Librairie Vuibert, Paris, p. 208 (Vol.1), and p. 340 (Vol. 2).

Ivanoff, A., 1978, *Introduction to Oceanography*, Mir, Moskva, p. 574 (in Russian).

Ivanov, A.P., 1975, *Physical Fundamentals of Hydrooptics*, Nauka i Tekh., Minsk, p. 503 (in Russian).

Ivlev, L.S. and Andreev, S.D., 1986, *Optical Properties of Atmospheric Aerosols*, Gidrometeoizdat, Leningrad (in Russian).

Jackson, G.A. and Burd, A.B., 1998, Aggregation in the marine environment, *Environ. Sci. Tech.*, **32**:2805–2814.

Jackson, J.D., 1975, *Classical Electrodynamics*, 2nd ed., John Wiley / Polish translation, PWN, Warszawa, 1982.

Jaeger, R.G., 1959, *Dosimetrie und Strahlenschutz:Physikalische und technische Daten*, Georg Thieme Verlag, Stuttgart.

Janca, A., Tereszchuk, K., Bernath, P.F., Zobov, N.F., Shirin, S.V., Polyansky, O.L., and Tennyson, J., 2003, Emission spectrum of hot HDO below 4000 cm^{-1}, *J. Mol. Spectrosc.* **219**:132–135.

Jeffrey, S.W. and Humphrey, G.F., 1975, New spectrophotometric equation for determining chlorophyll *a*, *b*, *c1* and *c2*, *Biochem. Physiol. Pfl.* **167**:194–204.

Jeffrey, S.W., 1980a, Algal pigment systems, in: *Primary Productivity in the Sea*, P.G. Falkowski (Ed.), Plenum, New York, pp. 33–58.

Jeffrey, S.W., 1980b, Cultivating unicellular marine plants, CSIRO Fish. and Oceanogr. Rep. 1977–1979, (Cronulla), *Res. Rep. Div. Fish. Oceanogr. CSIRO*, pp. 22–43.

Jeffrey, S.W. and Vesk, M., 1997, Introduction to marine phytoplankton and their pigment signatures, in: *Phytoplankton Pigments in Oceanography: Guidelines to Modern Methods,* S.W. Jeffrey, R.F.C. Mantoura, and S.W. Wright (Eds.), 1997, Unesco, Paris, pp. 37–84.

Jeffrey, S.W., Mantoura, R.F.C., and Wright, S.W. (Eds.), 1997, *Phytoplankton Pigments in Oceanography: Guidelines to Modern Methods,* Unesco, Paris, p. 661.

Jenkins, F.A. and White, H.E., 1957, *Fundamentals of Optics,* 3rd ed., McGraw-Hill, Tokyo.

Jerlov, N.G., 1968, *Optical Oceanography,* Elsevier, Amsterdam, p. 194.

Jerlov, N.G., 1975, Long period changes of the optical properties of the Baltic, *J. Cons. Int. Explor. Mer.* **36(**2):188–190.

Jerlov, N.G., 1976, *Marine Optics*, Elsevier, Amsterdam, p. 231.

Jerlov, N.G., 1978, The optical classification of sea water in the euphotic zone, *Rep. Inst. Fys. Oceanogr. Københavns Univ.* **36**.

Jiracek, G.R., *Radio Sounding of Antarctic Ice*, 1967, Res. Rep. 67–1, Geophys. Polar Res. Centr, U, Wisconsin.
Johari, G.P., Charette, P.A., and Glaciol, J., 1975, The permittivity and attenuation in polycrystalline and single-crystal ice Ih at 30 and 60 Mhz, *J. Glaciol.* **14**(71):293–303.
Johnson, K., 2000, "Water buckyballs" chemical, catalytic and cosmic implications, *Infin. Energy* **6**:29–32.
Jonasz, M. and Zalewski, M.S., 1978, Stability of the shape of particle size distribution in the Baltic, *Tellus* **30**:569–572.
Jonasz, M., 1980, *Characterization of physical properties of marine particles using light scattering* (*Wykorzystanie zjawiska rozpraszania swiatla do wyznaczenia wlasnosci fizycznych zawiesin morskich*; in Polish). Ph.D. Thesis, Institute of Oceanology, Polish Academy of Sciences, Sopot, Poland, p. 214.
Jonasz, M., 1983, Particle size distribution in the Baltic, *Tellus B* **35**:346–358.
Jonasz, M., 1986, Role of nonsphericity of marine particles in light scattering and comparison of results of its determination using SEM and two types of particle couters, *SPIE, Vol. 637, Ocean Optics VIII,* 148–154.
Jonasz, M., 1987, Nonsphericity of suspended marine particles and its influence on light scattering. *Limnol. Oceanogr*. **32**:1059–1065.
Jonasz, M. and Fournier, G., 1996, Approximation of the size distribution of marine particles by a sum of log-normal functions, *Limnol. Oceanogr*. **41**:744–754.
Jonasz, M., Fournier, G.F., and Stramski, D., 1997, Photometric immersion refractometry: A method for determining the refractive index of marine microbial particles from beam attenuation, *Appl. Opt*., **36**:4214.
Jonasz, M. and Fournier, G.F., 1999, Approximation of the size distribution of marine particles by a sum of log-normal functions (errata: corrections and additional results), *Limnol. Oceanogr*. **44**(5):1358.
Junge, C.E., 1963, *Air Chemistry and Radioactivity*, Academic Press, New York, p. 382.
Kalle, K., 1949, Fluoreszenz und Gelbstoff im Bottnischen und Finnischen Meerbusen, *Deutch. Hydrograph. Z*. **2**:117–124.
Kalle, K., 1961, What do we know about the "Gelbstoff," *IUGG Monogr*. **10**:59–62.
Kalle, K., 1962, Uber die gelosten organischen Komponenten im Meerwasser, *Kieler Meeresforsch*. **18**:128–131.
Kalle, K., 1966, The problem of Gelbstoff in the sea, *Oceanogr. Mar. Biol. Ann. Rev.* **4**:91–104.
Karabashev, G.S. and Agatova, A.I., 1984, Relationship of fluorescence intensity to concentration of dissolved organic substances in ocean water, *Oceanology* **24**(6):680–682.
Karabashev, G.S., 1987, *Fluorescence in the Ocean*, Gidrometeoizdat, Leningrad, p. 200 (in Russian).
Karabashev, G.S, Evdoshenko, M., and Sherberstov, S., 2002, Penetration of coastal waters into the Eastern Mediterranean Sea using SeaWiFS data, *Oceanologica Acta* **25**: 31–38.
Karentz, D., McEuen, F. S., Land, M.C., and Dunlap, W.C., 1991, Survey of mycosporinelike aminoacid compounds in Antarctic marine organisms: Potential protection from ultraviolet exposure, *Mar. Biol.* **108**(1):157–166.
Karl, D.M. and Björkman, K.M., 2002, *Dynamics of DOP*, in: D.A. Hansell, and C.A. Carlson (Eds.), 2002 (op. cit.), pp. 249–366.
Kęcki, Z., 1992, *Fundamentals of Molecular Spectroscopy*, PWN, Warszawa, p. 337 (in Polish).
Kent, G.S., Yue, G.K., Farrukh, U.O., and Deepak, A., 1983, Modeling atmospheric aerosol backscatter at CO_2 laser wavelengths. 1: Aerosol properties, modeling technics, and associated problems, *Appl. Opt*. **22**:1655–1665.
Kerr, P.F., 1977, *Optical Mineralogy*, McGraw-Hill, New York.

Ketchum, B.H. and Redfield, A.C., 1949, Some physical and chemical characteristics of algae growth in mass culture, *J. Cell. Comp. Physiol* **33**:281–299.

Ketelaar, J.A.A. and Hanson, E.A., 1937, Elementary cell and space group of ethylchlorophyllide. *Nature* **140**:196.

Kiefert, L., McTainish, G.H., and Nickling, W.G., 1996, Sedimentological characteristics of Saharan and Australian dusts, in: *The Impact of Desert Dust Across the Mediterranean*, Guerzoni S. and Chester R. (Eds), pp. 183–190, Kluwer Academic, Norwell, MA.

Kirk, J.T.O., 1976, Yellow substance (Gelbstoff) and its contribution to the attenuation of photosinthetically active radiation in some inland and coastal south-eastern Australian waters, *Aust. J .Mar. Fresh. Res.* **27**:61–71.

Kirk, J.T.O., 1980, Spectral absorption properties of natural waters: Contribution of the soluble and particulate fractions to light absorption in some inland waters of south-eastern Australia, *Aust. J. Mar. Freshwater Res.* **31**:287–296.

Kirk, J.T.O., 1994, *Light and Photosynthesis in Aquatic Ecosystems*, Cambridge University Press, London-New York, p. 509.

Kishino, M., Sugihara, S., and Okami, N., 1984, Estimation of quantum yield of chlorophyll *a* fluorescence from the upward irradiance spectrum in the sea. *La Mer* **22**:233–240.

Kishino, M., Takahashi, M., Okami, N., and Ichimura, S., 1985, Estimation of the spectral absorption coefficients of phytoplankton in the sea. *Bull. Mar. Sci.(made in United States of America)*, **37**(2):634–642.

Kishino, M., Okami, N., Takahashi, M., and Ichimura, S., 1986, Light utilization efficiency and qantum yield of phytoplankton in athermally stratified sea, *Limnol. Oceanogr*. **31**(3):557–566.

Kobayashi, T. (Ed.), 1987, *Primary Processes in Photobiology*, Springer Verlag, Berlin, p. 243.

Koblentz-Mishke, O.J., 1971, Some ecological and physiological properties of phytoplankton, in: *Functioning of Pelagic Communites in the Tropical Regions of the Ocean,* M.E. Vinogradov (Ed.), Nauka, Moskva, pp.183–208.

Koblentz-Mishke, O.I., 1977, Primary production, in: *Biology of the Ocean*, M. E. Vinogradov (Ed.), Vol. 1, Nauka, Moskva (in Russian), pp. 62–64.

Koblentz-Mishke, O.J. and Vedernikov, V.I., 1977, Primary production, in: *Biology of the Ocean*, M.E. Vinogradov (Ed.), Vol. 2, Nauka, Moskva (in Russian), pp. 183–208.

Koblentz-Mishke, O.I., Woźniak, B., and Ochakovskiy, Y.E., 1985, *Utilisation of Solar Energy in the Photosynthesis of the Baltic and Black Sea Phytoplankton*, Izd. Inst. Okeanol., AN SSSR, Moskva, p. 336 (in Russian).

Koblentz-Mishke, O.I., Woźniak, B., Kaczmarek, S., and Konovalov, B.V., 1995, The assimilation of light energy by marine phytoplankton. Part 1. The light absorption capacity of the Baltic and Black Sea phytoplankton (methods; relation to chlorophyll concentration), *Oceanologia* **37**(2):145–169.

Koike, I., Hara, S., Terauchi, K., and Kogure, K., 1990, Role of sub-micrometre particles in the ocean. *Nature* **345**:242–244.

Kolber, Z. and Falkowski, P.G., 1993, Use of active fluorescence to estimate phytoplankton photosynthesis 'in situ,' *Limnol. Oceanogr*. **38**(8):1646–1665.

Kondratev, K.J., 1969, *Radiation in the Atmosphere*, Academic Press, New York, p. 920.

Kondratev, K.J. and Pozdniakov, D.V., 1988, *The Optical Properties of Natural Waters and the Remote Sensing of Phytoplanton,* Nauka, Leningrad, p. 181 (in Russian).

Kondratev, K.J., Pozdniakov, D.V., and Isakov, V.J., 1990, *Radiation and Hydro-optical Experiments in Lakes,* Nauka, Leningrad, p. 115 (in Russian).

Konovalov, B.V., 1979, Some special features of spectral absorption by suspensions in sea water, in: *Optical Methods of Investigation of Oceans and Internal Waters*, G.I. Galasij, K.S. Shifrin, and P.P. Sherstyankin (Eds.), Nauka, Novosibirsk (in Russian), pp. 58–65.

Konovalov, B.V., 1985, The light-absorbing ability of phytoplankton pigments, in: *The Assimilation of Solar Energy in the Photosynthetic Process in Phytoplankton of the Black and Baltic Seas*, O.J. Koblentz-Mishke, B. Woźniak, and Y.E. Ochakovskiy (Eds.), Izd. Inst. Okeanol. AN SSSR, Moskva, pp. 59–67 (in Russian).

Konovalov, B.V., Belyayeva, G.A., Bekasowa, O.D., and Kosakowska, A., 1990, Light-absorbing capacity of phytoplankton in the Gulf of Gdańsk in May, 1987, *Oceanologia* **28**:25–37.

Konovalov, B.V., 1992, Determination of light absorption coefficient by seston components using absorption spectrum on the membrane filter, *Okeanologiya* **32**:588–593 (in Russian).

Kopelevitch, O.V., Rusanov, S.J., and Nosienko, N.M., 1974, Light absorption by sea water in: hydrophysical and hydrooptical investigation in the Atlantic and Pacific Oceans in view of results of the 5th cruise of r/v "Dimitry Mendeleyev" (*Gidrofizicheskiye i gidroopticheskiye issledovaniya v Atlanticheskom i Tikhom okieanakh po riezultatam issledovaniy v 5-tom rieysie m/s "Dimitry Mendeleyev"*), Nauka, Moskva, pp. 107–112 (in Russian).

Kopelevitch, O.V., 1976, Optical properties of pure water in the 250–600 nm range, *Opt. Spectrosc.* **41**:391–392.

Kopelevitch, O.W. and Burenkov, V.I., 1977, On the relationship between spectral light attenuation coefficients of sea water, phytoplankton pigments and yellow substances (in Russian), *Okeanologya* **17**: 427–433.

Kopelevitch, O.V. and Shifrin, K.S., 1981, A contemporary conception of the optical properties of sea water, in: *The Optics of the Sea and Atmopshere*, Nauka, Moskva (in Russian), pp. 4–55.

Kopelevitch, O.V., 1983, Factors defining the optical properties of sea water (Chapter 6), Experimental data on the optical properties of sea water (Chapter 7), A few-parameter model of the optical properties of sea water (Chapter 7*)*, in: *The Optics of the Ocean*, A.S. Monin (Ed.), Vol. 1, Nauka, Moskva, pp. 150–234 (in Russian).

Kopelevitch, O.V., Lyutsarev, S.V., and Rodionov, V.V., 1988, The spectral absorption of light by yellow substance in ocean water, in: *The Optics of the Sea and Atmopshere*, K.S. Shifrin (Ed.), AN SSSR, Leningrad, pp. 158–159 (in Russian).

Kou, L., 1993, *Refractive Indices of Water and Ice in the 0.65–2.5 μm Spectral Range*, Dalhousie University, Canada.

Kou, L., Labrie, D., and Chylek, P., 1993, Refractive indices of water and ice in the 0.65-2.5μm spectral range, *Appl. Optics* **32**:3531–3540.

Koukouzas, N., Vassilatos, C., and Glarakis, I., 2005, Mixture of lignite fly ash in concrete: Physical and mineralogical characterization – case study from Ptolemais, Northern Greece, *World of Coal Ash (WOCA) Conference,* Lexington, Kentucky, USA, April 11–15 (under consideration), www.lignite.gr/abstracts/c_Mixture LigniteFlyAshPhysicalMineralogical.html.

Kowalczuk, P., 1999, Seasonal variability of yellow substances absorption in the surface layer of the Baltic Sea, *J. Geophys. Res.* **104**:30,047–30,058.

Kowalczyk, P., 2000, *Particle Physics,* PWN, Warszawa, p. 206 (in Polish).

Krey, J., 1971, *Primary Production in the Indian Ocean*, Kiel **31**(N4):29–32.

Krey, J., 1973, Primary production in the Indian Ocean, in: *The Biology of the Indian Ocean, Ecol. Stud.* **3**:115–126.

Krey, J. and Babenerd, B., 1976, *Phytoplankton Production – Atlas of the International Indian Ocean Expedition*, Inst. Meereskunde-Kiel University, p. 70.

Krężel, A., 1997, *Identification of Mesoscale Hydrophysical Anomalies in a Shallow Sea with Broad-band Remote Sensing*, Wydaw. Uniw. Gdańsk., Gdańsk, p. 173 (in Polish).

Krężel, A., Kozłowski, Ł., Szymanek, L., and Szymelfenig, M., 2005, Influence of coastal upwelling on chlorophyll-like pigments concentration in the surface water along Polish coast of the Baltic Sea, *Oceanologia* **47**(4):433–452.

Krischok, S., Höfft, O., Günster, J., Stultz, J., Goodman, D.W., and Kempter, V., 2001, H_2O interaction with bare and Li-precovered TiO_2: Studies with electron spectroscopies (MIES and UPS (He I and II)), *Surf. Sci.* **495**:8–18.

Król, T., 1985, A model for calculating Mie coefficients for spherical absorbing and scattering particles, *Stud. Mater. Oceanol.* (Published by Polish SCOR), **49**:43–62 (in Polish).

Król, T., 1991, The effect of the size of dispersed substances on absorption and scattering properties of dispersed media, *Stud. Mater. Oceanol.* (Published by Polish SCOR), **59**:175–181.

Król, T., 1998, *Light scattering by phytoplankton* (in Polish: *Rozpraszanie swiatla przez fitoplankton*); Diss. and monogr., No **9**, Inst. Oceanol. PAS, Sopot, p. 147.

Kuhn H., 1949, A quantum–mechanical theory of light absorption of organic dyes and similar compounds, *J. Chem. Phys.* **17**(12):1198–1212.

Kuhn, H., 1951, Elektronengasmodell zur quantitativen Deutung der Lichtabsorption von organischen Farbstoffen, 11. Teil B: Störung des Elektronengases durch Heteroatomel, *Helv. Chem. Acta* **34**:2371–2402.

Kullenberg, G., 1974, Observed and computed scattering functions, in: *Optical Aspects of Oceanography*, N. G. Jerlov, and E. Steeman-Nielsen, eds., Acad. Press, London, New York, pp. 25–49.

Kurtz, V. and Salib, S., 1993, Scattering and absorption of electromagnetic radiation by spheroidally shaped particles: Computation of the scattering properties. *J. Imaging Sci. Technol.* **37**:43–60.

Lancelot, C. and Billen, G.,1985, Carbon-nitrogen relationships in nutrient metabolism of coastal ecosystems. *Adv. Aquatic Microbiol.* **3**:262–321.

Laurion, I., Vincent, W.F., and Lean, R.S., 1997, Underwater ultraviolet radiation: Development of spectral models for northern high latitude lakes, *Photochem. Photobiol.* **65**(1):107–114.

Laurion, I., Ventura, M., Catalan, J., Psenner, R., and Sommaruga, R., 2000, Attenuation of ultraviolet radiation in mountain lakes: Factors controling the among- and within-lake variability, *Limnol. Oceanogr.* **45**:1274–1288.

Lazzara, L., Bricaud, A., and Claustre, H., 1996, Spectral absorption and fluorescence excitation properties of phytoplanktonic populations at a mesotrophic and an oligotrophic site in the tropical North Atlantic (EUMELI program), *Deep-Sea Res. Pt I* **43**(8):1215–1240.

Lee, C., 2005, Biogeochemical C flux studies: Past, present and future, *EUR-OCEAN Kick-Off Meeting,* Paris, 14–15 April 2005.

Lee, Z.P., Carder, K.L., Hawes, S.K., Steward, R.G., Peacock, T.G., and Davis, C.O., 1994, Model for the interpretation of hyperspectral remote-sensing reflectance, *Appl. Optics* **33**(24):5721–5732.

Le Grand, Y., 1939, La pénétration de la lumière dans la mer, *Ann. de l'Inst. Océanogr.* **19,** 393.

Lemus, R., 2004,Vibrational excitations in H_2O in the framework of a local model, *J. Mol. Spectrosc.* **225**:73–92.

Lenoble, J. and Saint-Guilly, B., 1955, Sur l'absorption du rayonnement par l'eau distillée, *C. R. Acad. Sci. Paris* **240**:954–955.

Lenoble, J., 1956, Sur le role des principaux sels dans l'absorption ultraviolette de l'eau de Mer, *Comp. Rend.* **242**:806–808.

Leppanan, J.-M., 1989, Cycle of carbon and nitrogen in the sea with special emphasis on sedimentation, in: P. Wassmann and A.-S. Heiskanan (Eds.), 1989 (op. cit.), pp. 78–99.

Levin, J.C., 1962a, Silicification, in: *Physiology and Biochemistry of Algae*, R.A. Levin (Ed.), Academic Press, New York, pp. 445–455.

Levin, J.C., 1962b, Calcification, in: *Physiology and Biochemistry of Algae*, R.A.Levin (Ed), Academic Press, New York, pp. 457–465.

Levin, Z., Joseph, J.H., and Mekler, Y., 1980, Properties of Sharav (Khamsin) dust-comparison of optical and direct sampling data, *J. Atmos. Sci.* **37**:882–891.

Lewis, M.R., Cullen, J.J., and Platt, T., 1983, Phytoplankton and thermal structure in the upper ocean: consequences of non-uniformity in the chlorophyll profile, *J. Geophys. Res.* **88**:2565–2570.

Lias, S.G., 2005, Ionization Energy Evaluation, in: *NIST Chemistry WebBook*, P.J. Linstrom and W.G. Mallard (Eds.), NIST Standard Reference Database Number 69, National Institute of Standards and Technology, Gaithersburg, MD, 20899. Available from: <http://webbook.nist.gov>.

Libes, S., 1992, *An Introduction to Marine Biogeochemistry*, Wiley, New York, p. 733.

Lide, D.R. (Ed.), 1997, *Physical and Optical Properties of Minerals*, in: *CRC Handbook of Chemistry and Physics*, 77th ed., pp. 4.130–4.136, CRC Press, Boca Raton, FL.

Lieth, H. and Whittaker, R.H., 1975, *Primary productivity of the Biosphere*, Springer Verlag, Berlin-Heidelberg-New York, p. 339.

Light, B., Maykut, G.A., and Grenfell, T.C., 2004, A temperature-dependent, structural-optical model of first-year sea ice, *J. Geophys. Res.* **109**:C06013, doi: 10.1029/2003JC002164, 1–16.

Light, B., Maykut, G.A., and Grenfell, T.C., 2003, A two-dimensional Monte Carlo model of radiative transfer in sea ice, *J. Geophys. Res.* **108**(C7):3219, doi:10.1029/2002JC001513, 12-1–12-18.

Lin, M.Y., Lindsay, H.M., Weitz, D.A., Ball, R.C., Klein, R., and Meakin, P., 1989, Universality in colloid aggregation, *Nature* **339**:360–362.

Lindberg, J.D. and Laude, L.S., 1974, Measurement of the absorption coefficient of atmospheric dust, *Appl. Optics* **13**:1923–1927.

Lindberg, J.D., Gillespie, J.B., and Hinds, B.D., 1976, Measurements of imaginary refractive indices of atmospheric particulate matter from a variety of geographic locations. *Proc. Int. Symp. on Radiation in the Atmosphere, Garmisch-Partenkirchen*, H.J. Bolle (Ed.), Science Press, New York.

Linne, M.A., 2002, *Spectroscopic Measurement. An Introduction to the Fundamentals*, Elsevier, p. 268.

Lisitzin, A.P., 1959, New data on distribytion and composition of suspended matter in seas and oceans related to some geological questions (in Russian), Doklady AN SSSR, **126**(4):863–866.

Lisitzin, A.P. and Bogdanov, J.A., 1968, Granulometric composition of Pacific Ocean hydrosol (in Russian), *Okeanol. Issled.* **18**:53–74.

Litvin, F.F., Belyaeva, O.B., Gulyaev, B.A., Karneeva, H.V., Sineshchekov, V.A., Stadnichuk, I.N., and Shubin, V.V., 1974, Systems of the native forms of chlorophyll, its role in primary processes of photosynthesis and development in process of the plant leafs greening, in: *Chlorophyll*, A. A. Shlyk (Ed.), Nauka i Tekh., Minsk (in Russian), pp. 215–231.

Litvin, F.F. and Sineshchekov, V.A., 1975, Molecular organization of Chlorophyll and energetics of the initial stages in photosynthesis, in: Govindjee (Ed.), 1975 (op. cit.), pp. 619–661.

Lorentzen, C.J., 1968, Determination of chlorophyll and pheopigments spectrophotometric equation, *Contribut. SCRIPPS Inst. Oceanogr.* **38**:977–980.

Lorenzen, C.J. and Jeffrey, S.W., 1980, Determination of chlorophyll in seawater (Report of intercalibration tests sponsored by SCOR and carried out in September–October 1978), *Unesco Tech. Pap. Mar. Sci.* **35**:1–20.

Love, A.E.H., 1899, The scattering of electric waves by dielectric sphere, *Proc. London Math. Soc.* **30**:308–321.

Love, J.P. and Peterson, K., 2005, *Quantum Chemistry,* 3rd ed., Elsevier, New York, p. 728.

Lundgren, B. 1976, *Spectral transmittance measurements of the Baltic*, University of Copenhagen, *Inst. Phys. Oceanogr.* Rep. 30.

Lyendekkers, J., 1967, Fluorophors and light-absorbing substances in natural waters, *Konink. Nederl. Meteorol. Inst. (De Bilt), Wetenschappelijk Rapp.* **67**(1):IV–4.

Majchrowski, R. and Ostrowska, M., 1999, Modified relationships between the occurrence of photoprotecting carotenoids of phytoplankton and Potentially Destructive Radiation in the sea, *Oceanologia* **41**(4):589–599.

Majchrowski, R. and Ostrowska, M., 2000, Influence of photo- and chromatic acclimation on pigment composition in the sea, *Oceanologia* **42**(2):157–175.

Majchrowski, R., Woźniak, B., Dera, J., Ficek, D., Kaczmarek, S., Ostrowska, M., and Koblentz-Mishke, O.I., 2000, Model of the 'in vivo' spectral absorption of algal pigments. Part 2. Practical applications of the model, *Oceanologia* **42**(2):191–202.

Majchrowski, R., 2001, *Influence of Irradiance on the Light Absorption Characteristics of Marine Phytoplankton*, Rozpr. i monogr., Nr **1**, Pom. Akad. Pedagogiczna, Słupsk, p. 131 (in Polish).

Malkiel, E., Alquaddoomi, O., and Katz, J., 1999, Measurements of plankton distribution in the ocean using submersible holography, *Meas. Sci. Technol.* **10**:1142–1152.

Mantoura, R.F.C. and Llewellyn, C.A., 1983, The rapid determination of algal chlorophyll and carotenoid pigments and their breakdown products in natural waters by reversephase high performance liquid chromatography, *Anal. Chim. Acta* **151**:297–314.

Margalef, R., 1967, Some concepts relative to the organization of plankton, *Oceanogr. Mar. Biol. Ann. Rev.* **5**:257–289.

Maring, H., Savoie, D.L., Izaguirre, M.A., and Custals, L., 2003, Mineral dust aerosol size distribution change during atmospheric transport, *J. Geophys. Res.* **108** (D19): 8592, doi: 10.1029/2002JD002536, p.6.

Markager, S. and Vincent, W.F., 2000, Spectral light attenuation and the absorption of UV and blue light in natural waters, *Limnol. Oceanogr.* **45**(3):642–650.

Marra, A.C., Blanco, A., Fonti, S., Jurewicz, A., and Orofino, V., 2005, Fine hematite particles of Martian interest: absorption spectra and optical constants, *J. Physics: Conference Series* **6**:132–138.

Mashke, H. and Haardt, H., 1987, Quantitative in vivo absorption spectra of phytoplankton: Detrital absorption and comparsion with fluorescence excitation spectra, *Limnol. Oceanogr.* **32**:620–633.

Massel, S.R., 1999, *Fluid Mechanics for Marine Ecologists*, Springer, Heidelberg, p. 566.

Matsunaga, T., Burgess, J.G., Yamada, N., Komatsu, K., Yoshida, S., and Wachi, Y, 1993, An ultraviolet (UV-A) absorbing biopterin glucoside from the marine planktonic cyanobacterium *Oscillatoria sp.*, *Appl. Microbiol. Biotech.* **39**(2):250–253.

Maureliss, A. and Tennyson J., 2003, The climatic effect of water vapour, *Physics World, May 2003, physicswep. org*, 1–5.

McAllister, C.D., Parson, T.R., Stephens, K., and Strickland, L.H.D., 1961, Measurements of primary production in coastal sea water using a large-volume plastic sphere. *Limnol. Oceanogr.* **6**:237–259.

McCave, I.N., 1983, Particulate size spectra behavior and origin of nepheloid layers over the Nova Scotian continental Rise, *J. Geophys. Res.* **88(C)**:7647–7666.

McClain, C.R, Feldman, G.C, and Hooker, S.B, 2004, An overview of the SeaWiFS project and strategies for producing a climate research quality global ocean bio-optical time series, *Deep-Sea Res. Pt II* **51**(1–3):5–42.

McCrone, W.C., Draftz, R.G., and Delly, J.G., 1967, *The Particles Atlas,* Ann Arbor Science, Ann Arbor, MI.

McQuarrie, D.A., 1983, *Quantum Chemistry*, University Science, Mill Valley, CA.

Meyers-Schulte, K. and Hedges, J., 1986, Molecular evidence for a terrestrial component of organic matter dissolved in ocean water, *Nature* **321**:61–63.

Mie, G., 1908, A contribution to the optics of turbid media: Special colloidal metal solutions (in German: *Beitrage zur Optik trüber Medien speziell kolloidaler Metallösungen*). *Ann. Phys.* **25**:377–445.

Miller, N. and Carpentier, R., 1991, Energy dissipation and photoprotection mechanisms during chlorophyll photobleaching in thylakoid membranes, *Photochem. Photobio.* **54**:465–472.

Minczewski, J. and Marczenko, Z., 1985, *Analytical Chemistry*, Vol. 3, PWN, Warszawa, p. 521 (in Polish).

Mitchell, B.G. and Kieffer, D.A., 1988, Chlorophyll *a* specyfic absorption and fluorescence excitation spectra for light-limited phytoplankton, *Deep-Sea Res.* **35**(5A): 639–663.

Mitchell, B.G. and Kieffer, D.A., 1984, Determination of absorption and fluorescence excitation spectra for phytoplankton, in: *Marine Phytoplankton and Productivity*, O. Holm-Hansen, L. Bolis, and R. Gilles (Eds.), Springer, Berlin, pp. 157–169.

Mobley, C.D., 1994, *Light and Water; Radiative Transfer in Natural Waters*, Academic Press, San Diego, p. 592.

Mobley C.D. and Stramski, D., 1997, Effect of microbial particles on oceanic optics: Methodology for radiative transfer modeling and example simulations, *Limnol. Oceanogr.* **42**:550–560.

Mobley, C.D., Cota, G., Grenfell, T.C., Maffione, R.A., Pegau, W.S., and Perovich, D.K., 1998, Modeling light propagation in sea ice. *IEEE Trans. Geosci. Remote Sens.* **36**:1743–1749.

Moisan, T.A. and Mitchell, B.G., 2001, UV absorption by mycosporine-like amino acids in Phaeocystis antarctica (Karsten) induced by photosynthetically available radiation, *Mar. Biol.* **138(1)**:217–227.

Moiseev, P.A., 1969, *Biologitcheskie Resursy Mirovogo Okeana*, Pischtchepromizd., Moskva, p. 40.

Moore, G.F., Aiken, J., and Lavender, S.J., 1999, The atmospheric correction of water colour and the quantitative retrival of suspended particulate matter in Case II waters: Application to MARIS, *Int. J. Remote Sens.* **20**:1713–1733.

Mopper, K. and Kieber, D.J., 2002, Photochemistry and the cycling of carbon, sulfur nitrogen and phosforus, in: D.A. Hansell, and C.A. Carlson (Eds.), 2002 (op. cit.), pp. 455–507.

Mordasova, N.V., 1974a, *Chlorophyll in Peruvian Coastal Waters of the Pacific Ocean in 1972*, OI, Ser. **7**: Premyslovaya Okeanologiya, CNIITEIRH SSSR, Moskva (in Russian), pp. 11–21.

Mordasova, N.V., 1974b, The distribution of chlorophyll a in the surface waters of the Pacific and Atlantic Oceans, in: *Bonitation of the World Ocean*, P.A. Moiseev et al. (Ed.), Part 2, Vol. 98, VNIRO, Moscow, pp. 91–97 (in Russian).

Mordasova, N.V., 1976, *The Distribution of Chlorophyll a in the World Ocean*, OI, Ser. **9**(3), CNIITEIRH SSSR, Moskva, p. 49 (in Russian).

Morel A., 1974, Optical properties of pure water and pure sea water, in: *Optical Aspects of Oceanography,* N.G. Jerlov and E. Steemann-Nielsen (Eds.), 1974 (op. cit.), pp. 1–24.

Morel, A. and Prieur, L., 1977, Analysis of variations in ocean color, *Limnol. Oceanogr.* **22**(4):709–722.

Morel, A., 1978, Available, usable and stored radiant energy in relation to marine photosynthesis, *Deep-Sea Res.* **25**(8):673–688.

Morel, A. and Bricaud, A., 1981, Theoretical results concerning light absorption in a discrete medium and application to specyfic absorption of phytoplankton, *Deep Sea Res.* **28(**A11):1375–1393.

Morel, A. and Bricaud, A., 1986, Inherent optical properties of algal cells including picoplankton: theoretical and experimental results, in: *Photosynthetic Picoplankton*, T. Platt, and W.K.W. Li (Eds.), *Can. Bull. Fish. Aquat. Sci.* **214**:521–559.

Morel, A., 1988, Optical modelling of the upper ocean in relation to its biogenous matter content (case 1 waters*)*, *J. Geophys. Res.* **93**(C10):10749–10768.

Morel, A. and Berthon, J.F., 1989, Surface pigments, algal biomass profiles and potential production of the euphotic layer: relationships re-investigated in view of remote sensing applications, *Limnol. Oceanogr.* **34**(8):1545–1562.
Morel, A. and Ahn, Y.H., 1990, Optical efficiency factors of free-living marine bacteria: Influence of bacterioplankton upon the optical properties and particular organic carbon in oceanic waters, *J. Mar. Res.* **48**:145–175.
Morel, A., 1991, Light and marine photosynthesis: A spectral model with geochemical and climatological implications, *Prog. Oceanogr.* **26**:263–306.
Morel, A., Ahn, Y.H., Partensky, F., Vaulot, D., and Claustre, H., 1993, Prochlorococcus and Synechococcus: A comparative study of their optical properties in relation to their size and pigmentation, *J. Mar. Res.* **51**:617–649.
Morrow, J.H., Chamberlain, W.S., and Kiefer, D.A., 1989, A two component description of spectra absorption by marine particulates, *Limnol. Oceanogr.* **34**:1500–1509.
Mota, R., Parafita, R., Giuliani, A., Hubin-Franskin, M.J., Lourenço, J.M.C., Garcia, G., Hoffmann, S.V, Mason, N.J., Ribeiro, P.A., Raposo, M., and Limão-Vieira, P., 2005, Water VUV electronic state spectroscopy by synchrotron radiation, *Chem. Phys. Lett.* **416**:152–159.
Mulliken, R.S. and Person, W.B., 1962, Donor-acceptor complexes, *Ann. Rev. Phys. Chem.* **13**:107–126.
Murrel, J.N., 1963, *The Theory of the Electronic Spectra of Organic Molecules*, London.
Murugavel P., Pawar, S.D., and Kamra A.K., 2001, Size distribution of submicron aerosol particles over the Indian Ocean during IFP-99 of INDOEX, *Current Science* **80**:123–127.
Nawell, G.F. and Nawell, R.C., 1966, *Marine Plankton*, Hutchinson Educational, London (revised edition) p. 221.
Nelson, N.B. and Siegel, D.A., 2002, Chromophoric DOM in the open ocean, in: D.A. Hansell, and C.A. Carlson (Eds.), 2002 (op. cit.), pp. 547–578.
Niiler, P.P. and Kraus, E.B., 1977, One-dimensional models of the upper ocean, pages 143–172 in: Kraus, E.B. (ed), *Modelling and Prediction of the Upper Layers of the Ocean*. Pergammon Press, Oxford, U.K.
Nilsson, A., Ogasawara, H., Cavalleri, M., Nordlund, D., Nyberg, M., Wernet, P., and Pettersson, L.G.M., 2005, The hydrogen bond in ice probed by soft x-ray spectroscopy and density functional theory, *J. Chem. Phys.* **122**:154505.
Nyquist, G., 1979, *Investigation of some optical properties of sea water with special reference to lignin sulfonates and humic substances*, PhD thesis, Dept. Analytical and Marine Chemistry, Göteborg University, Göteborg, Sweden, p. 203.
Oden, S., 1919, The humic acids, studies in their chemistry, physics and soil science, *Kolloidchem. Beih.* **11**:75–260.
Ogura, N. and Hanya, T., 1967, UV absorption of organic and inorganic matter, *Int. J. Oceanol. Limnol.* **1**:91–102.
Okami, N., Kishino, M., Sugihara, S., Takematsu, N., and Unokil, S, 1982, Analysis of ocean color spectra. 3. Measurements of optical properties of sea water. *J. Oceanogr. Soc. Jpn.* **38**:362–372.
Okamura, M.Y., Feher, G., and Nelson, N., 1982, Reaction centers, in: *Photosynthesis*, Vol. 1, *Energy Conversion by Plants and Bacteria*, Govindjee (Ed.), Academic Press, New York, pp. 195–272.
Olaizola, M., 1996, High performance liquid chromatography, in: M.C. Ooms (Ed.) (op. cit.), pp. 363–382.
Olander, D.S. and Rice, S.A., 1972, Preparation of amorphous solid water, *Proc. Nat. Acad. Sci. USA* **69**:98–100.
Olmo, F.J., Quirantes, A., Alcántara, A., Lyamani, H., and Alados-Arboleda, L., 2006, Preliminary results of a non-spherical aerosol method for the retrieval of the atmospheric aerosol optical properties. *J. Quantitative Spectroscopy and Radiative Transfer* **100**:305–314.

Olson, R.J., Vaulot, D., and Chisholm, S.W., 1985, Marine phytoplankton distributions measured using shipboard flow cytometry, *Deep-Sea Res.* **32**:1273–1280.

Olszewski, J. and Darecki, M., 1999, Derivation of remote sensing reflectance of Baltic waters from above-surface measurements, *Oceanologia* **41**(1):1–13.

Oncley J.L., Scatchard, G., and Brown, A., 1947, Physical-chemical characteristics of the proteins of normal human plasma. *J. Phys. Chem.* **51**:184–198.

Ooms, M. (Ed.), 1996, *ULISSE (Underwater Light Seatruth Satellite Experiment)*, Europ. Commiss. Joint Res. Centre, Ispra, Italy, Spec. Publ., 1.96.29, pp. 506.

Ort, D.R. and Govindjee, 1987, A general survey of energy conversion during photosynthesis, in: *Photosynthesis*, Vol. 1, Govindjee (Ed.), Mir, Moskva (in Russian), pp. 316–402.

Ostrowska, M., 1990, Fluorescence 'in situ' method for the determination of chlorophyll *a* concentration in sea, *Oceanologia* **29**:175–202.

Ostrowska, M. and Woźniak, B., 1991a, Fluorescence properties of marine phytoplankton, *Stud. Mater. Oceanol.* (Published by Polish SCOR), **59**:159–174.

Ostrowska, M. and Woźniak, B., 1991b, An introduction to fluorescence methods of marine photosynthesis studies, *Stud. i Mater. Oceanol.* (Published by Polish SCOR), **59**:97–136.

Ostrowska, M., Majchrowski, R., Matorin, D.N., and Woźniak, B., 2000a, Variability of the specific fluorescence of chlorophyll in the ocean. Part 1. Theory of classical 'in situ' chlorophyll fluorometry, *Oceanologia* **42**(2):203–219.

Ostrowska, M., Matorin, D.N., and Ficek, D., 2000b, Variability of the specific fluorescence of chlorophyll in the ocean. Part 2. Fluorometric method of chlorophyll *a* determination, *Oceanologia* **42**(2):221–229.

Ostrowska, M., 2001, *Using the Fluorometric Method for Marine Photosynthesis Investigations in the Baltic*, Rozpr. i monogr., Nr **15**, Inst. Oceanol. PAN, Sopot, p. 194 (in Polish).

Otremba, Z., 2000, The impact on the reflectance in VIS of a type of crude oil film floating on the water surface, *Optics Express* **7**:129–134.

Otremba, Z. and Piskozub, J., 2001, Modelling of the optical contrast of an oil film on a sea surface, *Optics Express* **9**:411–416.

Otremba, Z. and Król, T., 2002, Modeling of the crude oil suspension impact on inherent optical properties of coastal seawater, *Polish J. Environ. Studies* **11**(4):407–411.

Otremba, Z., 2004, Modeling the bidirectional reflectance distribution functions (BRDF) of sea areas polluted by oil, *Oceanologia* **46**(4):505–518.

Padró, J.A. and Marti, J., 2003, An interpretation of the low-frequency spectrum of liquid water, *J. Chem. Phys.* **118**:452–453.

Palmer, K.F. and Williams, D., 1974, Optical properties of water in the near infrared, *J. Opt. Soc. Am.* **64**:1107–1110.

Parke, M. and Dixon, P.S., 1968, Check-list of British Marine algae-sesond revision. *J. Mar. Biol. Ass. U.K.* **48**:783–832.

Parson, V.V. and Ke, B., 1987, Primary photochemical reactions, in: *Photosynthesis*, Vol. 1, Govindjee (Ed.), Mir, Moskva, pp. 472–539 (in Russian).

Parsons, T.R., Stephens, K., and Strickland, J.H.D., 1961, On the chemical composition of eleven species of marine phytoplankton, *Fish. Res. Board. Can.* **18**:1001–1016.

Parsons, T.R., 1963, Suspended organic matter in sea water, *Prog. Oceanogr.* **1**:205–239.

Parsons, T.R., 1975, Particulate organic carbon in the sea, in: *Chemical Oceanography*, J.P. Riley and G. Skirrow (Eds.), Vol. 2, Academic Press, New York, pp. 365–385.

Parsons, T.R., Takahashi M., and Hargrave B., 1977, *Biological Oceanographic Processes*, 2nd ed., Pergamon Press, Oxford, p. 332.

Patterson, E.M., Gillette, D.A., and Stockton, B. H., 1977, Complex index of refraction between 300 and 700 nm for Saharan dust, *J. Geophys. Res.* **82**:3153–3160.

Patterson, E.M., Kiang, C.S., Delany, A.C., Wartburg, A.F., Leslie, A.C.D., and Huebert, B.J., 1980, Global measurements of aerosols in remote continental and

marine regions – concentrations, size distribution, and optical properties, *J. Geophys. Res. – Oceans and Atmospheres* **85** (NC12):7361–7376.
Pauling, L., 1959, The structure of water, in: *Hydrogen Bonding*, D. Hadzi, and H.W. Thompson (Eds.), Pergamon Press, London, pp. 1–6.
Payne, R.E., 1979, Albedo of the sea surface, *J. Atmos. Sci.* **29**:959–970.
Pelevin, V.N. and Rutkovskaya V.A., 1978, A review of photosynthetic radiation in the waters of the Pacific Ocean, *Okeanologiya* **18**(4):619–626 (in Russian).
Pelevin, V.N. and Rutkovskaya, V.A., 1980, The attenuation of the solar energy flux with depth in the waters of the Indian Ocean, in: *Light Fields in the Ocean*, V.N. Pelevin and M.V. Kozlaninov (Eds.), Moskva, pp. 73–85 (in Russian).
Pelevin, V.N. and Rostovtseva, V.V., 2001, Sea water scattering and absorption models development using classification of ocean waters on base of contact measurements data, *Proc. Int. Conf. on Current Problems in Optics of Natural Waters* (St. Petersburg), pp. 377–382.
Pempkowiak, J., 1977, The complex-forming properties of humus substances in the bottom sediments of the Baltic Sea, *Stud. Mater Oceanol.* (Published by Polish SCOR), **19**:136–142 (in Polish).
Pempkowiak, J. and Kupryszewski, G., 1980, The input of organic matter to the Baltic from the Vistula river, *Oceanologia* **12**:79–95.
Pempkowiak, J. and Pocklington, R., 1983, Phenolic aldehyds as indicators of the origin of humic substances in the marine environment, in: *Aquatic and Terrestrial Humic Materials*, R.F. Christman and E.T. Gjessing (Eds.), Ann Arbor Science, Ann Arbor, MI, pp. 371–386.
Pempkowiak, J., 1984, The origin of humic substances in the Baltic Sea evaluated on the basis of their physical and chemical properties, *Proc. XIV Conf. Baltic Oceanogr. (Gdynia)*, pp. 674–689.
Pempkowiak, J., 1989, *The Distribution, Origin and Properties of Humus Acids in the Baltic Sea,* Ser. Prace habilitacyjne Inst. Oceanol. PAN (1989), Ossolineum, Wrocław, p. 146 (in Polish).
Pempkowiak, J., 1997, *An Outline of Marine Geochemistry*, Wydaw. Uniw. Gdańsk., Gdańsk, p. 171 (in Polish).
Perovich, D.K., Maykut, G.A., and Grenfell, T.C., 1986, Optical properties of ice and snow in polar oceans, I: Observations, *Ocean Optics VIII, Proc. SPIE* **637**: 232–241.
Perovich, D.K., 1990, Theroretical estimates of light reflection and transmission by spatially complex and temperally varying sea ice covers, *J. Geophys. Res.* **95**:9557–9567.
Perovich, D.K. and Govoni, J.W., 1991, Absorption coefficients of ice from 250 to 400 nm, *Geophys. Res. Lett.* **18**(7):1233–1235.
Perovich, D.K., 1994, Light reflection from sea ice during the onset of melt, *J. Geophys. Res.* **99**:3351–3359.
Peterson, J.T. and Weinman, J.A., 1969, Optical properties of quartz dust particles at infrared wavelengths, *J. Geophys. Res.* **74**(28):6947–6952.
Petke, J.D., Maggiora, G.M., Shipman, L., and Christoffersen, R.E., 1979, Stereoelectronic properties of photosynthetic and related systems – V. AB initio configuration interaction calculations on the ground and lower excited singlet and triplet states of ethyl chlorophyllide *a* and ethyl pheophorbide *a*, *Photochem. Photobiol.* **30**:203–223.
Phillips, C.S.G. and Williams, R.J.P., 1966, *Inorganic Chemistry*, Vol. 2, Clarendon Press, Oxford.
Pinkley, L. and Williams, D., 1976, Optical properties of sea water in the infrared, *J. Opt. Soc. Amer.* **66(**9):554–558.
Pinkley, W.P., Sethna, P.P., and Williams, D., 1977, Optical constants of water in the infrared: Influence of temperature, *J. Opt. Soc. Am*er. **67(4)**:494–499.

Platt, T. and Sathyendranath, S., 1993a, Estimators of primary production for interpretation of remotely-sensed data on ocean color, *J. Geophys. Res. –Oceans* **98**: 14561–14576.
Platt, T. and Sathyendranath, S., 1993b, The remote sensing of ocean primary productivity – use of a new data compilation to test satellite algorithms – comment, *J. Geophys. Res.–Oceans*. **98**:16583–16584.
Platt, T., Sathyendranath, S., Cavarhill, C.M., and Lewis, M.R., 1988, Ocean primary production and available light: Further algorithms for remote sensing, *Deep-Sea Res*. **35**(6):855–879.
Platt, T., Sathyendranath, S., and Longhurst, A., 1995, Remote-sensing of primary production in the ocean – promise and fulfillment, *Philos. Trans. Roy. Soc. London, Series B* **348**:191–201.
Plaza, R.C., Durán, J.D.G., Quirantes, A., Ariza, M.J., and Delgado, A.V., 1997, Surface chemical analysis and electrokinetic properties of spherical hematite particles coated with Yttrium compounds. *J. Colloid and Interface Sci.* **194**: 398–407.
Plaza, R.C., Quirantes, A., and Delgado, A.V., 2002, Stability of dispersions of colloidal hematite/yttrium oxide core-shell particles. *J. Colloid and Interface Sci.* **252**: 102–108.
Polyansky, O.L., Császár, A.G., Shirin, S.V., Zobov, N.F., Berletta, P., Tennyson, J., Schwenke, D.W., and Knowles, P.J., 2003, High-accuracy ab initio rotation-vibration transitions for water, *Science* **299**:539–542.
Pope, R.M., 1993, *Optical Absorption of Pure Water and Sea Water Using the Integrating Cavity Absorption Meter*, Texas A&M University.
Pope, R.M. and Fry, E.S., 1997, Absorption spectrum (380–700 nm) of pure water, II. Integrating cavity measurements, *Appl. Optics* **36**(33):8710–8723.
Popov, N.I., Fyedorov, K.N., and Orlov, V.M., 1979, *Sea Water*, Nauka, Moskva, p. 327 (in Russian).
Postma, H., 1954, Hydrography of the Dutch Wadden Sea. *Archives Néerlandaises de Zoologie* **10**:405–511.
Preisendorfer, R.W., 1961, Application of radioactive transfer theory of light measurements in the sea. A Symposium on Radiant Energy in the Sea (Helsinki), *IUGG Monogr*. **10**:11–30.
Prieur, L. and Sathyendranath, S., 1981, An optical classification of coastal and oceanic waters based on the specific spectral absorption curves of phytoplankton pigments dissolved organic matter, and other particulate materials, *Limnol. Oceanogr*. **26**:671–689.
Prochorow, J., 1983, The electronic spectra of molecules, in: *Encyclopedia of Contemporary Physics*, PWN, Warszawa (in Polish).
Prochorow, J. and Siegoczyński, R., 1969, Radiative and radiationless processes in charge-transfer complexes, *Chem. Phys. Letters* **3**:635–639.
Pye, K., 1987, *Aeolian Dust and Dust Deposits*, Academic Press, San Diego, p. 334.
Quan, X. and Fry. E.S., 1995, Empirical equation for the index of refraction of seawater, *Appl. Optics* **34**:3477–3480.
Querry, M.R., Cary, P.G., and Waring, R.C., 1978, Split-pulse laser method for measuring attenuation coefficients of transparent liquids: application to deionized filtered water in the visible region, *Appl. Optics* **17**:3587–3592.
Querry, M.R., Osborn, G., Lies, K., Jordon R., and Coveney, R.M., 1978, Complex refractive index of limestone in the visible and infrared, *Appl. Optics* **17**:353–356.
Querry, M.R., 1987, Optical constants of minerals and other materials from the millimeter to the UV, *U.S. Army Aberdeen, MD* (Ref. after Sokolik and Toon, 1999, op.cit.).
Quickenden, T.I. and Irvin, J.A., 1980, The ultraviolet absorption spectrum of liquid water, *J. Chem. Phys*. **72**:4416–4428.
Quirantes, A. and Delgado, Á., 2003, Cross section calculations of randomly oriented bispheres in the small particle regime. *J. Quant. Spectros. Radiative Transfer* **78**:179–186.

Quirantes, A. and Bernard, S., 2004, Light scattering by marine algae: d two-layer spherical and nonspherical models. *J. Quant. Spectros. Radiative Transfer* **89**:311–321.
Quirantes, A., 2005, A T-matrix method and computer code for randomly oriented, axially symmetric coated scatterers. *J. Quant. Spectros. Radiative Transfer* **92**:373–381.
Quirantes, A. and Bernard, S., 2006, Light scattering models for modelling algal particles as a collection of coated and/or nonspherical scatterers. *J. Quant. Spectros. Radiative Transfer* **100**:315–324.
Raes, F., Van Dingenen, R., Cuevas, E., Van Velthoven, P.F.J., and Prospero, J.M., 1997, Observations of aerosols in the free troposphere and marine boundary layer of the subtropical Northeast Atlantic: Discussion of processes determining their size distribution, *J. Geophys, Res.* **102(**D17), 21,315– 21,328.
Raichlin, Y., Millo, A., and Katzir, A., 2004, Investigation of the structure of water using mid-IR fiberoptic evanescent wave spectroscopy, *Phys. Rev. Lett.* **93**:85703.
Ramming, H.G. and Kowalik, Z., 1980, *Numerical Modelling of Marine Hydrodynamics. Application to Dynamic Physical Processes*, Elsevier Oceanogr. Ser., No. **26**, Amsterdam, p. 368.
Rao, C.N.R., 1981, *Electron Spectroscopy of Organic Compounds*, PWN, Warszawa (in Polish).
Rashid, M.A., 1985, *Geochemistry of Marine Humic Compounds*, Springer Verlag, New York, p. 300.
Raspletina, G.F., Yliyanova, D.N., and Sherman, E.E., 1967, The hydrochemistry of Lake Ladoga, in: *Hydrochemistry and Hydro-optics of Lake Ladoga*, Gidrometeoizdat, Leningrad, pp. 60–122 (in Russian).
Reid, J.S., Jonsson, H.H., Maring, H.B., Smirnov, A., Savoie, D.L., Cliff, S.S., Reid, E.A., Livingston, J.M., Meier, M.M., Dubovik, O., and Tsay, S-C., 2003, Comparison of size and morphological measurements of coarse mode dust particles from Africa, *J. Geophys. Res.* **108**(D19):8593, doi: 10.1029/2002JD002485, p. 28.
Renk, H., 1973, Primary production in the waters of the southern Baltic, *Stud. Mater. MIR (Gdynia)*, A /**12**:5–126 (in Polish).
Reuter, R., Diabel-Langohr, D., Doerffer, R., Dorre, F., Haardt, H., and Hengstermann, T., 1986, The influence of Gelbstoff on remote sensing of seawater constituents from space, in: Grassl et al. (Eds.), 1986b (op. cit.), Vol. 2, Appendix 3, p. 58.
Richards, F.A., 1952, The estimation and characterisation of phytoplankton population by pigment analysis (part 1), *J. Mar. Res.* **11**:147–155.
Richards, F.A. and Thompson, T.G., 1952, The estimation and charakterisation of phytoplankton population by pigment analysis (part 2), *J. Mar. Res.* **11**:156–172.
Riley, J.P. and Skirrow G. (Eds.), 1965, *Chemical Oceanography*, Vol. 1, Academic Press, London, p. 712.
Riley, J.P. and Chester, R., 1971, *Introduction to Marine Chemistry*, Academic Press, London, p. 465.
Risovic, D., 1993, Two-component model of sea particle size distribution, *Deep-Sea Res. Pt I* **40**:1459–1473.
Robinson, G.W., Zhu, S.B., Singh, S., and Evans, M.W., 1996, *Water in Biology, Chemistry and Physics: Experimental Overviews and Computational Methodologies,* World Scientific, Singapore.
Roesler, C.S., Perry, M.J., and Carder, K.L., 1989, Modeling in situ phytoplankton absorption from total absorption spectra in productive inland marine waters, *Limnol. Oceanogr.* **34**(8):1510–1523.
Romankevich, E.A., 1977, *The Geochemistry of Organic Substances in the Ocean*, Nauka, Moskva, p. 256 (in Russian).
Romankevich, E.A., 1979, The composition of organic substances, in: *Oceanology: The Chemistry of Ocean Waters*, O.K. Bordovskii and V.N. Ivanenkov (Eds.) (in Russian).
Romanovsky, V., 1966, *Physique de l'océan*, Seuil, Paris, p. 189.

Ross, K.F.A., 1967, *Phase Contrast and lntelference Microscopy Jor Cell Biologists,* Arnold, London.

Rubin, A.B. and Samuilov, V.D. (Eds.), 1973, *Problems of Biophotochemistry*, Nauka, Moskov (in Russian).

Rubin, A.B., Kononenko, A.A., Shaitan, K.V., Paschenko, V.Z., and Riznichenko, G.Y., 1994, Electron transport in photosynthesis, *Biophysics* **39**(2):173–195.

Rubin, A.B., 1995, *Principles of Organisation and Regulation of Primary Processes of Photosynthesis*, OHTI PHC RAN, Moskva, **33**, p. 38 (in Russian).

Ruddick, K.G., Ovidio, F., and Rijkeboer, M., 2000, Atmospheric correction of SeaWiFS imagery for turbid coastal and inland waters, *Appl. Optics* **39**(6):897–912.

Samuła-Koszałka, T. and Woźniak, B., 1979, The share of particular sea water components in the light attenuation and analysis of the absorption spectra of yellow substances taking Gdansk Bay as an example, *Stud. Mater. Oceanol.* (Published by Polish SCOR), **26**:203–216 (in Polish with English abstract).

Sano, A., 1960, Studies on organic reagents for inorganic analysis, 6, Absorption spectrum of phenylfluorone-metal chelate and the nature of the chelating bond, *Bull. Chem. Soc. Japan* **3**:286–290.

Sarpal, R.S., Mopper, K., and Kieber, D.J., 1995, Absorbance properties of dissolved organic matter in Antarctic seawater, *Antarct. J. US* **30**:139–140.

Sathyendranath, S., Platt, T., Cavarhill, C.M., Warnock, R.E., and Lewis, M.R., 1989, Remote sensing of oceanic primary production: Computations using a spectral model, *Deep-Sea Res.* **36**(3):431–453.

Sathyendranath, S., Hoge, F.E., Platt, T., and Swift, R.N., 1994, Detection of phytoplankton pigments from ocean color – improved algorithms, *Appl. Optics* **33**: 1081–1089.

Sathyendranath, S. (ed.), 2000, *Remote sensing of ocean colour in coastal, and other optically-complex, waters*, IOCCG Rep. No 3, IOCCG Project Office, Dartmouth, Nova Scotia, pp. 140.

Sathyendranath, S., Platt, T., and Stuart, V., 2000, Remote sensing of ocean colour: Recent advances, exciting possibilities and unanswered questions, *Proc. Fifth Pacific Ocean Remote Sensing Conf. (PORSEC), 5–8 Dec. 2000*, Vol. **1**, p. 6.

Sathyendranath, S., Cota, G., Stuart, V., Maass, H., and Platt, T., 2001, Remote sensing of phytoplankton pigments: A comparison of empirical and theoretical approaches, *Int. J. Rem. Sens.* **22**:249–273.

Scheer, H. (Ed.), 1991, *Chlorophylls*, CRC Press, Boca Raton, FL.

Schiebener, P., Straub, J., Sengers, J.M.H.L., and Gallagher, J.S., 1990, Refractive index of water and steam as function of wavelength, temperature and density, *J. Phys. Chem. Res.* **19**:677–717.

Schneider, S.H., 1992, Introduction to climate modeling, pp. 3–26, in: *Modelling Climate System*, Cambridge University Press, Cambridge, pp. 788.

Schnitzer, H. and Khan, S., 1972, *Humic Substances in the Environment*, Marcel Dekker, New York, p. 327.

Scott, A.I., 1964, *Interpretation of the Ultraviolet Spectra of Natural Products*, Pergamon Press, Oxford, p. 443.

Segelstein, D.J., 1981, *The Complex Refractive Index of Water*, University of Missouri-Kansas City.

Segtnan, V.H., Šašic, Š., Isaksson, T., and Ozaki, Y., 2001, Studies on the structure of water using two-dimensional near-infrared correlation spectroscopy and principal component analysis, *Anal. Chem.* **73**:3153–3161.

Seki, M., Kobayashi, K., and Nakahara, J., 1981, Optical –spectra of hexagonal ice, *J. Phys. Soc. Jpn.* **50**(8):2643–2648.

Semina, G.I., 1957, Factors influencing the vertical distribution of phytoplankton in the sea, *Tr. Vsekhsoyuz. Gidrobiol. Obshch.* **7**:119–129 (in Russian).

Semina, G.I. (Ed.), 1985, *Study of Oceanic Phytoplankton*, Acad. Sci. USSR, P.P. Shirshov Inst. Oceanol., Moskow (in Russian).
Shapiro, J., 1957, Chemical and biological studies on the yellow organic acids of lake water, *Limnol. Oceanogr.* **2**(3):161–179.
Sharif, S., 1995, Chemical and mineral composition of dust and its effect on the dielectric constant, *IEEE Trans. Geosci. Remote Sens.* **33**:353–359.
Sharp, J.H., 2002, *Analitical Methods for Total DOM Pools*, in: D.A. Hansell, and C.A. Carlson (Eds.), 2002 (op. cit.), pp. 35–58.
Sheldon, R.W. and Parsons, T.R., 1967, A continuous size spectrum for particulate matter in the sea, *J. Fish. Res. Board of Canada* **24**:909–915.
Sheldon, R.W., Prakash, A., and Sutcliffe, W.H., Jr., 1972, The size distribution of particles in the ocean, *Limnol. Oceanogr.* **17**:327–340.
Shibaguchi, T., Onuki, H., and Onaka, R., 1977, Electronic structures of water ice, *J. Phys. Soc. Jap.* **42**(1):152–158.
Shifrin, K.S., 1951, *Light Scattering in a Turbid Medium*, Gostekhteorizdat, Leningrad, p. 288 (in Russian).
Shifrin, K.S., 1952, *Light Scattering by Two-layered Particles*, AN SSSR, Ser. Geofiz. No. **2**, pp. 15–20 (in Russian).
Shifrin, K.S., 1957, The optical investigation of charged particles in clouds, in: *Investigation of Clouds, Precipitation and Thundercloud Electricity,* Gidrometeoizdat, Leningrad, p. 19 (in Russian).
Shifrin, K.S., Kopelevitch, O.V., Burenkov, V.I., Mashtakov, Y.L., The light scattering function and the structure of the sea hydrosol, 1974, *Izv. Ak. Nauk SSSR, Fizika Atmosfere i Okeana* **10**:25–35.
Shifrin, K.S., 1983a, Theory of absorption and scattering of light in sea water, in: *Optics of the Ocean*, Vol. 1, *Physical Optics of the Ocean*, A.S. Monin (Ed.), Nauka, Moskva, pp. 18–54 (in Russian).
Shifrin, K.S., 1983b, *Introduction to Ocean Optics* (in Russian), Gidrometeoizdat, Leningrad, p. 278.
Shifrin, K.S., 1988, *Physical Optics of Ocean Water,* American Institute of Physics, New York.
Shifrin, K.S., 1995, Chemical and mineral composition of dust and its effect on the dielectric constant, *IEEE Trans. Geosci. Remote Sens.* **33**:353–359.
Shipman, L.L., 1982, Electronic structure and function of chlorophyls and their pheophytins, in: *Photosynthesis*, Vol. 1, Govindjee (Ed.), Academic Press, New York, pp. 275–291.
Shipman, L.L., 1987, Electronic structure and function of chlorophyls and their pheophytins, in: *Photosynthesis*, Vol. 1, Govindjee (Ed.), Mir, Moskva, pp. 403–420 (in Russian).
Shirazi, M.A., Boersma, L., and Johnson, C.B., 2001, Particle-size distribution: Comparing texture systems, adding rock, and predicting soil properties, *Soil Sci. Soc. Am. J.* **63**:300–310.
Shlyk, A.A., 1965, *Metabolism of Chlorophyll in Green Plant*, Nauka i Tekh., Minsk (in Russian).
Shlyk, A.A. (Ed.), 1974, *The Chlorophyll*, Nauka i Tekh., Minsk, p. 416 (in Russian).
Shuleykin, V.V., 1959, *A Short Course of Marine Physics*, Leningrad (in Russian).
Shuleykin, V.V., 1968, *Marine Physics*, 4th ed., Nauka, Moskva.
Sieburth, J. and Jensen, A., 1968, Studies an algal substances in the sea. I. Gelbstoff (humic material) in terrestrial and marine waters., *J. Exp. Mar. Biol. Ecol.* **2**:174–179.
Siegel, D.A., Iturriaga, R., Bidigare, R.R., Smit, R.C., Pak H, Dickey, T.D., Marra, J., and Baker, K.S., 1990, Meridional variations of the springtime phytoplankton community in the Sargasso Sea, *J. Mar. Res.* **48**:379–412.

Siegoczyński, R.M., Jędrzejewski, J., and Kawski, A., 1975, Optical properties and transfer of electronic excitation energy in poly-N-vinylcarbazole layers, *Acta Physica Polonica* **A47**:707–716.

Siegoczyński, R.M., Jędrzejewski, J., and Kawski, A., 1978, The excited triple complex of the poly-N-vinylcarbazole:benzanthrone system, *J. Molecular Structure* **45**: 445–449.

Siegoczyński, R.M. and Ejchart, W., 2004, Quenching of exciplex and charge-transfer complex fluorescence of poly-N-vinylcarbazole-benzanthrone system, *Macromolecular Symposia* **212**:575–580.

Silver, M.W., Pilskaln, C.H., and Steinberg, D., 1991, The biologists' view of sediment trap collections: Problems of marine snow and living organisms, in: P. Wassmann et al. (Eds.) (op. cit.), pp. 76–93.

Simson, J.P., 1976, *Photochemistry and Sectroscopy*, PWN, Warszawa (Polish translation: *Fotochemia i Spektroskopia*).

Skopintsev, B.A., 1950, Organic substances in natural waters, *Tr. GOIN*, **32**:3–32 (in Russian).

Skopintsev, B.A., 1971, Contemporary achievements in the study of organic substances in ocean waters, *Okeanologiya* **11(6)** (in Russian).

Skopintsev, B.A., Bordovski, O.K., and Ivanenkov, V.N., 1979, Carbon of the dissolved organic matter, in: *Chemistry of the ocean*, A.C., Monin, (Ed.), Nauka, Moscow, Vol. **1**, pp. 251–259 (in Russian).

Smekot-Wensierski, W., Woźniak, B., Grassl, H., and Doerffer, R., 1992, *Die Absorptionseigenschaften des marinen Phytoplanktons*, GKSS 92/E/105, GKSS-Forschungszentrum Geesthacht GMBH, Geesthacht, p.104.

Smirnov, A.V. and Shifrin, K.S., 1987, Light absorption by the disporse system, *Kolloid. Zh., (AN SSSR, Moskva)* **4**:809–811 (in Russian).

Smith, R.C. and Baker, K.S., 1978, Optical classification of natural waters, *Limnol. Oceanogr*. **23**(2):260–267.

Smith, R.C. and Baker, K.S., 1981, Optical properties of the clearest natural waters (200–800 nm), *Appl. Optics* **20**(2):177–184.

Smoluchowski, M., 1908, *Molekular-kinetiche Theorie der Opaleszenz von Gasen im Kritischen Zustande, sowie Einiger Verwandter Erscheinungen*, *Ann. Phys*. **25**:205.

Sogandares, F.M. and Fry, E.S., 1997, Absorption spectrum (340–640 nm) of pure water. I. Photothermal measurements, *Appl. Optics* **36**:8699–8709.

Sokolik, I.N., Andronova, A.W., and Johnson, T.C., 1993, Complex refractive index of atmospheric dust aerosols, *Atmos. Environ*. **27A**(16):2495–2502.

Sokolik, I.N., Toon, O.B., and Bergstrom, R.W., 1998, Modeling the relative characteristics of airborne mineral aerosol at infrared wavelengths, *J. Geophys. Res*. **103**: 8813–8826.

Sokolik, I.N. and Toon, O.B., 1999, Incorporation of mineralogical composition into models of the radiative properties of mineral aerosol from UV to IR wavelengths, *J. Geophys. Res*. **104**(D8):9423–9444.

SooHoo, J.B., Kiefer, D.A., Collins, D.J., and McDermid, I.S., 1986, *In vivo* fluorescence excitation spectra of marine phytoplankton. I. Taxonomic characteristics and responses to photoadaptation. *J. Plankton Res*. **8**:197–214.

Soper, A.K., 2000, The radial distribution functions of water and ice from 220 to 673 K and at pressures up to 400 MPa, *Chem. Phys*. **258**:121–137.

Sosik, H.M. and Mitchell, B.G., 1995, Light absorption by phytoplankton, photosynthetic pigments and detritus in the California Current System. *Deep Sea-Res*. **42**(10):1717–1748.

Sournia, A., 1965, Mesere de l'absorption de l'ultraviolet dans les eaux cosieres de Nossi-Be (Madagascar), *Bull. Inst. Oceanogr. (Monaco)* **65**:1348.

Spinard, R.W., Carder, K.L., and Perry, M.J., 1994, *Ocean Optics*, Oxford University Press, New York; Clarendon Press, Oxford, p. 283.

Spitzy, A. and Ittekkot, V., 1986, Gelbstoff: An uncharachterized fraction of dissolved organic carbon, in: Grassl et al. (Eds.), 1986b (op. cit.), Vol. 2, Appendix 1, p. 31.
Staniszewski, A., Lejman, A., and Pempkowiak, J., 2001, Horizontal and vertical distribution of lignin in surface sediments of the Gdańsk Bay, *Oceanologia* **43**(4): 421–439.
Starmach, K., 1966, Anabaena circinalis Rabenhorst and A.flos aquae Brebisson, ex Bornet et Flahault, *Flora Slodkowodna Polski* **10**:227–263.
Stecher, P.G. (Ed.), 1968, *The Merck Index,* 8th ed. Merck, Rahway, NJ.
Stedmon, C.A., Markager, S., and Kaas, H., 2000, Optical properties and signatures of chromophoric dissolved organic matter (CDOM) in Danish coastal waters, *Estuarine Coastal Shelf Sci.* **51**:267–278.
Steele, J.H., 1962, Environmental control of photosynthesis in the sea, *Limnol. Oceanogr.* **7**:137–150.
Steele, J.H., 1964, A study of the production in the Gulf of Mexico, *J. Mar. Res.* **22**: 211–220.
Steemann Nielsen, E., 1975, *Marine Photosynthesis with Special Emphasis on the Ecological Aspect*, Elsevier, Amsterdam, p. 141.
Stevenson, F., 1982, *Humus Chemistry. Genesis, Composition, Reactions*, Wiley, New York, p. 443.
Stramski, D., Morel, A., and Bricaud, A., 1988, Modeling the light attenuation and scattering by spherical phytoplanktonic cells: a retrieval of the bulk refractive index. *Applied Optics* **27**:3954–3956.
Stramski, D. and Kieffer D.A., 1990, Optical properties of marine bacteria, in: R.W. Spinard (Ed.), *Ocean Optics X, Proc. SPIE (Bellingham)* **1302**:250–268.
Stramski, D. and Kiefer, D.A., 1991, Light scattering by microorganisms in the open ocean, *Progr. Oceanogr.* **28**:343–383.
Stramski, D., Rossoulzadegan, F., and Kiefer, D.A., 1992, Changes in optical properties of a particle suspension coused by protist grazing, *J. Plankton Res.* **14**:961–977.
Stramski D. and Reynolds R.A., 1993, Diel variations in the optical properties of a marine diatom. *Limnol. Oceanogr.* **38**:1347–1364.
Stramski, D., Rosenberg, G., and Legendre, L., 1993, Photosynthetic and optical properties of the marine chlorophyte Dunaliella tertiolecta grown under fluctuating light caused by surface wave focusing, *Marine Biol.* **115**:363–372.
Stramski, D., Shalapyonok, A., and Reynolds, R.A., 1995, Optical characterization of the oceanic unicellular cyanobacterium *Synechococcus* grown under a day–night cycle in natural irradiance. *J. Geophys. Res.* **100**:13295–13307.
Stramski, D., 1999, Refractive index of planktonic cells as a measure of cellular carbon and chlorophyll *a* content, *Deep-Sea Res.* **46**:335–351.
Stramski, D., Bricaud, A., and Morel, A., 2001, Modeling the inherent optical properties of the ocean based on the detailed composition of the planktonic community, *Appl. Optics* **40**:2929–2945.
Stramski, D., Babin, M., and Woźniak, S.B., 2002, Variation in the absorption coefficient of mineral particles caused by changes in the particle size distribution and refractive index, *CD ROM - Ocean Optics* XVI, *Proc.* 10 pp., Santa Fe.
Stramski, D., Sciandra, A., and Claustre, H., 2002, Effects of temperature, nitrogen, and light limitation on the optical properties of the marine diatoms Thalassiosira pseudonana, *Limnol. Oceanogr.* **47**:392–403.
Stramski, D., Woźniak, S.B., and Flatau, P.J., 2004, Optical properties of Asian mineral dust suspended in seawater, *Limnol. Oceanogr.* **49**:749–755.
Stramski, D. and Woźniak, S.B., 2005, On the role of colloidal particles in light scattering in the sea, *Limnol. Oceanogr.* **50**(5):1581–1591.
Strickland, J.D.H. and Parsons, T.R., 1968, A practical handbook of seawater analysis. Pigment analysis, *Bull. Fish. Res. Bd. Can.* **167**:1–311.
Stuermer, D.H. and Harvey, G.R., 1974, Humic substances from sea water, *Nature* **250**:480–481.

Stuermer, D.H., 1975, *The characterization of humic substances in sea water*, Ph.D. Thesis, Mass. Inst. Technol., Woods Hole Oceanogr. Inst., p. 163.

Sullivan, S.A., 1963, Experimental study of the absorption indistilled water, artifical water and heavy water in the visible region of the spectrum, *J. Opt. Soc. Am.* **53**: 962–967.

Summerhayes, C.P., Emeis, K.C., Angel, M.V., Smith R.L., and Zeitzschel, B. (Eds.), 1995, *Upwelling in the Ocean: Modern Processes and Ancient Records*, John Wiley, Chichester, p. 422.

Symons, M.C.R., 2001, Water structure, unique but not anomalous, *Phil. Trans. R. Soc. Lond. A* **359**:1631–1646.

Takahashi, M., Nagai, H., Yamagushi, Y., and Yohimura, S., 1974, *J. Oceanogr. Soc. Jap*. **30**(3):37–150.

Takematsu, N., Kishino, M., and Okami, N., 1981, The quantum yield of phytoplankton photosynthesis in Lake Fukami-ike, *La Mer* **19**:132–138.

Tam, A.C. and Patel, C.K.N., 1979, Optical absorption of light and heavy water by laser optoacoustic spectroscopy, *Appl. Optics* **18**:3348–3358.

Tassan, S. and Ferrari, G.M., 1995, An alternative approach to absorption measurements of aquatic particles retained on filters, *Limnol. Oceanogr*. **40**(8):1347–1357.

Tcherkasov, A.Y., 1957, Optics and spectroscopy, *Opt. Spectrosc.-USSR,* **38**:827 (in Russian).

Thurman, E.M., 1985, *Organic Geochemistry of Natural Waters*, Niehoff and Junk, Dordrecht.

Tomczak, M. and Stuart, J.G., 1995, *Regional Oceanography: An Introduction*, Pergamon Press, Oxford, UK, p. 422.

Treiber, E., 1955, Die Chemie del′ Zellwand, in: W. Ruhland (Ed.), *Handbuch der Pflanzenphysiologie*. Vol. I. Springer-Verlag, Berlin, pp. 668–721.

Trenberth, K.E. (Ed.), 1992, *Climate System Modelling*, Cambridge University Press, Cambridge, UK, p. 788.

Tsai, K.H. and Wu, T.M., 2005, Local structural effects on low-frequency vibrational spectrum of liquid water: The instantaneous-normal-mode analysis, *Chem. Phys. Lett*. **417**:390–395.

Tunved, P., Nilsson, E.D., Hansson, H.-C., Ström, J., Kulmala, M., Aalto, P., and Viisanen, Y., 2005, Aerosol characteristics of air masses in northen Europe: Influence of location, transport, sinks, and sources, *J. Geophys. Res*. **110**, D07201, doi:10.1029/2004JD005085, p. 13.

Twardowski, M.S., Boss, E., Macdonald, J.B., Pegau, W.S., Barnard, A.H., and Zaneveld, J.R.V., 2001, A model for estimating bulk refractive index from the optical backscattering ratio and the implication for understanding particle composition in case I and case II waters, *J. Geophys. Res*. **106**(C7):14,129–14,142.

Underwood, J. and Wittig, C., 2004, Two-photon photodissociation of H_2O via the B state, *Chem. Phys. Lett*. **386**:190–195.

Vaccaro, R.F., 1965, Inorganic nitrogen in sea water, in: *Chemical Oceanography*, Academic Press, London.

Vaida, V., Daniel, J.S., Kjaergaard, H.G., Goss, L.M., and Tuck, A.F., 2001, Atmospheric absorption near infrared and visible solar radiation by the hydrogen bonded water dimer, *Q. Soc. J. Roy. Meteor*. 1627–1663.

Venyaminov, S.Y. and Prendergast, F.G., 1997, Water (H_2O and D_2O) molar absorptivity in the 1000–4000 cm^{-1} range and quantitative infrared spectroscopy of aqueous solutions, *Anal. Biochem*. **248**:234–245.

Vernet, M. and Whitehead, K., 1996, Release of ultraviolet-absorbing compounds by the red-tide dinoflagellate Lingulodinium polyedra, *Mar. Biol*. **127**(1):35–44.

Vij, K., Simpson, D.R.J., and Panarina O.E., 2003, Far infrared spectroscopy of water at different temperatures: GHz to THz dielectric spectroscopy of water, *J. Mol. Liq*. **112**:125–135.

Vinogradov, A.P., 1968, *The Chemistry of the Ocean*, Nauka, Moskva (in Russian).

Vinogradov, M.E., (Red.), 1977, *Biology of the ocean*, Vol. 1, p. 398, Vol. 2, p. 399, Nauka, Moscow (in Russian).
Vinogradov, M.E. and Shushkina, E.A., 1987, *The Functioning of Plankton Associations in the Epipelagial of the Ocean*, Nauka, Moskva, p. 240 (in Russian).
Vinogradov, M.E., Tsejtlin, V.B., and Sapozhnikov, V.V., 1992, Primary production in the ocean (in Russian: Pervichnaya produktsiya v okeane), Zh. Obshch. Biol. **53**(3): 314–327.
Visser, M.P., 1967, Shipboard laboratory measurements of light transmittance of Sargasso Sea in the visible and near infrared part of the spectrum, *NATO Subcomm. Oceanogr. Res. Tech. Rep*. **30**, p. 24.
Vodacek, A., Blough, N.V., DeGrandpre, M.D., Peltzer, E.T., and Nelson, R.K., 1977, Seasonal variation of CDOM and DOC in the Middle Atlantic Bight: Terrestrial inputs and photooxidation, *Limnol. Oceanogr*. **42**(4):674–686.
Vodacek, A., Hoge, F.E., Swift, R.N., Yungel, J.K., Peltzer, E.T., and Blough, N.V., 1995, The use of in situ airborne fluorescence measurements to determine UV absorption coefficient and DOC concentrations in surface waters, *Limnol. Oceanogr*. **40**(2):411–415.
Vogel, H.-J., Huffmann, H., and Roth, K., 2004, Studies of crack dynamics in clay soil; I. Experimental methods, results, and morphological quantification, *Geoderma* **125**:203–211 (www.elsevier.com/locate/geoderma).
Vogel, H.-J., Hoffmann, H., and Roth, K., 2005, Studies of crack dynamics in clay soil: I. Experimental methods, results, and morphological quantification, *Geoderma* **125**:203–211 [www.sciencedirect.com].
Volten, H., Muñoz, O., Rol, E., de Haan, J.F., Vassen, W., Hovenier, J.W., Muinonen, H., and Nousiainen, T., 2001, Scattering matrices of aerosols at 441.6 nm and 632.8 nm, J. Geophys. Res. **106**(D15):17,375–17,401.
Volz, F.E., 1973, Infrared constants of ammonium sulfate, Sahara dust, volcanic pumice and flyash, *Appl. Optics* **12**:564–568.
Voshchinnikov, N.V. and Farafonov, V.G., 1993, Optical properties of spheroidal particles, *Astrophys. Space Sci*. **204**:19–86.
Walrafen, G.E., 2004, Raman H-bond pair volume for water, *J. Chem. Phys*., **121**: 2729–2736.
Warren, S.G., 1982, Optical properties of snow, *Rev. Geophys*. **20**:67–89.
Warren, S.G., 1984, Optical constants of ice from the ultraviolet to the microwave, *Appl. Optics* **23**:1026–1225.
Wassmann, P. and Heiskanen, A.-S. (Eds.), 1988, *Sediment trap studies in the Nordic Countries,* 1, Workshop Proceedings, (Tvarminne Zoological Station, Hanko, Finland, 24–28 February 1988), Yliopistopaino, Helsinki, p. 207.
Wassmann, P., Heiskanen, A.-S., and Lindahl O. (Eds.), 1991, *Sediments trap studies in the Nordic Countries,* 2, Symposium Proceedings (Kristineberg Marine Biological Station, Sweden, 21–23 November 1990), NurmiPrint Oy, Nurmijarvi, p. 309.
Weast, R.C. (Ed.), 1981, *CRC Handbook of Chemistry and Physics,* 62nd ed., CRC, Cleveland.
Wells, M.L. and Goldberg, E.D., 1994, The distribution of colloids in the North Atlantic and Southern Oceans, *Limnol. Oceanogr*. **39(2)**:286–302.
Wells, M.L., 2002, Marine colloids and trace metals, in: D.A. Hansell and C.A. Carlson, (Eds.), 2002 (op.cit.), pp. 367–404.
Whitehead, K. and Hedges, J.I., 2003, Electrospray ionization tandem mass spectrometric and electron impact mass spectrometric characterization of mycosporinelike amino acids, *Rapid Commun. Mass Sp*. **17(18)**:2133–2138.
Whittaker, D.H. and Likens, G.E., 1975, The biosphere and man, in: *Primary Productivity of the Biosphere*, H. Lieth and R.H. Wittaker (Eds.), Springer, Berlin, pp. 305–328.

Wieliczka, D.M., Weng, S., and Querry, M.R., 1989, Wedge shaped cell for highly absorbent liquids: infrared optical constants of water, *Appl. Optics* **28**:1714–1719.

Williams, D.H. and Fleming, I., 1980, *Spectroscopic Methods in Organic Chemistry*, McGraw-Hill, New York.

Willson, R.C., 1993, Solar irradiance, in: *Atlas of Satellite Observations Related to Global Change*, R.J. Gurney, J.L. Foster, and C.L. Parkinson (Eds.), Cambridge University Press, Cambridge, UK, pp. 5–18.

Winter, B., Weber, R., Widdra, W., Dittmar, M., Faubel, M., and Hertel, I.V., 2004, Full valence band photoemission from liquid water using EUV synchrotron radiation, *J. Phys. Chem. A* **108**:2625–2632.

Wishart, J.F. and Nocera, D.G. (Eds.), 1998, *Photochemistry and Radiation Chemistry*, Oxford University Press, Oxford, p. 448.

Wolken, J.J., 1975, *Photoprocesses, Photoreceptors, and Evolution*, Academic Press, New York, p. 317.

Woods, K.N. and Wiedemann, H., 2004, The relationship between dynamics and structure in the far infrared absorption spectrum of liquid water, *Chem. Phys. Lett.* **393**:159–165.

Worley, J.D. and Klotz, I.M., 1966, Near-infrared spectra of H_2O-D_2O solutions, *J. Chem. Phys.* **45**:2868–2871.

Woźniak, B., 1973, An investigation of the influence of the components of seawater on the light field in the sea, *Stud. Mater. Oceanol.* (Published by Polish SCOR), **6**:69–132 (in Polish).

Woźniak, B., 1977, New results of studies into the proportions of absorption and scattering in the total attenuation of light in Baltic Sea waters, *Stud. i Mater, Oceanol.* (Published by Polish SCOR), **19**:90–96.

Woźniak B, Hapter R., and Dera J., 1989, Light curves of marine plankton photosynthesis in the Baltic, *Oceanologia* **27**:61–78.

Woźniak, B., 1990a, Statistical relations between photosynthesis and abiotic conditions of the marine environment; an initial prognosis of the World Ocean productivity ensuing from warming up of the Earth, *Oceanologia* **29**:147–174.

Woźniak, B., 1990b, *The energetics of marine photosynthesis*, Vols. 1–3, unpublished dissertation, Inst. Oceanol. PAS, (in Polish).

Woźniak, B. and Ostrowska, M., 1990a, Optical absorption properties of phytoplankton in various seas, *Oceanologia* **29**:117–146.

Woźniak, B. and Ostrowska, M., 1990b, Composition and resources of photosynthetic pigments of the sea phytoplankton, *Oceanologia* **29**:91–115.

Woźniak, B. and Ostrowska, M., 1991, Photosynthesis pigments: Their individual optical (absorption and fluorescence) properties, *Stud. Mater Oceanol.* (Published by Polish SCOR), **59**:137–158.

Woźniak, B. and Pelevin, V.N., 1991, Optical classifications of the seas in relation to phytoplankton characteristics, *Oceanologia* **31**:25–55.

Woźniak, B., Dera, J., and Koblentz-Mishke, O.I., 1992a, Bio-optical relationships for estimating primary production in the Ocean, *Oceanologia* **33**:5–38.

Woźniak, B., Dera, J., and Koblentz-Mishke, O.I., 1992b, Modelling the relationship between primary production, optical properties, and nutrients in the sea, *Ocean Optics 11, Proc. SPIE* **1750**:246–275.

Woźniak, B., Dera, J., Semovski, S., Hapter, R., Ostrowska, M., and Kaczmarek, S., 1995, Algorithm for estimating primary production in the Baltic by remote sensing, *Stud. Mater Oceanol.* (Published by Polish SCOR), **68(**8):91–123.

Woźniak, B., Dera, J., Ficek, D., Majchrowski, R., Kaczmarek, S., Ostrowska, and M., Koblentz-Mishke, O.J., 1998, Modelling the influence of photo- and achromatic-acclimation on the absorption properties of marine phytoplankton, *CD ROM -Ocean Optics XIV*, (Kailua-Kona Hawaii), 1998, 8.

Woźniak, B., Dera J., Ficek, D., Majchrowski, R., Kaczmarek, S., Ostrowska, M., and Koblentz-Mishke, O.J., 1999, Modelling the influence of acclimation on the absorption properties of marine phytoplankton, *Oceanologia* **41**(2):187–210.

Woźniak, B., 2000, Algorithm of Gaussian analysis of marine phytoplankton absorption spectra set – for using in BIOCOLOR programme, *Słupsk. Pr. Mat.-Przyrod.* **13a**:329–341.

Woźniak, B., Dera, J., Ficek, D., Majchrowski, R., Kaczmarek, S., Ostrowska, M., and Koblentz-Mishke, O.I., 2000, Model of the 'in vivo' spectral absorption of algal pigments. Part 1. Mathematical apparatus, *Oceanologia* **42**(2):177–190.

Woźniak B., Dera J., Ficek D., Majchrowski R., Ostrowska M., and Kaczmarek S., 2003, Modelling light and photosynthesis in the marine environment, *Oceanologia* **45**(2):171–245.

Woźniak, B., Krężel, A., and Dera J., 2004, Development of a satellite method for Baltic ecosystem monitoring (DESAMBEM) -an ongoing project in Poland, *Oceanologia* **46**(3):445–455.

Woźniak, B., Woźniak, S.B., Tyszka, K., and Dera, J., 2005a, Modelling the light absorption properties of particulate matter forming organic particles suspended in seawater. Part 1. Model description, classification of organic particles, and example spectra of the light absorption coefficient and the imaginary part of the refractive index of particulate matter for phytoplankton cells and phytoplankton-like particles, *Oceanologia* **47**(2):129–164.

Woźniak, B., Woźniak, S B., Tyszka, K., Ostrowska, M., Majchrowski, R., Ficek, D., and Dera, J., 2005b, Modelling the light absorption properties of particulate matter forming organic particles suspended in seawater. Part 2. Modeling results, *Oceanologia* **47**(4):621–662.

Woźniak, B., Woźniak, S.B., Tyszka, K., Ostrowska, M., Majchrowski, R.. Ficek, D, and Dera J., 2006, Modelling the light absorption properties of particulate matter forming organic particles suspended in seawater. Part 3. Practical application. *Oceanologia* **48**(4):479–507.

Woźniak, S.B and Stramski, D., 2004, Modeling the optical properties of mineral particles suspended in seawater and their influence on ocean reflectance and chlorophyll estimation from remote sensing algorithms, *Appl. Optics* **43**(17): 3489– 3503.

Wozniak, S.B., Miksic, E.Y., Stramski, D., and Stramska, M., 2006, Relationships between the inherent optical properties and seawater constituents in the Southern California coastal waters at Imperial Beach. *EOS Trans. AGU*, **87**(36), Ocean Sci. Meet. Suppl., Abstract OS11F–05.

Wriedt, T., 1998, A review of elastic light scattering theories, *Part. Syst. Charact.* **15**: 67–74.

Yakamura, G., Glazer, A.N., and Williams, R.C., 1980, Molecular architecture of a light-harvesting antenna. Comparison of wild type and mutant *Synechococcus* 6301 phycobilisomes. *J.Biol.Chem.* **255**:11004–11010.

Yakovenko, A.A., Yashin, V.A., Kovalev, A.E., and Fesenko, E.E., 2002, Structure of the vibrational absorption spectra of water in the visible region, *Biophysics* **47**:891–895.

Yamamoto, G. and Onishi, G., 1952, Absorption of solar radiation by water vapor in the atmosphere, *J. Meteorology* **9**:415–421.

Yamasaki, A., Fukuda, H., Fukuda, R., Miyajima, T., Nagata, T., Ogawa, H., and Koike, I., 1998, Submicrometer particles in northwest Pacific coastal environments: Abundance, size distribution and biological origins, *Limnol. Oceanogr.* **43**(3):536–542.

Yentsch, C.S., 1960, The influence of phytoplankton pigments on the colour of the sea, *Deep Sea Res.* **7**:1–9.

Yentsch, C.S., 1962, Measurement of visible light absorption by particulate matter in the Ocean, *Limnol. Oceanogr.* **7**(2):207–217.

Yentsch, C.S., 1965, Distribution of chlorophyll and pheophitin in the open sea, *Deep–Sea Res.* **12**(5):653–666.

Zakrzewski, W., 1982, *Ices on Seas* (*Lody na morzach*) Wyd. Morskie, Gdańsk, p. 313 (in Polish).

Zalewski, S.M., 1977, *Analysis of size distributions of the particles suspended in sea water*, Doctoral dissertation, Institute of Geophysics of the Polish Academy of Sciences, Warszawa (in. Polish).

Zaneveld, J.R.V., Barnard, A.H., Boss, E., 2005a, Theoretical derivation of the depth average of remotely sensed optical parameters, *Opt. Express* **13**(22):9052–9061.

Zaneveld, J.R.V., Twardowski, M.S., Lewis, M., and Barnard, A., 2005b, Radiative transfer and remote sensing, in: *Remote Sensing of Coastal Aquatic Waters*, R. Miller and C.E. Del-Castillo (Eds.), Springer-Kluwer, 1–20.

Zelsmann, H.R., 1995, Temperature dependence of the optical constants for liquid H_2O and D_2O in the far IR region, *J. Mol. Struct.* **350**:95–114.

Zepp, R. and Schlotzhauer, P.F., 1981, Comparison of photochemical behaviour of various humic substances in water: 3. Spectroscopic properties of humic substances, *Chemosphere* **10**:479–486.

Zieliński, A., Król, T., and Gędziorowska, D., 1987, The influence of the Chlorella vulgaris cell on the light scattering properties, *Bull. Pol. Acad. Sci., Earth Sci.* **32**(2): 119–125.

Zoha, S.J., Ramnarain, S., and Allnutt, F.C.T., 1998, Ultrasensitive Direct Fluorescent Immunoassay for Thyroid Stimulating Hormone, *Clinical Chemistry* **44**(9):2045–2046.

Zolotarev, V.M., Mikhilov, B.A., Alperovich, L.L., and Popov, S.I., 1969, Dispersion and absorption of liquid water in the infrared and radio regions of the spectrum, *Opt. Spectrosc.* **27**:430–432.

Zuev, V.E. and Krekov, G.M., 1986, *Atmospheric Optical Models*, Gidrometeoizdat, Leningrad (in Russian).

Zvalinsky, V.I., 1986, *Principles of Influence of Light Intensity and Spectral Composition on Marine Phytoplankton*, Thesis, 2nd Doctoral degree dissertation, Akad. Nauk Belarus, SSR, Minsk, p. 45 (in Russian).

Subject Index

D

M

List of Symbols and Abbreviations

Symbol	Denotes	Units
A1	Aromatic amino acids (a chemical class of SPM)	
A2	Mycosporinelike amino acids (a chemical class of SPM)	
A3	Natural proteins (a chemical class of SPM)	
AGP	Aggregated particles	
AO	Atomic orbital	
A_M	Molecular mass	
C	Concentration of solution	$kg \cdot m^{-3}$ or $M \equiv mol \cdot dm^{-3}$
C_a	Concentration of chlorophyll *a*, i.e., sum of chlorophyll *a* + pheo. or total chlorophyll (chlorophyll *a* + divinyl chlorophyll *a*) concentration in sea water, used as the trophicity index of waters	mg tot. chl $a \cdot m^{-3}$
C_c	Intracellular concentration of carbon	$kg \cdot m^{-3}$
C_{chl}	Intracellular concentration of chlorophyll *a*	kg tot. chl $a \cdot m^{-3}$
C_{DOM}	Concentration of dissolved organic matter	$kg \cdot m^{-3}$
$C_a(0)$	Concentration of total chlorophyll *a* in a surface layer of the sea (surface chlorophyll), used as the trophicity index of sea basen	mg tot. chl $a \cdot m^{-3}$
C_{org}	Concentration of organic carbon	$kg \cdot m^{-3}$ or $g \cdot m^{-3}$
Chl. *a*	Chlorophyll *a*	
Chl. *b*	Chlorophyll *b*	

(*Continued*)

Symbol	Denotes	Units
Chl. *c*	Chlorophyll *c*	
C_j	Concentration of the *j*th group of pigments (including chlorophyll *a*)	mg of the *j*th pigment · m^{-3}
$C_{org,pl}$	Organic carbon content in phytoplankton	kg · m^{-3}
C_{PPC}	Concentration of the photoprotecting carotenoids	mg of the carotenoids · m^{-3}
C_{PSC}	Concentration of the photosynthetic carotenoids	mg of the carotenoids · m^{-3}
CZCS	Coastal Zone Colour Scanner	
DOC	Dissolved organic carbon	
DOM	Dissolved organic matter	
D	Equivalent spherical diameter of suspended particle (also phytoplanktoncell diameter)	μm
DNA	Deoxyribonucleic acids	
CDOM	Colored dissolved organic matter	
D1	Oceanic DOM-like particles (a chemical class of SPM)	
D2	Baltic DOM-like particles (a chemical class of SPM)	
E	Eutrophic basin	
E1, E2, E3, ...	Subtypes of eutrophic basin	
E_d	Downward irradiance	W · m^{-2} or Ein · m^{-2} · s^{-1}
E_0	Scalar irradiance	W · m^{-2} or Ein · m^{-2} · s^{-1}
E_v	Energy of a photon	J or eV
E_E	Energy of a electron in a molecule	J or eV
E_M	Energy of a molecule	J or eV
E_n	Energy of an electron	J or eV
E_{ROT}	Rotational energy of a molecule	J or eV
E_{TR}	Translational energy of a molecule	J or eV
E_{VIB}	Vibrational energy of a molecule	J or eV
FA	Fulvic acids	
F_j	Spectral fitting functions for the *j*th pigment,	Dimensionless
FLP	Flakes particles	
FOP	Fragment of organisms	
GOM	Gulf of Mexico	
H1, H2, H3, ...H13	Chemical classes of particles consistingof organic humus matter (see Table 5.16)	

Symbol	Denotes	Units
HOMO	Highest-energy occupied molecular orbital	
HA	Humic acids	
I	Transitional meso–eutrophic basin (intermediate)	
Ih	A type of ice crystals	
IR	Infrared radiation	
IRP	Irregular particles	
IO PAS	Institute of Oceanology Polish Academy of Sciences	
J	Rotational quantum number	Dimensionless
K_d	Coefficient of downward irradiance attenuation	m^{-1}
L	Radiance of light	$W \cdot m^{-2} \cdot sr^{-1}$
L_u	Upward radiance	$W \cdot m^{-2} \cdot sr^{-1}$
L	Libration bend of absorption	μm
L	Lignins (a chemical class of SPM)	
LO	Living Organisms	
LUMO	Lowest unoccupied molecular orbital	
M	Mesotrophic basin	
M	Concentration	$mol \cdot dm^{-3}$ or M
M	Multiplicity of electron state	Dimensionless
MAAs	Mycosporinelike amino acids	
MD	Microdetritus	
MO	Molecular orbital	
MCM	Multicomponent Complex Model	
N	Purine and pyridine compounds (a chemical class of SPM)	
N	Numerical concentration of suspended particles (the density number)	$m^{-3} \cdot \mu m^{-1}$
N	Number of links in a molecular chain	Dimensionless
N_{inorg}	Total concentration of inorganic nitrogen: this includes the nitrogen contained in nitrates, nitrites and ammonia dissolved in sea water ($N_{inorg} = N(NO_3) + N(NO_2) + N(NH_4)$), but not dissolved nitrogen gas N_2	$\mu M \equiv 10^{-6}\ mol \cdot dm^{-3}$
N_{tot}	Total numerical concentration of particles in the medium	m^{-3}
NAP	Nonalgal particles	
NAP	Nonalgal Particles concentration	$g \cdot m^{-3}$ (dry mass)
ND	Nano detritus	

(*Continued*)

Symbol	Denotes	Units
NPP	Nonphotosynthetic pigments index $NPP = C_{PPC}/(C_{PPP}+C_{PSP})$	Dimensionless
NUV	Near Ultraviolet Radiation	
O	Oligotrophic basin	
Od	Organic Detritus	
Od	Organic Detritus concentration	$g \cdot m^{-3}$ (dry mass)
OM	Organic matter	
O1, O2, O3, ...	Subtypes of oligotrophic basins	
PAR	Photosynthetically Available Radiation, i.e., a radiation in the spectral range from c. 400 nm to 700 nm	
PAR	Downward irradiance in the Photosynthetically Available Radiation spectral range (from c. 400 nm to 700 nm)	$W \cdot m^{-2}$ or $Ein \cdot m^{-2} \cdot s^{-1}$
$PAR(0^+)$	Downward irradiance in the photosynthetically available radiation spectral range (from c. 400 nm to 700 nm) just below the water surface (surface PAR)	$W \cdot m^{-2}$ or $Ein \cdot m^{-2} \cdot s^{-1}$
P1	Ocean phytoplankton pigments (a chemical class of SPM)	
P2	Baltic phytoplankton pigments (a chemical class of SPM)	
P	Probability	%
Ph1	ocean Phytoplankton and phytoplanktonlike particles (a chemical class of SPM)	
Ph2	Baltic phytoplankton and phytoplanktonlike particles (a chemical class of SPM)	
PhM	Polar phytoplankton and phytoplanktonlike particles (a chemical class of SPM)	
P_i	Pigment index of phytoplankton *in vivo*, $P_i = a_{pl}(\lambda = 441 \text{ nm}) / a_{pl}(\lambda = 675 \text{ nm})$	Dimensionless
$P_{i,extr}$	Pigment index of acetone extracts of pigments $P_{i,extr} = a_{pl,e}(\lambda = 433 \text{ nm}) / a_{pl,e}(\lambda = 661 \text{ nm})$	Dimensionless
PDR	Potentially destructive radiation	$\mu Ein \cdot m^{-2} \cdot s^{-1}$
*PDR**	Potentially destructive radiation per unit of the chlorophyll *a* mass	$\mu Ein \cdot (mg\ chl\ a)^{-1} \cdot s^{-1}$

Symbol	Denotes	Units
PIM	Particulate inorganic matter	
PIM	Particulate inorganic matter concentration	$g \cdot m^{-3}$ (dry mass)
PL	Phytoplankton concentration	$g \cdot m^{-3}$ (dry mass)
POC	Particulate organic carbon	
POC	Particulate organic carbon concentration	$g \cdot m^{-3}$
POM	Particulate organic matter	
POM	Particulate organic matter concentration	$g \cdot m^{-3}$ (dry mass)
POS	Particulate organic matter in skeletons	
PPC	Photoprotecting carotenoids	
PPP	Photoprotecting pigments	
PSC	Photosynthetic carotenoids	
PSP	Photosynthetic pigments	
PSU	Practical salinity units	Dimensionless
Q^*	Absorption deficiency function due to the pigment packing in the phytoplankton cells, so-called packing (packaging, package) function	Dimensionless
Q_a	Relative cross-sections (optical efficiency factors) for the absorption of light	Dimensionless
Q_c	Relative cross-sections (optical efficiency factors) for the attenuation, of light	Dimensionless
Q_s	Relative cross-sections (optical efficiency factors) for the scattering of light	Dimensionless
R_{CC}	Length of the bond C–C or C=C	m
RNA	Ribonucleic acids	
R_{rs}	Remote reflectance of irradiance	sr^{-1}
R_y	Rydberg constant = 109677.5810	cm^{-1}
S	Spin of electron	Dimensionless
$\boldsymbol{S}(\theta, \varphi)$	Scattering matrix	
S_g	Geometrical cross-section of the particle	m^2
S_H	Slope of visible light absorption spectra of yellow substances or humus	nm^{-1}
S_0	Ground singlet state of molecule	
$S_1, S_2, \ldots$	Excited singlet states of molecule	
SD	Standard Deviation	
SDF	Size distribution functions of particles	

(*Continued*)

Symbol	Denotes	Units
SLP	Spherelike particles	
SOC	Suspended Organic Carbon	
SOM	Suspended organic matter	
SPM	Suspended particulate matter	
T	Transmittance of light	Dimensionless
T	Period of vibrations	s
T_B	Intermolecular translation bend	
T_S	Intermolecular stretch bend	
T_1, T_2, T_3, ...	Triplet states of molecule	
U	Potential energy of electron	eV
UV	Ultraviolet light	
VIS	Visible light	
V	Volume	m^3
X	Roentgen radiation (X radiation)	
Z	Zooplankton and zooplankton and/ornekton-like particles (a chemical class of SPM)	
a	Cross-section of a molecule	m^2
a	Coefficient of light absorption	m^{-1}
a_{DOM}	Coefficient of light absorption by the Dissolved Organic Matter in sea water	m^{-1}
$a_{H'}$	Coefficient of light absorption by the humus	m^{-1}
a_h	Coefficient of light absorption by the humic acids	m^{-1}
a_L	Coefficient of light absorption by the lignins	m^{-1}
a_{OM}	Coefficient of light absorption by organic matter	m^{-1}
a_{pm}	Coefficient of light absorption by the particulate material	m^{-1}
a_{pl}	Coefficient of light absorption by the monodispersant, uniform medium of a suspension of phytoplankton in water (often identified with the total absorption coefficient of all the phytoplankton pigments)	m^{-1}
$a_{pl,e}$	Coefficient of light absorption by the acetone extracts of pigments	m^{-1}
a_{POM}	Coefficient of light absorption by the Particulate Organic Matter in sea water	m^{-1}
a_s	Coefficient of light absorption by the sea salt in sea water	m^{-1}

Symbol	Denotes	Units
a_{sol}	Coefficients of light absorption by dissolved substances	m^{-1}
a_{SM}	Coefficient of light absorption by suspended minerals in sea water	m^{-1}
a_w	Coefficient of light absorption by pure liquid water	m^{-1}
$a_{w,\ ice}$	Coefficient of light absorption by the ice	m^{-1}
$a_{w,\ v}$	Coefficient of light absorption by the water vapor	m^{-1}
a_y	Coefficient of light absorption by yellow substances	m^{-1}
$\bar{a}_{pl}$	Mean coefficients of light absorption in the 400–700 nm spectral range for the monodispersant, uniform medium of a suspension of phytoplankton in water	m^{-1}
$\bar{a}_{pl,\ PSP}$	Mean coefficients of light absorption in the 400–700 nm spectral range by the phytoplankton photosynthetic pigments in vivo	m^{-1}
a^*	Mass-specific absorption	$m^2 \cdot g^{-1}$
a_a^*	Mass-specific coefficients of light absorption by the chlorophyll *a* (in solvent, i.e., in the unpackaged dispersed state)	m^2 (mg tot. chl *a*)$^{-1}$
a_b^*	Mass-specific coefficients of light absorption by the chlorophyll *b* (in solvent, i.e., in the unpackaged dispersed state)	m^2 (mg chl *b*)$^{-1}$
a_c^*	Mass-specific coefficients of light absorption by the chlorophyll *c* (in solvent, i.e., in the unpackaged dispersed state)	m^2 (mg chl *c*)$^{-1}$
a_f^*	Mass-specific coefficients of light absorption by fulvic acids	m^2 (g fulvic acids)$^{-1}$
a_H^*	Mass-specific absorption coefficient of light by the humus substances	m^2 (g humus)$^{-1}$
a_h^*	Mass-specific absorption coefficient of light by the humic acids	m^2 (g humic acids)$^{-1}$
a_L^*	Mass-specific absorption coefficient of light by the lignins	m^2 (g lignins)$^{-1}$

(*Continued*)

Symbol	Denotes	Units
a_j^*	Mass-specific coefficient of light absorption by the *j*th phytoplankton pigments group in vivo	m^2 (mg of the *j*th pigment)$^{-1}$
a^*_{OM}	Mass-specific absorption coefficient of light by the organic matter	$m^2\ g^{-1}$
a^*_{OC}	Carbon-specific coefficient of light absorption by organic matter	m^2 (g C)$^{-1}$
$a^{*\ chl}_{NAP}$	Chlorophyll-specific coefficient of light absorption by nonalgal particles	m^2 (mg tot. chl *a*)$^{-1}$
a^*_{pl} or $a^{*\ chl}_{pl}$	Chlorophyll-specific coefficient of light absorption by phytoplankton in vivo ($a^*_{pl} = a_{pl}(\lambda)/C_a$)	m^2 (mg tot. chl *a*)$^{-1}$
$a^{*\ chl}_{p}$	Chlorophyll-specific coefficient of light absorption by suspended particles	m^2 (mg tot. chl *a*)$^{-1}$
$a^*_{pl,s}$, $a^*_{pl,sol}$	Chlorophyll-specific coefficient of light absorption by phytoplankton pigments (in solvent, i.e., in the unpackaged dispersed state)	m^2 (mg tot. chl *a*)$^{-1}$
a^*_{PPC}	Mass-specific coefficients of light absorption by the photoprotrcting carotenoids (in solvent, i.e., in the unpackaged dispersed state)	m^2 (mg PPC)$^{-1}$
a^*_{PSC}	Mass-specific coefficients of light absorption by the photosynthetic carotenoids (in solvent, i.e., in the u npackaged dispersed state)	m^2 (mg PSC)$^{-1}$
a^*_{PSP}	Mass-specific coefficients of light absorption by the photosynthetic pigments (in solvent, i.e., in the unpackaged dispersed state)	m^2 (mg PSP)$^{-1}$
$a^*_{PPP,s}$	Chlorophyll specific coefficients of light absorption by the 'unpackaged' photoprotecting pigments (in solvent, i.e., in the unpackaged dispersed state)	m^2 (mg tot. chl *a*)$^{-1}$
$a^*_{PSP,s}$	Chlorophyll specific coefficients of light absorption by the 'unpackaged' photosynthetic pigments (in solvent, i.e., in the unpackaged dispersed state)	m^2 (mg tot. chl *a*)$^{-1}$
a^*_{sol}	Mass specific absorption coefficients of dissolved substances	$m^2\ mg^{-1}$
b	Light scattering coefficient	m^{-1}
b_b	Light backscattering coefficient	m^{-1}
c	Velocity of light in vacuum	$m \cdot s^{-1}$

Symbol	Denotes	Units
c	Total volumetric coefficient of light attenuation	m^{-1}
chl *a*	Chlorophyll *a*	
d	Cell diameter	μm
d	Grain width of suspended particle	μm
f	Degree of fulvization	Dimensionless
h	Planck's constant $h = 6.62517 \cdot 10^{-34}$	J · s
j	Usually denote the light absorbing component in question	
k	Wave number	m^{-1} or cm^{-1}
l	Orbital electron quantum number	Dimensionless
m	Complex refractive indices	Dimensionless
m	Mass	g
m	Magnetic electron quantum number	Dimensionless
n	A type of nonbonding molecular orbital; a type of valence electron	
n	Principal electron quantum number	Dimensionless
n	Real part of the absolute complex refractive index of light	Dimensionless
n	Numerical concentration of suspended particles (relative number of particles)	$m^{-3}\mu m^{-1}$
n'	Imaginary part of the absolute complex refractive index (the full expression is $i\,n'$, where $i = \sqrt{-1}$)	Dimensionless
n'_p	Imaginary part of the absolute refractiveindex of light of particulate material	Dimensionless
$n'_{p,dry}$,	Imaginary part of the absolute refractive index of light of pure particulate organic matter, without any interstitial water (dry material)	Dimensionless
$n'_{p,r}$	Imaginary part of the relative (to water)refractive index of light of particulate material	Dimensionless
$n'_{p,VIS}$	Imaginary part of the absolute reflective index of light in the VIS spectral range, for particulate material	Dimensionless
r_o	Particle radius	m
r	Distance	m
s	Spin	Dimensionless
temp	Temperature	°C
x	Particle size-parameter	Dimensionless
z	Depth in the sea	m

(*Continued*)

Symbol	Denotes	Units
$z = 0^-$	Location just above the sea surface (at the surface)	0 m
$z = 0^+$	Location just below the sea surface	0 m
Δ_a	Absorption defect due to the packaging effect	Dimensionless
$\varphi_e(x, y, z, s)$	Single-electron wave function (spin orbital)	Dimensionless
Λ	Electronic quantum number	Dimensionless
Ψ_e	Configurational wave function of single-electron	Dimensionless
$\xi(s)$	Spin function	Dimensionless
α	Angle	deg
$\beta(\theta)$	Volume scattering function of light	$m^{-1} \cdot sr^{-1}$
$\beta_p(\theta)$	Volume scattering function of light by suspended particles	$m^{-1} \cdot sr^{-1}$
ε	Molar decimal extinction coefficient	$dm^3 \cdot (mol \cdot cm)^{-1}$
ε_{max}	Spectral maximum of the molar, decimal extinction coefficient	$dm^3 \cdot (mol \cdot cm)^{-1}$
γ	Gamma radiation	
λ	Wavelength	m (often μm or nm)
λ_{max}	Center of the spectral band maximum i.e. Wavelength of maximum of a spectral band	m (often μm or nm)
ν	Frequency	s^{-1}
ν^*	Wave number ($\nu^* = 1/\lambda$)	cm^{-1}
π	A type of bonding molecular orbital; a type of valence electron	
π^*	A type of antibonding molecular orbital; a type of valence electron	
θ	Angle (of light scattering)	deg
ρ	Density	$kg \cdot m^{-3}$
ρ'	Optical parameter of a particle dimensions	Dimensionless
σ	A type of bonding molecular orbital; a type of valence electron	
σ_j	Optical cross-section of a particle	m^2
σ_a	Cross-section for absorption of light	m^2
σ_c	Cross-section for attenuation of light	m^2
σ_s	Cross-section for scattering of light	m^2
σ^*	A type of antibonding molecular orbital; a type of valence electron	
τ	Optical depth	Dimensionless
υ	Vibrational quantum number	Dimensionless

ATMOSPHERIC AND OCEANOGRAPHIC SCIENCES LIBRARY

1. F.T.M. Nieuwstadt and H. van Dop (eds.): *Atmospheric Turbulence and Air Pollution Modelling.* 1982; rev. ed. 1984 ISBN 90-277-1365-6; Pb (1984) 90-277-1807-5
2. L.T. Matveev: *Cloud Dynamics.* Translated from Russian. 1984 ISBN 90-277-1737-0
3. H. Flohn and R. Fantechi (eds.): *The Climate of Europe: Past, Present and Future.* Natural and Man-Induced Climate Changes: A European Perspective. 1984 ISBN 90-277-1745-1
4. V.E. Zuev, A.A. Zemlyanov, Yu.D. Kopytin, and A.V. Kuzikovskii: *High-Power Laser Radiation in Atmospheric Aerosols.* Nonlinear Optics of Aerodispersed Media. Translated from Russian. 1985 ISBN 90-277-1736-2
5. G. Brasseur and S. Solomon: *Aeronomy of the Middle Atmosphere.* Chemistry and Physics of the Stratosphere and Mesosphere. 1984; rev. ed. 1986 ISBN (1986) 90-277-2343-5; Pb 90-277-2344-3
6. E.M. Feigelson (ed.): *Radiation in a Cloudy Atmosphere.* Translated from Russian. 1984 ISBN 90-277-1803-2
7. A.S. Monin: *An Introduction to the Theory of Climate.* Translated from Russian. 1986 ISBN 90-277-1935-7
8. S. Hastenrath: *Climate Dynamics of the Tropics.* Updated Edition from *Climate and Circulation of the Tropics.* 1985; rev. ed. 1991 ISBN 0-7923-1213-9; Pb 0-7923-1346-1
9. M.I. Budyko: *The Evolution of the Biosphere.* Translated from Russian. 1986 ISBN 90-277-2140-8
10. R.S. Bortkovskii: *Air-Sea Exchange of Heat and Moisture During Storms.* Translated from Russian, rev. ed. 1987 ISBN 90-277-2346-X
11. V.E. Zuev and V.S. Komarov: *Statistical Models of the Temperature and Gaseous Components of the Atmosphere.* Translated from Russian. 1987 ISBN 90-277-2466-0
12. H. Volland: *Atmospheric Tidal and Planetary Waves.* 1988 ISBN 90-277-2630-2
13. R.B. Stull: *An Introduction to Boundary Layer Meteorology.* 1988 ISBN 90-277-2768-6; Pb 90-277-2769-4
14. M.E. Berlyand: *Prediction and Regulation of Air Pollution.* Translated from Russian, rev. ed. 1991 ISBN 0-7923-1000-4
15. F. Baer, N.L. Canfield and J.M. Mitchell (eds.): *Climate in Human Perspective.* A tribute to Helmut E. Landsberg (1906–1985). 1991 ISBN 0-7923-1072-1
16. Ding Yihui: *Monsoons over China.* 1994 ISBN 0-7923-1757-2
17. A. Henderson-Sellers and A.-M. Hansen: *Climate Change Atlas.* Greenhouse Simulations from the Model Evaluation Consortium for Climate Assessment. 1995 ISBN 0-7923-3465-5
18. H.R. Pruppacher and J.D. Klett: *Microphysics of Clouds and Precipitation*, 2nd rev. ed. 1997 ISBN 0-7923-4211-9; Pb 0-7923-4409-X
19. R.L. Kagan: *Averaging of Meteorological Fields.* 1997 ISBN 0-7923-4801-X
20. G.L. Geernaert (ed.): *Air-Sea Exchange: Physics, Chemistry and Dynamics.* 1999 ISBN 0-7923-5937-2
21. G.L. Hammer, N. Nicholls and C. Mitchell (eds.): *Applications of Seasonal Climate Forecasting in Agricultural and Natural Ecosystems.* 2000 ISBN 0-7923-6270-5
22. H.A. Dijkstra: *Nonlinear Physical Oceanography.* A Dynamical Systems Approach to the Large Scale Ocean Circulation and El Niño. 2000 ISBN 0-7923-6522-4
23. Y. Shao: *Physics and Modelling of Wind Erosion.* 2000 ISBN 0-7923-6657-3
24. Yu.Z. Miropol'sky: *Dynamics of Internal Gravity Waves in the Ocean.* Edited by O.D. Shishkina. 2001 ISBN 0-7923-6935-1
25. R. Przybylak: *Variability of Air Temperature and Atmospheric Precipitation during a Period of Instrumental Observations in the Arctic.* 2002 ISBN 1-4020-0952-6
26. R. Przybylak: *The Climate of the Arctic.* 2003 ISBN 1-4020-1134-2
27. S. Raghavan: *Radar Meteorology.* 2003 ISBN 1-4020-1604-2
28. H.A. Dijkstra: *Nonlinear Physical Oceanography.* A Dynamical Systems Approach to the Large Scale Ocean Circulation and El Niño. 2nd Revised and Enlarged Edition. 2005 ISBN 1-4020-2272-7 Pb; 1-4020-2262-X
29. X. Lee, W. Massman and B. Law (eds.): *Handbook of Micrometeorology.* A Guide for Surface Flux Measurement and Analysis. 2004 ISBN 1-4020-2264-6
30. A. Gelencsér: *Carbonaceous Aerosol.* 2005 ISBN 1-4020-2886-5

31. A. Soloviev and L. Roger: *The Near-Surface Layer of the Ocean. Structure, Dynamics and Applications.* 2006 ISBN 1-4020-4052-0
32. G.P. Brasseur and S. Solomon: *Aeronomy of the Middle Atmosphere. Chemistry and Physics of the Stratosphere and Mesosphere.* 2005
ISBN 1-4020-3284-6; Pb 1-4020-3285-4
33. B. Woźniak and J. Dera: *Light Absorption in Sea Water.* 2007
ISBN 0-387-30753-2
34. A. Kokhanovsky: *Cloud Optics.* 2005 ISBN 1-4020-3955-7
35. T.N. Krishnamurti, H.S. Bedi, V.M. Hardiker, and L. Ramaswamy: *An Introduction to Global Spectral Modeling, 2nd Revised and Enlarged Edition.* 2006
ISBN 0-387-30254-9
36. L. Pratt and J. Whitehead: *Rotating Hydraulics.* 2007 ISBN 0-387-36639-5
37. Y. Shao: *Physics and Modelling of Wind Erosion.* 2007
38. S. Massel: *Ocean Waves Breaking and Marine Aerosol Fluxes.* 2007
ISBN 0-387-36638-8
39. H.M. van Aken: *The Oceanic Thermohaline Circulation.* 2006
ISBN 0-387-36637-1

MIX
Papier aus verantwortungsvollen Quellen
Paper from responsible sources
FSC® C105338

If you have any concerns about our products,
you can contact us on
ProductSafety@springernature.com

In case Publisher is established outside the EU,
the EU authorized representative is:
Springer Nature Customer Service Center GmbH
Europaplatz 3, 69115 Heidelberg, Germany

Printed by Libri Plureos GmbH
in Hamburg, Germany